Progress in Mathematics
Vol. 36

Edited by
J. Coates and
S. Helgason

Springer Science+Business Media, LLC

Arithmetic and Geometry

Papers Dedicated to I.R. Shafarevich on the Occasion of His Sixtieth Birthday

Volume II Geometry

Michael Artin,
John Tate,
editors

1983 **Springer Science+Business Media, LLC**

Editors:
Michael Artin
Mathematics Department
Massachusetts Institute of Technology
Cambridge, MA 02139

John Tate
Mathematics Department
Harvard University
Cambridge, MA 02138

This book was typeset at Stanford University using the TEX document preparation system and **computer modern** type fonts by Y. Kitajima. Special thanks go to Donald E. Knuth for the use of this system and for his personal attention in the development of additional fonts required for these volumes. In addition, we extend thanks to the contributors and editors for their patience and gracious help with implementing this system.

Library of Congress Cataloging in Publication Data
Main entry under title:

Arithmetic and geometry.

(Progress in mathematics ; v. 35-36)
Contents: v. 1. Arithmetic — v. 2. Geometry.
1. Algebra—Addresses, essays, lectures. 2. Geometry, Algebraic—Addresses, essays, lectures. 3. Geometry—Addresses, essays, lectures. 4. Shafarevich, I. R. (Igor' Rostislavovich), 1923-
I. Shafarevich, I. R. (Igor' Rostislavovich), 1923-
II. Artin, Michael. III. Tate, John Torrence, 1925- . IV. Series: Progress in mathematics (Cambridge, Mass.) ; v. 35-36.
QA7.A67 1983 513'.132 83-7124

ISBN 978-0-8176-3133-8

CIP-Kurztitelaufnahme der Deutschen Bibliothek

Geometry / Michael Artin ; John Tate, ed. -
(Arithmetic and geometry ; Vol. 2) (Progress in mathematics ; Vol. 36)
ISBN 978-0-8176-3133-8 ISBN 978-1-4757-9286-7 (eBook)
DOI 10.1007/978-1-4757-9286-7

NE: Artin, Michael (Hrsg.); 2. GT

Originally published by Birkhäuser Boston in 1983

ISBN 978-0-8176-3133-8

Igor Rostislavovich Shafarevich has made outstanding contributions in number theory, algebra, and algebraic geometry. The flourishing of these fields in Moscow since World War II owes much to his influence. We hope these papers, collected for his sixtieth birthday, will indicate to him the great respect and admiration which mathematicians throughout the world have for him.

Michael Artin
Igor Dolgachev
John Tate
A.N. Todorov

Spencer Bloch
Barry Mazur
Arthur Ogus
Steve Lichtenbaum
V. Arnold
Tetsuji Shioda
Michael Artin
Mark Spivakovsky
C. Musili
Hyman Bass
Oscar Zariski
Dale Peterson
Поздравляем с днём рождения!
A. P. Ogg
Nicholas M Katz
Miles Reid
Victor Kac
J S Milne
Andrew Pressley
John Tate
Robert Bryant
John Coates

Volume II Geometry

Some Algebro-Geometrical Aspects of the Newton Attraction Theory*

V.I. Arnold

To I. R. Shafarevich

According to the Zeldovich theory[1], the observed large scale structure of the universe (the drastically non-uniform distribution of galaxy clusters) is explained by the geometry of caustics of a mapping of a Lagrange submanifold of the symplectic total space of the cotangent bundle to its base space. This Lagrange submanifold is formed by the particle velocities. Contemporary theory of the hot universe predicts a smooth potential velocity field at an early stage (when the universe was about 1000 times "smaller" than now). At this stage the Lagrange manifold is a cotangent bundle section. Then it evolves according to Hamiltonian equations of motion, and hence continues to be Lagrangian. However, it does not need to be a section at all times. The set of critical values of its projection on the base space is called the caustic. At the caustic the particle density becomes infinite (mathematically); the caustic is the place where clustering occurs (generation of galaxies and so on).

The singularities of caustics and their metamorphoses are classified in a way usual for the Lagrangian singularity theory: $A_k, D_k, E_k, \ldots$. This is true for non-interacting particles or for particles in any potential field. But in cases where the field is generated by the particles, a new difficulty occurs. After the caustic has been formed the force field is no longer smooth (because of density singularities). Hence our Lagrange manifold may acquire singularities. Thus we are led to the problem: to generalize the Lagrange mappings singularity theory to the case where the Lagrange

*from a letter to M. Artin, May 10, 1982

[1] Arnold, V.I., Shandarin, S.F., Zeldovich, Ya. B., the Large Scale Structure of the Universe I, Geophys, Astrophys. Fluid Dynamics 1982.

Arnold, V.I., Surgery of singularities in potential collisionless media and caustics metamorphoses in 3-space, Trudy Seminary, 8, (1982), 22–58.

source manifold becomes a Lagrange variety.

[Lagrange varieties occur also in other situations, for instance in the study of the shortest length function on a manifold with boundary (see V.I. Arnold, Lagrange varieties singularities, asympototical rays and the open swallowtail, Funkt. Anal. 15 (1982) 1–14). In this case typical singularities of Lagrange varieties are those of the set of odd degree polynomials having a root of multiplicity greater than one half of the degree. (The symplectic structure of the space on polynomials is inherited from the invariant symplectic structure on the space of binary forms.) But Lagrange variety singularities arising from the Newton attraction theory seem to be different (and in any case they are not known).]

The first step in the study of such Lagrange variety singularities is the study of the singularities of force fields, generated by clustering of free particles. Among typical clusterings in the plane, elliptic and hyperbolic D_4 singularities occur at some (exceptional) moments in time. At these singularities, the particle density is inversely proportional to a quadratic form (in suitable local coordinates). Thus constant density lines are homothetic ellipses or hyperbolas. We are led to the problem of Newtonian attraction by an elliptic or hyperbolic layer.

The theory of attraction by elliptic layers is classic. The first results are due to Newton: a uniform spherical layer does not attract interior points, and attracts exterior ones as if the mass were concentrated at its center. These results were extended to the case of ellipsoids by Ivory, but the hyperbolic case seems not to be settled by classical authors.

Two generalizations of the Newton interior points theorem are formulated below. Let us consider a level hypersurface of a real polynomial in a euclidean space of dimension h, and a point not on the hypersurface. We call natural the density inversely proportional to the gradient length. Let us distribute a Coulomb charge (force inversely proportional to the $(n-1)$st power of the distance) along the hypersurface, with natural density but with a sign, depending on the chosen point: + for points of the hypersurface which one can see from the chosen one, − for those obstructed once, + for those obstructed twice, and so on. We call such a charge "associated" to the chosen point. Then the following generalized Newton theorem holds:

(1) The charge on a second degree hypersurface, associated to a point, does not attract this point.

For instance, let us consider a plane hyperbola with natural density and component charges of different signs. Such a hyperbola does not attract points of the convex parts of the plane it bounds.

Now let us consider any real algebraic hypersurface. We call a point p hyperbolic with respect to the hypersurface if all real lines through p meet the hypersurface at real points only. The following generalization of Newton's theorem holds:

(2) The charge associated to a hyperbolic point does not attract this point.

For instance, let us consider a plane algebraic curve of even degree, consisting of a sequence of ovals one inside other (their number being equal to one half of the degree). Let the charges change signs from each oval to the next one and let the density be natural. Such a curve does not attract points inside the inner oval.

The theorem on the attraction of exterior points leads to results in the geometry of confocal quadrics which seem to be new. The generalization to nonconnected hyperboloids is direct; for instance, the attraction of points between the components of the hyperbola charged as above does not change if we substitute for the hyperbola a larger confocal one. For quadratic forms of other signatures one needs to consider differential forms of appropriate degree instead of charges and instead of the Newton-Coulomb attraction law one needs to consider the Biot-Savart law and its higher dimensional analogues.

Received June 2, 1982

Professor V. I. Arnold
Mathematics Department
Moscow State University
Moscow 117 234
USSR

Smoothing of a Ring Homomorphism Along a Section

M. Artin and J. Denef

To I.R. Shafarevich

Introduction: This paper studies the problem of smoothing a homomorphism of commutative rings along a section. The data needed to pose the problem make up a commutative diagram

(0.1)

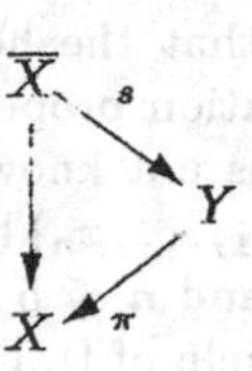

of affine schemes, such that Y is finitely presented over X. Our standard notation is that X, $\overline{X}$, Y are the spectra of A, $\overline{A}$, B respectively, and that B is a finitely presented A-algebra. (In the body of the text, we work primarily with the rings rather than with their spectra. This reverses the arrows.) The problem is to embed the commutative diagram (0.1) into a larger one,

(0.2)

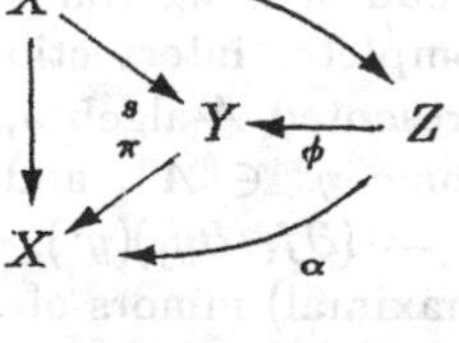

such that

(i) α is smooth, and

(ii) ϕ is smooth wherever possible — roughly speaking, except above the singular (nonsmooth) locus of π.

It is probable that a diagram (0.2) exists under fairly general conditions, perhaps whenever the map $\overline{X} \to X$ is regular. Special cases have been known for some time (Néron [13], Elkik [9], Pfister [11]). We prove its existence for normal X in two special cases: when $X = \overline{X}$ (section 2), and when A, $\overline{A}$ are henselian local rings with the same completion, and π is smooth at every point of $s(\overline{X})$ except the closed point — the isolated singularity case (section 3).

Section 4 contains a version of Néron's p-desingularization [13], and in section 5 we apply the results to henselian local rings of dimension 2.

The subject of this paper was motivated by the following observation. Let A, $\mathfrak{a}$ be a henselian pair, $\hat{A}$ the completion of A, $X = \operatorname{Spec} A$, and $\overline{X} = \operatorname{Spec} \hat{A}$. Suppose that every diagram (0.1) can be embedded in a diagram (0.2) satisfying (i) and (ii). Then A has the approximation property, i.e., every system of polynomial equations over A with a solution in $\hat{A}$ has a solution in A. It is known [2] that the henselization of $R[x_1, \ldots, x_n]$ at $(\mathfrak{p}, x_1, \ldots, x_n)$ has the approximation property if R is an excellent discrete valuation ring with prime $\mathfrak{p}$. It is not known whether the henselization of $k[[x_1, \ldots, x_r]][x_{r+1}, \ldots, x_n]$ at $(x_1, \ldots, x_n)$ has the approximation property, though this is true when $r = 2$ and $n \leq 5$ [14, 4].

We want to acknowledge the help of D. Popescu. Part of the work in this paper was done in collaboration with him, and is announced in [3, see also 17]. We would also like to thank D. Eisenbud for explaining the resolution (1.11) to us.

1. The Complete Intersection Case

This section refines a method of Tougeron [18, 2], which can be applied when B is a relative complete intersection over A. As always, $B = A[y]/(f)$ is a finitely presented A-algebra, with $y = (y_1, \ldots, y_n)$ and $f = (f_1, \ldots, f_r)^t$. We fix some $y^\circ \in A^n$, and suppose $r \leq n$. Let the Jacobian matrix at y° be $J = (\partial f_i / \partial y_j)(y^\circ)$, and let D denote the ideal generated by the r-rowed (maximal) minors of J.

Theorem (1.1). *Assume that* $f(y^\circ) \equiv 0$ (modulo $D^2\mathfrak{a}$), *where* $\mathfrak{a}$ *is some ideal of* A. *Thus* $y = y^\circ$ *defines an* A*-homomorphism*

$$s^\circ : B \longrightarrow A/D^2\mathfrak{a}.$$

There is a finitely presented B-algebra C and an A-homomorphism $\sigma^\circ: C \longrightarrow A/\mathfrak{a}$:

$$A \xrightarrow{\pi} B \xrightarrow{\phi} C \xrightarrow{\sigma^\circ} A/\mathfrak{a}, \tag{1.2}$$

such that

(i) $\phi\pi = \alpha$ *is smooth.*
(ii) *Let* $\mathfrak{p} \in \operatorname{Spec}(A/\mathfrak{a}) - V(D)$. *Then* ϕ *is smooth at* $\sigma^{\circ -1}(\mathfrak{p}) \in \operatorname{Spec} C$.
(iii) $\sigma^\circ\phi \equiv s^\circ$ (modulo $\mathfrak{a}$).

Note that if the pair $(A, \mathfrak{a})$ is henselian, the homomorphism σ° can be extended to a section $\sigma: C \longrightarrow A$.

Proof of Theorem (1.1). We first consider the homogeneous linear case, with $y^\circ = 0$, $\mathfrak{a} = 0$, and with $f(y)$ replaced by

$$f'(y) = Jy. \tag{1.3}$$

To avoid confusion, we denote the ring defined by these equations by B'. In this case the scheme $Y' = \operatorname{Spec} B'$ represents, functorially, the kernel of the linear map defined by J, i.e., there is an exact sequence of group schemes over $X = \operatorname{Spec} A$:

$$0 \to Y' \to \mathbb{G}^n_{aX} \xrightarrow{J} \mathbb{G}^r_{aX}. \tag{1.4}$$

An obvious choice of smoothing of Y' can be obtained by mapping a free module to $\mathcal{Y}' = \ker(A^n \xrightarrow{J} A^r)$:

$$A^N \xrightarrow{P} A^n \xrightarrow{J} A^r. \tag{1.5}$$

If $JP = 0$, the corresponding map

$$\mathbb{G}^N_{aX} \xrightarrow{P} \mathbb{G}^n_{aX} \tag{1.6}$$

will send $\mathbb{G}^N_{aX}$ to Y'. So we may take $C = A[z_1, \dots, z_N]$, and $\operatorname{Spec} C = \mathbb{G}^N_{aX}$. Conditions (i), (iii) of the theorem are trivially verified.

Lemma (1.7). *With the above notation, condition (ii) of theorem (1.1) holds if the sequence (1.5) is exact at all points $\mathfrak{p}$ of X at which J has maximal rank, i.e., $\mathfrak{p} \notin V(D)$.*

Proof of Lemma (1.7). Condition (ii) only concerns points of $Z = \operatorname{Spec} C$ lying over such $\mathfrak{p}$. So, by localizing, we may assume that J has maximal rank. Then the maps (1.4), (1.5) determined by J are surjective, $\mathcal{Y}'$ is a projective A-module, and π is smooth. Clearly $\mathbb{G}^N_{aX} \xrightarrow{P} Y'$ will be smooth if and only if $A^N \xrightarrow{P} \mathcal{Y}'$ is surjective. Namely, if P is surjective then it splits because $\mathcal{Y}'$ is projective, and so $A^N \approx Z' \oplus \mathcal{Y}'$ with Z' projective. Similarly, $\mathbb{G}^N_{aX} \approx Z' \times_X Y'$. Since Z' is projective, Z' is smooth over X, hence P is also smooth. Conversely, if $\mathbb{G}^N_{aX} \to \mathcal{Y}'$ is smooth, one proves surjectivity by the Nakayama Lemma.

It follows from this lemma that we can get a solution in the homogeneous linear case by choosing P so that the sequence (1.5) is exact. However, we prefer to make a canonical choice, one which corresponds to a resolution of the A-module $\mathcal{M} = A^r/J$ when J is a generic matrix. The resolution in that case is treated in [8].

Let a_{ij} denote the entries of J. An r-rowed minor of J is determined by an r-multi-index $\rho = (j_1, \ldots, j_r)$, with $1 \le j_1 < \cdots < j_r \le n$. We denote this minor by d_ρ:

$$d_\rho = \det(a_{ij_\nu}), \quad 1 \le i,\ \nu \le r. \tag{1.8}$$

Thus D is the ideal generated by the set $\{d_\rho\}$. Let $d = (d_\rho)$ denote the row vector obtained by some ordering of the multi-indices ρ.

There are natural relations among $\{a_{ij}\}$ and $\{d_\rho\}$, as follows: Let σ denote an $(r+1)$-multi-index $\sigma = (j_1, \ldots, j_{r+1})$ with

$$1 \le j_1 < \cdots < j_{r+1} \le n,$$

and let σ_ν denote the r-multi-index obtained by deleting j_ν:

$$\sigma_\nu = (j_1, \ldots, \hat{j}_\nu, \ldots, j_{r+1}). \tag{1.9}$$

Expansion by minors leads to the relations

$$\sum_{\nu=1}^{r+1} (-1)^\nu\, a_{ij_\nu} d_{\sigma_\nu} = 0. \tag{1.10}$$

These relations determine the resolution for a generic matrix J, which can be written as

$$\Lambda^{r+1}A^n \xrightarrow{P} A^n \xrightarrow{J} A^r. \tag{1.11}$$

Here P is contraction by the vector $d^t \in \Lambda^r A^{n*}$. If $w = (w_\sigma)$ represents an element of $\mathcal{W} = \Lambda^{r+1}A^n$ and $y = (y_j)$ a vector in A^n, then P is given as

$$y_j = \sum_{\substack{\sigma, \\ j=j_\nu\in\sigma}} (-1)^\nu w_\sigma d_{\sigma_\nu}. \tag{1.12}$$

The relations (1.10) imply that this is a complex, and it is proved in [8] that the complex is exact when J is a generic matrix. Therefore, for generic J, it is split exact (i.e., J has a section) locally at any point at which J has maximal rank (is surjective), and by pull-back, it is also split exact at such a point for any J. By Lemma (1.7), the associated map (1.5) satisfies the requirements of Theorem (1.1).

For the nonlinear case, we will need some extra variables to absorb higher order terms, and so we embed $\mathcal{W}$ into $A^n \otimes \Lambda^r A^n$ by the rule $w \to v = (v_{j\rho})$, with

$$\begin{aligned} v_{j\rho} &= 0 \qquad \text{if } j \in \rho, \\ v_{j\rho} &= \pm w_\sigma \quad \text{if } j \notin \rho, \end{aligned} \tag{1.13}$$

where $\sigma = \{j\} \cup \rho$, arranged in increasing order, and where the sign is $(-1)^\nu$, if $j = j_\nu$ in σ. Thus $\mathcal{W}$ is defined in $A^n \otimes \Lambda^r A^n$ by the equations L_v:

$$\begin{aligned} &v_{j\rho} = 0 \quad \text{if } j \in \rho, \\ -v_{j_1\sigma_1} = \cdots &= (-1)^\nu v_{j_\nu\sigma_\nu} = \cdots = (-1)^{r+1} v_{j_{r+1}\sigma_{r+1}}, \end{aligned} \tag{1.14}$$

for all $(r+1)$-multi-indices σ. The map P is induced by

$$\begin{aligned} A^n \otimes \Lambda^r A^n &\xrightarrow{P'} A^n, \\ y_j &= \sum_\rho v_{j\rho} d_\rho, \text{ or} \\ y &= v d^t. \end{aligned} \tag{1.15}$$

With this notation, the ring required by Theorem (1.1) is presented as $C_v = A[v]/(L_v)$, where L_v is the system (1.14) of linear equations, and the map ϕ': $B' \longrightarrow C_v$ is defined by the substitution (1.15). The linear equation has thus become

$$Jy = Jvd^t = 0, \tag{1.16}$$

which is implied by the system L_v (1.14).

If one wants to avoid the reference to [8], one can check the requirements of Lemma (1.7) for the system L_v directly, as follows: Suppose that some minor $d = d_{\rho_0}$ is invertible, say $\rho^o = (1, \ldots, r)$. Then equations (1.3) can be used to establish an isomorphism of $\mathcal{Y}'$ with the free module with coordinates $\{y_{r+1}, \ldots, y_n\}$. Let $\mathcal{Y}''$ be a free module with coordinates $\{y'_{r+1}, \ldots, y'_n\}$, and define a homomorphism S': $\mathcal{Y}'' \longrightarrow \mathcal{W}$ by

$$v_{j\rho} = \begin{cases} d^{-1} y'_j, & \text{if } \rho = \rho^o, \quad j > r \\ 0, & \text{if } \rho \neq \rho^o, \quad j > r. \end{cases} \tag{1.17}$$

The relations L_v (1.14) determine $v_{j\rho}$ uniquely in terms of y'_j, if $j \leq r$. The map $P'S'$: $\mathcal{Y}'' \longrightarrow \mathcal{Y}'$ is

$$y_j = \begin{cases} \sum_\rho v_{j\rho} d_\rho = y'_j, & \text{if } j > r \\ \quad * \quad , & \text{if } j \leq r. \end{cases} \tag{1.18}$$

Therefore $P'S'$ is surjective, hence P' is surjective, as required. This completes our discussion of the linear case.

Now consider the general case. Since $f(y^o) \in D^2\mathfrak{a}$, we may write

$$f_\nu(y^o) = \sum_{\rho', \rho} d_{\rho'} d_\rho \epsilon_{\rho' \rho \nu} \tag{1.19}$$

for some elements $\epsilon_{\rho'\rho\nu}$ of $\mathfrak{a}$. Substitute

$$\begin{aligned} y_j &= y^o_j + \sum_\rho u_{j\rho} d_\rho, \text{ or} \\ y &= y^o + ud^t, \end{aligned} \tag{1.20}$$

where $u = (u_{j\rho})$ and $d = (d_\rho)$. Taylor's expansion gives us

$$\begin{aligned} f(y) &= f(y^o) + Jud^t + \sum_{\rho', \rho} d_{\rho'} d_\rho Q_{\rho'\rho} \\ &= Jud^t + \sum_{\rho', \rho} d_{\rho'} d_\rho (\epsilon_{\rho'\rho} + Q_{\rho'\rho}), \end{aligned} \tag{1.21}$$

where $Q_{\rho'\rho\nu}$ are polynomials in $\{u_{j\rho}\}$ having no term of degree < 2.

Since d_ρ is a minor of J, there is a matrix N_ρ with

$$d_\rho I = J N_\rho \quad (I \text{ the } r \times r \text{ identity}). \tag{1.22}$$

This allows us to factor out J on the left from the right-hand term of (1.21), using the index ρ'. We can also collect the terms d_ρ into a factor d^t on the right.

$$f(y) = J u d^t + J\Big(\sum_{\rho'} N_{\rho'}(\epsilon_{\rho'} + Q_{\rho'})\Big) d^t, \tag{1.23}$$

where $\epsilon_{\rho'}$ is the matrix $(\epsilon_{\rho'\rho\nu})_{\nu,\rho}$, etc. Let

$$q = \sum_{\rho'} N_{\rho'}(\epsilon_{\rho'} + Q_{\rho'}). \tag{1.24}$$

This is a matrix whose row and column indices are (j, ρ). Thus

$$f(y) = J(u + q)d^t = J v d^t, \tag{1.25}$$

where $v = (v_{j\rho})$, and

$$v_{j\rho} = u_{j\rho} + q_{j\rho}, \quad \text{or} \quad v = u + q. \tag{1.26}$$

In this way the original equation is related to the homogeneous linear system (1.3).

Let C_u denote the quotient of $A[u]$ by the ideal defined by the linear relations (1.14) with $v = u + q$ (1.26):

$$C_u = A[u]/(L_{u+q}). \tag{1.27}$$

Then C_u is a B-algebra, via (1.20), because of (1.16) and (1.25). The substitution $u = 0$ implies $v = u + q \equiv 0(\text{modulo}\, \mathfrak{a})$, by (1.24) and (1.19), hence determines an A-homomorphism σ_u: $C_u \longrightarrow A/\mathfrak{a}$. Note that the Jacobian matrix of the system (1.26) is congruent to the identity $(\text{modulo}(u, \mathfrak{a}))$. Let γ denote the Jacobian of this system. Let

$$C = C_u[\gamma^{-1}]. \tag{1.28}$$

The homomorphism σ_u extends to give the map $\sigma^0\colon C \longrightarrow A/\mathfrak{a}$ required by (1.1)(iii). Also, C is étale over $C_v = A[v]/(L_v)$, and C_v is smooth over A, hence C is smooth over A, as required by (1.1)(i). Finally, the tangent map (relative to $\operatorname{Spec} A$) of $\operatorname{Spec} C \xrightarrow{\phi} \operatorname{Spec} B$ at $\sigma^{0^{-1}}(\mathfrak{p})$ is the same as the tangent map of $\operatorname{Spec} C_v \xrightarrow{\phi'} \operatorname{Spec} B'$ at $\sigma^{0'^{-1}}(\mathfrak{p})$, for all $\mathfrak{p} \in \operatorname{Spec}(A/\mathfrak{a})$. Thus if the second map is surjective, so is the first, and so condition (1.1)(ii) carries over from ϕ' to ϕ.

2. Smoothing Along a Section

The data we consider here are a finitely presented A-algebra $B = A[y]/(f)$, $(f) = (f_1, \ldots, f_N)^t$, and a section s:

(2.1)
$$\begin{array}{c} B \\ s \downarrow \uparrow \pi \\ A \end{array} \qquad s\pi = id_A,$$

and we denote $X = \operatorname{Spec} A$, $Y = \operatorname{Spec} B$. Let $H = H_{B/A}$ be the ideal of B generated by elements of the form cd, where for some integer r, d has the form

(2.2)
$$d = \det(\partial g_i/\partial y_{j_\nu}), \quad 1 \le i, \nu \le r, \quad g_i \in (f),$$

and where $c \in (g){:}(f)$, i.e., $c(f) \subset (g)$. Thus $V(H)$ is the singular (non-smooth) locus of π in Y. Also, let sH denote the ideal of A induced by H via the section s. If A is not noetherian, the ideal H need not be finitely generated; in fact the singular locus of π is not always the set of zeros of a finitely generated ideal. An auxiliary, finitely generated ideal $I \subset sH$ is introduced in the following theorem because of this.

Theorem (2.3). *Let A be a normal domain, not necessarily noetherian, and suppose given a diagram (2.1). Let $I \subset sH$ be a finitely generated ideal of A. There is a finitely presented B-algebra C and a homomorphism $\sigma\colon C \longrightarrow A$ such that the diagram*

(2.4)

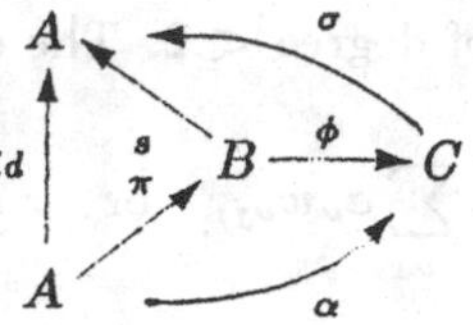

commutes, and that moreover

(i) *α is smooth, and*

(ii) *ϕ is smooth except on $V(\alpha I)$ in $Z = \operatorname{Spec} C$.*

The data (2.1), $I \subset s\mathfrak{H}$ of this theorem descend to some normal subring A_0 of A which is finitely generated over $\mathbb{Z}$, and it suffices to prove the theorem for the descended data — the required diagram (2.4) for A is then obtained by tensor product $A \otimes_{A_0} \cdot$. We may therefore assume that A is noetherian, which we do from now on.

Lemma (2.5). *To prove Theorem (2.3), it is permissible to replace (ii) by the weaker condition*

(ii′) *Let $\mathfrak{p} \in X$. If $\mathfrak{p} \notin V(I)$, then ϕ is smooth at $\sigma^{-1}\mathfrak{p} \in Z$.*

Proof of Lemma (2.5). Assume given a diagram (2.4) satisfying (i) and (ii′). Say that C is presented in the form $A[z]/J$, and let C' be the symmetric algebra of $P = J/J^2$ over C, i.e., $C' = C \oplus P \oplus P^2 \oplus \cdots$. By Elkik's Lemma 3 [9], there is an embedding of $\operatorname{Spec} C'$ into affine space over A whose conormal bundle is trivial. We may replace C by C'. Then since the conormal bundle is trivial, C admits a presentation of the form

$$C = \left(A[z_1, \ldots, z_m]/(g_1, \ldots, g_r)\right)[p^{-1}], \tag{2.6}$$

where g_i, $p \in A[z]$ and where C is of constant relative dimension $m - r$ over A. Let $\mathfrak{a} = (a_1, \ldots, a_l)$ be an ideal of C such that $V(\mathfrak{a})$ is the singular locus of ϕ. Then $\sigma\mathfrak{a}$ contains some power of I. We may also assume that the section σ is defined by $z = 0$.

Write

$$g_i(z) = \sum_j b_{ij} z_j + G_i(z), \tag{2.7}$$

where $G_i(z)$ has no terms of degree < 2. The change of coordinates

$$(2.8) \qquad z_j = \sum_{\nu} a_\nu w_{\nu j}, \quad \text{or} \quad z = aw,$$

gives us

$$g_i(aw) = \sum_{\nu,j} a_\nu b_{ij} w_{\nu j} + G_i(aw).$$

Collecting terms suitably, we may write

$$(2.9) \qquad \begin{aligned} g_i(aw) &= \sum_{\nu} a_\nu h_{i\nu}(w), \\ h_{i\nu}(w) &= \sum_{j} b_{ij} w_{\nu j} + H_{i\nu}(w), \end{aligned}$$

where $H_{i\nu}(w)$ has no terms of degree < 2, and does not involve variables $w_{\lambda j}$ with $\lambda > \nu$. Define

$$(2.10) \qquad R = C[w]/K,$$

where K is the ideal generated by $\{h_{i\nu}\}$ and by the linear relations (2.8), say

$$(2.11) \qquad d_j(z, w) = z_j - \sum_{\nu} a_\nu w_{\nu j}.$$

Let $C \xrightarrow{\tau} R$ denote the canonical map. We claim that if we replace C by a suitable localization of R, and α, ϕ by $\tau\alpha$, $\tau\phi$, we will obtain a diagram (2.4) satisfying (i), (ii). To show this, it suffices to verify (i), (ii) for R in a neighborhood of the zero section $z = w = 0$.

As A-algebra, R is obtained by inverting p (2.6) in $A[z, w]/(g, h, d)$, where the elements a_ν are represented by polynomials in $A[z]$ in the computations of h, d. By (2.8), (2.9), the elements $g_i(z)$ are not needed to generate the ideal (g, h, d), so $R = A[z, w]/(h, d)[p^{-1}]$. The Jacobian matrix of this presentation at $z = w = 0$ is

$$(2.12) \qquad \frac{\partial(h,d)}{\partial(z,w)}(0) = \begin{array}{c|cccc|} & z. & w_1. & \cdots & w_l. \\ \hline h._1 & 0 & J & & 0 \\ \vdots & \vdots & & \ddots & \\ h._l & 0 & 0 & & J \\ d. & I & -a_1^\circ I & \cdots & -a_l^\circ I \\ \hline \end{array}$$

where $J = (b_{ij}) = \partial g/\partial z(0)$, I is the $m \times m$ identity, and $a_\nu^\circ = a_\nu(0)$. Since J has maximal rank r, the above matrix has maximal rank $lr + m$. Thus R is smooth over A along the zero section, as required by (i).

To verify (ii), consider the Jacobian of the presentation (2.10):

$$(2.13) \qquad \frac{\partial(h,d)}{\partial w} = \begin{array}{|cccc|} \hline h'_1 & & & 0 \\ & \ddots & & \\ & * & \ddots & \\ & & & h'_l \\ -a_1 I & \cdots & \cdots & -a_l I \\ \hline \end{array}$$

where

$$h'_\nu = \left(\frac{\partial h._\nu}{\partial w_\nu.}\right).$$

Since $h'_\nu(0) = J$ and J has maximal rank, there is a neighborhood U of the zero section in Spec R such that each h'_ν has maximal rank on U. Now if $\mathfrak{p} \in U$ and some a_ν is not zero at $\mathfrak{p}$, the rank of the matrix (2.13) is at least $(l-1)r + m$. Since the relative dimensions of C/A and R/A are $(m-r)$ and $(m-r)l$ respectively, τ is smooth at $\mathfrak{p}$, of relative dimension $(m-r)(l-1)$. Also ϕ is smooth at $\tau^{-1}\mathfrak{p}$ by hypothesis on $\mathfrak{a}$. Thus $\tau\phi$ is smooth at $\mathfrak{p}$. If on the other hand $\mathfrak{p} \in U$ and $a_\nu = 0$ at $\mathfrak{p}$ for all ν, then equations (2.8) imply $z = 0$ at $\tau^{-1}\mathfrak{p}$, hence that $\tau^{-1}\mathfrak{p}$ lies on the section $\sigma^{-1}X \subset Z$. Since $\sigma\mathfrak{a}$ contains a power of I, it follows that $(\tau\alpha)^{-1}\mathfrak{p} \in V(I)$, as required by (ii). This completes the proof of Lemma (2.5).

Proof of Theorem (2.3). We may replace I be a larger finitely generated ideal contained in sH, hence we may assume $I = s(c_1d_1, \ldots, c_md_m)$, where

c_i, d_i are as above (2.2). If $I = (0)$, we take $m = 1$ and $c_1 = d_1 = 0$. The proof is by induction on m.

Lemma (2.14). *The theorem is true if* $m = 1$.

Proof. Say that $I = s(cd)$. Let B_1 be the ring $A[y]/(g)$, where g is in (2.2). We apply Theorem (1.1) and Lemma (2.5), with B_1 replacing B and $\mathfrak{a} = 0$, to find a diagram as required. Note that Y is a closed subscheme of $Y_1 = \operatorname{Spec} B_1$. Since A is a normal domain and C is smooth over A, C is also normal. We can take C to be a domain as well.

Let $K = \ker(B_1 \to B)$. By definition, c annihilates K. If c is not zero in C then, since C is a domain, the image of K in C is zero, and $B_1 \to C$ factors through B. Replacing B_1 by B gives us the required diagram. If c maps to zero in C, then c also maps to zero via the section $B_1 \to A$, because this section extends to C. In that case, condition (ii) of the theorem becomes vacuous, and the required diagram is obtained by taking $C = A$.

Now for the induction step, we assume the theorem true for $I_{m-1} = s(c_1 d_1, \ldots, c_{m-1} d_{m-1})$, and we have to prove it for $I_m = s(c_1 d_1, \ldots, c_m d_m)$, where $m > 1$. We relabel the triple (C, σ, ϕ) already constructed for I_{m-1} as (A', σ, ϕ'):

(2.15)

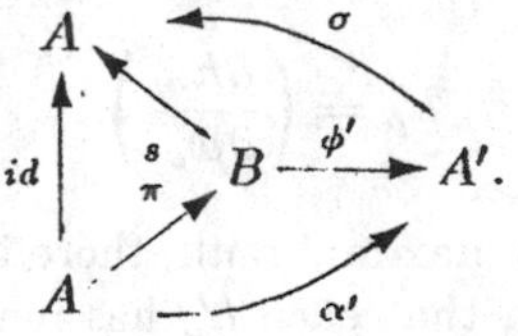

Thus α' is smooth, and ϕ' is smooth except on $V(\alpha' I_{m-1}) \subset X' = \operatorname{Spec} A'$. Also, we consider the ring $B' = A' \otimes_A B$:

(2.16)

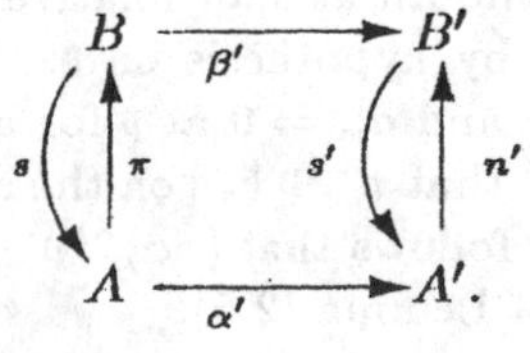

Finally, let t' denote the map

(2.17) $$t' = (id_{A'}, \phi') : A' \otimes_A B =: B' \to A'.$$

It is a section for π'.

Now we apply Lemma (2.14) to the map π' and section t', with the ideal $I' \subset A'$ generated by the single element $\phi(c_m d_m)$, to obtain a diagram which we label as

(2.18)

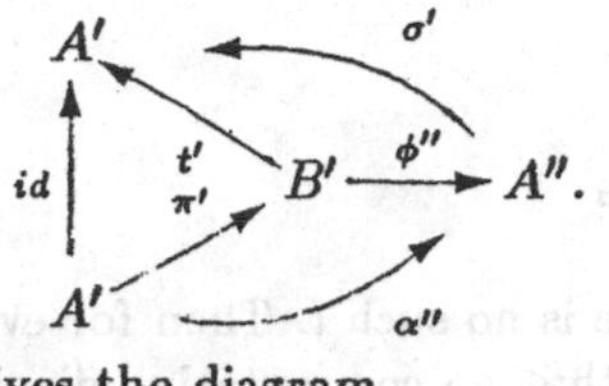

Composing with (2.16) gives the diagram

(2.19)

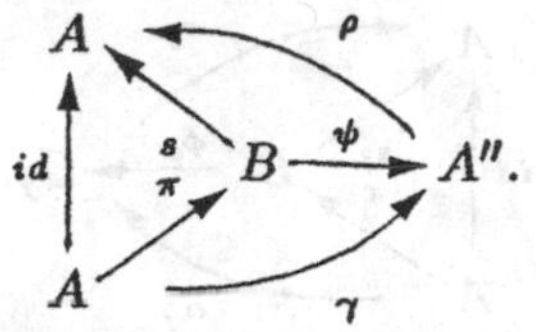

where $\rho = \sigma\sigma'$, $\psi = \phi''\beta'$, and $\gamma = \psi\pi$. We claim that this diagram commutes and has properties (i), (ii') for the ideal I_m. This will prove the theorem, by Lemma (2.5).

The diagram commutes because $\rho\psi = \sigma\sigma'\phi''\beta' = \sigma t'\beta' = \sigma\phi' = s$. Since α' and α'' are smooth, and $\alpha''\alpha' = \phi''\pi'\alpha' = \phi''\beta'\pi = \gamma$, it follows that γ is smooth. Finally, to verify (ii'), let $\mathfrak{p} \in X$ be a point not in $V(I_m)$. If $s(c_m d_m) \neq 0$ at $\mathfrak{p}$, then by construction ϕ'' is smooth at $\rho^{-1}\mathfrak{p}$. Also, β' is smooth because α' is (2.15). Therefore $\psi = \phi''\beta'$ is smooth at $\rho^{-1}\mathfrak{p}$, as required.

Suppose $\mathfrak{p} \notin V(I_m)$, but $s(c_m d_m) = 0$ at $\mathfrak{p}$. Then $\mathfrak{p} \notin V(I_{m-1})$. We need to show that ψ is smooth at $\rho^{-1}\mathfrak{p}$ in this case as well. By Lemma (2.20) below, it suffices to show that $\sigma'\psi$ is smooth at $\sigma^{-1}\mathfrak{p}$. We have $\sigma'\psi = \sigma'\phi''\beta' = t'\beta' = \phi'$, and ϕ' is smooth at $\sigma^{-1}\mathfrak{p}$ by construction. This completes the proof of the induction step and of Theorem (2.3).

Lemma (2.20). *Let $\phi' : U' \to V$ be a map of schemes over a scheme X, and let U be a closed subscheme of U'. Assume that U', U, V are all smooth over X. If the restriction ϕ of ϕ' to U is smooth at a point $\mathfrak{p} \in U$, then ϕ' is smooth at $\mathfrak{p}$.*

Proof. Since all the schemes are smooth over X, smoothness of ϕ' at $\mathfrak{p}$ is equivalent with surjectivity of the induced map of relative tangent spaces [10]. This surjectivity obviously carries over from U to U'.

Theorem (2.21). *(a uniform version of (2.3)). Let A be a noetherian normal domain, and let B be a finitely presented A-algebra. There is an integer l with the following property: For every section $s : B \to A$ there exists a commutative diagram (2.4) such that*

(i) α *is smooth*

(ii) $\alpha s H^l_{B/A} \subset H_{C/B}$.

Proof. Suppose there is no such l. Then for every $j \in \mathbb{N}$, there exists a section $s_j\colon B \longrightarrow A$ such that no commutative diagram

(2.22) 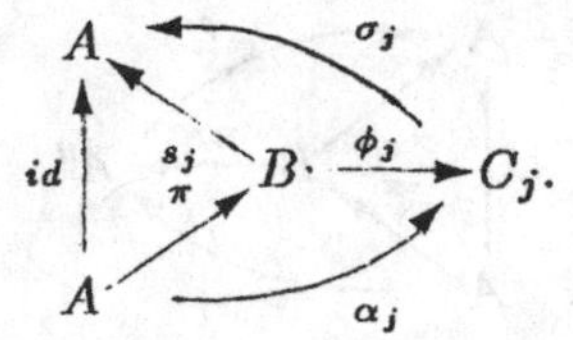

with α_j smooth and $\alpha_j s_j (H_{B/A})^j \subset H_{C_j/B}$ exists. We will obtain a contradiction by proving that for infinitely many of these s_j there nevertheless exists such a diagram. For this we use the ultraproduct construction (see e.g. [7] or [5]).

Let D be a nonprincipal ultrafilter on $\mathbb{N}$, and let

$$A^* = \Big(\prod_{j\in\mathbb{N}} A\Big)/D$$

be the ultrapower of A with respect to D. Then A^* is a normal domain, not necessarily noetherian. The sequence of maps, $s_j\colon B \longrightarrow A$, defines a homomorphism

$$s^*\colon B \longrightarrow A^*,$$

which sends $b \in B$ to the equivalence class in A^* of the sequence $\big(s_j(b)\big)_{j\in\mathbb{N}}$. Let $\overline{B} = B \otimes_A A^*$, and let $\overline{s}$ denote the map

$$\overline{s} = (s^*, id_{A^*})\colon \quad B \otimes_A A^* := \overline{B} \to A^*.$$

Thus we have a commutative diagram

(2.23) 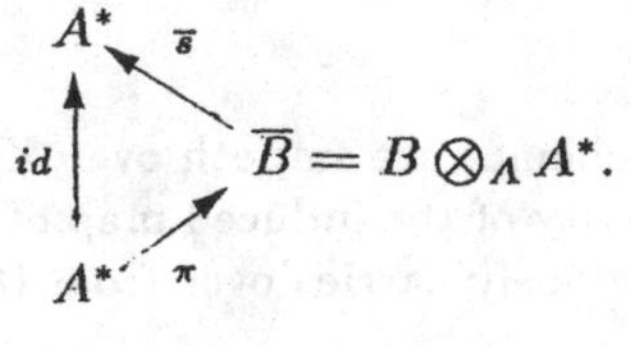

Let I be the ideal of A^* generated by $s^*(H_{B/A})$. The ideal I is finitely generated, since B is noetherian. By base change, we have $I \subset \overline{s}(H_{\overline{B}/A^*})$.

Now we apply Theorem (2.3) to the diagram (2.23) and the ideal I, to obtain a commutative diagram which we label as

(2.24)

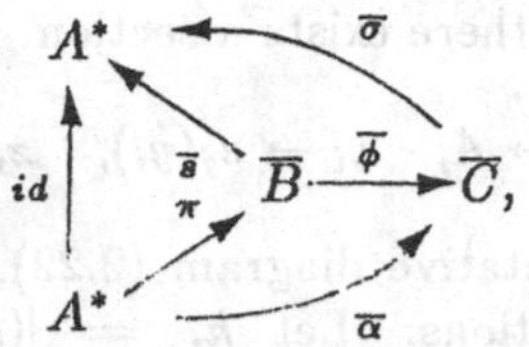

with $\overline{\alpha}$ smooth, and $\overline{\phi}$ smooth except on $V(\overline{\alpha}I)$. Thus there exists an $e \in \mathbb{N}$ such that $\overline{\alpha}I^e \subset H_{\overline{C}/\overline{B}}$. Combining this with the definition of I, we get

(2.25) $$\overline{\alpha}s^*(H_{B/A})^e \subset H_{\overline{C}/\overline{B}}.$$

We will now proceed to construct from (2.24), for infinitely many j, a commutative diagram (2.22) with α_j smooth and

(2.26) $$\alpha_j s_j(H_{B/A})^e \subset H_{C_j/B}.$$

This will establish the desired contradiction.

First we need some notation. If $(a_j)_{j\in\mathbb{N}}$ is a sequence of elements of A, then we will denote the equivalence class in A^* of this sequence by $[(a_j)_{j\in\mathbb{N}}]$. Let $(g_j)_{j\in\mathbb{N}}$ be a sequence of polynomials of $A[y,z]$, of *bounded* degree. Such a sequence defines in a natural way a polynomial over A^*, which we will denote by $[(g_j)_{j\in\mathbb{N}}]$. The coefficients of this polynomial are the equivalence classes in A^* of the sequences of coefficients of the g_j. When we use this notation, we will always suppose without mention that the g_j, $j = 0,1,2,\ldots$, have bounded degree. We have the presentations $B = A[y]/(f)$ and $\overline{C} = A^*[y,z]/(f,g)$, with $z = (z_1,\ldots,z_m)$ and $g = (g_1,\ldots,g_M)$. Write

$$g_i = [(g_{ij})_{j\in\mathbb{N}}], \quad i = 1,\ldots,M,$$

with $g_{ij} \in A[y,z]$, and write

$$\overline{\sigma}(z_i) = [(a_{ij})_{j\in\mathbb{N}}], \quad i = 1,\ldots,m,$$

with $a_{ij} \in A$. For all $j \in \mathbb{N}$, define

$$C_j = A[y,z]/(f, g_{1j}, g_{2j}, \ldots, g_{Mj}).$$

For almost all $j \in \mathbb{N}$ (almost is meant with respect to the ultrafilter D; thus for all j in some set $S \in D$) we have

$$g_{ij}(s_j(y_1), \ldots, s_j(y_n), a_{1j}, \ldots, a_{mj}) = 0.$$

Thus for almost all $j \in \mathbb{N}$ there exists a section

$$\sigma_j \colon C_j \to A, \quad y_i \to s_j(y_i), \quad z_i \to z_{ij},$$

and there exists a commutative diagram (2.22). To finish the proof, we first make some observations: Let $h_i = [(h_{ij})_{j \in \mathbb{N}}] \in A^*[y, z]$, for $i = 1, 2, \ldots, r$, and $h = [(h_j)_{j \in \mathbb{N}}] \in A^*[y, z]$. Let $\theta_j \colon A[y, z] \longrightarrow C_j$ and $\overline{\theta} \colon A^*[y, z] \to \overline{C}$ be the natural projections.

(2.27) (i) *If* $h \in (h_1, \ldots, h_r)A^*[y, z]$, *then* $h_j \in (h_{1j}, \ldots, h_{rj})A[y, z]$, *for almost all* j.
(ii) *If* $\overline{\theta}(h) \in H_{\overline{C}/A^*}$, *then* $\theta_j(h_j) \in H_{C_j/A}$, *for almost all* j.
(iii) *If* $\overline{\theta}(h) \in H_{\overline{C}/\overline{B}}$, *then* $\theta_j(h_j) \in H_{C_j/B}$, *for almost all* j.

Since $\overline{\alpha}$ is smooth, $1 \in H_{\overline{C}/A^*}$, and observation (ii) implies that α_j is smooth for almost all j. Finally from (2.25), observation (iii), and the fact that $H_{B/A}$ is finitely generated, (2.26) follows. This completes the proof of Theorem (2.21).

3. Smoothing an Isolated Singularity

In this section, A denotes a noetherian normal, local ring, with maximal ideal $\mathfrak{m}$ and residue field $k = A/\mathfrak{m}$. Let $\overline{A}$ be a noetherian, henselian local, flat A-algebra with maximal ideal $\overline{\mathfrak{m}} = \mathfrak{m}A$ and with the same residue field $k = \overline{A}/\overline{\mathfrak{m}}$. Thus the completions of A and $\overline{A}$ are isomorphic. Let $B = A[y]/(f)$ be a finitely presented A-algebra with a section into $\overline{A}$, i.e., a commutative diagram

(3.1)

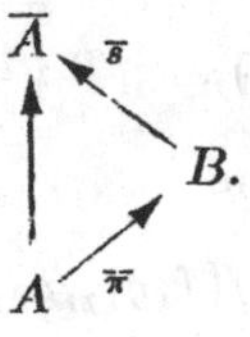

We denote the spectra of $A, \overline{A}, B$ by $X, \overline{X}, Y$, respectively, and the pointed spectra by $X^{\cdot} = X - \{\mathfrak{m}\}$, $\overline{X}^{\cdot} = \overline{X} - \{\overline{\mathfrak{m}}\}$.

Theorem (3.2). *With the above notation, assume that π is smooth at every point of $\overline{s}^{-1}(\overline{X}^{\cdot})$. There exists a finitely presented B-algebra C and a homomorphism $\overline{\sigma}$ such that the diagram*

(3.3)

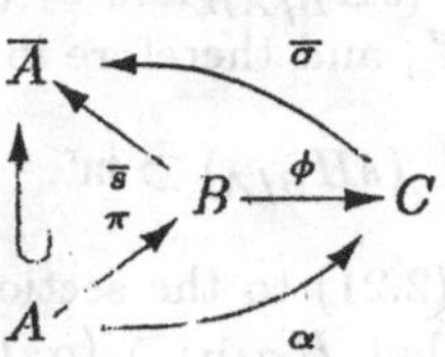

commutes, and that moreover

(i) *α is smooth,*

(ii) *ϕ is smooth except on $V(\alpha\mathfrak{m})$ in $Z = \operatorname{Spec} C$.*

Proof. We may replace A by its henselization A^h and B by $A^h \otimes_A B$, thereby reducing to the case that A is henselian.

Since π is smooth on $\overline{s}^{-1}(X^{\cdot})$, some power of $\overline{\mathfrak{m}}$ is contained in $(\overline{s}H_{B/A})$, say

(3.4)
$$(\overline{s}H_{B/A}) \supset \overline{\mathfrak{m}}^r,$$

where $H_{B/A}$ is as in section 2, and $(\overline{s}H_{B/A})$ denotes the ideal of $\overline{A}$ generated by its image in $\overline{A}$. Choose the integer l as in Theorem (2.21).

We apply Elkik's theorem [9, Thm. 2] to the ring $\overline{A}$, with $h = lr$: There are integers n_0, k so that for any integer $n \geq n_0$, any finitely presented $\overline{A}$-algebra $\overline{R}$, and any approximate section

$$\rho^{\circ}: \overline{R} \longrightarrow \overline{A}/\overline{\mathfrak{m}}^n, \text{ with}$$

$$\overline{\mathfrak{m}}^k/\overline{\mathfrak{m}}^n \subset \rho^{\circ}(H_{\overline{R}/\overline{A}}),$$

there exists a section

(3.5)
$$\overline{\rho}: \overline{R} \longrightarrow \overline{A}, \text{ with}$$
$$\overline{\rho} \equiv \rho^{\circ} \text{ (modulo } \overline{\mathfrak{m}}^{n-k}).$$

We may choose n so that $n \geq n_0$ and $n > r$.

It also follows from Elkik's theorem, since $\overline{A}/\overline{\mathfrak{m}}^n \approx A/\mathfrak{m}^n$, that the section $\overline{s}$ can be approximated by a section $s : B \to A$, i.e.,

$$s \equiv \overline{s} \quad (\text{modulo } \overline{\mathfrak{m}}^n). \tag{3.6}$$

Therefore $(sH_{B/A})\overline{A} + \overline{\mathfrak{m}}^n \supset (\overline{s}H_{B/A})$, and by (3.4), $(sH_{B/A})\overline{A} + \overline{\mathfrak{m}}^n \supset \overline{\mathfrak{m}}^r$. Since $n > r$, $(sH_{B/A})\overline{A} \supset \overline{\mathfrak{m}}^r$, and therefore in A,

$$(sH_{B/A}) \supset \mathfrak{m}^r. \tag{3.7}$$

We now apply Theorem (2.21) to the section s, to find a B-algebra C and a diagram (2.4), such that $H_{C/B} \supset (\alpha s H^l_{B/A})$. We claim that this algebra will work, and since conditions (i), (ii) are satisfied, it remains to find a homomorphism $\overline{\sigma} : C \to \overline{A}$ so that $\overline{\sigma}\phi = \overline{s}$. Since $\sigma\alpha = id_A$, it follows from the above inclusion that $(\sigma H_{C/B}) \supset (sH^l_{B/A})$, hence by (3.7) that

$$(\sigma H_{C/B}) \supset \mathfrak{m}^{lr}. \tag{3.8}$$

We consider $\overline{A}$ as B-algebra by the map $\overline{s}$. Let $\overline{R} = C \otimes_B \overline{A}$, and let the canonical maps be

$$\begin{array}{ccc} \overline{A} & \xrightarrow{\overline{\psi}} & \overline{R} \\ {\scriptstyle \overline{s}}\uparrow & & \uparrow{\scriptstyle \overline{t}} \\ B & \xrightarrow{\phi} & C. \end{array} \tag{3.9}$$

By base change,

$$H_{\overline{R}/\overline{A}} \supset (\overline{t}H_{C/B}). \tag{3.10}$$

Denote the projections from A, $\overline{A}$ to $A/\mathfrak{m}^n \approx \overline{A}/\overline{\mathfrak{m}}^n$ by ϵ_n, $\overline{\epsilon}_n$ respectively. Then $\overline{\epsilon}_n$ and $\epsilon_n\sigma$ determine a homomorphism $\rho^\circ : \overline{R} \to A/\mathfrak{m}^n$ such that $\rho^\circ\overline{\psi} = \overline{\epsilon}_n$ and $\rho^\circ\overline{t} = \epsilon_n\sigma$. By (3.10) and (3.8),

$$(\rho^\circ H_{\overline{R}/\overline{A}}) \supset (\rho^\circ\overline{t}H_{C/B}) = (\epsilon_n\sigma H_{C/B}) \supset (\epsilon_n\mathfrak{m}^{lr}) = \mathfrak{m}^{lr}/\mathfrak{m}^n.$$

Therefore Elkik's theorem applies, and there is a section $\overline{\rho}$ of $\overline{R}/\overline{A}$, as in (3.5). We define

$$\overline{\sigma} = \overline{\rho}\overline{t}. \tag{3.11}$$

The diagram (3.3) commutes because $\overline{\sigma}\phi = \overline{\rho}\overline{\iota}\phi = \overline{\rho}\overline{\psi}\overline{s} = \overline{s}$. This completes the proof.

4. Smoothing Singularities in Codimension 1

This section contains an extension of Néron's p-desingularization [13,2]. The first such extension was made by Pfister [11] for certain two-dimensional rings, and has been generalized in various ways by several authors, especially Popescu [16] and Brown [6].

Theorem (4.1). *Let $A \subset \overline{A}$ be normal noetherian domains. Assume that for every prime ideal $\overline{\mathfrak{p}}$ of $\overline{A}$ of height ≤ 1, the height of $\mathfrak{p} = \overline{\mathfrak{p}} \cap A$ is at most 1, and the map of local rings $A_\mathfrak{p} \to A_{\overline{\mathfrak{p}}}$ is a regular homomorphism [12, p.249]. Let a commutative diagram (3.1) be given, in which B is a finitely generated A-algebra. There exists a finitely generated B-algebra C and a commutative diagram (3.3), such that*

(i) *α is smooth at $\overline{\sigma}^{-1}(\overline{\mathfrak{p}})$ for every height 1 prime $\overline{\mathfrak{p}}$ of $\overline{A}$.*
(ii) *If $\overline{\mathfrak{p}}$ has height 1 and π is smooth at $\overline{s}^{-1}\overline{\mathfrak{p}}$, then ϕ is smooth at $\overline{\sigma}^{-1}\overline{\mathfrak{p}}$.*

Note that the assumption when $\overline{\mathfrak{p}} = (0)$ is just that the field extension $\text{Fract}(A) \subset \text{Fract}(\overline{A})$ is separable. If $\overline{\mathfrak{p}}$ and $\mathfrak{p}$ both have height 1, the assumption says that $A_\mathfrak{p} \subset \overline{A}_{\overline{\mathfrak{p}}}$ is *unramified* (i.e., $\mathfrak{p}\overline{A}_{\overline{\mathfrak{p}}} = \overline{\mathfrak{p}}$), and the residue field extension $\text{Fract}(A/\mathfrak{p}) \subset \text{Fract}(\overline{A}/\overline{\mathfrak{p}})$ is separable. It is harder to pin the assumption down when $\overline{\mathfrak{p}}$ has height 1 and $\mathfrak{p} = (0)$, but it is implied by the stronger condition that $\text{Fract}(A) \subset \text{Fract}(\overline{A}/\overline{\mathfrak{p}})$ is separable.

If π is not smooth at $\overline{s}^{-1}(0)$, then $\overline{s}^{-1}(0) \neq 0$ because $\text{Fract}(\overline{A})$ is separable over $\text{Fract}(A)$. In this case it is permissible to replace B by $B/\overline{s}^{-1}(0)$. So we may assume π smooth at $\overline{s}^{-1}(0)$. The theorem follows by induction from this lemma:

Lemma (4.2). *Let $A \subset \overline{A}$ be normal noetherian domains, let $\overline{\mathfrak{p}}$ be a fixed height 1 prime of $\overline{A}$, and let $\mathfrak{p} = \overline{\mathfrak{p}} \cap A$. Suppose that the homomorphism $A_\mathfrak{p} \to \overline{A}_{\overline{\mathfrak{p}}}$ is regular. Let a commutative diagram (3.1) be given, in which B is a finitely presented A-algebra, smooth over A at $\overline{s}^{-1}(0)$. There* *exists a finitely generated B-algebra C and a commutative diagram (3.3), such that*

(i) *α is smooth at $\overline{\sigma}^{-1}\overline{\mathfrak{p}}$,*
(ii) *ϕ is smooth at $\overline{\sigma}^{-1}\overline{\mathfrak{q}}$ for every height 1 prime $\overline{\mathfrak{q}}$ of $\overline{A}$ except $\overline{\mathfrak{p}}$.*

Proof. Assume first that $\overline{A} = \overline{A}_{\overline{\mathfrak{p}}}$, i.e., that $\overline{A}$ is a discrete valuation ring. Then the lemma follows from Néron's p-desingularization [13], which we will review briefly in order to establish notation. Say that B is presented as usual, in the form $A[y_1, \ldots, y_n]/(f_1, \ldots, f_r)$, and that the relative dimension of $Y = \operatorname{Spec} B$ over $X = \operatorname{Spec} A$ at $\overline{s}^{-1}(0)$ is d. Néron's measure of the singularity of Y over X at $\overline{s}^{-1}\overline{\mathfrak{p}}$ is

$$l = l(B/A, \overline{\mathfrak{p}}) = \inf_M v(\det \overline{s}(M)), \tag{4.3}$$

where M is a minor of $J = (\partial f/\partial y)$ of rank $n - d$, and v is the $\overline{\mathfrak{p}}$-adic valuation of $\overline{A}$ [13]. Thus $l = 0$ if and only if B is smooth over A at $\overline{\mathfrak{p}}$. Note that l is defined for any diagram (3.1) in which $\overline{A}_{\overline{\mathfrak{p}}}$ is a discrete valuation ring and B/A is smooth at $\overline{s}^{-1}(0)$. One need not assume the map $A \to \overline{A}_{\overline{\mathfrak{p}}}$ regular.

We omit the proof of the following proposition:

Proposition (4.4). *Given a smooth map $B \to B_1$ and an extension $\overline{s}_1 \colon B_1 \longrightarrow \overline{A}$ of $\overline{s}$, we have*

$$l(B/A, \overline{\mathfrak{p}}) = l(B_1/A, \overline{\mathfrak{p}}).$$

Let $Y' \to Y = \operatorname{Spec} B$ be the blowing up of the ideal $P = \overline{s}^{-1}\overline{\mathfrak{p}}$ of B. Let $b_1, \ldots, b_m$ be generators for the ideal P. Then Y' is covered by the affine open sets $\operatorname{Spec} B[b_1/b_i, \ldots, b_m/b_i]$, $i = 1, \ldots, m$. Since $\overline{A}$ is a discrete valuation ring, $\overline{s}$ extends to a homomorphism $\overline{s}' : B' \to \overline{A}$, where B' is the coordinate ring of a suitable affine open set of Y'. Notice that $l(B'/A, \overline{\mathfrak{p}})$ does not depend on the choice of B'.

Theorem (4.5). *Consider a commutative diagram (3.1), in which $\overline{A}$ is a discrete valuation ring, B is a finitely generated A-algebra, and A is a noetherian subring of $\overline{A}$. Let $\mathfrak{p} = \overline{\mathfrak{p}} \cap A$ and $P = \overline{s}^{-1}\overline{\mathfrak{p}}$. Assume that B/A is smooth at $\overline{s}^{-1}(0)$, and that one of the following conditions holds:*

(i) *height* $\mathfrak{p} = 1$, *the map* $A_{\mathfrak{p}} \to \overline{A}$ *is unramified, and the residue field extension* $\mathrm{Fract}(A/\mathfrak{p}) \subset \mathrm{Fract}(B/P)$ *is separable,*

(ii) $\mathfrak{p} = (0)$, *and the field extension* $\mathrm{Fract}(A) = k \subset \mathrm{Fract}(B/P)$ *is separable,*

(iii) $\mathfrak{p} = (0)$, *and the map* $\mathrm{Fract}(A) = k \to \overline{A}$ *is a regular homomorphism.*

Then $l(B'/A, \overline{\mathfrak{p}}) \leq l(B/A, \overline{\mathfrak{p}})$, *with equality if and only if* $l(B/A, \overline{\mathfrak{p}}) = 0$, *i.e., B is smooth over A at P.*

We will discuss the proof of this theorem at the end of the section. Lemma (4.2) follows from it by induction, when $\overline{A}$ is a discrete valuation ring.

If $\overline{A}$ is not a discrete valuation ring, we can still define the integer $l(B/A, \overline{\mathfrak{p}})$ by (4.3), and (4.4) holds. However, the homomorphism $\overline{s}$ may not extend to the blowing up, and so a substitute has to be found. Once this is done, it will be possible to apply Theorem (4.5) again, to complete the proof. We may assume $l(B/A, \overline{\mathfrak{p}}) > 0$.

Choose elements $\overline{p}, \overline{\pi} \in \overline{A}$ so that both are local parameters at $\overline{\mathfrak{p}}$, and that

(4.6) $\overline{\mathfrak{p}}$ is the only height 1 prime of $\overline{A}$ containing both $\overline{p}$ and $\overline{\pi}$.

We may also choose an element $\overline{a} \in \overline{A}$ such that

$$v_{\overline{\mathfrak{p}}}(\overline{a}) = 0, \quad \text{and} \quad v_{\overline{\mathfrak{q}}}(\overline{a}) \geq v_{\overline{\mathfrak{q}}}(\overline{p}), \tag{4.7}$$

if $v_{\overline{\mathfrak{q}}}$ is the valuation at any height 1 prime $\overline{\mathfrak{q}} \neq \overline{\mathfrak{p}}$. Since $\overline{A}$ is normal, it is the intersection of its localizations $\overline{A}_{\overline{\mathfrak{q}}}$ at height 1 primes $\overline{\mathfrak{q}}$. Therefore (4.7) shows that $\overline{p}$ divides $\overline{a}\,\overline{\pi}$, say

$$\overline{p}\,\overline{w} = \overline{a}\,\overline{\pi}.$$

Let a, π, p, w be variables, and let

$$B_0 = B[a, \pi, p, w]/(pw - a\pi). \tag{4.8}$$

Extend $\overline{s}$ to $\overline{s}_0 \colon B_0 \longrightarrow \overline{A}$, by $a \to \overline{a}$, etc ... Let $\overline{\mathfrak{q}}$ be a height 1 prime of $\overline{A}$. Then $\overline{\mathfrak{q}}$ does not contain $\{\overline{a}, \overline{\pi}, \overline{p}\}$. Therefore $Q_0 = \overline{s}_0^{-1}(\overline{\mathfrak{q}})$ does not contain

$\{a, \pi, p\}$, which implies, by the defining equation, that B_0 is smooth over B at Q_0.

Let $\{p, b_1, \ldots, b_m\}$ be generators for $P_0 = \bar{s}_0^{-1}\bar{\mathfrak{p}}$, and let $\bar{b}_i = \bar{s}_0(b_i)$. Define B_1 as the ring obtained by killing p-torsion in

$$(4.10) \qquad B_0[z_1, \ldots, z_m]/(pz_i - ab_i).$$

It follows from (4.7) that $\bar{p}$ divides $\bar{a}\bar{b}_i$ in $\bar{A}$, hence that $\bar{s}_0$ extends to $\bar{s}_1 : B_1 \to \bar{A}$, by $z_i \to \bar{a}\bar{b}_i/\bar{p}$.

Note that $B_1 \subset B_0[p^{-1}]$. Also, p is not a zero divisor in $B_0[\pi^{-1}]$. Using this, it follows from the presentation (4.10) that $B_1 \subset B_0[\pi^{-1}]$ too. Thus by (4.6), B_1 is smooth over B_0 at $\bar{s}_1^{-1}\bar{\mathfrak{q}}$, for every height 1 prime $\bar{\mathfrak{q}} \neq \bar{\mathfrak{p}}$ of $\bar{A}$, and so B_1 is smooth over B at $\bar{s}^{-1}\bar{\mathfrak{q}}$ too.

The ring B_1 is our substitute for Néron's blowing up. If we show that $l(B_1/A, \bar{\mathfrak{p}}) < l(B/A, \bar{\mathfrak{p}})$, then we will be done by induction on l. To show this, we may localize, replacing $A, \bar{A}$ by $A_{\mathfrak{p}}, \bar{A}_{\bar{\mathfrak{p}}}$ and B, B_0, B_1 by $A_{\mathfrak{p}} \otimes_A B$, $A_{\mathfrak{p}} \otimes_A B_0$, $A_{\mathfrak{p}} \otimes_A B_1$ respectively. Then $\bar{p}$, $\bar{\pi}$ are local parameters in $\bar{A}$, and $\bar{a}$ is a unit there. So, $\bar{s}_0$ extends to $B_0[a^{-1}] = B[a, p, w, a^{-1}]$. By Proposition (4.4),

$$(4.11) \qquad l(B/A, \bar{\mathfrak{p}}) = l(B_0[a^{-1}]/A, \bar{\mathfrak{p}}) = l(B_0/A, \bar{\mathfrak{p}}).$$

Note that the elements $\{p, ab_1, \ldots, ab_m\}$ generate $\bar{s}_0^{-1}\bar{\mathfrak{p}}$ in $B_0[a^{-1}]$. Therefore by (4.10), $B_1[a^{-1}]$ is Néron's blowing up of $B_0[a^{-1}]$, and by (4.4), (4.5), (4.11),

$$l(B_1/A, \bar{\mathfrak{p}}) = l(B_1[a^{-1}]/A, \bar{\mathfrak{p}}) < l(B_0[a^{-1}]/A, \bar{\mathfrak{p}}) = l(B/A, \bar{\mathfrak{p}}),$$

as required. This completes the proof of Lemma (4.2).

It remains to consider Theorem (4.5). Case (i) of that theorem is precisely Néron's form [13, 2, Thm. 4.5]. The ring described in [2, (4.3)] is an affine coordinate ring of the blowing up. The proof of Case (ii) parallels the argument of [2, pp. 40-42] closely, and so we omit it.

Case (iii) is not as clear. We will show how to reduce it to case (ii). It is permissible to localize, replacing A by $k = \text{Fract}(A)$, and B by $B \otimes_A k$. Thus $\bar{A}$ is a geometrically regular k-algebra. Since $L = \text{Fract}(B/P)$ is a finitely generated field extension of k, there is a finite purely inseparable field extension $k \to k_1$ so that the compositum $L_1 = Lk_1$ of L and k_1 is separable over k_1. Given any finite extension k_1 of k, $\overline{A}\otimes_k k_1$ is a regular semi-local ring of dimension 1. Let $(\overline{A}_1, \bar{\mathfrak{p}}_1)$ be the localization at one of its maximal ideals, and let $B_1 = B\otimes_k k_1$. Case (iii) of Theorem (4.5) is reduced to Case (ii) by

Lemma (4.12). *Let k_1 be a finite normal field extension of k. If Theorem (4.5) is true for k_1-algebras B_1 with residue field $L_1 = Lk_1$, then it is true for k-algebras B with residue field L.*

Proof. We extend notation from $(k, B, \overline{A})$ to $(k_1, B_1, \overline{A}_1)$ by the subscript 1. It suffices to treat the cases: (a) k_1/k is separable, and (b) char $k = p$, and $k_1 = k[\alpha]$, where $\alpha^p = a \in k$. In the first case, one sees immediately that l is preserved by the extension of scalars to k_1, i.e., $l(B/k, \bar{\mathfrak{p}}) = l(B_1/k_1, \bar{\mathfrak{p}}_1)$ and $l(B'/k, \bar{\mathfrak{p}}) = l(B'_1/k_1, \bar{\mathfrak{p}}_1)$, from which the lemma follows. In case (b) there are two possibilities. Either $L\otimes_k k_1 = L_1$ is a field, or else $L\otimes_k k_1 \approx L[\epsilon]/(\epsilon^p)$ and $L \approx L_1$. If $L\otimes_k k_1 = L_1$, then $PB_1 = P_1$, and the blowing up of B_1 at P_1 is $B'\otimes_k k_1$. It may happen that $\overline{A} \subset \overline{A}_1$ is ramified, and then l changes. But since the change is uniform, it is of no importance. Thus the lemma follows as in case (a).

Assume that $L\otimes_k k_1 \approx L[\epsilon]/(\epsilon^p)$. Then the valuation changes to $v_1 = pv$. Next, note that since k_1/k is purely inseparable, $\overline{A}\otimes_k k_1$ is local, hence equal to $\overline{A}_1$. We have

$$(4.13)\ \overline{A}_1/P\overline{A}_1 \approx (\overline{A}/P\overline{A})\otimes_k k_1 \approx (\overline{A}/P\overline{A})\otimes_L L\otimes_k k_1 \approx (\overline{A}/P\overline{A})[\epsilon]/(\epsilon^p).$$

Since $\overline{A}_1$ is a discrete valuation ring, the maximal ideal of $\overline{A}_1/P\overline{A}_1$ is generated by one element. From the right-hand term of (4.13), it follows that $\overline{A}/P\overline{A}$ is a field, i.e., that $P\overline{A} = \bar{\mathfrak{p}}$.

Let $z \in P_1$ be a lifting of $\epsilon \in L\otimes_k k_1$ (We may suppose $\epsilon \in B/P\otimes_k k_1$). Then, again from (4.13), z generates $\bar{\mathfrak{p}}_1$, and $z^p \in PB_1$. Let $\{u_0, \ldots, u_m\}$ generate P in B. Then P_1 is generated locally by $\{z, u_0, \ldots, u_m\}$. Thus after a suitable localization of B we may suppose that $P_1 = (z, u_0, \ldots, u_m)B_1$. Moreover,

$$(4.14)\qquad z^p = b_0u_0 + \cdots + b_mu_m, \quad \text{for some } b_\nu \in B_1.$$

Since z generates $\bar{\mathfrak{p}}_1$, we have $v_1(z) = 1$, while $v_1(u_i) = pv(u_i) \geq p$. Since $P\overline{A} = \bar{\mathfrak{p}}$, we may suppose that $v(u_0) = 1$. Then u_0/z^p is a unit at P_1. An affine ring B' of the blowing up of P in B, such that $\bar{s}$ extends to B', is obtained by adjoining the ratios u_i/u_0. These ratios also generate the affine ring $B' \otimes_k k_1$ of the blowing up of PB_1 in B_1. On the other hand, locally at P_1, the ratios u_i/u_0 differ by unit factor from the elements u_i/z^p, and $B_1[u_0/z^p, \dots, u_m/z^p]$ is the result of blowing B_1 up p times in succession, beginning with the ideal P_1. The ν-th blowing up is $B_1^{(\nu)} = B_1[u_i/z^\nu]$, and the relevant prime ideal is $P_1^{(\nu)} = (z, u_0/z^\nu, \dots, u_m/z^\nu)B_1^{(\nu)}$, if $\nu < p$. Thus $\mathrm{Fract}(B_1^{(\nu)}/P_1^{(\nu)}) \approx L_1$ for each $\nu < p$. Assuming theorem (4.5) for k_1-algebras B_1 with residue field L_1, each of these operations will lower l, which implies (4.5) for $(k, B, \overline{A})$, and proves Lemma (4.12).

5. Application to 2-dimensional Local Rings

Using the result of section 3 and 4, we can prove that two-dimensional local rings are limits of smooth algebras in certain cases. Roughly, the method of Pfister (section 4) allows us to smooth singularities of codimension 1, after which Theorem (3.2) can be applied. In this way we obtain

Theorem (5.1). *Let $A \to \overline{A}$ be a regular map [12] of excellent local rings with residue field k. Assume that $\overline{A}$ is normal, henselian, and of dimension 2. Then $\overline{A}$ is a filtering direct limit of smooth A-algebras.*

The connection of this theorem with the result of section 3 is this:

Lemma (5.2). *Theorem (5.1) is equivalent with the following assertion: Given a finitely presented A-algebra B, and an A-homomorphism $\bar{s}$: $B \longrightarrow \overline{A}$, i.e., a diagram (3.1), there exists a finitely presented B-algebra C and a commutative diagram (3.3), such that α is smooth.*

Proof. If the theorem holds, say $\overline{A} = \varinjlim S_i$ where S_i are smooth A-algebras, then the map $\bar{s}$ of (5.2) will factor through S_i for large enough i. Then $C = S_i$ is the required algebra. In any case, any algebra is the limit of finitely presented algebras. Let I denote the category of pairs $(B, \bar{s})$ of finitely presented A-algebras B and A-homomorphisms $\bar{s} : B \to \overline{A}$. This category is filtering, and $\overline{A} \approx \varinjlim_{(B,\bar{s})} B$. If the assertion of Lemma (5.2) holds, then the subcategory of I consisting of pairs with B smooth over A

is cofinal. Therefore $\overline{A}$ is the limit over this subcategory.

Proof of Theorem (5.1). The first step is to replace A by a larger ring, smooth over A, so that the maximal ideal of A generates that of $\overline{A}$. We choose elements $x = \{x_i\} \in \overline{A}$ which reduce to a system of parameters for the regular ring $\overline{A}/\mathfrak{m}\overline{A}$ ($\mathfrak{m} = \max A$), and replace A by a localization of $A[x]$. This is permissible, by the following lemma.

Lemma (5.3). *Let $A \to \overline{A}$ be a regular local homomorphism of noetherian local rings. Let $\mathfrak{m}$ and $\overline{\mathfrak{m}}$ be the maximal ideals of A and $\overline{A}$. Assume that A is excellent and that $\overline{A}/\overline{\mathfrak{m}}$ is separable over $A/\mathfrak{m}$. Let $x = \{x_i\} \in \overline{A}$ induce a system of parameters in $\overline{A}/\mathfrak{m}\overline{A}$. Then the map $A[X] \to \overline{A}$ sending $X_i \to x_i$ is regular.*

Proof. Let $A' = A[X]_{(\mathfrak{m},X)}$ be the localization of the polynomial ring $A[X]$. It suffices to show that $\theta : A' \to \overline{A}$ is regular. By [1], we only have to prove θ formally smooth for the adic topologies. Thus we have to show that θ is flat, and that the map $A/\mathfrak{m} \approx A'/\mathfrak{m}' \to \overline{A}/\mathfrak{m}'\overline{A} = \overline{A}/\overline{\mathfrak{m}}$ is formally smooth, which is true because it is a separable field extension. To prove θ flat, we apply [12, p. 152 (20.G)] to the maps $A \to A' \xrightarrow{\theta} \overline{A}$. Since $\overline{A}$ is flat over A, it suffices to show $\overline{A}/\mathfrak{m}\overline{A}$ flat over $A'/\mathfrak{m}A'$. This follows from [12, p. 150 (20.D)].

The next step in the proof of (5.1) is to apply Theorem (4.1) to the given diagram (3.1). We replace B by the ring C of (4.1). Then since $\overline{A}$ has dimension 2, B will be smooth over A at all points of $\overline{X} = \operatorname{Spec} \overline{A}$ except the closed point. At this point Theorem (3.2) applies, to complete the proof.

References

[1] M. André, Localisation de la lissité formelle, Manuscr. Math. 13 (1974) 297–307.

[2] M. Artin, Algebraic approximation of structures over complete local rings, Pub. Math. Inst. Hautes Études Sci. 36 (1969) 23–58.

[3] M. Artin, Algebraic structure of power series rings Contemp. Math. 13, Amer. Math. Soc., Providence (1982) 223–227.

[4] J. Becker, J. Denef, and L. Lipshitz, The approximation property for some 5-dimensional henselian rings, Trans. Amer. Math. Soc. (to appear).

[5] J. Becker, J. Denef, L. Lipshitz, and L. van den Dries, Ultraproducts and approximation in local rings-I, Invent. Math. 51 (1979) 189–203.

[6] M. Brown, Artin's approximation property, Thesis, Cambridge, 1979.

[7] C. C. Chang and H. J. Keisler, Model theory, North Holland, Amsterdam 1973.

[8] J. A. Eagon and D. G. Northcott, Generically acyclic complexes and generically perfect ideals, Proc. Royal Soc. London 269A (1967) 147–172.

[9] R. Elkik, Solutions d'équations à coefficients dans un anneau hensélien, Ann. Sci. École Norm. Sup. 4e sér. 6 (1973) 553–604.

[10] A. Grothendieck and J. Dieudonné, Éléments de géométrie algébrique IV, Pub. Math. Inst. Hautes Etudes Sci. 32 (1967).

[11] H. Kurke, T. Mostowski, G. Pfister, D. Popescu, and M. Roczen, Die Approximationseigenschaft lokaler Ringe, Lec. Notes in Math. 634, Springer Verlag, Berlin 1978.

[12] H. Matsumura, Commutative algebra, Benjamin, New York, 1970.

[13] A. Néron, Modèles minimaux des variétés abéliennes sur les corps locaux et globaux, Pub. Math. Inst. Hautes Études Sci. 21 (1964).

[14] G. Pfister and D. Popescu, On three-dimensional local rings with the property of approximation, Rev. Roum. Math. Pures Appl. 26 (1981) 301–307.

[15] A Płoski, Note on a theorem of M. Artin, Bull. Acad. Pol. Sci. 22 (1974) 1107–1110.

[16] D. Popescu, Global form of Néron's p-desingularization and approximation, Teubner Texte Bd 40, Leipzig 1981.

[17] D. Popescu, Higher dimensional Néron desingularization and approximation, (preprint).

[18] J. C. Tougeron, Idéaux de fonctions différentiables, thèse, Rennes 1967.

Received October 1, 1982
Supported by N.S.F.

Professor Michael Artin
Department of Mathematics
Massachusetts Institute of Technology
Cambridge, Massachusetts 02139

Professor Jan Denef
Departement Wiskunde
Universiteit van Leuven
Celestijnenlaan 200 B
3030 Heverlee, Belgium

Convexity and Loop Groups

M. F. Atiyah and A. N. Pressley

To I.R. Shafarevich

1. Introduction

The purpose of this paper is to extend certain convexity results associated with compact Lie groups to an infinite-dimensional setting, in which the Lie group is replaced by the corresponding loop group. To recall the finite-dimensional results which we shall generalize let G be a simply connected, compact Lie group, T a maximal torus of G and W its Weyl group. Consider the adjoint action of G on its Lie algebra $L(G)$ and fix a G-invariant metric on $L(G)$ so that we can define orthogonal projection. A result of Kostant [8] describes the images of the G-orbits in $L(G)$ under the orthogonal projection onto $L(T)$. To state it, recall that such G-orbits correspond to W-orbits in $L(T)$. Then Kostant's result is:

(1.1). *The orthogonal projection of a G-orbit onto $L(T)$ coincides with the convex hull of the corresponding W-orbit.*

Recall that any adjoint orbit of G has a natural cell decomposition, the so called Bruhat decomposition. Atiyah [1] and, independently, Heckman [4] have refined (1.1) by showing that *the images of the closures of the Bruhat cells are convex sub-polyhedra of the image of the corresponding G-orbit.*

It is well known from various points of view (Kac [6], Segal [13]) that the space ΩG of smooth loops on G (based at the identity) has many properties which are analogous to those of G and its adjoint orbits. In particular, it can essentially be thought of as a Lie group whose rank is one greater than that of G, and it also has a Bruhat decomposition (to be precise, this is a decomposition of a dense subgroup of ΩG, but we will ignore this for the moment). It is, therefore, natural to ask if (1.1) has an analogue for the loop space. Our purpose is to show that this is indeed the case.

In fact, there is a natural smooth mapping from ΩG to $L(T) \oplus \mathbf{R}$, analogous to the orthogonal projection in (1.1), whose image will be a convex polyhedron. More precisely, define "Energy" and "Momentum" functions $E\colon \Omega G \longrightarrow \mathbf{R}$ and $p\colon \Omega G \longrightarrow L(T)$ by

$$
\begin{aligned}
E(f) &= \frac{1}{4\pi}\int_0^{2\pi} \|f(\theta)^{-1}f'(\theta)\|^2\,d\theta \\
p(f) &= pr_{L(T)}\left(\frac{1}{2\pi}\int_0^{2\pi} f(\theta)^{-1}f'(\theta)\,d\theta\right)
\end{aligned}
\tag{1.2}
$$

where $f \in \Omega G$ and $pr_{L(T)}$ is the orthogonal projection of $L(G)$ onto $L(T)$. The norm is that corresponding to the G-invariant metric on $L(G)$. Then the required mapping is $f \to \big(p(f), E(f)\big)$, and will be called the *moment map*, for reasons to be explained later.

Our main result is then:

Theorem 1. *The image of ΩG under the moment map is the convex hull of the graph of the real valued function $\frac{1}{2}\|\ \ \|^2$ restricted to the integer lattice in $L(T)$.*

Moreover, we will show that the image of the closure of each Bruhat cell in ΩG is a compact convex sub-polyhedron of the image ΩG. In fact, theorem 1 can also be refined in another direction. It is known that ΩG admits a natural decomposition into finite-codimensional submanifolds, the so called Birkhoff decomposition [10]. We will show that the image of the closure of each stratum of the Birkhoff decomposition is a (non-compact) convex subpolyhedron of the image of ΩG. The analogue of the Birkhoff decomposition in the finite dimensional situation (1.1) is just a different Bruhat decomposition (corresponding to a different choice of positive roots) and so yields no new information.

Remarks. 1. The assumption that the loops $f\colon S^1 \longrightarrow G$ in theorem 1 be smooth is, of course, unnecessarily restrictive. In fact, for the moment map (1.2) to be defined, it is only necessary to assume that f lies in the Sobolev class H^1. Apart from being more natural in this context, the space Ω^1 of H^1 loops on G has the advantage of being an honest Hilbert manifold. For these reasons, we shall work with Ω^1 rather than the smooth loops ΩG in the proof of theorem 1. This will not affect the results, because the images of ΩG and Ω^1 under the moment map are actually the same.

2. Recall that the integer lattice Λ is the inverse image of the identity

under the exponential map $L(T) \to T$. The set of homomorphisms from the circle group S^1 into T is naturally parameterised by Λ, and its image under the moment map is precisely the graph of the function $\frac{1}{2}\|\ \ \|^2$ restricted to Λ. Thus, theorem 1 says that the image of ΩG under the moment map is a convex polyhedron whose vertices are the images of the homomorphisms $S^1 \to T$.

3. By the Cauchy-Schwarz inequality, the image of ΩG lies inside the paraboloid

$$E = \frac{1}{2}\|p\|^2$$

in $L(T)\oplus\mathbf{R}$. The homomorphisms $S^1 \to T$ are mapped onto its surface, but other loops in G are mapped strictly inside the paraboloid. Thus, theorem 1 can be viewed as a strengthening of the Cauchy-Schwarz inequality. Its origin lies in the fact that ΩG consists of *closed* paths in G — no such strengthening is possible for arbitrary smooth maps of an interval into G.

This phenomenon has an analogue in the finite dimensional situation. Since $L(G)$ has a G-invariant inner product, each adjoint orbit is contained in a sphere centered at the origin in $L(G)$. The projection of the orbit onto $L(T)$ therefore lies in a disc of the same radius, and the points in its boundary correspond to the intersection of $L(T)$ with the orbit, which is a finite set of points.

4. As a first example we illustrate in fig. 1 the image of ΩG in the simplest non-trivial case $G = SU(2)$. Here, $L(G)$ consists of the 2×2 skew-Hermitian, trace zero matrices. It is convenient to write a general element of $L(G)$ in the form

$$2\pi\begin{pmatrix} ix & y+iz \\ -y+iz & -ix \end{pmatrix}$$

where $x, y, z \in \mathbf{R}$. Then $L(T)$ consists of the diagonal matrices

$$\begin{pmatrix} 2\pi ix & 0 \\ 0 & -2\pi ix \end{pmatrix},$$

and the integer lattice consists of those elements of $L(T)$ for which $x \in \mathbf{Z}$. The invariant norm can be taken to be $x^2 + y^2 + z^2$ — it is unique up to multiplication by a positive real number.

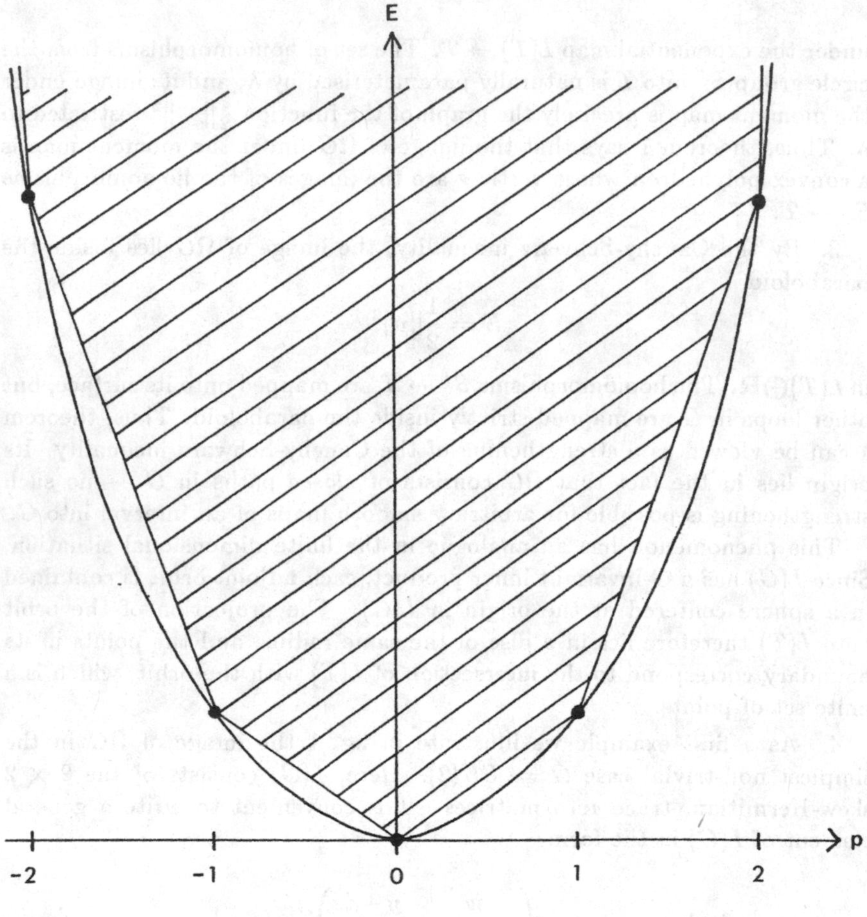

Fig. 1 The case $G = SU(2)$

5. We give one further example to show what happens if we weaken the requirement that G be simply-connected to its being semi-simple. For this, we consider the case $G = SO(3)$. The loop space $\Omega SO(3)$ has two components. The identity component is naturally identified with $\Omega SU(2)$ (via the double covering $SU(2) \to SO(3)$) and the moment map is the same as that in the $SU(2)$ case. Thus, the image of the identity component of $\Omega SO(3)$ is the convex polyhedron shown in fig. 1. The remaining vertices, corresponding to the homomorphisms $S^1 \to T$ lying in the non-identity

component, are the points on the parabola $E = \frac{1}{2}p^2$ in fig. 1 corresponding to the half odd integer values of p. The image of the non-identity component is, in fact, the convex hull of these extra vertices, as shown in fig. 2. In particular, the image of $\Omega SO(3)$ is *not* convex.

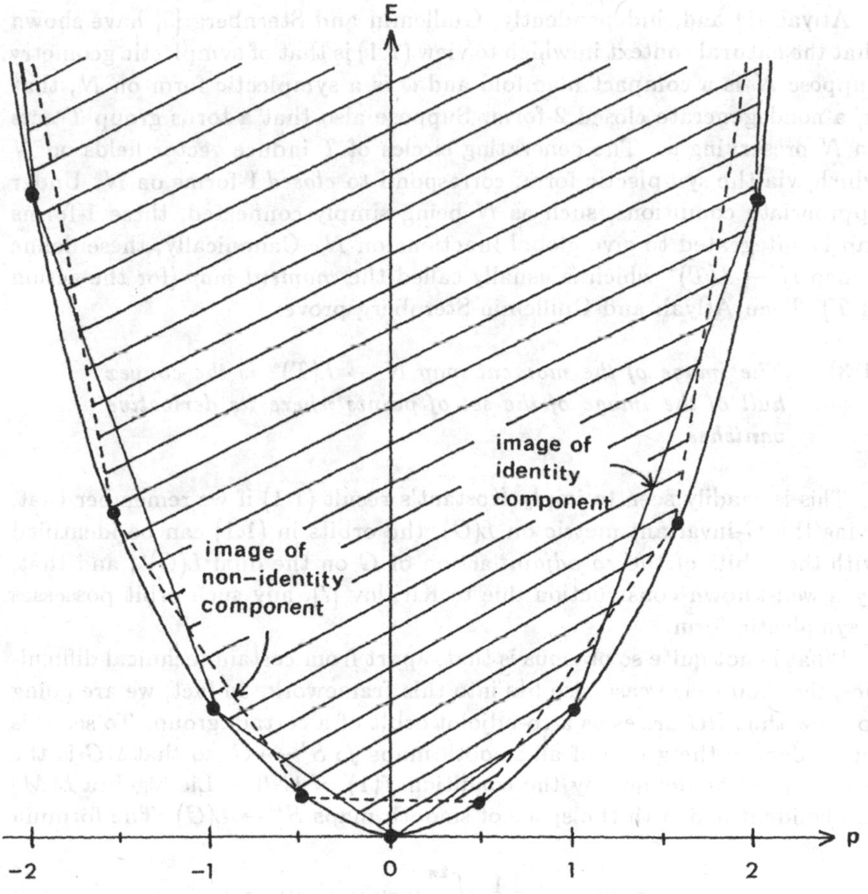

Fig. 2 The case $G = SO(3)$

Nevertheless, it *is* true in this example that the image of each component of the loop space is a convex polyhedron. This situation is typical. The methods of this paper allow one to prove that for a compact semi-simple Lie group G, the image of *each component* of ΩG is the convex hull of the graph of the function $\frac{1}{2}\|\ \|^2$ restricted to the part of the integer lattice corresponding to the homomorphisms $S^1 \to T$ lying in that particular component.

Atiyah [1] and, independently, Guillemin and Sternberg [3] have shown that the natural context in which to view (1.1) is that of symplectic geometry. Suppose N is a compact manifold and ω is a symplectic form on N, that is, a nondegenerate closed 2-form. Suppose also that a torus group T acts on N preserving ω. The generating circles of T induce vector fields on N which, via the symplectic form, correspond to *closed* 1-forms on N. Under appropriate conditions, such as N being simply-connected, these 1-forms can be integrated to give global functions on N. Canonically, these define a map $N \to L(T)^*$ which is usually called the *moment map* (for the action of T). Then Atiyah and Guillemin-Sternberg prove:

(1.3) *The image of the moment map $N \to L(T)^*$ is the convex hull of the image of the set of points where its derivative vanishes.*

This is readily seen to imply Kostant's result (1.1) if we remember that, using the G-invariant metric on $L(G)$, the orbits in (1.1) can be identified with the orbits of the *co-adjoint* action of G on the dual $L(G)^*$, and that, by a well-known construction due to Kirillov [7], any such orbit possesses a symplectic form.

What is not quite so obvious is that, apart from certain technical difficulties, the loop space case also fits into this framework. In fact, we are going to show that ΩG arises as a co-adjoint orbit of a certain group. To see this let M denote the group of all smooth maps $f\colon S^1 \longrightarrow G$, so that ΩG is the subgroup of M defined by the condition $f(1) = 1$. The Lie algebra $L(M)$ can be identified with the space of smooth maps $S^1 \to L(G)$. The formula

$$S(X,Y) = \frac{1}{2\pi}\int_0^{2\pi} \langle X'(\theta), Y(\theta)\rangle\, d\theta, \tag{1.4}$$

where $X, Y \in L(M)$ and $\langle\ ,\ \rangle$ is the G-invariant metric on $L(G)$, defines a 2-cocycle on $L(M)$ and hence a central extension of Lie algebras

$$0 \to \mathbf{R} \to \hat{L} \to L(M) \to 0.$$

In [13] Segal shows that this can be 'exponentiated' to a central extension of M by the circle group S^1,

$$1 \to S^1 \to \hat{M} \to M \to 1$$

where $L(\hat{M}) = \hat{L}$. We do not need the explicit construction of the group $\hat{M}$, but we should point out that the definition of the extension on the group level is non-trivial. Indeed, the circle bundle $\hat{M} \to M$ is non-trivial (its first Chern class is non-zero) so there cannot be an extension

$$0 \to \mathbf{R} \to F \to M \to 1$$

with $\hat{M} = F/\mathbf{Z}$.

The co-adjoint representation of $\hat{M}$ takes place inside $L(\hat{M})^*$, which is a space of distributions on the circle. But we need only consider the action on the 'smooth' part of the dual, which can be identified with $\mathbf{R} \oplus L(M)$, using the invariant inner product on $L(G)$. The action of $\hat{M}$ preserves the affine hyperplane $1 \oplus L(M)$ and its restriction to it can be thought of as an affine action of M on $L(M)$. In [13] this is computed to be

$$X \cdot f = f^{-1}f' + f^{-1}Xf.$$

The orbit of $0 \in L(M)$ is $M/G \cong \Omega G$, so ΩG is a co-adjoint orbit of $\hat{M}$. In particular, ΩG will be a symplectic manifold; in fact, formula (1.4) restricted to $L(\Omega G) \subset L(M)$ defines a non-degenerate skew form, from which the required symplectic form can be obtained by left translation.

The maximal torus T of G acts on G, and hence on ΩG, by conjugation. The circle group S^1 acts on ΩG by "rotating the loop": if $f \in \Omega G$, then $e^{it} \in S^1$ acts on f by

$$(e^{it} \cdot f)(\theta) = f(\theta + t)f(t)^{-1}.$$

These two actions obviously commute, and so define an action of $T \times S^1$ on ΩG. It is easy to see that this action preserves the symplectic form, so there is a corresponding moment map

$$\Omega G \to L(T \times S^1)^* \cong L(T) \oplus \mathbf{R}$$

(using the metric on $L(T)$ to identify $L(T)$ with $L(T)^*$); note that ΩG is simply-connected. It is not hard to verify that this is precisely the map defined in (1.2).

Theorem 1 could therefore be proved by extending the method of proof of (1.3) so as to cover our infinite dimensional situation. This would involve using the techniques of Morse theory on infinite dimensional manifolds and, although there seems to be no reason in principle why this proof could not be carried out, we prefer to follow a different course: we shall reduce our problem to a finite dimensional one.

To do this we use the Bruhat decomposition mentioned earlier, but now we must describe it a little more precisely. For this we must introduce the subgroup Ω^{alg} of ΩG consisting of the *algebraic maps* $S^1 \to G$. For example, if $G = SU(n)$, these are the maps given by $n \times n$ matrices whose entries are finite trigonometric polynomials. It is Ω^{alg}, and *not* ΩG, which has a Bruhat decomposition. In fact,

$$\Omega^{alg} = \bigcup_{\lambda \in \Lambda} C_\lambda .$$

where the C_λ are finite dimensional cells, indexed by the integer lattice Λ. We will prove that the action of $T \times S^1$ on ΩG preserves each cell C_λ. We should then be able to use (1.3) to determine the image of the closure $\overline{C}_\lambda$ of C_λ in Ω^{alg}. Then the image Ω^{alg} will be the union of the images of the $\overline{C}_\lambda$. Finally, we will prove that Ω^{alg} is dense in ΩG. Since the image of Ω^{alg} is, in fact, already closed, it is the same as the image of ΩG.

The difficulty with this approach is that the closed cells $\overline{C}_\lambda$ are manifolds *with singularities*, so (1.3) does not apply. To overcome this, we use the more precise results of [1] which apply when the symplectic manifold N in (1.3) is actually a Kähler manifold and the torus action preserves the Kähler form. As N is then complex, the torus action extends to an action of the corresponding complex torus, and theorem 2 of [1] describes the image, under the moment map, of the closure of any orbit of the complex torus. Note, in particular, that such a closed orbit may have singularities.

To apply this to our situation, we will show that there is an increasing family of complex Grassmann manifolds and algebraic embeddings

$$\mathcal{Y}_1 \subset \mathcal{Y}_2 \subset \mathcal{Y}_3 \subset \cdots$$

and that each Bruhat cell C_λ is naturally embedded in $\mathcal{Y}_m$, for m sufficiently large. Moreover, the action of $T \times S^1$ extends to each $\mathcal{Y}_m$. Finally, we will find a point of C_λ whose orbit under the complex torus $T_c \times \mathbf{C}^*$ is dense in C_λ. Then theorem 2 of [1] immediately tells us the image of $\overline{C}_\lambda$ under the moment map. It is the convex hull of the graph of $\frac{1}{2}\|\quad\|^2$ restricted to a finite subset of Λ. Every integer lattice point occurs so it is obvious that the image of Ω^{alg} is contained in the polyhedron described in theorem 1. To prove equality, it is enough to prove that the image is convex, and this follows from a simple fact about the Bruhat decomposition, namely that for any two closed Bruhat cells there is another which contains them both.

This method allows us to obtain, as a bonus, the image of the closure (in ΩG) of each stratum of the Birkhoff decomposition. The Birkhoff strata are infinite dimensional submanifolds of ΩG and we again reduce to a finite dimensional situation by considering the intersections of a given Birkhoff stratum with each Bruhat cell. The above proof goes through without any essential change, and proves that the image of the closure of each Birkhoff stratum is a sub-polyhedron of the image of ΩG, but this time using all but a finite set of its vertices.

2. Geometry of the Loop Space

In this section we review certain aspects of the geometry of the space of loops on G, leaving most of the proofs to the references. We begin with

Definition 2.1. M^{alg} is the group of all maps $f: S^1 \longrightarrow G$ which are restrictions of morphisms of complex varieties from $\mathbf{C}^*$, the non-zero complex numbers, to the complexification G_c of G. Ω^{alg} is the subgroup of M^{alg} consisting of the basepoint preserving maps: $f(1) = 1$.

For example, if $G = SU(n)$, M^{alg} consists precisely of those maps $f: S^1 \longrightarrow SU(n)$ which are given by a finite trigonometric polynomial:

$$f(\theta) = \sum_{k=-m}^{m} A_k e^{ik\theta}$$

where the A_k are $n \times n$ complex matrices.

The loop space Ω^{alg} can be thought of as an infinite dimensional variety. More precisely:

Proposition 2.2.[10]. *For each $m = 0, 1, 2, \ldots$ there is a closed subvariety Ω_m of a (finite dimensional) complex Grassmannian, together* *with algebraic embeddings*

$$\Omega_0 \subset \Omega_1 \subset \Omega_2 \subset \cdots$$

such that Ω^{alg} *is the union* $\bigcup_{m=0}^{\infty} \Omega_m$.

To construct such a filtration, consider first the case $G = SU(n)$. Let H be the Hilbert space of L^2 maps $S^1 \to \mathbf{C}^n$ and let H_0 be the closed subspace of H consisting of those maps which extend holomorphically inside the unit circle. Multiplication by $e^{i\theta} \in S^1$ is a unitary map $H \to H$ and, more generally, Ω^{alg} acts on H by unitary transformations. Now H has a filtration

$$\cdots \subset H_2 \subset H_1 \subset H_0 \subset H_{-1} \subset H_{-2} \subset \cdots \subset H$$

where $H_m = e^{im\theta} . H_0$. Let $X_m (m = 0, 1, 2, \ldots)$ denote the space of subspaces V of H such that

$$(a) \quad e^{i\theta} . V \subseteq V,$$
$$(b) \quad H_m \subseteq V \subseteq H_{-m}.$$

It is not hard to see that X_m is a closed subvariety of the Grassmannian $\mathcal{Y}_m$ of nm-dimensional subspaces of the $2nm$-dimensional complex vector space H_{-m}/H_m. Moreover,

$$X_0 \subset X_1 \subset X_2 \subset \cdots \subset X = \bigcup_{m=0}^{\infty} X_m.$$

The action of Ω^{alg} on H induces an action on X which is strictly transitive — thus the map $f \to f . H_0$ is a bijection $\Omega^{alg} \cong X$. Moreover, Ω_m corresponds to X_m.

In the case where G is arbitrary, the corresonding varieties Ω_m can be defined using an embedding $G \hookrightarrow SU(n)$.

Note that the group M_c^{alg} of all algebraic maps $\mathbf{C}^* \to G_c$ acts on H and that the stabiliser of $H_0 \in X$ is precisely the subgroup P consisting of the maps which are regular at the origin. Thus, we have an identification

$$\Omega_{alg} \cong M_c^{alg}/P.$$

Choose a Borel subgroup of G_c containing the maximal torus T, and let B be the subgroup of P consisting of those maps $f \in P$ for which $f(0)$ lies in this Borel subgroup. The natural left action of B on M_c^{alg} induces an action on Ω^{alg}. One can show [11] that each B-orbit on Ω^{alg} is topologically a finite dimensional cell, and that the orbits are indexed naturally by the integer lattice Λ of G. Thus, Ω^{alg} has a cell decomposition

$$\Omega^{alg} = \bigcup_{\lambda \in \Lambda} C_\lambda. \tag{2.3}$$

In fact, M_c^{alg} is an algebraic group over the ring $\mathbf{C}[z, z^{-1}]$ of Laurent polynomials and P is a parabolic subgroup, so it follows that M_c^{alg}/P has a Bruhat decomposition in the sense of Tits[2]. The induced decomposition of Ω^{alg} is precisely (2.3).

In the proof of theorem 1, it will be convenient to have available the coarser decomposition of Ω^{alg} into the orbits of the group P. This time the orbits are indexed by the set of orbits of the Weyl group W on Λ:

$$\Omega^{alg} = \bigcup_{\lambda \in W\backslash\Lambda} F_\lambda.$$

By abuse of notation, we label the P-orbits by elements $\lambda \in \Lambda$, but it is understood that λ is only determined up to the action of W. The F_λ are not cells, but each F_λ is the union of the Bruhat cells corresponding to the points of the W-orbit of λ.

There is a close relationship between the Bruhat decomposition of Ω^{alg} and the classical Bruhat decomposition of flag manifolds. To see this, let $\Gamma_\lambda(\lambda \in \Lambda)$ be the set of conjugates, by elements of G, of the homomorphism $S^1 \to T$ given by $\theta \to \exp(\lambda\theta)$ (note that Γ_λ depends only on the W-orbit of λ). Γ_λ is a complex flag manifold, and it can be shown [11] that F_λ has a natural structure of holomorphic vector bundle over Γ_λ. Then the Bruhat cells C_μ, for μ lying in the W-orbit of λ, are precisely the parts of F_λ lying over the Bruhat cells in Γ_λ.

As we mentioned in the introduction, the natural context in which to study the moment map is the space of loops in the Sobolev class H^1. We now turn to a discussion of this space.

Definition 2.4. M^1 is the group of maps $f: S^1 \longrightarrow G$ which lie in the Sobolev class H^1 (i.e., if Φ is any chart of G, $\Phi \circ f$ is in the class H^1 as a map $S^1 \to \mathbf{R}^{\dim G}$). Ω^1 is the subgroup of M^1 consisting of the basepoint preserving maps.

The exponential map $L(G) \to G$ induces an exponential map

$$L(M^1) \to M^1$$

which is a local diffeomorphism onto a neighbourhood of the identity in M^1; here $L(M^1)$ is the Hilbert space $H^1(S^1, L(G))$ of H^1 maps $S^1 \to L(G)$. This can be used to define an atlas for M^1 and to prove further that M^1 is a Hilbert Lie group with $L(M^1)$ as its Lie algebra. Moreover, Ω^1 is a closed (Lie) subgroup of M^1, whose Lie algebra $L(\Omega^1)$ consists of those H^1 maps $X: S^1 \longrightarrow L(G)$ satisfying $X(1) = 0$.

Notice that Ω^1 can be identified, as a space, with the quotient M^1/G, where G is thought of as the constant maps in M^1. Moreover, $L(\Omega^1)$ can be identified, as vector space, with the quotient $L(M^1)/L(G)$. This observation will be used repeatedly.

The relationship between the algebraic and H^1 loops is given by the following result, the proof of which is due to G. B. Segal and will be given in §5.:

Theorem 2. *Ω^{alg} is dense in Ω^1.*

This result, together with the fact that Ω^{alg} is a complex variety, suggests that Ω^1 should be a complex manifold. This is indeed the case, as we will now show.

We begin by defining an almost complex structure J on Ω^1. The construction works for *any* Lie group G. First consider the case where G is the circle group S^1. Any element $u \in H^1(S^1, \mathbf{R})/\mathbf{R}$ has a Fourier expansion

$$u(\theta) = \sum_{n \in \mathbf{Z}\setminus\{0\}} u_n e^{in\theta}$$

where the complex numbers u_n satisfy the reality condition $u_{-n} = \overline{u}_n$. Then let

$$(Ju)(\theta) = \sum_{n \in \mathbf{Z}\setminus\{0\}} i \operatorname{sign}(n) u_n e^{in\theta}.$$

It is clear that J is a unitary operator on $H^1(S^1, \mathbf{R})/\mathbf{R}$ satisfying $J^2 = -1$. For a general G, just observe that $L(M^1)/L(G) \cong (H^1(S^1, \mathbf{R})/\mathbf{R}) \otimes L(G)$. Extending by left translation gives a smooth almost complex structure J on Ω^1.

With more effort one can show that J is integrable (see [12]), so that Ω^1 is a complex Hilbert manifold. It is worth emphasising that this is true even though the underlying Lie group G is not complex.

The first piece of information we require concerning the complex geometry of the loop space is the existence of a Kähler form on Ω^1. First observe that the formula

$$S(X,Y) = \frac{1}{2\pi}\int_0^{2\pi} \langle X'(\theta), Y(\theta)\rangle \, d\theta$$

defines a skew-symmetric bilinear form on $L(\Omega^1)$, and S is clearly non-degenerate as we are dealing with basepoint preserving maps. Extending by left translation gives a left invariant 2-form S on Ω^1, and it is straightforward to verify that S is closed (for details, see [12]). Thus, *S is a symplectic form on Ω^1*.

To prove that S and J fit together to give a Kähler form on Ω^1, of which S is the imaginary part, we have to verify that the bilinear form on $L(\Omega^1)$ given by

$$R(X,Y) = S(X,JY)$$

is positive-definite and symmetric — then R will be the real part of the Kähler form. But this is easy to check: in terms of the Fourier series of X and Y, one finds

$$R(X,Y) = \sum_{n\in\mathbf{Z}} |n|\langle X_n, Y_n\rangle.$$

Notice that R is not the same as the H^1 metric used to define the topology of Ω^1. In fact, R defines a metric which makes Ω^1 into an incomplete Riemannian manifold, whereas in the H^1 metric Ω^1 is complete. Nevertheless, it is the metric R which will be used when, in the next two sections, we construct the gradient vector fields of various real-valued functions on Ω^1. This is in contrast to previous treatments of Morse theory on Ω^1 (see, for example, [9]) where the H^1 metric is used to define the gradient fields.

Finally, we must discuss the Birkhoff decomposition (see [10] for proofs and further details). For this it is convenient to introduce the group M_c^1 of H^1 maps $S^1 \to G_c$, together with two closed subgroups $P_\pm$ of M_c. P_+ consists of the maps in M_c which extend continuously to *holomorphic* maps of the interior of the unit disc into G_c. Similarly, P_- consists of the maps which extend holomorphically to the exterior of the unit disc in the Riemann sphere. The group P_- acts in an obvious way on the coset space M_c^1/P_+, and the orbits are naturally indexed by the set of orbits of the Weyl group W on Λ. In fact, one can take as orbit representatives the homomorphisms $\theta \to \exp(\lambda\theta)$, where λ runs through a set of representative of the W-orbits on Λ. Thus, M_c^1/P_+ has a decomposition indexed by $W\backslash\Lambda$. Using the natural map $\Omega^1 \to M_c^1/P_+$, which is easily seen to be injective, one obtains the induced decomposition of Ω^1:

$$\Omega^1 = \bigcup_{\lambda \in W\backslash\Lambda} \mathcal{F}^\lambda$$

where, abusing notation as before, $\mathcal{F}^\lambda$ depends only on the W-orbit of λ. The $\mathcal{F}^\lambda$ are submanifolds of Ω^1 of finite codimension, and are called the *Birkhoff manifolds*.

There is a close relationship between the Birkhoff manifolds $\mathcal{F}^\lambda$ and the Bruhat manifolds F_λ which we will exploit in the proof of theorem 1 and its refinements. To explain this, we need to introduce a partial ordering on $W\backslash\Lambda$. This can be defined in terms of the closures of the Bruhat manifolds (in Ω^{alg}) as follows:

Definition 2.5. If $\lambda, \mu \in W\backslash\Lambda$, we write $\lambda \le \mu$ if $F_\lambda \subseteq \overline{F}_\mu$.

Then one can establish the following facts relating the two decompositions:

Proposition 2.6 [11].
(1) $\mathcal{F}^\mu \subseteq \overline{\mathcal{F}^\lambda}$ *(closure in* Ω^1*) if and only if* $\lambda \le \mu$:
(2) $\mathcal{F}^\lambda$ *meets* F_μ *if and only if* $\lambda \le \mu$:
(3) *in particular,* $\mathcal{F}^\lambda \cap F_\lambda = \Gamma_\lambda$, *the set of homomorphisms* $S^1 \to G$ *conjugate to* $\theta \to \exp(\lambda\theta)$.

3. The Torus Action and its Moment Map

The maximal torus T of G acts on G, and hence on Ω^1, by conjugation. The circle group S^1 acts on Ω^1 by "rotating the loop": if $e^{it} \in S^1$, $f \in \Omega^1$,

$$(e^{it} . f)(\theta) = f(t+\theta)f(t)^{-1}.$$

These two actions obviously commute and so define an action of $T \times S^1$ on Ω^1. Using the almost complex structure J on Ω^1, we obtain an action of the complexification $T_c \times \mathbf{C}^*$ on Ω^1.

We can make this action explicit on the algebraic loops, if we use the identification $\Omega^{alg} = M_c^{alg}/P$. In fact, $\mathbf{C}^*$ acts on itself by multiplication, and T_c acts on G_c by conjugation. Thus, both groups act naturally on M_c^{alg} and the two actions commute and preserve P. It is now clear from the description of the Bruhat decomposition that:

Lemma 3.1. *The action of $T_c \times \mathbf{C}^*$ on Ω^1 preserves each Bruhat cell C_λ. It therefore also preserves the Bruhat manifolds F_λ.*

To obtain the corresponding facts for the Birkhoff decomposition, observe first that from their description in §2, it is clear that the Birkhoff manifolds $\mathcal{F}^\lambda$ are preserved by the action of the real torus $T \times S^1$. It is therefore enough to show that the tangent bundle of each $\mathcal{F}^\lambda$ is preserved by the almost complex structure J. But this follows immediately from the double coset description of $\mathcal{F}^\lambda$ and the fact that, for all $X \in L(M^1)$,

$$J\,.\,X + i\,.\,X \in L(P_+).$$

This establishes :

Lemma 3.2. *The action of $T_c \times \mathbf{C}^*$ on Ω^1 preserves the Birkhoff manifolds $\mathcal{F}^\lambda$.*

The action of the real torus $T \times S^1$ (but not that of the complex torus $T_c \times \mathbf{C}^*$) preserves the symplectic form S on Ω^1. This is essentially because the inner product on $L(G)$ is invariant under the adjoint action of T (but not that of T_c). Since Ω^1 is simply connected, the action of $T \times S^1$ must be generated by $l+1$ Poisson-commuting real-valued functions on Ω^1, where $l = \dim T = \operatorname{rank} G$. In fact, define the *energy function*

$$E(f) = \frac{1}{4\pi} \int_0^{2\pi} \|f(\theta)^{-1} f'(\theta)\|^2 \, d\theta$$

and the *momentum function*

$$p(f) = pr_{L(T)} \frac{1}{2\pi} \int_0^{2\pi} f(\theta)^{-1} f'(\theta) \, d\theta,$$

where $pr_{L(T)}$ is the orthogonal projection $L(G) \to L(T)$. We claim that the l components of p, together with the energy function E, are the required functions.

To verify this, let $X \in L(\Omega^1)$ and compute

$$\begin{aligned} dE(f)(f\,.\,X) &= \frac{1}{2\pi}\int_0^{2\pi} \langle -f^{-1}\,.\,fX\,.\,f^{-1}\,.\,f' + f^{-1}\,.\,(fX)', f^{-1}f'\rangle\, d\theta \\ &= \frac{1}{2\pi}\int_0^{2\pi} \langle X' + [f^{-1}f', X], f^{-1}f'\rangle\, d\theta \\ &= \frac{1}{2\pi}\int_0^{2\pi} \langle X', f^{-1}f'\rangle\, d\theta \end{aligned}$$

since, by the G-invariance of $\langle\, ,\rangle$,

$$\langle [f^{-1}f', X], f^{-1}f'\rangle = \langle [f^{-1}f', f^{-1}f'], X\rangle = 0.$$

Further, since $X(0) = X(2\pi)$, we can write

$$\begin{aligned} dE(f)(f\,.\,X) &= \frac{1}{2\pi}\int_0^{2\pi} \langle X', f^{-1}f' - f'(0)\rangle\, d\theta \\ &= S(X, f^{-1}f' - f'(0)). \end{aligned}$$

Of course, the term $f'(0)$ has been introduced to make the map $\theta \to f(\theta)^{-1}f'(\theta) - f'(0)$ basepoint preserving. This last result shows that the Hamiltonian vector field on Ω^1 corresponding to the energy function is given, at $f \in \Omega^1$, by $f' - f\,.\,f'(0)$. The corresponding flow on Ω^1 is precisely the rotation flow. For

$$\frac{d}{dt} f(t+\theta)f(t)^{-1} = f'(t+\theta)^{-1} - f(t+\theta)f(t)^{-1}f'(t)f(t)^{-1};$$

evaluating at $t = 0$ and using the fact that $f(0) = 1$ gives

$$\left.\frac{d}{dt}\right|_{t=0} f(t+\theta)f(t)^{-1} = f'(\theta) - f(\theta)f'(0)$$

as claimed. Similarly, one checks that the function p generates the conjugation action of T.

In the language of symplectic geometry we have proved:

Lemma 3.3. *The rotation action of the circle group S^1 and the conjugation action of T define a symplectic action of $T \times S^1$ on Ω^1. Its moment map is the function $(p, E) \colon \Omega^1 \to L(T) \oplus \mathbf{R}$.*

Finally we need to determine the common critical points of the functions E and (the components of) p. Equivalently we want the fixed points of the torus $T \times S^1$. Now a loop $f \in \Omega^1$ is fixed by the circle group if

$$f(\theta + t) = f(\theta) f(t) \quad \text{for all } t, \theta \in [0, 2\pi],$$

in other words, if f is a homomorphism $S^1 \to G$. Moreover, f is fixed by T precisely when the image of f lies in the centraliser of T, which is just T. Hence

Lemma 3.4. *The fixed points of the action of $T \times S^1$ on Ω^1 are precisely the homomorphisms $S^1 \to T$.*

Remark. The space $L(T) \oplus \mathbf{R}$, in which the image of the moment map lies, should be thought of as an *affine space*, with no preferred origin. In fact, the affine Weyl group W_{aff} of G acts on $L(T) \oplus \mathbf{R}$ by affine transformations, and the action preserves the image of Ω^1.

To see this, recall that W_{aff} is the semi-direct product $W \tilde{\times} \Lambda$, where W acts on $\Lambda \subset L(T)$ in the natural way; explicitly,

$$(w, \lambda) . (w', \lambda'), = (ww', \lambda + w . \lambda').$$

Then one easily checks that the formula

$$(w, \lambda) . (t, x) = (w . t + \lambda, x + \langle \lambda, w . t \rangle + \tfrac{1}{2} \|\lambda\|^2),$$

where $t \in L(T)$, $x \in \mathbf{R}$, defines an action of W_{aff} on $L(T) \oplus \mathbf{R}$ by affine transformations. That it preserves the image of Ω^1 under the moment map follows from the formula

$$(w, \lambda) . (p(f), E(f)) = (p(nfn^{-1} \exp(\lambda\theta)), E(nfn^{-1} \exp(\lambda\theta))),$$

where $f \in \Omega^1$ and n is any element of $N(T)$ representing $w \in W = N(T)/T$. In other words, the action of W corresponds to the conjugation action of

$N(T)$ on Ω^1, while that of Λ corresponds to the right multiplication by the homomorphisms $S^1 \to T$.

It is clear from this discussion that the action of W_{aff} preserves the set of vertices of the image of Ω^1 and that any two vertices are equivalent under this action. In fact, this is already true for the subgroup $\Lambda \subset W_{aff}$.

It is worth comparing this with the finite dimensional situation (1.1). There the image of an adjoint orbit in $L(G)$ under orthogonal projection onto $L(T)$ is a *compact* convex polyhedron, and its centre of mass can therefore be taken as a preferred origin. Notice also that the moment map in the finite dimensional case, namely orthogonal projection onto $L(T)$, is analogous to the momentum function p. There is no finite dimensional analogue of the energy function E because the length function on $L(G)$, being invariant under the adjoint action G, is constant on each adjoint orbit. Ultimately, therefore, the difference between the finite and infinite dimensional situations is explained by (or, perhaps, explains) the absence of a bi-invariant metric on Ω^1.

4. Proof of the Main Theorem

In this section we will determine the image of the loop space Ω^1 under the moment map by carrying out the programme outlined in §1. We begin with

Proposition 4.1. *The image of the closure of the Bruhat cell C_λ is the convex hull of the images of the homomorphisms $S^1 \to T$ which lie in the closure $\overline{C}_\lambda$.*

Note. The closures here are taken in Ω^{alg}, which is topologised as a subspace of Ω^1.

Proof. To prove that the image of $\overline{C}_\lambda$ is contained in the convex set described in the statement of (4.1), consider a generic linear combination

$$m = \langle p, \alpha \rangle + \beta E$$

where $\alpha \in L(T)$, $\beta \in \mathbf{R}$. Then the Hamiltonian vector field corresponding to m generates the action of $T \times S^1$ on Ω^1 (this is the meaning of "generic") and the critical set of m is the set of common critical points of E and (the components of) p, namely the set of homomorphisms $S^1 \to T$ (3.4). From elementary properties of vector fields (see lemma (3.1) of [1]), it follows that the linear form

$$(t, x) \to \langle t, \alpha \rangle + \beta x$$

on $L(T) \oplus \mathbf{R}$, when restricted to the image of $\overline{C}_\lambda$, takes its maximum at the images of the homomorphisms $S^1 \to T$ contained in $\overline{C}_\lambda$, and the result follows.

To prove the reverse inclusion, we will apply theorem 2 of [1]. To do this we must embed the singular variety $\overline{C}_\lambda$ in a (non-singular) compact Kähler manifold to which the action of $T_c \times \mathbf{C}^*$ extends. To this end, we recall from §2 that the space of algebraic loops Ω^{alg} has a filtration

$$\Omega_0 \subset \Omega_1 \subset \Omega_2 \subset \cdots \subset \Omega^{alg} = \bigcup_{m=0}^{\infty} \Omega_m$$

by finite dimensional (singular) varieties, and that each Ω_m embeds in a finite dimensional complex Grassmannian $\mathcal{Y}_m$, where $\mathcal{Y}_0 \subset \mathcal{Y}_1 \subset \cdots$. Moreover, the action of $T_c \times \mathbf{C}^*$ on Ω^{alg} preserves each Ω_m and extends in an obvious way to $\mathcal{Y}_m$.

We shall need to make the action of $\mathbf{C}^*$ on $\mathcal{Y}_m$ explicit. To do this, recall from §2 that, after fixing an embedding $G \to SU(n)$, $\mathcal{Y}_m$ consists of the nm-dimensional subspaces of the $2nm$-dimensional complex vector space H_{-m}/H_m, and that the embedding $\Omega_m \hookrightarrow \mathcal{Y}_m$ is given explicitly by $f \to fH_0/H_m$. If $\{v_1, \ldots, v_n\}$ is the usual basis of $\mathbf{C}^n$, then

$$\{e^{-ik\theta} v_j \mid 1 \leq j \leq n, \quad -m < k \leq m\}$$

is a basis of H_{-m}/H_m, and the rotation action of $e^{it} \in S^1$ on $\mathcal{Y}_m$ is that induced from the linear action on H_{-m}/H_m given by multiplication by e^{-ikt} on the basis vector $e^{-ik\theta}v_j$. Order the basis $\{e^{-ik\theta}v_j\}$ lexicographically by (k, j) and call it $\{w_1, \ldots, w_{2nm}\}$. In the corresponding Plücker coordinates $(Z_1, \ldots, Z_N)$ on $\mathcal{Y}_m$ (where $N = \binom{2nm}{nm}$) the action of $e^{it} \in S^1$ is still linear, namely

$$e^{it} . (Z_1, \ldots, Z_N) = (e^{ip_1 t} Z_1, \ldots, e^{ip_N t} Z_N)$$

where the p_i are certain integers. All we need to know about the p_i is that, if $Z_1 = w_1 \wedge w_2 \wedge \cdots \wedge w_{nm}$, then $p_1 > p_j$ for all $j > 1$.

The flow ϕ_t on $\mathcal{Y}_m$ generated by the radial part of the $\mathbf{C}^*$ action is thus given by

$$\phi_t(Z_1, \ldots, Z_N) = (e^{p_1 t} Z_1, \ldots, e^{p_N t} Z_N).$$

This is the flow corresponding to a real valued function on $\mathcal{Y}_m$ which extends the energy function on Ω_m. It is clear that the absolute minimum of this function is attained precisely at the single point with Plücker coordinates $(1, 0, \ldots, 0)$. These are the coordinates of the constant loop $1 \in \Omega_m \subset \mathcal{Y}_m$. Thus, the absolute minimum of the function on $\mathcal{Y}_m$ is actually attained at a point of Ω_m (see fig. 3). Finally, the formula for ϕ_t shows that all points

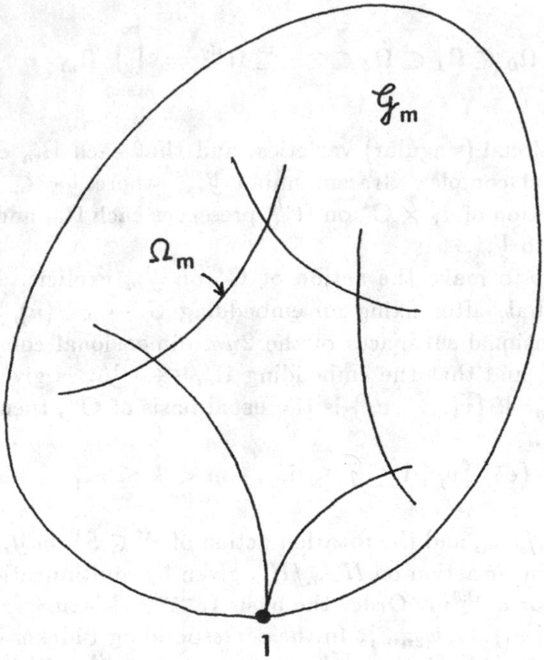

Fig. 3

of $\mathcal{Y}_m$ whose Z_1 coordinate is non-zero converge to the constant loop under the flow ϕ_t as $t \to \infty$. In particular, these points form a *Zariski open subset of* $\mathcal{Y}_m$.

We can now complete the proof of (4.1). Fix $\lambda \in \Lambda$ and choose m so large that $\exp(-\mu\theta) . C_\lambda \subset \mathcal{Y}_m$ for each μ in the finite subset $\Lambda_\lambda \subset \Lambda$ defined by the condition $\exp(\mu\theta) \in \overline{C}_\lambda$. If $\mu \in \Lambda_\lambda$, then $1 \in \exp(-\mu\theta) . \overline{C}_\lambda$ so all points lying in a Zariski open subset of $\exp(-\mu\theta) . \overline{C}_\lambda$ converge to 1 under the flow ϕ_t as $t \to \infty$. It follows that there is a dense open subset $C_\lambda(\mu)$ of C_λ which converges to $\exp(\mu\theta)$ under the flow ψ_t, where ψ arises from ϕ by action of μ. Then the finite intersection

$$\bigcap_{\mu \in \Lambda_\lambda} C_\lambda(\mu)$$

is non-empty, and if f is any point in this intersection, the closure of the $T_c \times \mathbf{C}^*$ orbit Y of f contains $\exp(\mu\theta)$ for each $\mu \in \Lambda_\lambda$. Then by theorem 2 of [1], the image of $\overline{Y}$ under the moment map is precisely the convex hull of the images of the points $\exp(\mu\theta)$ for $\mu \in \Lambda_\lambda$. Hence, the image of $\overline{C}_\lambda$ contains this convex set, as was to be proved.

As a corollary of (4.1) we can determine the image of the closure of the Bruhat manifolds F_λ. In a sense, this is trivial, since F_λ is a union of Bruhat cells, but a more precise result is possible using the results of §2. There we saw that F_λ fibres over the flag manifold Γ_λ, and that the Bruhat cells in F_λ are precisely the inverse images of the classical Bruhat cells in Γ_λ. Now it is well-known that one of the Bruhat cells of Γ_λ is dense in Γ_λ, and it follows that one Bruhat cell of Ω^{alg} is dense in F_λ. Hence, the image of $\overline{F}_\lambda$ coincides with that of the closure of this particular Bruhat cell. In particular,

Corollary 4.2. *The image of the closure of the Bruhat manifold F_λ is the convex hull of the images of the homomorphisms $S^1 \to T$ which lie in $\overline{F}_\lambda$.*

We can now complete the proof of theorem 1. Since

$$\Omega^{alg} = \bigcup_{\lambda \in \Lambda} C_\lambda$$

it follows from (4.1) that the image of Ω^{alg} under the moment map is contained in the convex hull of the images of the homomorphisms $S^1 \to T$.

To show that the image of Ω_{alg} coincides with this convex hull, it is enough to show that the image is convex, and this follows from:

Lemma 4.3. *For any $\mu, \mu' \in \Lambda$, there is a $\nu \in \Lambda$ such that*

$$\overline{F}_\mu \cup \overline{F}_{\mu'} \subseteq \overline{F}_\nu.$$

Proof. Since μ, μ', ν are only determined up to the action of W, we can assume they all lie inside a fixed Weyl chamber in $L(T)$. Then the partial ordering on Λ has the following characterisation (see [5]) :

$$\mu \leq \nu \text{ if and only if} \left\{ \begin{matrix} \text{convex hull} \\ \text{of } W.\mu \end{matrix} \right\} \subseteq \left\{ \begin{matrix} \text{convex hull} \\ \text{of } W.\nu \end{matrix} \right\}.$$

Take $\nu = \mu + \mu'$ and use (2.5).

Remark. If G is not simply-connected, F_μ and $F_{\mu'}$ may lie in different components of the loop space, so (4.3) is no longer true. The proof fails because in this case μ and μ' are unrelated by the partial ordering. However, the result remains true if F_μ and $F_{\mu'}$ lie in the same component of the loop space, and this together with (4.2) proves that, *if G is semi-simple, the image of each component of the loop space is a convex polyhedron.*

Finally, by theorem 2, the space Ω^1 of H^1 loops on G has the same image as Ω^{alg}, and since the smooth loops ΩG satisfy $\Omega^{alg} \subset \Omega G \subset \Omega^1$, theorem 1 follows.

We have now shown that the image of Ω^1 under the moment map is a convex polyhedron, the images of the Bruhat manifolds being convex sub-polyhedra. To complete the picture, we will determine the images of the Birkhoff manifolds $\mathcal{F}^\lambda$:

Proposition 4.4. *The image under the moment map of the closure of the Birkhoff manifold $\mathcal{F}^\lambda$ in Ω^1 is the convex hull of the images of the homomorphisms $S^1 \to T$ which lie in $\overline{\mathcal{F}^\lambda}$.*

Proof. As the $\mathcal{F}^\lambda$ are themselves infinite dimensional manifolds, we need to reduce to a finite dimensional problem using the same technique we used for the whole loop space Ω^1. We therefore consider the finite dimensional varieties $F_\mu \cap \mathcal{F}^\lambda$, and assume $\lambda \leq \mu$ as the intersection is empty otherwise (2.6). The first part of the argument used to prove (4.1) shows that the image of $\overline{F_\mu \cap \mathcal{F}^\lambda}$ (the closure of F_μ being taken in Ω^{alg}) is contained in the convex hull of the images of the homomorphisms $S^1 \to T$ which lie in $\overline{F_\mu \cap \mathcal{F}^\lambda}$, while the second part shows that the image of $\overline{F_\mu \cap \mathcal{F}^\lambda}$ (closure in Ω^{alg}) contains this convex set. To complete the proof of (4.4) we only need the following analogue of (4.3): for any μ, $\mu' \geq \lambda$, there is a $\nu \in \Lambda$ for which

$$(\overline{F_\mu \cap \mathcal{F}^\lambda}) \cup (\overline{F_{\mu'} \cap \mathcal{F}^\lambda}) \subseteq \overline{F_\nu \cap \mathcal{F}^\lambda}.$$

But this is an immediate consequence of (4.3) itself.

Example : $G = SU(3)$

To illustrate theorem 1 and its refinements (4.1), (4.2), and (4.4) we describe the case $G = SU(3)$ in some detail.

Here

$$L(T) = \left\{ \begin{pmatrix} 2\pi i x_1 & & 0 \\ & 2\pi i x_2 & \\ 0 & & 2\pi i x_3 \end{pmatrix} \text{where } x_i \in \mathbf{R} \text{ and } \sum_{i=1}^{3} x_i = 0 \right\}$$

with norm $\sum_{i=1}^3 x_i^2$, the integer lattice consists of the points of $L(T)$ for which the $x_i \in \mathbf{Z}$, and the Weyl group W is the symmetric group on 3 letters, which acts on $L(T)$ by permuting the coordinates.

In fig. 4 we show the part of the image of the loop space $\Omega SU(3)$ under the moment map lying below the hyperplane $E = 3$, together with the circumscribing paraboloid $E = \frac{1}{2}\|p\|^2$.

The closures of the first few Bruhat cells are the convex polyhedra with vertices as shown:

$\overline{C}_{(0,0,0)}$	0
$\overline{C}_{(1,0,-1)}$	$0, A_1$
$\overline{C}_{(0,1,-1)}$	$0, A_1, A_2$
$\overline{C}_{(1,-1,0)}$	$0, A_1, A_6$
$\overline{C}_{(0,-1,1)}$	$0, A_2, A_1, A_6, A_5$
$\overline{C}_{(-1,1,0)}$	$0, A_3, A_2, A_1, A_6$
$\overline{C}_{(-1,0,1)}$	$0, A_1, A_2, A_3, A_4, A_5, A_6$

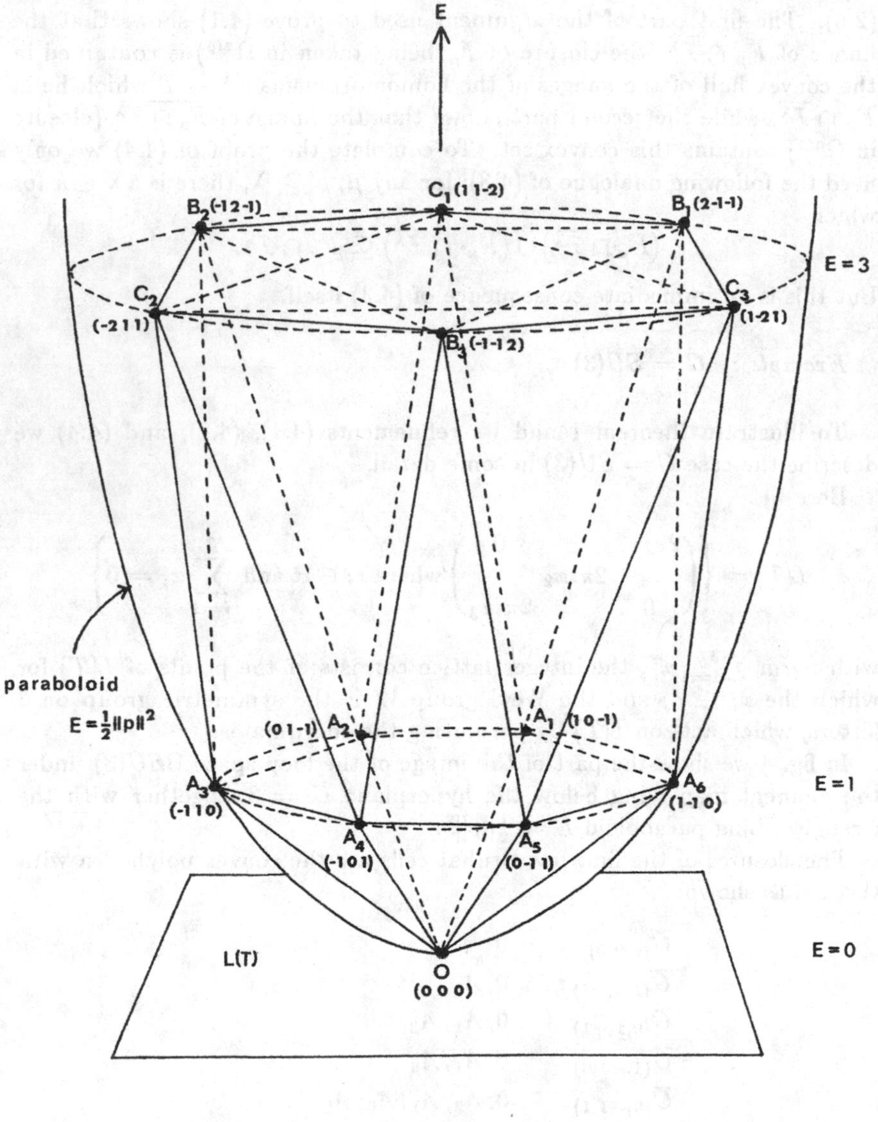

Fig. 4 The case $G = SU(3)$

For the closures of the Bruhat manifolds we have:

$\overline{F}_{(0,0,0)}$	0
$\overline{F}_{(1,0,-1)}$	$0, A_1, \ldots, A_6$
$\overline{F}_{(2,-1,-1)}$	$0, A_1, \ldots, A_6, B_1, B_2, B_3$
$\overline{F}_{(1,1,-2)}$	$0, A_1, \ldots, A_6, C_1, C_2, C_3$

Needless to say, it is difficult to show the images of the Birkhoff manifolds in this diagram. We content ourselves with the observation that the closure of $\mathcal{F}^{(0,0,0)}$ is the entire loop space Ω^1 (by (2.6)(1)), while the image of $\overline{\mathcal{F}^{(1,0,-1)}}$ is the part of the image of Ω^1 lying above the hyperplane $E = 1$.

Remarks 1. The same method can be used to find the images under the moment map of other varieties in the loop space. Consider, for example, the conjugacy classes Γ_λ. Any loop $f \in \Gamma_\lambda$ is of the form $f(\theta) = \exp(\mu\theta)$ where $\mu \in \Lambda$ lies in the orbit of $\lambda \in \Lambda$ under the adjoint action of G. Thus,

$$E(f) = \frac{1}{4\pi}\int_0^{2\pi} \|f(\theta)^{-1}f'(\theta)\|^2 d\theta = \frac{1}{2}\|\mu\|^2 = \frac{1}{2}\|\lambda\|^2,$$

a constant, as the norm $\|\ \|$ on $L(G)$ is G-invariant. On the other hand,

$$p(f) = pr_{L(T)}\left(\frac{1}{2\pi}\int_0^{2\pi} f(\theta)^{-1}f'(0)\, d\theta\right) = pr_{L(T)}(\mu),$$

so identifying Γ_λ with the adjoint orbit of λ, we see that $p\colon \Gamma_\lambda \longrightarrow L(T)$ is precisely the function considered in [1]. Thus, the image of Γ_λ under the moment map is a compact convex polyhedron lying inside the 'horizontal' hyperplane $E = \frac{1}{2}\|\lambda\|^2$.

The images of the conjugacy classes Γ_λ in the case $G = SU(3)$ are shown in fig. 4 for $\lambda = (0,0,0), (1,0,-1), (1,1,-2), (2,-1,-1)$. Their vertices are given by:

$$\begin{array}{lll}
\Gamma_{(0,0,0)} \cong \text{point} & & 0 \\
\Gamma_{(1,0,-1)} \cong \{\text{full flags in } \mathbf{C}^3\} & & A_1, A_2, A_3, A_4, A_5, A_6 \\
\Gamma_{(2,-1,-1)} \cong P_2(\mathbf{C}) & & B_1, B_2, B_3 \\
\Gamma_{(1,1,-2)} \cong P_2(\mathbf{C}) & & C_1, C_2, C_3 \quad .
\end{array}$$

Notice that more than one conjugacy class can lie at the same energy level ($\Gamma_{(2,-1,-1)}$ and $\Gamma_{(1,1,-2)}$ in fig.4.). This means that the part of the image of the loop space lying at such an energy level is not simply the union of the images of the corresponding conjugacy classes, but rather its convex hull.

This discussion shows that the study of the moment map on the loop space includes that for all the complex flag manifolds as considered in [1].

2. Using the symmetry of the image of Ω^1 under the affine Weyl group of G, as discussed at the end of §3, we can give a rather more precise description of the image than that afforded by theorem 1. We restrict ourselves to the case $G = SU(l+1)$, though the results are presumably true in general.

Consider the finite set of integer lattice points which are closest to the origin in $L(T)$, other than the origin itself. They form a single Weyl group orbit. Their convex hull can be expressed uniquely as the union of $|W|$ l-simplices, any two having at most a face in common. The images of these simplices under the action of W_{aff} form a tiling of $L(T)$ by l-simplices. We "lift" each l-simplex in $L(T)$ to one in $L(T) \oplus \mathbf{R}$ by taking the convex hull of the graph of the function $\frac{1}{2}\|\ \|^2$ restricted to its $(l+1)$ vertices. We claim that these l-simplices in $L(T) \oplus \mathbf{R}$ are precisely the faces of the image of Ω^1. To prove this it is sufficient to show that they bound a *convex* polyhedron in $L(T) \oplus \mathbf{R}$. This in turn reduces to the following assertion: Choose one of the original l-simplices in $L(T)$, with vertices $0, \lambda_1, \ldots, \lambda_l$, say, and consider the neighbouring simplex with vertices $\lambda_1, \ldots, \lambda_l, \lambda_{l+1}$, say. Then the lifts of these two simplices make an acute angle with each other. (Note that any l-simplex in $L(T) \oplus \mathbf{R}$ lies in a codimension 1 (affine) hyperplane. The angle between two such hyperplanes is that between their normals, where the orientation of the normal is such that it has a positive component in the "$\mathbf{R}$"-direction).

In the case of $SU(l+1)$ we have the following:

(i) $\lambda_{l+1} = \frac{2}{l}(\lambda_1 + \cdots + \lambda_l)$;
(ii) $\| \lambda_i \|^2 = 2, \langle \lambda_i, \lambda_j \rangle = 1$ if $i \neq j$.

Thus, the assertion to be proved is

$$\left\| \frac{2}{l} \sum_{i=1}^{l} \lambda_i \right\|^2 \geq \frac{2}{l} \sum_{i=1}^{l} \|\lambda_i\|^2,$$

i.e., $\frac{4}{l^2}(2l + l(l-1)) \geq \frac{2}{l} \cdot 2l$ which is true for $l \geq 0$.

Fig. 5 illustrates the situation when $l = 2$.

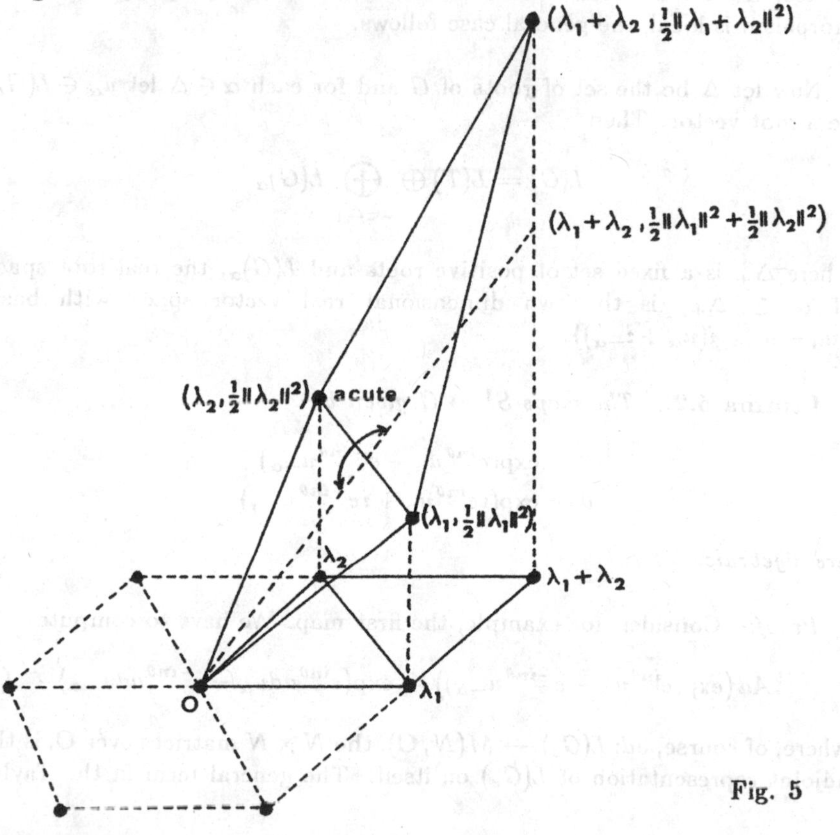

Fig. 5

5. Density of the Algebraic Loops

In this section we give the proof of theorem 2. Recall that an algebraic loop in G is a map $S^1 \to G$ which extends to a morphism of algebraic varieties $\mathbf{C}^* \to G_c$. We begin with a characterisation of such loops.

Lemma 5.1. *A map $f\colon S^1 \longrightarrow G$ is algebraic if and only if the composite map $S^1 \xrightarrow{f} G \xrightarrow{Ad} U(N)$ has a finite Laurent expansion. Here, Ad is the adjoint representation of G on $L(G)$, and $N = \dim G$.*

Proof. Only the 'if' part requires proof, and this would follow from general properties of algebraic varieties if the adjoint representation $Ad\colon G_c \longrightarrow GL(N,\mathbf{C})$ were an embedding. However, as G is simply connected, G_c has at most a finite centre, and since the property of being a morphism is local, the general case follows.

Now let Δ be the set of roots of G and for each $\alpha \in \Delta$ let $u_\alpha \in L(G_c)$ be a root vector. Then

$$L(G) = L(T) \oplus \bigoplus_{\alpha\in\Delta_+} L(G)_\alpha$$

where Δ_+ is a fixed set of positive roots and $L(G)_\alpha$, the real root space of $\alpha \in \Delta_+$, is the two dimensional real vector space with basis $\{u_\alpha - u_{-\alpha}, i(u_\alpha + i_{-\alpha})\}$.

Lemma 5.2. *The maps $S^1 \to G$ given by*

$$\theta \to \exp(e^{in\theta}u_\alpha - e^{-in\theta}u_{-\alpha})$$
$$\theta \to \exp(ie^{in\theta}u_\alpha + ie^{-in\theta}u_{-\alpha})$$

are algebraic.

Proof. Consider, for example, the first map. We have to compute

$$Ad\left(\exp(e^{in\theta}u_\alpha - e^{-in\theta}u_{-\alpha})\right) = \exp(e^{in\theta}\,ad u_\alpha - e^{-in\theta}\,ad u_{-\alpha}) \qquad (*)$$

where, of course, $ad\colon L(G_c) \longrightarrow M(N,\mathbf{C})$, the $N\times N$ matrices over $\mathbf{C}$, is the adjoint representation of $L(G_c)$ on itself. The general term in the Taylor expansion of $(*)$ is of the form

$$e^{in(r-s)\theta} \times (\text{a product of } r \; ad u_\alpha\text{'s and } s \; ad u_{-\alpha}\text{'s in some order})$$

It suffices to show that any such product vanishes if $|r-s|$ is large. If not, then either

(i) (product). $u_\beta \neq 0$ for some u_β, or
(ii) (product). $t \neq 0$ for some $t \in L(T_c)$.

This implies the existence in $L(G_c)$ of an element of weight (for the adjoint representation)

(i) $(r-s)\alpha + \beta$, or
(ii) $(r-s)\alpha$

and this is impossible if $|r-s|$ is very large.

Now the elements

$$\{e^{in\theta}u_\alpha - e^{-in\theta}u_{-\alpha'}, \quad ie^{in\theta}u_\alpha + ie^{-in\theta}u_{-\alpha} \mid \alpha \in \Delta_+, \quad n \in \mathbf{Z}\}$$

span the subspace

$$V = H^1\Big(S^1, \bigoplus_{\alpha\in\Delta_+} L(G)_\alpha\Big) \subset H^1(S^1, L(G)) = L(M^1).$$

But, $L(\overline{M^{alg}})$ (where the closure is in M^1) is obviously invariant under the adjoint action of G, and since G is semi-simple, $G.V = L(M^1)$. Hence, $L(\overline{M^{alg}}) = L(M^1)$.

That M^{alg} is dense in M^1, and hence theorem 2 now follows from

Lemma 5.3. *Let $\mathcal{H}$ be a closed subgroup of a Banach Lie group $\mathcal{G}$. Then the elements $\xi \in L(\mathcal{G})$ such that the 1-parameter group $t \to \exp(t\xi)$ lies in $\mathcal{H}$ form a closed subspace $L(\mathcal{H})$ of $L(\mathcal{G})$.*

Proof. This is presumably well-known. In any case, it suffices to prove that $L(\mathcal{H})$ is closed under addition, and this follows from the formula

$$\exp(\xi + \eta) = \lim_{n\to\infty}\left(\exp\frac{\xi}{n}\exp\frac{\eta}{n}\right)^n.$$

Remarks 1. Theorem 2 is true for any compact semi-simple Lie group G, but not for a general compact Lie group. For example, if $G = U(n)$, any map $f \in M^{alg}$ has determinant det $f\colon S^1 \longrightarrow S^1$ which is an algebraic map, that is a monomial. For a similar reason, (5.1) is false for $G = U(n)$: it would imply that any map $S^1 \to S^1$ is algebraic.

2. It has been pointed out to us by Segal that a version of theorem 2 holds for maps of the 2-sphere $S^2 \to G$.

References

[1] M. F. Atiyah, Convexity and commuting Hamiltonians, Bull. London Math. Soc. *14* (1982), 1–15.

[2] N. Bourbaki, Groupes et algèbres de Lie, Ch. 4–6, Hermann (Paris) 1968.

[3] V. Guillemin & S. Sternberg, Convexity properties of the moment mapping, Invent. Math. *67*, 491–513 (1982).

[4] G. Heckman, Thesis, Leiden (1980).

[5] N. Iwahori & H. Matsumoto, On some Bruhat decompositions and the structure of Hecke rings of p-adic Chevalley groups, Publ. Math. I.H.E.S. (Paris) *25* (1965), 5–48.

[6] V. G. Kac, Simple irreducible graded Lie algebras of finite growth, Izv. Akad. Nauk. *32* (1968), 1271–1311.

[7] A. A. Kirillov, Elements of the theory of representations, Springer-Verlag 1978.

[8] B. Kostant, On convexity, the Weyl group and the Iwasawa decomposition, Ann. Sci. Éc. Norm. Sup. *6* (1973), 413–455.

[9] R. S. Palais, Morse theory on Hilbert manifolds, Topology *2* (1962), 299–340.

[10] A. N. Pressley, Decompositions of the space of loops on a Lie group, Topology *19* (1980), 65–79.

[11] A. N. Pressley, Thesis, University of Oxford (1980).

[12] A. N. Pressley, The energy flow on the loop space of a compact Lie group, J. London Math. Soc. (to appear).

[13] G. B. Segal, Unitary representations of some infinite dimensional groups, Commun. Math. Phys. *80* (1981), 301–342.

Received May 7, 1982

Professor Michael Francis Atiyah
Mathematical Institute
24-29 St. Giles
Oxford OX1 3LB, England

Dr. Andrew Nicholas Pressley
Mathematical Institute
24-29 St. Giles
Oxford OX1 3LB, England

The Jacobian Conjecture and Inverse Degrees

Hyman Bass

Dedicated to I. R. Shafarevich

1. Introduction

If k is a commutative ring and n is a positive integer we have the polynomial algebra $k^{[n]} = k[x_1, \ldots, x_n]$ and the affine space $\mathbb{A}_k^n = \operatorname{Spec}(k^{[n]})$. A k-endomorphism F of $\mathbb{A}_k^n$ will be identified with its sequence $F = (F_1, \ldots, F_n)$ of coordinate functions $F_i \in k^{[n]}$ $(i = 1, \ldots, n)$. Its Jacobian matrix is $J(F) = (\partial F_i / \partial X_j)$. From the chain rule,

$$J(G(F)) = J(G)(F) \cdot J(F),$$

we see that, if F is invertible then $J(F)$ is invertible, i.e., $\det J(F) \in (k^{[n]})^\times$. (If k is reduced, i.e., has no nonzero nilpotent elements, then $(k^{[n]})^\times = k^\times$.) The Jacobian Conjecture asserts the converse when k is a field of characteristic zero: If $J(F)$ is invertible then F is invertible.

We here study this conjecture in terms of

$$\deg(F) = \max \deg(F_i).$$

Specifically, consider the following assertion, in which Λ denotes a class of commutative rings.

$JC(\Lambda, n, d)$: *If $k \in \Lambda$ and F is a k-endomorphism of $\mathbb{A}_k^n$ of degree $\leq d$ with $J(F)$ invertible, then F is invertible.*

We shall relate $JC(\Lambda, n, d)$ to the following "bounded inverse" assertion:

$BI(\Lambda, n, d)$: *There is a constant $c = c(\Lambda, n, d)$ such that if $k \in \Lambda$ and F is a k-automorphism of $\mathbb{A}_k^n$ of degree $\leq d$ with* $\det J(F) \in k^\times$, *then* $\deg(F^{-1}) \leq c$.

If k is a commutative ring we shall, in the following results, write "k-alg" for the class of k-algebras, and "k-field" for the subclass of those that are fields.

The main conclusion of the results presented here is the following. (This is (b) $\Rightarrow$ (a) of Theorem (2.2) below.)

Theorem (1.1). *Let k be a commutative noetherian ring. Then*

$$BI(k\text{-}alg, n, d) \Rightarrow JC(k\text{-}alg, n, d)$$

for all $n, d \geq 1$.

The following converse was proved in [BWC].

Proposition (1.2). ([BCW],Ch. I, Prop. (1.2).) *Let k be a commutative noetherian ring. Then*

$$JC(k\text{-}field, n, d) \Rightarrow BI(k\text{-}alg, n, d)$$

for all $n, d \geq 1$.

This was stated in [BCW] for a subring k of $\mathbf{Q}$, but the proof applies unchanged in the above generality.

The validity of the Jacobian Conjecture for quadratic maps ($d = 2$) can be stated as follows.

Theorem (1.3). (S. Wang [S]; see [BCW], Ch. II, Theorem (2.4.)) *Let $k = \mathbf{Z}[1/2]$. Then $JC(k\text{-}alg, n, 2)$, and therefore also $BI(k\text{-}alg, n, 2)$ (by (1.2)), hold for all $n \geq 1$.*

The following result was communicated by O. Gabber, and will play a basic role in what follows.

Theorem (1.4). ([BCW], Ch. I, Cor. (1.4)) *$BI(\mathbf{Z}\text{-}field, n, d)$ holds for all $n, d \geq 1$, with $c = d^{n-1}$, where "$\mathbf{Z}$-field" denotes the class of all fields.*

In fact one can easily deduce the same result with "Z-field," replaced by the class of all *reduced* commutative rings. However, in the presence of nilpotent elements we can only obtain the following result. (See Proposition (2.8) below.)

Proposition (1.5). *Let Λ_e denote the class of commutative rings whose nil radicals have $(e+1)^{st}$ power equal to zero. Then $BI(\Lambda_e, n, d)$ holds for all $n, d \geq 1$, with a constant $c = c(e, n, d)$ depending on e.*

In this optic, using Theorem (1.1), we can deduce that the Jacobian Conjecture is equivalent to making c in (1.5) independent of e. Since the Jacobian Conjecture is false in characteristic $p > 0$, we can only hope to make c independent of e for $\mathbf{Q}$-algebras.

In the final section we note (Remark (3.4)) how the considerations here are related to Shafarevich's study [Sh 1] and [Sh 2] of $\mathrm{Aut}(\mathbf{A}^n)$ as an infinite dimensional algebraic group.

This paper was stimulated by a conversation about Shafarevich's papers with T. Kambayashi, to whom I am grateful. Kambayashi's observation is explained in detail in Remark (3.5) below.

2. The Main Theorem

We fix an integer $n \geq 1$.

Consider the following monoid valued functors defined for commutative rings k:

$$E(k) \supset J(k) \supset G(k).$$

$$E(k) = \{F \in \mathrm{End}_k(\mathbf{A}_k^n) \mid F(0) = 0, dF(0) = \text{identity}\}$$

Explicitly $F \in E(k)$ iff $F = X + F'$ where $X = (X_1, \ldots, X_n)$ (the identity endormophism) and each F'_i involves only monomials of degree ≥ 2.

$$J(k) = \{F \in E(k) \mid \det J(F) = 1\}$$

$$G(k) = \{F \in J(k) \mid F \quad \text{is invertible}\}.$$

Thus the Jacobian Conjecture asserts that $G(k) = J(k)$ when k is a field of characteristic zero. (Cf. [BCW], Ch. I, Remark (1.1)1.)

The above objects are "infinite dimensional" over k, but they are stratified by "finite dimensional" layers,

$$E_d(k) = \{F \in E(k) \mid \deg(F) \leq d\}$$
$$J_d(k) = J(k) \cap E_d(k)$$
$$G_d(k) = G(k) \cap E_d(k).$$

Clearly E_d can be identified with the functor of points of an affine space $\mathbf{A}^{N_d}$ with coordinates $c_{i,M}$ indexed by $i = 1, \ldots, n$ and M varying over monomials in $x_1, \ldots, x_n$ of degrees $2, \ldots, d$. We shall thus interpret E_d as an affine scheme (defined over $\mathbf{Z}$). Moreover, J_d is evidently a closed subscheme of E_d, defined by the system of equations arising from the condition, "$\det J(F) = 1$." It is not however clear whether G_d corresponds to a closed subscheme of J_d. Indeed we shall see (Theorem (2.2) below) that the Jacobian Conjecture is equivalent to G_d being a closed subscheme of J_d over $\mathbf{Q}$.

In any case, however, we can express G_d as an ascending union of closed subschemes.

$$G_d = \bigcup_{r=1}^{\infty} G_{d,r}$$

where

$$G_{d,r}(k) = \{F \in G_d(k) \mid \deg(F^{-1}) \leq r\}.$$

Lemma (2.1). *Each $G_{d,r}$ defines a closed subscheme of E_d.*

Indeed, each $F \in E_d(k)$ has a formal inverse

$$H = (H_1, \ldots, H_n) : H_i \in k[[X_1, \ldots, X_n]] \text{ and } H_i(F) = X_i (i = 1, \ldots, n).$$

(See [BCW], Ch. III, Cor. (2.2) for an explicit formula for H.) Let $c_{i,M}(H)$ denote the coefficient of the monomial M in H_i. Then $c_{i,M}(H)$ is given by a $\mathbf{Z}$-polynomial $f_{i,M}(F)$ in the coefficients of F. Now we see that $G_{d,r}$ is defined by the (infinite) system of equations, $f_{i,M}(F) = 0$ whenever $\deg(M) > r$.

We can now state our main theorem.

Theorem (2.2). *Let k be a commutative noetherian ring and let d be an integer ≥ 1. The following conditions are equivalent.*

(a) *$G_d(k') = J_d(k')$ for all k-algebras k'. (Jacobian Conjecture in degree d over k.)*

(b) *For some $c \geq 1, G_d(k') = G_{d,c}(k')$ for all k-algebras k'.*

(c) *G_d defines a closed subscheme of J_d over k.*

(d) *For all k-algebras k' which are algebraically closed fields, $G_d(k')$ contains a Zariski neighborhood of X (the identity endomorphism) in $J_d(k')$.*

The implication (a) ⇒ (b) follows from Proposition (1.2) above, proved in [BCW]. The implication (b) ⇒ (c) results from Lemma (2.1). The implication (c) ⇒ (d) is a consequence of Lemma (2.4) below. Following that we shall prove (d) ⇒ (a).

Lemma (2.3). *Let C be a commutative ring, N a nil ideal of C, and $\overline{C} = C/N$. If $\overline{F} \in G_d(\overline{C})$ then $G_d(C)$ contains the fiber over $\overline{F}$ of $J_{(d)}(C) \to J_{(d)}(\overline{C})$.*

Given $F \in J_d(C)$ such that $\overline{C} \otimes_C F = \overline{F}$ is invertible, we must show that F is invertible. After replacing C by a finitely generated subring C_0 containing the coefficients of F, and so that $C_0/C_0 \cap N$ contains the coefficients of $\overline{F}^{-1}$, we can reduce to the case when C is noetherian and hence N is nilpotent. Then the assertion follows from Nakayama lemma argument, as in [BCW], Ch. I, Remark (1.1)6. Explicitly, put $D = \mathrm{Coker}(C^{[n]} \xrightarrow{F} C^{[n]})$, where F here represents the C-algebra endomorphism of $C^{[n]}$ sending X_i to $F_i (i = 1, \ldots, n)$. Since $C/N \otimes_C F = \overline{F}$ is an automorphism, we have $C/N \otimes_C D = 0$. Thus $D = ND = N^2 D = \cdots$. Since N is nilpotent, it follows that $D = 0$. Now since $C^{[n]}$ is noetherian, the surjective endomorphism F is an isomorphism; otherwise $\mathrm{Ker}(F^n)$ is a strictly ascending chain of ideals.

Remark. The same proof shows that if $F \in \mathrm{End}_C(C^{[n]})$ and if $\overline{C} \otimes_C F$ is invertible, then F is invertible.

Lemma (2.4). *Suppose that G_d defines a closed subscheme of J_d over k. Let $\varphi: A \longrightarrow B$ denote the surjection of affine k-algebras corresponding to the inclusion of G_d in J_d. Then the localization $\varphi_X: A_X \longrightarrow B_X$ at the identity X is an isomorphism.*

Let e be a positive integer. Consider the commutative diagram

$$\begin{array}{ccc} A & \xrightarrow{p} & A/m_X^e \\ \varphi \downarrow & & \downarrow q \\ B & \xrightarrow[X]{\delta} & k = A/m_X \end{array}$$

where m_X is the maximal ideal of X in A, p and q are canonical projections, and $X \in G_d(k) = \mathrm{Hom}(B, k)$. Since $p \in \mathrm{Hom}(A, A/m_X^e) = J_d(A/m_X^e)$ lies over $X \in J_d(A/m_X) = J_d(k)$, it follows from Lemma (2.3) that there is a δ: $B \longrightarrow A/m_X^e$ leaving the diagram commutative. It follows that $\mathrm{Ker}(\varphi) \subset m_X^e$. This being so for all $e \geq 1$, the lemma follows.

The proof that (d) $\Rightarrow$ (a) will use the following observation.

Lemma (2.5). *There is a functorial action $F \mapsto F^t$ of the multiplicative monoid of k on the monoid $E_d(k)$ ($F \in E_d(k), t \in k$), defined as follows: Write $F = X + F_{(2)} + \cdots + F_{(d)}$ where, for each j, the components of $F_{(j)}$ are homogeneous polynomials of degree j; then $F^t = X + tF_{(2)} + \cdots + t^{d-1}F_{(d)}$. If F belongs to $J_d(k)$ or to $G_d(k)$ then the same is true of F^t for all $t \in k$. The orbit map $t \to F^t$ defines a morphism from the affine line into $E_d(k)$ joining $F = F^1$ to $X = F^0$. Hence J_d and G_d are connected.*

That the formula for F^t satisfies $F^1 = F$, $(F^s)^t = F^{st}$, and $(F \circ H)^t = F^t \circ H^t$ follows from [BCW2], Example (2.3). It is evidently a scheme theoretic action preserving G_d. Since

$$J(F)^t(X) = I + tJ(F_{(2)}) + \cdots + t^{d-1}J(F_{(d)}) = J(F)(tX),$$

the action likewise preserve J_d. The other assertions are now clear.

Remark. If $t \in k^\times$, then F^t is just the conjugate of F by the linear automorphism tX of $\mathbb{A}_k^n$.

As a first step to proving (d) $\Rightarrow$ (a) we show that (d) implies:

(d′) *For all k-algebras k' which are algebraically closed fields, $G_d(k') = J_d(k')$.*

Indeed, if k' is as in (d′) then, by (d), $G_d(k')$ contains a Zariski neighborhood V of X in $J_d(k')$. From Lemma (2.5) we see that V meets every orbit of the action of $k'^\times$ on $J_d(k')$. Since $V \subset G_d(k')$ and $G_d(k')$ is $k^\times$-invariant (Lemma (2.5) again), (d′) follows.

Finally, (d′) $\Rightarrow$ (a): For $r \geq 1$ let φ_r: $A \longrightarrow B_r$ denote the surjection of affine k-algebras corresponding to the inclusion of $G_{d,r}$ in J_d. Let N denote the nil radical of A. Since A is noetherian $N^{e+1} = 0$ for some $e \geq 0$. Choose r so that

$$r \geq d^{2^e n - 1}. \qquad (*)$$

Since $r \geq d^{n-1}$ it follows from Theorem (1.4) that $G_{d,r}(k') = G_d(k')$ for any algebraically closed field k'. Therefore, by (d'), J_d and $G_{d,r}$ have the same geometric points over any field. It follows then from the Nullstellensatz that $\mathrm{Ker}(\varphi_r) \subset N$. Let F denote the generic element of $J_d(A) = \mathrm{Hom}(A, A)$, corresponding to Id_A. Then $\overline{F} = B_r \otimes_A F \in J_d(B_r)$ is the generic element of $G_{d,r}(B_r) \subset J_d(B_r)$. It follows now from Lemma (2.3) that $F \in G_d(A)$, and so, from (*) and Proposition (2.8) below, $F \in G_{d,r}(A) = \mathrm{Hom}(B_r, A)$. This means that there is a $\psi: B_r \longrightarrow A$ such that $\psi \circ \varphi_r = Id_A$, and so the surjection φ_r is an isomorphism.

This completes the proof of (d) $\Rightarrow$ (a), and so of Theorem (2.1), modulo Proposition (2.8) below.

Let $GA_n(k)$ denote the full group $\mathrm{Aut}_k(\mathbf{A}_k^n)$.

Lemma (2.6). *Let C be a commutative ring and N an ideal of C of square zero. Let Γ denote the additive group of n-tuples $F' = (F'_1, \ldots, F'_n)$ where each F'_i is a polynomial of $C^{[n]}$ with coefficients in N. Then*

$$1 \to \Gamma \overset{\epsilon}{\to} GA_n(C) \overset{\pi}{\to} GA_n(C/N) \to 1,$$

where $\epsilon(F') = X + F'$, is an exact sequence of groups.

It follows from the Remark after Lemma (2.3) that π is surjective with kernel $\epsilon(\Gamma)$. It remains to show that ϵ is a homomorphism. Let $F = X + F'$ and $H = X + H'$ with $F', H' \in \Gamma$. We must show that $F(H) = X + H' + F'$. Since $F(H) = H + F'(H) = X + H' + F'(H)$, we must show that $F'_i(H) = F'_i (i = 1, \ldots, n)$. Since both terms are additive in F'_i it suffices to show this when F'_i is a monomial $aX^p = aX_1^{p_1} \ldots X_n^{p_n}$ with $a \in N$. Then

$$\begin{aligned} F'_i(H) &= a(X_1 + H'_1)^{p_1} \ldots (X_n + H'_n)^{p_n} \\ &= a \sum_{\substack{0 \leq r_i \leq p_i \\ i=1,\ldots,n}} \binom{p_n}{r_1} \cdots \binom{p_n}{r_n} X^{p-r} H'^{r}. \end{aligned}$$

Since $N^2 = 0$ we have $aH'^r = 0$ unless $r = (0, \ldots, 0)$, and so $F'_i(H) = aX^p = F'_i$, as claimed.

Lemma (2.7). *Let C be a commutative ring and N an ideal of C such that $N^{e+1} = 0$. Let $F \in J_d(C)$. Suppose that $C/N \otimes_C F \in J_d(C/N)$ belongs to $G_{d,r}(C/N)$. Then $F \in G_{d,s}(C)$ where $s = r^{2^e} d^{2^e - 1}$.*

We argue by induction on $e \geq 0$, the case $e = 0$ being trivial. Consider the case $e = 1$, when $N^2 = 0$. Choose a lifting $H \in E(C)$ of $(C/N \otimes_C F)^{-1}$ so that $\deg(H) \leq r$. By the Remark after Lemma (2.3) H is invertible. Let $L = H(F) = X + L' \in \operatorname{Ker}(GA_n(C) \to GA_n(C/N))$. Lemma (2.6) implies that $L^{-1} = X - L'$, and $F^{-1} = L^{-1}(H)$. Now $\deg(F^{-1}) \leq \deg(L^{-1}) \cdot \deg(H)$. By construction, $\deg(H) \leq r$. Moreover, $\deg(L^{-1}) = \deg(L) = \deg(H(F)) \leq \deg(H) \cdot \deg(F) \leq r \cdot d$. Thus $\deg(F^{-1}) \leq r^2 d$, as claimed.

In the general case put $\overline{C} = C/N^2$ and $\overline{F} = \overline{C} \otimes_C F \in G_d(\overline{C})$. The case $e = 1$ implies that $\overline{F}$ is invertible with $\deg(\overline{F}^{-1}) \leq \overline{r} = r^2 d$. Since $(N^2)^e = 0$, it follows by induction on e that F is invertible with $\deg(F^{-1}) \leq \overline{r}^{2^{e-1}} d^{2^{e-1}-1} = (r^2 d)^{2^{e-1}} d^{2^{e-1}-1} = r^{2^e} d^{2^e-1}$.

Remark. Evidently the bound for $\deg(F^{-1})$ given in (2.7) is far from optimal.

Proposition (2.8). *Let C be a commutative ring with nil radical N. Suppose that $N^{e+1} = 0$. If $F \in G_d(C)$ then* $\deg(F^{-1}) \leq (d^{n-1})^{2^e} d^{2^e-1} = d^{2^e n-1}$. *In other words, $G_d(C) = G_{d,r}(C)$ for $r \geq d^{2^e n-1}$.*

Let $\overline{C} = C/N$ and $\overline{F} = \overline{C} \otimes_C F \in G_d(\overline{C})$. In view of Lemma (2.7), it suffices to show that $\deg(\overline{F}^{-1}) \leq d^{n-1}$. When $\overline{C}$ is a field this follows from Theorem (1.4) (proved in [BCW]). In general it follows since $\overline{C}$ can be embedded in a cartesian product of fields.

Proposition (2.8) gives an explicit form of Proposition (1.5).

3. Concluding Remarks

We shall discuss matters over $\mathbf{C}$. Fix n and $d \geq 0$. For each $r \geq 1$ let

$$\varphi_r \colon A \longrightarrow B_r$$

denote the surjection of affine $\mathbf{C}$-algebras corresponding to the inclusion of $G_{d,r}$ in J_d, and put $Q_r = \operatorname{Ker}(\varphi_r)$. We have $Q_r \supset Q_{r+1}$ for all $r \geq 1$.

Remark (3.1). If $r \geq d^{n-1}$ then, by Theorem (1.4), $G_{d,r}(\mathbf{C}) = G_d(\mathbf{C})$. Therefore (Nullstellensatz), for $r \geq d^{n-1}$, $B_{r+1} \to B_r$ has a nilpotent kernel, i.e., the radicals $\sqrt{Q_r}$ and $\sqrt{Q_{r+1}}$ coincide. Put $Q = \sqrt{Q_r}$ for $r \geq d^{n-1}$, and $\varphi: A \longrightarrow B = A/Q$.

Remark (3.2). Let m be a maximal ideal of A containing Q. Let e be an integer ≥ 1. Suupose that $r \geq d^{2^e n-1}$. Then we claim that $Q_r \subset m^e$, and so $\bigcap_r Q_r \subset \bigcap_e m^e$. To see this, consider the commutative diagram

$$\begin{array}{ccc} A & \xrightarrow{p} & A/m^e \\ \varphi_r \downarrow & & \downarrow q \\ B_r & \xrightarrow[F]{} & A/m = \mathbf{C} \end{array}$$

where p and q are natural projections and $F \in G_{d,r}(\mathbf{C}) \subset J_d(\mathbf{C})$. Since $p \in J_d(A/m^e)$ lies over F, it follows from Lemma (2.3) that $p \in G_d(A/m^e)$. By Proposition (2.8) and the condition $r \geq d^{2^e n-1}$ we have

$$G_d(A/m^e) = G_{d,r}(A/m^e).$$

This means that $p = h \circ \varphi_r$ for some homomorphism $h: B_r \longrightarrow A/m^e$. It follows, as claimed, that $Q_r = \operatorname{Ker}(\varphi_r) \subset \operatorname{Ker}(p) = m^e$.

Remark (3.3). Let $F \in G_d(\mathbf{C})$ and consider the tangent space $T^F(J_d)$ to J_d at F. If $\mathbf{C}[\epsilon]$ denotes the ring of dual numbers, $\epsilon^2 = 0$, then $T_F(J_d)$ is the fiber over F of $J_d(\mathbf{C}[\epsilon]) \to J_d(\mathbf{C})$. By Lemma (2.3) this fiber is contained in $G_d(\mathbf{C}[\epsilon])$, and by Proposition (2.8) (with $e = 1$) $G_d(\mathbf{C}[\epsilon]) = G_{d,r}(\mathbf{C}[\epsilon])$ for $r \geq d^{2n-1}$. Thus, for $r \geq d^{2n-1}$, the inclusion $G_{d,r} \to J_d$ induces an isomorphism of tangent spaces at each point of $G_{d,r}(\mathbf{C})$. Unfortunately, since we don't know that the scheme $G_{d,r} = \operatorname{Spec}(B_r)$ is reduced, we are not entitled to the nice geometric consequences that we are tempted to derive from this. For example $G_{d,r}$ and J_d are isomorphic in a neighborhood of each simple point of $G_{d,r}$; however, we don't know that $G_{d,r}$ has any simple points if it is not reduced.

When $F = X \in G_d(\mathbf{C})$ and $X + F'\epsilon \in T_X(J_d)$, it follows from Lemma (2.6) that $(X + F'\epsilon)^{-1} = X - F'\epsilon$, which still has degree $\leq d$. Thus $T_X(G_{d,r}) \to T_X(J_d)$ is an isomorphism already for $r \geq d$.

Remark (3.4). In [Sh 1] Shafarevich defined an infinite dimensional variety V to be an ascending union of finite dimensional varieties V_d, with each $V_d \subset V_{d+1}$ a closed immersion. A morphism $f: V \longrightarrow W$ of such varieties is required to map each V_d into some $W_{e(d)}$, as an ordinary morphism. He thus defined infinite dimensional algebraic groups, and gave $GA_n(\mathbf{C}) = \mathrm{Aut}(\mathbf{A}^n_{\mathbf{C}})$ as an example.

Consider the subgroup $G(\mathbf{C}) = \{F \in GA_n(\mathbf{C}) \mid F(0) = 0, dF(0) = Id\}$ which we have studied above. Shafarevich first stratified G by the subsets G_d, viewed as contained in the affine space of all n-tuples of polynomials of degree $\leq d$. To show that G_d is a subvariety of the latter, Shafarevich invoked the Jacobian Conjecture, $G_d = J_d$, apparently then unaware that the published proofs of the Jacobian Conjecture were faulty. In fact it follows from Theorem (1.4) that $G_d(\mathbf{C}) = G_{d,r}(\mathbf{C})$ for $r \geq d^{n-1}$ is indeed a closed variety in $J_d(\mathbf{C})$. Indeed, if the inverse map in G is to be a morphism in Shafarevich's sense, then we must have $G_d = G_{d,r}$ for some r. Thus by Theorem (2.2), Shafarevich's first definition is legitimate scheme theoretically iff the Jacobian Conjecture is true.

In a second, more detailed, version of the paper [Sh 2], Shafarevich bypassed the above issue by stratifying G essentially by the subsets $G_{d,d}$.

Remark (3.5). T. Kambayashi observed that if X is a simple point of (the scheme) $G_{d,d}$, then the Jacobian Conjecture in the form $G_{d,d}(\mathbf{C}) = J_d(\mathbf{C})$, follows. Indeed we saw in Remark (3.3) above that $T_X(G_{d,d}) \to T_X(J_d)$ is an isomorphism. Hence, if X is simple on $G_{d,d}$, then $G_{d,d} \to J_d$ is an isomorphism in a neighborhood of X. Now one uses the action of $\mathbf{C}^\times$ (cf. Lemma (2.5)) to deduce, as in the proof of (d) $\Rightarrow$ (d′) in section 2, that $G_{d,d}(\mathbf{C}) = J_d(\mathbf{C})$. Unfortunately it seems unlikely that X is simple on $G_{d,r}(\mathbf{C})$ or on $J_d(\mathbf{C})$. The present paper grew out of an attempt to see what could be further recovered from Kambayashi's idea.

Remark (3.6). Since writing this note I received a preprint from Pekka Nousiainen, "The variety of automorphisms of affine space," which treats some of the same issues considered here. He gives an independent proof that, for k a field, $G_d(k)$ is Zariski closed in $J_d(k)$. (Notation as in section 2.) From this he deduces that $G_d(k) = G_{d,r}(k)$ for some r, but without obtaining the explicit bound $r = d^{n-1}$ given in Theorem (1.4) quoted above. He further derives our Proposition (1.5), just as we do, but again without an explicit constant $c(e, n, d)$.

References

[BCW] H. Bass, E. H. Connell, and D. Wright, The Jacobian Conjecture: Reduction of degree, and formal expansion of the inverse, Bull. Amer. Math. Soc. (1982).

[BCW 2] H. Bass, E. H. Connell, and D. Wright, Locally polynomial algebras are symmetric algebras, Inventiones math., 38 (1977) 279–299.

[Sh 1] I. R. Shafarevich, On some inifinite dimensional groups Rend. di Matematica E delle sue applicagioni, Ser. V, vol. XXV (1966) 208–212.

[Sh 2] I. R. Shafarevich, On some infinite dimensional groups II, Math. USSR Izvestiga, vol. 18 (1982) 214–226.

[W] S. Wang, A jacobian criterion for separability, Jour. Algebra 65 (1980) 453–494.

Received September 28, 1982

Partially supported by the National Science Foundation Grant No. MCS 82-02633

Professor Hyman Bass
Department of Mathematics
Columbia University
New York, New York 10027

Some Observations on the Infinitesimal Period Relations for Regular Threefolds with Trivial Canonical Bundle

Robert L. Bryant and Phillip A. Griffiths

To I.R. Shafarevich

1. Introduction

It is well-known that, aside from algebraic curves, abelian varieties, and a few other isolated cases such as K3 surfaces, the period matrices of a family of algebraic varieties satisfy non-trivial universal infinitesimal period relations. In this note we shall discuss some remarkable properties of *any* local solution to the differential system given by the infinitesimal period relation associated to polarized Hodge structures of weight three with Hodge number $h^{3,0} = 1$.

Motivation for the study of this special case arises from the following algebro-geometric considerations: As is well known, an interesting class of projective algebraic varieties consists of those smooth X satisfying

$$(1.1)\qquad \begin{cases} K_X \cong \mathcal{O}_X \\ q(X) = 0 \qquad \text{if } n = \dim X \geq 2. \end{cases}$$

Examples include elliptic curves, algebraic K3 surfaces, and smooth hypersurfaces $X \subset \mathbf{P}^{n+1}$ of degree $n+2$. For these varieties it is clear that the differential

$$(1.2)\qquad \kappa\colon H^1(X,\Theta) \longrightarrow \bigoplus_{p+q=n} \operatorname{Hom}(H^{p,q}(X), H^{p-1,q+1}(X))$$

of the period mapping is injective (cf. Chapter III in[6]). In fact, the "first piece"

$$(1.3)\qquad \kappa\colon H^1(X,\Theta) \longrightarrow \operatorname{Hom}(H^{n,0}(X), H^{n-1,1}(X))$$

is an isomorphism. This implies the infinitesimal Torelli theorem (Chapter VIII in loc. cit.), and it seems reasonable to expect a global Torelli theorem for X satisfying (1.1). For example, in the case of polarized K3 surfaces there is the theorem of Piateski-Shapiro and Shafarevich [5]. Interestingly, the smooth hypersurfaces of degree $n+2$ in $\mathbf{P}^{n+1}$ are exceptional cases in the recent generic global Torelli theorem of Donagi [4].

When $n = 3$ all of $H^3(X)$ is primitive, and in this paper we shall be interested in variations of Hodge structure that "look like" period matrices of such an X. More precisely, let us make the reasonable assumption that all of $H^1(X, \Theta)$ is unobstructed, so that the local moduli space (Kuranishi space) $\{X_s\}_{s\in S}$ of $X = X_0$ is smooth of dimension $m = h^1(X,\Theta) = h^{2,1}(X)$ with all Kodaira-Spencer mappings

$$\rho_s\colon T_s(S) \longrightarrow H^1(X_s, \Theta)$$

being isomorphisms. The period mapping for this family then gives an m-dimensional integral manifold

$$\varphi\colon S \longrightarrow D \subset \check{D}$$

for the differential system I on $\check{D}$ given by the infinitesimal period relations (this terminology is explained below). It turns out that m is the maximal dimension of integral manifolds of I, and we are able to put the differential system I on $\check{D}$ in local (in the *Zariski topology*) normal form. This in turn gives information on all integral manifolds of I, and from this we may draw several conclusions, two of which are:

(1.4) *Let $\alpha_0, \dots, \alpha_m; \beta_0, \dots, \beta_m \in H_3(X, \mathbf{Z})$ be a canonical homology basis (i.e., the intersection numbers $\alpha_i \cdot \alpha_j = \beta_i \cdot \beta_j = 0$, $\alpha_i \cdot \beta_j = \delta_{ij}$), $\omega(s) \in H^{3,0}(X_s)$ a non-zero generator, and set*

$$\begin{cases} A_i(s) = \int_{\alpha_i} \omega(s) & (A-\text{periods}) \\ B_i(s) = \int_{\beta_i} \omega(s) & (B-\text{periods}) \end{cases}$$

Then the complete Hodge structure $\{H^{p,q}(X_s)\}$ is determined by the functions $A_i(s)$.

More precisely, we shall show that the $B_i(s)$ are canonically and explicitly expressible in terms of the functions $A_i(s)$, $\partial A_i(s)/\partial s^j$; and that

$$H^{3,0}(X_s) \oplus H^{2,1}(X_s)$$

is then expressible in terms of the functions

$$A_i(s), \quad \frac{\partial A_i(s)}{\partial s^j}, \quad \frac{\partial^2 A_i(s)}{\partial s^i \partial s^j}.$$

By (2.2) below we then know the complete Hodge structure.

For the next result we shall use the concept of the infinitesimal variation of Hodge structure associated to X [3], which is given by the linear algebra data (1.2) in the manner expained at the end of this section.

(1.5) *Let $J(X)$ be the intermediate Jacobian of X. Then the infinitesimal variation of Hodge structure associated to X gives an $\otimes^2 H^{0,3}(X)$-valued cubic form Ξ on $T^*_{(0)}(J(X))$. Moreover, the algebraic invariants of this infinitesimal variation of this Hodge structure are uniquely determined by Ξ.*

It is our feeling that, in general, this cubic form uniquely determines X.

The precise statement of (1.5) is given in §2, and the cubic form is further discussed in §§6, 7. The precise formulation of (1.4) is given in §7 (cf. diagram (7.1) and the ensuing interpretation of it).

In concluding this section we want to make some remaks on terminology. Let M be a complex manifold and $\Omega^*_M = \oplus_{q\geq 0}\Omega^q_M$ the sheaf of exterior algebras given by all holomorphic forms. *By a sheaf of differential ideals* or *differential system* we shall mean a subsheaf of graded ideals $I \subset \Omega^*_M$ that is closed under exterior differentiation. An *integral manifold* of I will be given by a complex manifold S together with a holomorphic immersion

$$f: S \longrightarrow M \tag{1.6}$$

satisfying

$$f^*(I) = 0.$$

An *integral element* of I is given by a linear subspace $E \subset T_p(M)$ such that

$$\theta(p)\,|_E = 0 \text{ for all } \theta \in I.$$

Thus (1.6) is an integral manifold if, and only if,

$$f_*(T_s(S)) \subset T_{f(s)}(M)$$

is an integral element for all $s \in S$.

Given a sub-bundle $W \subset T^*(M)$, the holomorphic sections of W generate a differential ideal I (take I to be the ideal generated algebraically by the forms θ^α and $d\theta^\alpha$ where $\theta^1, \ldots, \theta^s$ is a local coframe for W). Given a sub-bundle $V \subset T(M)$, we may set $W = V^\perp$ and take the corresponding differential ideal.

Let D be a classifying space for polarized Hodge structures and $\check{D}$ the dual classifying space for weight n Hodge filtrations

$$F^n \subset F^{n-1} \subset \cdots \subset F^1 \subset F^0 = H$$

satisfying the 1st Hodge-Riemann bilinear relation (cf. Chapter I in [6]). A holomorphic mapping

$$\varphi\colon S \longrightarrow \check{D}$$

may be thought of as a holomorphically varying filtration $\{F^p(s)\}$ on H, and the *infinitesimal period relation*

$$dF^p(s) \subseteq F^{p-1}(s) \tag{1.7}$$

shall mean the following: For $s^1, \ldots, s^m$ local holomorphic coordinates on S and $v(s) \in F^p(s)$, all derivatives

$$\frac{\partial v(s)}{\partial s^i} \in F^{p-1}(s).$$

Thinking of tangent vectors as tangents to curves, (1.7) defines a sub-bundle $T_h(\check{D}) \subset T(\check{D})$, and the corresponding sheaf of differential ideals $I \subset \Omega^*_{\check{D}}$ will be called the *differential system giving the infinitesimal period relation.*

This paper was motivated by trying to see if studying I via the general theory of differential systems would yield any insight into period matrices of algebraic varieties. Our basic point is that in the case $n = 3$ and $h^{3,0} = 1$, I on $\check{D}$ is essentially the 1st prologation of another differential system that is birationally equivalent to the canonical contact system. More precisely, *in this paper we agree that integral manifolds for I in the case $n = 3, h^{3,0} = 1$ shall have the additional property that the map $\psi\colon S \longrightarrow \mathbf{P}H$ given by $\psi(s) = F^3(s) \subset H$ be an immersion,* and then the statement about prolongations is correct. In this way we are able to determine all local maximal integral manifolds of I and to draw the conclusions mentioned above.

One final terminology we shall use is that of an *infinitesimal variation of Hodge structure* $\{H_{\mathbf{Z}}, H^{p,q}, Q, T, \delta\}$. This is given by a polarized Hodge structure $\{H_{\mathbf{Z}}, H^{p,q}, Q\} = p \in D$ together with an injective linear mapping

$$\delta\colon T \longrightarrow T_p(D) \tag{1.8}$$

whose image $E = \delta(T)$ is an integral element for the differential system giving the infinitesimal period relation (1.7). More concretely, (1.8) is given by

$$\delta\colon T \longrightarrow \bigoplus_{p+q=n} \operatorname{Hom}(H^{p,q}, H^{p-1,q+1}) \tag{1.9}$$

satisfying the conditions

$$\begin{cases} (i) & \delta(\xi_1)\delta(\xi_2) = \delta(\xi_2)\delta(\xi_1) \qquad \xi_1, \xi_2 \in T \\ (ii) & Q(\delta(\xi)\eta, \psi) + Q(\eta, \delta(\xi)\psi) = 0 \qquad \xi \in T \end{cases} \tag{1.10}$$

where $\eta \in H^{p,q}, \psi \in H^{n-p+1,q-1}$. We remark that if $\theta^1, \ldots, \theta^s$ are local holomorphic 1-forms on $\check{D}$ such that $\theta^\alpha = 0$ defines $T_h(\check{D}) \subset T(D)$, then (1.9) is equivalent to $\theta^\alpha(p)|_E = 0$ and (i) in (1.10) is equivalent to $d\theta^\alpha(p)|_E = 0$. Condition (ii) in (1.10) is the obvious one that the infinitesimal variation preserve the 1st bilinear relation.

2. Variation of Hodge Structure Associated to Certain 3-Folds

Let $H_{\mathbf{Z}}$ be a lattice of rank $2(m+2)$ (thus $H_{\mathbf{Z}} \simeq \mathbf{Z}^{2m+2}$) having a non-degenerate alternating bilinear form

$$Q\colon H_{\mathbf{Z}} \otimes H_{\mathbf{Z}} \longrightarrow \mathbf{Z}.$$

We will be concerned with polarized Hodge structure $\{H_{\mathbf{Z}}, H^{p,q}, Q\}$ of weight three on $H = H_{\mathbf{Z}} \otimes \mathbf{C}$ having Hodge numbers

$$h^{3,0} = 1, \; h^{2,1} = m.$$

This is given either by a Hodge decomposition

$$\begin{cases} H = H^{3,0} \oplus H^{2,1} \oplus H^{1,2} \oplus H^{0,3} \\ H^{p,q} = \overline{H^{q,p}} \end{cases}$$

satisfying the 1st and 2nd Hodge-Riemann bilinear relations, or by the corresponding Hodge filtration

$$(0) \subset F^3 \subset F^2 \subset F^1 \subset F^0 = H \tag{2.1}$$

where $F^p = \bigoplus_{k\geq 0} H^{p+k,3-p-k}$. We remark that the 1st bilinear relation is

$$\begin{cases} F^1 = F^{3\perp} \\ F^2 = F^{2\perp} \end{cases} \tag{2.2}$$

where $\perp$ denotes the orthogonal complement with respect to Q.

Motivated by the study of variation of Hodge structure for threefolds satisfying (1.1), we want to study holomorphic mappings,

$$\varphi \colon S \longrightarrow \check{D} \tag{2.3}$$

where S is a complex manifold and where the following conditions are satisfied:

(i) $\dim S = m = h^{2,1}$ and the composed mapping

$$S \overset{\varphi}{\to} \check{D} \overset{\pi}{\to} \mathbf{P}H \simeq \mathbf{P}^{2m+1} \tag{2.4}$$

is an immersion, where $\pi\{F^p\} = \{F^3 \subset H\}$ (this is a reflection of (1.3));

(ii) the infinitesimal period relations

$$\begin{cases} (i) & dF^3 \subseteq F^2 \\ (ii) & dF^2 \subseteq F^1 \end{cases} \tag{2.5}$$

are satisfied (this notation is explained in §1 – cf. just below (1.7)).

We remark that, in this case, $(i) \Rightarrow (ii)$ in (2.5). To see this let $s^1, \ldots, s^m$ be local holomorphic coordinates on S and write $\varphi(s) = \{F^p(s)\}$. If $0 \neq z(s) \in F^3(s)$, then (i) in (2.5) gives

$$\frac{\partial z(s)}{\partial s^i} \in F^2(s) \qquad i = 1, \ldots, m.$$

By (2.2) this implies that

$$(2.6)\qquad \begin{cases} (i) & Q\left(z(s), \dfrac{\partial z(s)}{\partial s^j}\right) = 0 \\ (ii) & Q\left(\dfrac{\partial z(s)}{\partial s^i}, \dfrac{\partial z(s)}{\partial s^j}\right) = 0 \end{cases}$$

The derivatives of (i) give, using (ii), that

$$(2.7)\qquad Q\left(z(s), \frac{\partial^2 z(s)}{\partial s^i \partial s^j}\right) = 0$$

On the other hand, since (2.4) is an immersion

$$\dim \operatorname{span}\left\{z(s), \frac{\partial z(s)}{\partial s^1}, \ldots, \frac{\partial z(s)}{\partial s^m}\right\} = m+1.$$

This implies that

$$F^2(s) = \operatorname{span}\left\{z(s), \frac{\partial z(s)}{\partial s^1}, \ldots, \frac{\partial z(s)}{\partial s^m}\right\},$$

and then (2.2) and (2.7) give that

$$dF^2(s) \subseteq F^1(s).$$

Next we want to comment on an infinitesimal variation of Hodge structure $\{H_{\mathbf{Z}}, H^{p,q}, Q, T, \delta\}$ where $\{H_{\mathbf{Z}}, H^{p,q}, Q\}$ is a Hodge structure of the type we are considering and where

$$\delta\colon T \longrightarrow \operatorname{Hom}(H^{3,0}, H^{2,1})$$

is an isomorphism (here we may think of $T = T_s(S)$ and $\delta = \varphi_*$). Using the natural identifications

$$\begin{cases} H^{1,2} \simeq H^{2,1*} \\ H^{0,3} \simeq H^{3,0*} \end{cases}$$

the infinitesimal variation of Hodge structure induces maps

$$(2.8)\qquad \begin{cases} (i) & T \otimes H^{3,0} \xrightarrow{\sim} H^{2,1} \\ (ii) & T \otimes H^{2,1} \rightarrow H^{2,1*} \\ (iii) & T \otimes H^{2,1*} \xrightarrow{\sim} H^{3,0*} \end{cases}$$

where (i) is an isomorphism, (ii) is a symmetric map, and (iii) is the dual of (i). These three combine to induce a map

$$\delta^{(3)}: T \otimes T \otimes T \longrightarrow \otimes^2 H^{0,3} \qquad (\simeq \mathbf{C}), \tag{2.9}$$

and (i) in (1.10) implies that this mapping is *symmetric;* i.e., we have an induced map

$$\Xi: \mathrm{Sym}^3\, T \longrightarrow \otimes^2 H^{0,3} \tag{2.10}$$

Definition. We shall call (2.10) the *cubic form* Ξ associated to the infinitesimal variation of Hodge structure $\{H_{\mathbf{Z}}, H^{p,q}, Q, T, \delta\}$.

Given Ξ we may define $\delta^{(3)}$ in (2.9), and then if we set $H^{2,1} = T \otimes H^{0,3^*}$ we may define the maps (2.8) where (i), (iii) are the identity and (ii) is given by $\delta^{(3)}$. In this way, any algebraic invariant of the linear algebra data (1.2) (in this case) is uniquely determined by Ξ. This is our assertion (1.5).

For a mapping (2.3) satisfying the condition that (2.4) be an immersion, we have for each $s \in S$ a cubic form $\Xi(s)$ on $T_s(S)$ with values in $\otimes^2 H^{0,3}(s)$. To compute $\Xi(s)$ we let $s^1, \ldots, s^m$ be local holomorphic coordinates on S and $0 \neq z(s) \in F^3(s)$. For $\xi = \sum_i \xi^i \partial/\partial s^i \in T_s(S)$, it follows from (2.6) and (2.7) that

$$\Xi(s)(\xi) = Q\left(z(s), \sum_{i,j,k} \xi^i \xi^j \xi^k \frac{\partial^3 z(s)}{\partial s^i \partial s^j \partial s^k}\right) \tag{2.11}$$

is a well-defined function with values in $H^{0,3}(s)$, and this is the cubic form $\Xi(s)$.

For a geometrically given infinitesimal variation of Hodge structure (1.2) where X is a 3-fold satisfying (1.1), the cubic form is given as follows: Represent $H^1(X,\Theta)$ by Dolbeault cohomology $H^{0,1}_{\bar\partial}(X,\Theta)$ and define

$$\det: H^1(X,\Theta) \longrightarrow H^3(X, K^*_X) \tag{2.12}$$

by

$$\det\left(\sum_{i,j} \theta^i_{\bar j} \partial/\partial z^i \otimes d\bar z^j\right) = \det\|\theta^i_{\bar j}\| \cdot \wedge^i \partial/\partial z^i \otimes \wedge^j d\bar z^j .$$

Using the natural isomorphism

$$H^3(X, K_X^*) \simeq \otimes^2 H^{0,3}(X)$$

resulting from $K_X \simeq \mathcal{O}_X$, it follows from standard arguments (cf. [3]) that the cubic form is given by (2.12).

In the case of a smooth quintic hypersurface $X \subset \mathbf{P}^4$ given by $F(x) = 0$, we let

$$\begin{cases} S_d = \text{forms } P(x) \text{ of degree } d \\ J_{F,d} = \text{Jacobian ideal in degree } d. \end{cases}$$

Then (cf. [3]) the Grothendieck residue symbol gives an isomorphism

$$\text{Res} : S_{15}/J_{F,15} \simeq \mathbf{C},$$

and the cubic form is the natural mapping

$$\text{Sym}^3(S_5/J_{F,5}) \rightarrow S_{15}/J_{F,15} \simeq \mathbf{C}$$

indiced by multiplying polynomials (loc. cit.).

3. Contact Systems and Legendre Manifolds

A *contact manifold* is given by a complex manifold M of dimension $2m+1$ together with a holomorphic line sub-bundle $L \subset T^*(M)$ such that, if ω is a local generator of $\mathcal{O}(L) \subset \Omega^1_M$, then

$$\omega \wedge (d\omega)^m \neq 0. \tag{3.1}$$

It is easy to verify that this condition is independent of ω.

Given a contact manifold (M, L), a *Legendre manifold* is given by an m-dimensional complex manifold S together with an immersion

$$f \colon S \longrightarrow M \tag{3.2}$$

satisfying

$$f^*(\omega) = 0. \tag{3.3}$$

Put differently, let $I \subset \Omega_M^* = \oplus_{q \geq 0} \Omega_M^q$ be the sheaf of differential ideals generated over Ω_M^* by ω and $d\omega$. An integral manifold of I is given by a holomorphic immersion (3.2), where now S may be a complex manifold of any dimension, satisfying

$$f^*(I) = 0. \tag{3.4}$$

Since $f^*: \Omega_M^* \longrightarrow \Omega_S^*$ is a map of differential algebras, (3.3) and (3.4) are equivalent conditions, and so Legendre manifolds are simply m-dimensional integral manifolds of I.

Let $p \in M$ and $E \subset T_p(M)$ be an integral element of I. This is equivalent to

$$\omega(p)|_E = 0, \quad d\omega(p)|_E = 0.$$

It is well-known that any such integral element has $\dim E \leq m$. It follows that integral manifolds of I have dimension $\leq m$, and those of maximal dimension are exactly the Legendre manifolds.

Given a contact manifold (M, L) and local generator ω of $\mathcal{O}(L)$, by the well known theorem of Pfaff-Darboux (cf. [2]) we may choose local holomorphic coordinates $(x^1, \ldots, x^m, u, y_1, \ldots, y_m) = (x, u, y)$ for M such that

$$\omega = du - \sum_{i=1}^{m} y_i dx^i.$$

For a "general" Legendre manifold we will have $f^*(dx^1 \wedge \cdots \wedge dx^m) \neq 0$, and then locally the Legendre manifold is given parametrically by

$$x \to (x, u(x), y(x)).$$

The condition (3.3) is

$$y_i(x) = \frac{\partial u(x)}{\partial x^i},$$

so that the Legendre manifold is locally given by the 1-jet $(x, u(x), \partial u(x)/\partial x)$ of an arbitrary function $u(x)$.

In the next section we will give a global algebro-geometric version of this construction.

4. The Canonical Contact System and its Legendre Manifolds

We will describe a canonical example of a contact manifold and determine its Legendre manifolds. Let V be an $(m+2)$-dimensional vector space with coordinates $w^0, w^1, \ldots, w^{m+1}$; denote by $p_0, p_1, \ldots, p_{m+1}$ the dual coordinates in V^*. In $\mathbf{P}V \times \mathbf{P}V^*$ we consider the *incidence subvariety*

$$R \subset \mathbf{P}V \times \mathbf{P}V^*$$

defined by

$$\sum_{\alpha=0}^{m+1} p_\alpha w^\alpha = 0.$$

We may also write this as

$$\langle p, w\rangle = 0, \tag{4.1}$$

and we think of points in R as pairs (w,p) where $w \in \mathbf{P}V$ and $p \in \mathbf{P}V^*$ is a hyperplane in $\mathbf{P}V$ containing w.

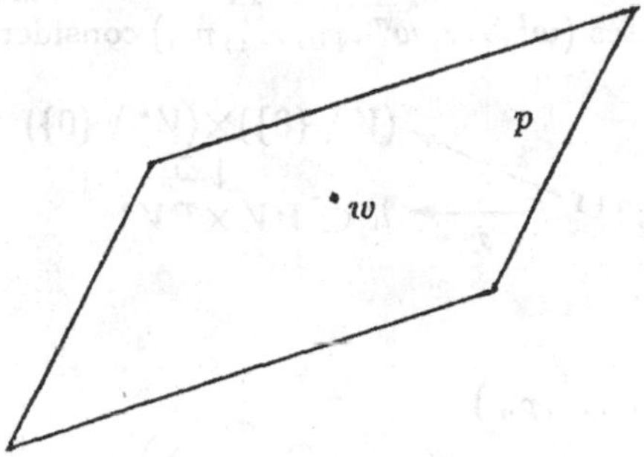

Such pairs (w,p) are sometimes called *contact elements.*

We shall give a canonical contact structure on R. For this we consider the standard projection

$$\begin{array}{c}(V\setminus\{0\})\times\ (V^*\setminus\{0\})\\ \downarrow \tilde{\omega}\\ \mathbf{P}V \quad \times \quad \mathbf{P}V^*,\end{array} \tag{4.2}$$

and on $V \times V^*$ we consider the tautological 1-form

$$\Omega = \langle p, dw \rangle = \sum_\alpha p_\alpha dw^\alpha. \tag{4.3}$$

For any holomorphic cross-section s of (4.2) over an open set $U \subset R$ we consider the 1-form on U given by

$$\omega = s^*\Omega = \langle p \circ s,\ d(w \circ s) \rangle. \tag{4.4}$$

Using (4.1) we may easily verify that ω is well-defined up to non-zero multiples.

We shall now check that

$$\omega \wedge (d\omega)^m \neq 0$$

by explicit computation. For this we will give a covering of R by open sets isomorphic to $\mathbf{C}^{2m+1}$ and over each open set a natural crosss-section s. Suppose that $(w, p) \in R$ and that $w^0 \neq 0$. Since $\langle p, w \rangle = 0$ it follows that one of $p_1, \dots, p_{m+1}$ must be non-zero. Suppose that $p_{m+1} \neq 0$, and for $\mathbf{C}_R^{2m+1}$ with coordinates $(w^1, \dots, w^m, p_0, \dots, p_m)$ consider the diagram.

$$\begin{array}{ccc} & \overset{s}{\nearrow} & (V \setminus \{0\}) \times (V^* \setminus \{0\}) \\ & & \downarrow \tilde{\omega} \\ \mathbf{C}_R^{2m+1} & \underset{j}{\hookrightarrow} & R \subset \mathbf{P}V \times \mathbf{P}V^* \end{array} \tag{4.5}$$

where

$$\begin{aligned} & s(w^1, \dots, w^m, p_0, \dots, p_m) \\ & \quad = \left(1, w^1, \dots, w^m, -\left(p_0 + \sum_{i=1}^m p_i w^i\right)\right) \times (p_0, \dots, p_m, 1). \end{aligned}$$

It is clear that $s(\mathbf{C}_R^{2m+1}) \subset \tilde{\omega}^{-1}(R)$ and that $j = \tilde{\omega} \circ s$ is one-to-one. By (4.4)

$$\begin{aligned} \omega &= \langle p \circ s,\ d(w \circ s) \rangle \\ &= \sum_i p_i dw^i - d\left(p_0 + \sum_i p_i w^i\right) \\ &= -\left(dp_0 + \sum_i w^i dp_i\right). \end{aligned}$$

Thus

$$d\omega = -\sum dw^i \wedge dp_i$$

and

$$\omega \wedge (d\omega)^m = (-1)^{m+1} m! \quad dp_0 \wedge dw^1 \wedge dp_1 \wedge \cdots \wedge dw^m \wedge dp_m.$$

Definition. We shall call ths construction the *canonical contact structure*, and shall denote it by (R, J).

Here, $J \subset T^*(R)$ is the line bundle determined by the 1-form (4.3) using the prescription (4.4).

We shall now determine the Legendre manifolds

$$f \colon S \longrightarrow R \tag{4.6}$$

for the canonical contact structure. Assuming that S is connected and has local holomorphic coordinates $s^1, \ldots, s^m$, we may locally lift (4.6) to $(V \setminus \{0\}) \times (V^* \setminus \{0\})$ and give f by a vector-valued holomorphic function

$$s \rightarrow (w(s), p(s))$$

where

$$\begin{cases} (i) & \sum_\alpha p_\alpha(s) w^\alpha(s) = 0 \\ (ii) & \sum_\alpha p_\alpha(s) dw^\alpha(s) = 0. \end{cases} \tag{4.7}$$

Consider the mapping

$$w \colon S \longrightarrow \mathbf{P}V$$

given by projecting f on the first factor, and suppose that this mapping has rank k at a general point. The image

$$X = w(S) \subset \mathbf{P}V$$

is then a k-dimensional piece of analytic variety. Denote by $X_{reg} \subset X$ the open dense set of smooth points, and for $w \in X_{reg}$ denote by $T_w(X) \subset \mathbf{P}V$ the tangent k-plane to X_{reg} at W.

Definition. We define the *Gauss correspondence*

$$\Gamma_X \subset R$$

to be the closure of the set

$$\{(w,p) : w \in X_{reg} \text{ and } T_w(X) \subseteq p\}.$$

It is clear that $\dim \Gamma_X = m$ and that we have a diagram

$$\begin{array}{ccc} & \Gamma_X \subset R \subset \mathbf{P}V \times \mathbf{P}V^* & \\ \quad \swarrow^{\pi_1} & & \searrow^{\pi_2} \quad \\ \mathbf{P}V \supset X & & X^* \subset \mathbf{P}V^* \end{array}$$

where X^* is the *dual variety* of tangent hyperplanes to X. Equations (i) and (ii) in (4.7) say exactly that:

(4.8) *The image of a Legendre manifold for the canonical contact structure is a Gauss correspondence.*

To have the converse we must relax the condition that (4.6) be an immersion, and simply require that f have maximal rank at a general point of each irreducible component of S. With this technicality being understood, using resolution of singularities we may say that:

(4.9) *The images of Legendre manifolds for the canonical contact structure are exactly the Gauss correspodences of analytic subvarieties $X \subset \mathbf{P}V$.*

Remark. Using (4.1) we have on $\tilde{\omega}^{-1}(R)$

$$\Omega = \langle p, dw \rangle = -\langle dp, w \rangle.$$

This suggests that in the construction of the canonical contact system the roles of w and p (i.e., of $\mathbf{P}V$ and $\mathbf{P}V^*$) may be interchanged. In fact this is obviously the case, and using (4.9) reflects the well-known fact that

$$(X^*)^* = X;$$

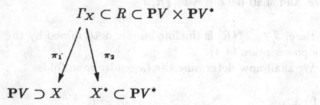

i.e., for subvarieties $X \subset \mathbf{P}V$ the dual of the dual is the original X (no matter what $\dim X$ is).

5. The Canonical Contact System Associated to an Alternating Bilinear from Q

We will now describe another example of an algebraic contact manifold that on the one hand has to do with variation of Hodge structure, and that on the other hand may be mapped by a birational contact transformation into the canonical contact system.

Let H be a complex vector space of dimension $2(m+1)$ and

$$Q: H \otimes H \longrightarrow \mathbf{C} \tag{5.1}$$

a non-degenerate alternating bilinear form. Setting

$$P = \mathbf{P}H \simeq \mathbf{P}^{2m+1}$$

we shall canonically associate to (H, Q) a contact manifold (P, L). For this we let z denote a typical point of H and consider the projection

$$\begin{array}{c} H \setminus \{0\} \\ \downarrow \pi \\ P \end{array} \tag{5.2}$$

Given a local holomorphic section s of (5.2) we set

$$\begin{aligned} \omega &= s^* Q(dz, z) \\ &= Q\big(d(z \circ s), z \circ s\big). \end{aligned}$$

Since $Q(z, z) = 0$, it follows that ω is well-defined up to non-zero multiples.

To show that this gives a contact structure, we consider the alternating bilinear form (5.1) as an element

$$Q \in \wedge^2 H^*$$

and choose coordinates $z^0, z^1, \ldots, z^{2m+1} \in H^*$ for H so that

$$Q = z^0 \wedge z^{m+1} + \cdots + z^m \wedge z^{2m+1}. \tag{5.3}$$

Then

$$Q(dz, z) = \sum_{\alpha=0}^{m} z^{m+1+\alpha} dz^{\alpha} - z^{\alpha} dz^{m+1+\alpha}.$$

For $\mathbf{C}_P^{2m+1}$ with coordinates $(x^1, \ldots, x^m, y_0, y_1, \ldots, y_m) = (x, y)$ we consider the diagram

$$\begin{array}{ccc} & \overset{s}{\nearrow} & H\backslash\{0\} \\ & & \downarrow \pi \\ \mathbf{C}_P^{2m+1} & \hookrightarrow & P \end{array}$$

where

$$s(x, y) = (1, x^1, \ldots, x^m, y_0, \ldots, y_m)$$

Then for this cross-section s,

$$\begin{aligned} \omega &= Q\big(ds(x, y), s(x, y)\big) \\ &= -dy_0 + \sum_{i=1}^{m} y_i dx^i - x^i dy_i, \end{aligned}$$

and

$$d\omega = 2 \sum_i dy_i \wedge dx^i.$$

Consequently

$$\omega \wedge (d\omega)^m = (-1)^{m+1} 2^m m! \quad dy_0 \wedge dx^1 \wedge dy_1 \wedge \ldots dx^m \wedge dy_m,$$

so that we have in this way determined a contact manifold (P, L).

Postponing the Hodge-theoretic discussion until the two next sections, we shall give a birational contact transformation of (P, L) to the canonical contact manifold (R, J). For this we recall the two open sets

$$\begin{cases} \mathbf{C}_R^{2m+1} \subset R \\ \mathbf{C}_P^{2m+1} \subset P \end{cases}$$

given in (4.5) and (5.4). We shall define a one-to-one rational holomorphic mapping

$$T\colon \mathbf{C}_R^{2m+1} \longrightarrow \mathbf{C}_P^{2m+1} \tag{5.5}$$

that preserves the contact forms. This mapping is given by the formulas

$$(5.6)\qquad \begin{cases} y_0 = p_0 + \frac{1}{2}\sum_{i=1}^{m} w^i p_i \\ y_i = \frac{1}{\sqrt{2}} p_i \\ x^i = \frac{1}{\sqrt{2}} w^i \end{cases}$$

Under this transformation

$$dy_0 + \sum_i (x^i dy_i - y_i dx^i) = dp_0 + \sum_i w^i dp_i,$$

so that (5.6) is indeed a birational contact transformation.

In concluding this section we want to discuss briefly the data needed to define (5.5). Given a filtration

$$F^3 \subset F^2 \subset F^1 \subset F^0 = H$$

with $\dim F^3 = 1, \dim F^2/F^1 = \dim F^1/F^2 = m$ and satisfying the 1st Hodge-Riemann bilinear relation

$$(5.7)\qquad \begin{cases} F^1 = F^{3\perp} \\ F^2 = F^{2\perp} \end{cases}$$

where $\perp$ is with respect to the bilinear form (5.1), we may choose coordinates $z^0, \ldots, z^{2m+1}$ so that (5.3) holds and where

$$(5.8)\qquad \begin{cases} F^3 = \{z^1 = \cdots = z^{2m+1} = 0\} \\ F^2 = \{z^{m+1} = \cdots = z^{2m+1} = 0\} \\ F^1 = \{z^{m+1} = 0\} \end{cases}$$

Having choosen such a coordinate system we may define T by (5.6). Of course such a coordinate system is not unique. In particular, a transformation

$$(5.9)\qquad \begin{cases} \tilde{z}^0 = z^0 \\ \tilde{z}^i = z^i + \sum_{j=1}^{m} q^{ij} z^{m+1+j}, & q^{ij} = q^{ji}, \\ \tilde{z}^{m+i+\alpha} = z^{m+i+\alpha}, & 0 \le \alpha \le m, \end{cases}$$

leaves invariant the form (5.3) and sets of equations (5.8). Under a transformation (5.9) we have

$$\begin{cases} \tilde{x}^i = x^i + \sum_j q^{ij} y_j, & q^{ij} = q^{ji} \\ \tilde{y}_\alpha = y_\alpha \end{cases}$$

and, using (5.6),

$$\begin{cases} \tilde{w}^i = w^i + \sum_j q^{ij} p_j \\ \tilde{p}_i = p_i \\ \tilde{p}_0 = p_0 - \sum_{i,j} q^{ij} p_i p_j \end{cases}$$

Remark. One of our motivations for studying this particular birational contact transformation is that the case $m = 1$ has already proved useful in another context, which we now explain. For more details consult the paper by the first author [1].

The celebrated "twistor" map of Penrose is a smooth fibration $\tau: \mathbf{P}^3 \rightarrow S^4$ where the fibres are linear $\mathbf{P}^1$'s. It yields a method of transforming Riemannian geometry problems in S^4 into complex analysis problems in $\mathbf{P}^3$. In particular, the complex 2-plane field on $\mathbf{P}^3$ orthogonal (in the Fubini-Study metric) to the fibers of τ is dual to a holomorphic line bundle $L \subset T^*\mathbf{P}^3$ which furnishes a contact structure on $\mathbf{P}^3$ with the following remarkable property: the holomorphic integral curves of this contact structure project via τ to be minimal surfaces in S^4.

Using the above birational transformation in the case $m = 1$, we are able to transform algebraic curves in $\mathbf{P}^2$ into minimal surfacs in S^4. From the fact that every compact Riemann surface occurs in $\mathbf{P}^2$ as an algebraic curve and from an elementary general position construction, the first author then concludes that every compact Riemann surface immerses minimally and conformally in S^4.

6. How Period Mappings Uniquely Arise as 1st Prolongations of the Legendre Manifolds Associated to Q

We retain the notations of the preceeding sections. Let S be a k-dimensional complex manifold and

(6.1) $$\varphi_0: S \longrightarrow P$$

an immersion. Denote by $\mathcal{F}(1,k)$ the manifold of all flags

$$\begin{cases} F^3 \subset F^2 \subset H & \text{where} \\ \dim F^3 = 1 \text{ and } \dim F^3/F^2 = k \end{cases}$$

We define the *1st prolongation* of (6.1) to be the map

$$\varphi_0^{(1)}: S \longrightarrow \mathcal{F}(1,k)$$

given for $s \in S$ by

$$\begin{cases} F^3(s) = \varphi_1(s) \\ F^2(s) = \text{projective tangent space to } \varphi_0(S) \text{ at } \varphi_0(s). \end{cases}$$

If locally φ_0 is given by

$$(s^1, \ldots, s^k) \to z(s^1, \ldots, s^k) \in H \setminus \{0\},$$

then

$$\begin{cases} F^3(s) = \text{span}\{z(s)\} \\ F^2(s) = \text{span}\left\{z(s), \dfrac{\partial z(s)}{\partial s^1}, \ldots, \dfrac{\partial z(s)}{\partial s^k}\right\} = \text{span}\left\{z(s), \dfrac{\partial z(s)}{\partial s^i}\right\}. \end{cases}$$

If also the linear subspace

$$F^1(s) = \text{span}\left\{z(s), \frac{\partial z(s)}{\partial s^i}, \frac{\partial^2 z(s)}{\partial s^i \partial s^j}\right\}$$

has constant dimension $k + l + 1$, then with the obvious notation we may define the *2nd prolongation*

$$\varphi_0^{(2)}: S \longrightarrow \mathcal{F}(1,k,l)$$

by

$$\varphi_0^{(2)}(s) = \{F^3(s) \subset F^2(s) \subset F^1(s)\}$$

(the reason for the indexing on the $F^p(s)$ will appear in a moment).

Remark 6.2. (i) In general, making suitable constant rank assumptions, we may define the kth *prolongation* to be the mapping given by the kth *osculating flag*

$$\operatorname{span}\{z(s)\} \subset \operatorname{span}\{z(s), \partial z(s)\} \subset \cdots \subset \operatorname{span}\{z(s), \ldots, \partial^k z(s)\}$$

where $\partial^k z(s)$ denotes all vectors $\partial^k z(s)/\partial s^{i_1} \cdots \partial s^{i_k}$.

(ii) Denoting tangent vectors to S by

$$\xi = \sum_i \xi^i \partial/\partial s^i \in T_s(S),$$

the kth *fundamental form* of (6.1) is by definition the symmetric k-linear function on $T(S)$ with values in $H/\operatorname{span}\{z(s), \ldots, \partial^{k-1}z(s)\}$ given by

$$\xi \to \sum_{i_1,\ldots,i_k} \xi^{i_1} \cdots \xi^{i_k} \frac{\partial^k z(s)}{\partial s^{i_1} \cdots \partial s^{i_k}} \in H/\operatorname{span}\{z(s), \ldots, \partial^{k-1} z(s)\}.$$

Equivalently, it is given by the linear function on

$$\left(H/\operatorname{span}\{z(s), \ldots, \partial^{k-1} z(s)\}\right)^*$$

defined for $\xi \in T_s(S)$ and $\lambda \in \left(H/\operatorname{span}\{z(s), \ldots, \partial^{k-1} z(s)\}\right)^*$ by

$$(\xi, \lambda) \to \langle \lambda, \sum_{i_1,\ldots,i_k} \xi^{i_1} \cdots \xi^{i_k} \frac{\partial^k z(s)}{\partial s^{i_1} \cdots \partial s^{i_k}} \rangle$$

(iii) We note the *expected dimension count*

$$\dim F^1(s) = k + 1 + k(k+1)/2 = (k+1)(k+2)/2.$$

To say that $\dim F^1(s) < (k+1)(k+2)/2$ means that the vector-valued function $z(s)$ satisfies a linear 2nd order PDE system.

Returning to the general discussion, we suppose that $k = m$ so that we have a canonical inclusion

$$\check{D} \subset \mathcal{F}(1, m)$$

(recall (cf. (2.2)) that $\{F^3 \subset F^2 \subset F^1 \subset H\} \in \check{D}$ is uniquely determined by $\{F^3 \subset F^2 \subset H\}$). We denote by $I \subset \Omega^*_{\check{D}}$ the sheaf of differential ideals given by the infinitesimal period relation (2.5). One of our main points is the following observation:

(6.3) *Let φ_0: $S \longrightarrow P$ be a Legendre manifold for the contact system on $P = \mathbf{P}H$ given by the alternating form Q. Then its 1st prolongation is a map*

$$\varphi_0^{(1)}: S \longrightarrow \check{D}$$

that is an m-dimensional integral manifold of the differential system I on $\Omega^*_{\check{D}}$. Coversely, any m-dimensional integral manifold of I is the 1st prolongation of such a Legendre manifold.

Proof. Let φ_0: $S \longrightarrow P$ be given locally by a vector-valued function

$$(s^1, \dots s^m) \to z(s) \in H \setminus \{0\}.$$

Then

$$\omega(s) = Q\big(dz(s), z(s)\big)$$

is the pullback under φ_0 of a local generator for the contact system on P. Using that $d\omega(s) = Q\big(dz(s), dz(s)\big)$, the conditions

$$\begin{cases} \omega(s) &= 0 \\ d\omega(s) &= 0 \end{cases}$$

give respectively

$$\begin{cases} (i)\; Q\left(\dfrac{\partial z(s)}{\partial s^i}, z(s)\right) = 0 \\ (ii)\; Q\left(\dfrac{\partial z(s)}{\partial s^i}, \dfrac{\partial z(s)}{\partial s^j}\right). \end{cases}$$

When combined, these imply that

$$\begin{cases} (i) & F^2(s) \subseteq F^2(s)^{\perp} \Rightarrow F^2(s) = F^2(s)^{\perp} \\ (ii) & dF^3(s) \subseteq F^2(s). \end{cases}$$

Condition (i) means that $\varphi_0^{(1)}(s) \in \check{D} \subset \mathcal{F}(1, m)$ (cf. (5.7)) while (ii) exactly means that $\varphi_0^{(1)}$ is an integral manifold of I.

To prove the converse we assume that $\varphi\colon S \longrightarrow \check{D}$ is an integral manifold of I. Writing $F^3(s)) = \text{span}\{z(s)\}$, from $dF^3(s) \subseteq F^2(s)$ and using our blanket assumption that $s \to \{F^3(s) \subset H\}$ be an immersion we have

$$F^2(s) = \text{span}\left\{z(s), \frac{\partial z(s)}{\partial s^i}\right\},$$

so that φ is the 1st prolongation of a mapping $\varphi_0\colon S \longrightarrow P$. The condition $F^2(s) = F^2(s)^\perp$ then implies that $\varphi_0\colon S \longrightarrow P$ is a Legendre manifold for the contact system on P given by Q.

Remarks 6.5. (i) Differentiation of (i) in (6.4) gives

$$Q\left(\frac{\partial^2 z(s)}{\partial s^i \partial s^j}, z(s)\right) = 0;$$

clearly this is equivalent to

$$dF^2(s) \subseteq F^3(s)^\perp.$$

It follows that

$$F^1(s) = F^3(s)^\perp = \text{span}\left\{z(s), \frac{\partial z(s)}{\partial s^i}, \frac{\partial^2 z(s)}{\partial s^i \partial s^j}\right\}$$

is the 2nd osculating space to $\varphi\colon S \longrightarrow P$. In particular,

$$\dim F^1(s) = 2m + 1 = (m+1)(m+2)/2 - m(m-1)/2$$

so that $z(s)$ satisfies a system of $m(m-1)/2$ linear 2nd order P.D.E.'s

$$\text{(6.6)} \qquad \sum_{i,j} q^{ij}(s) \frac{\partial^2 z(s)}{\partial s^i \partial s^j} \equiv 0 \text{ mod span } \left\{z(s), \frac{\partial z(s)}{\partial s^i}\right\}.$$

(ii) Referring to remark (ii) in (6.2) the 3rd fundamental form of $\varphi_0\colon S \longrightarrow P$ is the linear function on $(H/F^1(s))^* \simeq F^3$ given by the cubic form

$$\text{(6.7)} \qquad \xi \to Q\left(z(s), \sum_{i,j,k} \xi^i \xi^j \xi^k \frac{\partial^3 z(s)}{\partial s^i \partial s^j \partial s^k}\right)$$

in the tangent bundle $T(S)$. Referring to (2.11), this is also the cubic form canonically associated to the image tangent spaces $\varphi_*(T_s(S)) \subset T(\check{D})$ (these integral elements of I are *1st order* invariants of φ).

The P.D.E. system (6.6) has the following interpretation: Locally consider the cubic form (6.7) as a section

$$\Xi(s) = \sum_{i,j,k} \Xi_{ijk}(s) ds^i ds^j ds^k$$

of $\mathrm{Sym}^3 T^*(s)$. The quadrics

$$\frac{\partial \Xi(s)}{\partial(ds^k)} = \sum_{i,j} \Xi_{ijk}(s) ds^i ds^j$$

span a subspace $II(s) \subset \mathrm{Sym}^2 T^*(S)$, and we let $II(s)^{\perp} \subset \mathrm{Sym}^2(S)$ be the annihilator. Then the P.D.E. system (6.6) corresponds to operators $q = \sum q^{ij}(s)\partial^2/\partial s^i \partial s^j$ whose symbol belongs to $II(s)^{\perp}$.

7. How Period Mappings Arise as 2nd Prolongations of Analytic Subvarieties $X \subset \mathbf{P}H^{2,1}$

We are now ready to put everything together. The situation thus far may be summarized by the diagram

$$\text{(7.1)} \qquad \begin{array}{ccc} (\check{D}, I) & \overset{T^{(1)}}{\Leftrightarrow} & (R, J)^{(1)} \\ \downarrow & & \downarrow \\ (P, L) & \underset{T}{\Leftrightarrow} & (R, J) \\ & & \downarrow \\ & & PV \end{array}$$

that we may explain as follows:

(i) (P, L) is the contact manifold given by the alternating form Q on H (cf. §5);

(ii) $\check{D}$ is the dual classifying space for our particular Hodge structures, and I is the sheaf of differential ideals given by the infinitesimal period relation;

(iii) As explained in (6.3), $(\check{D}, I)$ may be considered as the 1st prolongation $(P, L)^{(1)}$ of the contact system (P, L) (Note: The general definition of the 1st prolongation of a sheaf of differential ideals is given, e.g., in [2]. All that we need to know here is that the integral manifolds of a differential system and of its 1st prolongation are in one-to-one correspondence.)

(iv) (R, J) is the canonical contact system as defined in §4. As explained there the Legendre manifolds for (R, J) are given by the Gauss correspondences of local analytic subvarieties $X \subset \mathbf{P}V$;

(v) $(R, J)^{(1)}$ is the 1st prolongation of (R, J) (loc. cit; we don't need to know explicitly what this is); and

(vi) T is the birational contact transformation given in §5), and $T^{(1)}$ is its 1st prolongation (we also don't need to know what this means).

Now let

$$\varphi\colon S \longrightarrow \check{D} \tag{7.2}$$

be a holomorphic immersion where $\dim S = m$ and $\varphi^*(I) = 0$, as might arise from the periods of a threefold satisfying (1.1). Then, by (6.3),

$$\varphi = \varphi_0^{(1)}$$

is the 1st prolongation of a Legendre manifold

$$\varphi_0\colon S \longrightarrow P. \tag{7.3}$$

Under the birational contact transformation T, which by the discussion at the end of §5) may be assumed to be well-defined in a *Zariski* neighborhood of a given point of P, we may consider (7.3) as a Legendre manifold

$$f\colon S \longrightarrow R \tag{7.4}$$

for the canonical contact system (R, J). Finally, (7.4) is given by the set of tangent hyperplanes to the image X of a holomorphic mapping

$$w\colon S \longrightarrow \mathbf{P}V \simeq \mathbf{P}^{m+1}. \tag{7.5}$$

Conclusion. *The assignment*

$$\varphi \to w$$

may be thought of as de-prolonging twice a variation of Hodge structure such as might arise from a family of 3-folds satisfying (1.1). The technical meaning of this statement has just been explained. *Intuitively, the period-type mapping (7.2) has been shown to arise from the 1st and 2nd derivatives* (2-jet) *of a holomorphic mapping (7.5).* In particular, referring to (5.3)-(5.6), in the case of the periods of a family $\{X_s\}_{s\in S}$ of 3-folds as discussed in §1 we may give (7.5) by the A-periods

$$w(s) = \left[\int_{\alpha_0} \omega(s), \int_{\alpha_1} \omega(s), \ldots, \int_{\alpha_m} \omega(s)\right], \tag{7.6}$$

and then the whole period mapping (7.2) is determined via (7.1) by the small piece (7.6). This establishes our assertion (1.4).

References

[1] R. Bryant, Every compact Riemann surface may be immersed conformally and minimally in S^4, to appear in Dec. (1982) issue of Jour. Diff. Geom.

[2] R. Bryant, S.S. Chern, and P. Griffiths, Exterior differential systems, Proc. Beijing Symposium, China Press (1982).

[3] J. Carlson and P. Griffiths, Infinitesimal variation of Hodge structure and the global Torelli problem, Journees de geometric algebrique d'Angers, Sijthoff and Nordhoff (1980), pp. 51-76.

[4] R. Donagi, The generic global Torelli theorem for most smooth hypersurfaces, to appear in Compositio Math.

[5] A. Piateski-Shapiro and I. Shafarevich, A. Torelli theorem for algebraic surfaces of type K3, Izv. Akad, Nauk., vol. 35 (1971), pp. 530-572.

[6] Transcendental topics in algebraic geometry (Proceedings of the IAS Seminar 1981/82), P. Griffiths editor, to apper as an Annals of Math Studies.

Received August 24, 1982
Supported in part by N.S.F. grant #MC580-03237

Professor Robert L. Bryant
Department of Mathematics
Rice University
Houston, Texas 77001

Supported in part by the Guggenheim Foundation and N.F.S. grant #MCS81-04249

Professor Phillip A. Griffiths
Department of Mathematics
Harvard University
Cambridge, Massachusetts 02138

On Nash Blowing-Up

Heisuke Hironaka

To I.R. Shafarevich

Let X be an algebraic variety, reduced and equidimensional, over the base field k of characteristic zero. Let us consider a sequence of transformations

$$X_0 = X \overset{\sigma_1}{\leftarrow} X_1 \overset{\sigma_2}{\leftarrow} X_2 \leftarrow \cdots$$

where $\sigma_i \colon X_i \longrightarrow X_{i-1}$ for each $i \geq 1$ is

(1) birational, i.e., proper and almost everywhere isomorphic, while X_i is reduced and equidimensional, and

(2) $\sigma_i^*(\Omega_{X_{i-1}})/(\text{its torsion})$ is locally free as $\mathcal{O}_{X_i}$-module. Here Ω denotes the sheaf of Kähler differentials on the variety and the torsion means the subsheaf consisting of those local sections whose supports are nowhere dense.

We are primarily interested in the cases in which

(a) σ_i is Nash blowing-up, namely, σ_i has a universal mapping property with respect to (1) and (2), i.e. if $\sigma' \colon X' \longrightarrow X_{i-1}$ is another having the same properties, then there exists a unique morphism $g \colon X' \longrightarrow X_i$ with $\sigma_i g = \sigma'$.

(b) σ_i is Nash blowing-up followed by normalization.

It is, however, of some geometric interest to consider a more general situation, including the sequence in which each σ_i is obtained by taking the graph of the birational correspondence between the Nash blowing-up of X_{i-1} and the monoidal transformation of X_{i-1} with a certain subvariety Y_{i-1}. This kind of σ_i is meaningful in regards to the question concerning Whitney conditions between $X_{i-1} \setminus Y_{i-1}$ and Y_{i-1}.

At any rate, given a sequence with the properties (1) and (2), we consider a sequence of points $\xi_i \in X_i$ such that $\sigma_i(\xi_i) = \xi_{i-1}$ for all i. Here, in general, the ξ_i are not required to be closed. In other words, the coordinates of the ξ_i may have transcendence over the base field k. We then have a natural homomorphism

$$\underline{O}_{X_{i-1},\xi_{i-1}} \to \underline{O}_{X_i,\xi_i}$$

by means of σ_i, and we consider the limit

$$R = \lim_{\substack{\rightarrow \\ i}} \underline{O}_{X_i,\xi_i}.$$

It is easy to see that R has the property

(*) If $\Omega^*_{R/k} = \Omega_{R/k}/(\text{its torsion})$, then every finitely generated R-submodule of $\Omega^*_{R/k}$ is contained in a free R-submodule of rank n of $\Omega^*_{R/k}$, where $n = \dim X$.

Let $K = k(X)$ be the ring of rational functions on X, which is a direct sum of the function fields of the irreducible components of X. The total ring of fractions of $\underline{O}_{x_i,\xi_i}$ is a direct summand of K, or more precisely, the sum of the function fields of those components of X_i which contain the point ξ_i. R has the same total ring of fractions as $\underline{O}_{X_i,\xi_i}$ for all $i \gg 0$.

Let V be a valuation ring of K/k, i.e., there exists a summand field K_λ of K such that V is a valuation ring of K_λ/k_λ where k_λ is the projection of k into the summand K_λ. We then have a unique point ξ_i of X_i for every i, determined by V in such a way that the canonical homomorphism of $\underline{O}_{X_i,\xi_i}$ into K_λ induces a map from $\underline{O}_{x_i,\xi_i}$ into V which sends the maximal ideal M_{X_i,ξ_i} into the maximal ideal M_V of V. The point ξ_i is called the center of V in X_i. The sequence $\{\xi_i\}$, so obtained, clearly has the property $\sigma_i(\xi_i) = \xi_{i-1}$ for all i as before.

The set of all valuation rings of K/k, denoted by $\underline{E}$, has Zariski topology. Namely, $\underline{E}$ is in a natural manner the inverse limit of all birational transformations of X, and the Zariski topology of $\underline{E}$ is the limit topology of Zariski topologies of those transforms. $\underline{E}$ has a dense subset consisting of divisorial valuation rings of K/k. Here a valuation ring V is said to be divisorial if its residue field V/M_V has transcendence degree $n-1$ over k_λ (i.e., one less than that of K_λ over k_λ). This condition is equivalent to saying that there exists a smooth algebraic variety X' over k such that $K = k(X')$ and $V = \underline{O}_{X',\xi'}$ where ξ' is a generic point of a hypersurface of X'.

If V is a valuation ring of K/k, then $\Omega_{V/k}$ has no torsion. If in particular V is divisorial, then $\Omega_{V/k}$ is a free V-module of rank n because $\Omega_{X'}$ is locally free of rank $= \dim X'$ for a smooth X'.

The purpose of this paper is to prove the following.

Theorem. *Assume that V is a divisorial valuation ring of K/k and ξ_i is its center in each X_i of the sequence having the properties (1) and (2) as in the beginning of this paper. Then there exists $i_0 \geq 0$ such that $\sigma_{i+1}\colon X_{i+1} \longrightarrow X_i$ is locally isomorphic at ξ_{i+1} and X_i is smooth at ξ_i for all $i \geq i_0$.*

Remark. We see, first of all, that the proof of the theorem can be reduced to the case in which $k = \mathbb{C}$ and ξ_i is a closed point of X_i (so that ξ_i has its coordinates in $\mathbb{C}$) for all i. In fact, let $x = (s_1, \ldots, x_s)$ be a system of non-zero divisors of R which induces a transcendence base of R/M over the base field k, where M denotes the maximal ideal of R. Replacing the given sequence $\{\sigma_i\}$ by a suitable subsequece, we may assume that the rational maps $f_i\colon X_i \longrightarrow k^s$, given by x, are well defined in some neighborhoods of the points ξ_i. Moreover, we may assume that $f_i(\xi_i)$ is the generic point of k^s. Then replace each X_i by the generic liber of f_i and ξ_i by its corresponding point in this fiber. By doing so, the assertion of the theorem is clearly unaffected. In other words, we may assume that ξ_i is a closed point of X_i for all i. Next, by an obvious descent of the base field, we may assume that k has a countable transcendence over $\mathbb{Q}$, because we are dealiug with only countably many algebraic varieties and points. Thus we have an injection $k \rightarrow \mathbb{C}$. Replace each X_i by its base field extension with respect to this $k \rightarrow \mathbb{C}$, and then replace V by any of its extensions to $\mathbb{C}(X)/\mathbb{C}$. Since the smoothness in characteristic zero is independent of the base field extensions, the assertion of the theorem is again unaffected.

The proof of the theorem will be completed after a sequence of lemmas as we see below.

Lemma 1. *Let $t = \mathrm{rank}_{R/M}\Omega^*/M\Omega^*$ where $\Omega^* = \Omega^*_{R/k}$. Then $t \leq n$.*

Proof. Let $\omega_1, \ldots, \omega_{n+1}$ be any system of elements of Ω^* and let $\overline{\omega}_1, \ldots, \overline{\omega}_{n+1}$ be their images modulo $M\Omega^*$. Since Ω^* has the property (*), there exists a free R-submodule of rank n, say L, of Ω^* which contains all the ω_i.

Therefore the $\overline{\omega}_i$ are contained in the image of L into $\Omega^*/M\Omega^*$, which is clearly generated by n elements. Thus the $\overline{\omega}_i$, $1 \le i \le n+1$, should be R/M-linearly dependent.

Remark. This lemma holds without the assumption that V is divisorial. In general, it is possible that $t < n$. In the lemmas that follow, we assume that V is divisorial.

Lemma 2. $t = n$.

Proof. Let us pick any system of elements $\omega_1, \dots, \omega_t$ of Ω^* which induces a free base of R/M-module $\Omega^*/M\Omega^*$. Let $E = \sum_{i=1}^t R\omega_i$. Then we have $\Omega^* = E + M\Omega^*$ and hence $\Omega^* = E + M^\nu\Omega^*$ for all $\nu > 0$. On the other hand, $\Omega = \Omega_{V/k}$ is a free V-module of rank n. Since $K_\lambda\Omega^* = \Omega_{K_\lambda/k} = K_\lambda\Omega$, there exists a non-zero element $b \in V$ such that $V\Omega^* \supset b\Omega$. Thus $\Omega \subset b^{-1}VE + b^{-1}M^\nu V\Omega^* \subset b^{-1}VE + (b^{-1}M^\nu V)\Omega$ for all $\nu > 0$. Since $b^{-1}VE$ is free of rank t and $\bigcap_{\nu=1}^\infty b^{-1}M^\nu V = (0)$, $\Omega \subset b^{-1}VE$ and rank $_V\Omega \le t$. Thus $n \le t$ and, by Lemma 1, $n = t$.

Lemma 3. *Ω^* is a free R-module of rank n.*

Proof. By Lemma 2, there exists a system of elements $\omega_1, \dots, \omega_n$ of Ω^* which induces a free base of $\Omega^*/M\Omega^*$. Let $E = \sum_{i=1}^n R\omega_i$. If ω is any element of Ω^*, then there exists a free submodule L of Ω^* which contains E and ω. Take a free base $\eta_1, \dots, \eta_n$ of L and write $\omega_i = \sum_{j=1}^n a_{ij}\eta_i$ with $a_{ij} \in R$. By considering these equalities modulo $M\Omega^*$, we see that $\det(a_{ij}) \ne 0$ modulo M, so that (a_{ij}) is invertible in the local ring R. Hence $\eta_j \in E$ for all j and hence $\omega \in E$.

Lemma 4. *Let $\hat{R}$ be the M-adic completion of R. Then the M-adic completion of $\Omega_{\hat{R}/k}$ has an $\hat{R}$-homomorphic image which is a free $\hat{R}$-module of rank n.*

Proof. The M-adic completion $\hat{\Omega}$ of $\Omega_{\hat{R}/k}$ is

$$\varprojlim_{\nu} \Omega_{\hat{R}/k}/M^\nu\Omega_{\hat{R}/k}$$

which is canonically isomorphic to

$$\varprojlim_{\nu} \Omega_{\hat{R}/k}/\hat{R}d(M^{\nu+1}\hat{R}) = \varprojlim_{\nu} \Omega_{R_\nu/k}$$

where $R_\nu = R/M^{\nu+1}$. This last limit is isomorpic to

$$\varprojlim_{\nu} \Omega_{R/k}/M^{\nu+1}\Omega_{R/k}.$$

Therefore we have a canonical homomorphism

$$h: \hat{\Omega} \longrightarrow \varprojlim_{\nu} \Omega^*/M^{\nu+1}\Omega^*$$

induced by $\Omega_{R/k} \to \Omega^*$. Since the target of h is generated as an $\hat{R}$-module by the images of finitely many elements of $\Omega_{R/k}$, h is clearly surjective. The image of h is a free R-module of rank n by Lemma 3.

Lemma 5. *Let S be a complete local $\mathbb{C}$-algebra whose residue field is isomorphic to $\mathbb{C}$. Let N be the maximal ideal of S and let $\hat{\Omega}$ be the N-adic completion of $\Omega_{S/\mathbb{C}}$. Assume that $\hat{\Omega}$ has a homomorphic image which is free of rank t. Then there exists a $\mathbb{C}$-subalgebra S_0 of S such that S is isomorphic to a formal power series ring of t variables over S_0.*

Proof. The idea is classical, and here we follow the method of Zariski. Let $h: \hat{\Omega} \longrightarrow E$ be the given epimorphism, where E is a free S-module of rank t. Since $\hat{\Omega}$ is generated by dx, $x \in N$, and S is local, we can find $x_i \in N$ such that $h(dx_i)$ form a free base of E, where $1 \leq i \leq t$. Take a projection $E \to S$ which sends $h(dx_1)$ to 1. Composed with h, this gives rise to a derivation $\partial: S \longrightarrow S$ such that $\partial x_1 = 1$. For simplicity, write x for x_1. Now the lemma will be proved in the following steps.

(1) Let $T(z)$ be an entire function of a complex variable z. Then $T(x\partial)$ is convergent as a $\mathbb{C}$-linear map $S \to S$. In fact, we can write

$$(x\partial)^l = \sum_{i=1}^{l} a_{l,i} x^i \partial^i \qquad \text{for } l \geq 1$$

because $\partial \cdot x = x\partial + 1$ as operators. Then we get

$$a_{l,i-1} + i a_{l,i} = a_{l+1,i}$$

so that $|a_{l,i}| \leq (2i)^{l-1}$ for all i and l. Since $T(z)$ is entire, we can write

$$T(x\partial) = \sum_{i=0}^{\infty} c_i x^i \partial^i.$$

As S is N-adically complete, $T(x\partial)$ is well defined.

(2) If $T(z) = e^z$ then $T(x\partial)$ is bijective. In fact, $T(-x\partial)$ is its inverse.

(3) x is not a zero-divisor in S. In fact, if $x^l g = 0$ with $l > 0$ and $g \in S$, then $0 = \partial^l(x^l g) = (l!)g + xg'$ with $g' \in S$. Hence $g = xg_1$ with $g_1 \in S$ and we get $x^{l+1}g_1 = 0$. Since S is N-adically complete and $x \in N$, no element but 0 is divisible by all powers of x. Thus $g = 0$.

(4) $e^{x\partial} - 1$ maps S into xS. (Obvious.)

(5) $\mathrm{Ker}(e^{x\partial} - 1) = \mathrm{Ker}(\partial)$, which will be denoted by S_1. In fact, write $e^z - 1 = U(z)z$ with an entire function $U(z)$. Then for every $g \in S$ and $l \geq 1$, we have

$$U(x\partial) \cdot x^l g = \frac{e^l - 1}{l} x^l g + x^{l+1} g'$$

with some $g' \in S$. It follows that the $\mathbb{C}$-linear map $U(x\partial)\colon xS \longrightarrow xS$ is bijective. Therefore $\mathrm{Ker}(e^{x\partial} - 1) = \mathrm{Ker}(x\partial) = \mathrm{Ker}(\partial)$, where the last equality is by (3).

(6) S_1 is a $\mathbb{C}$-subalgebra of S, because ∂ is a derivation.

(7) $e^{x\partial} - 1\colon xS \longrightarrow xS$ is bijective. In view of (5), it is enough that $x\partial\colon xS \longrightarrow xS$ bijective. But this is clear by $(x\partial)x^l g = l \cdot x^l g + x^{l+1}(\partial g)$.

(8) By (4) and (7), every element $g \in S$ is uniquely written as $g_0 + xg_1$ with $g_0 \in S_1$ and $g_1 \in S$.

(9) We have a canonical isomorphism $S_1[[x]] \xrightarrow{\sim} S$. This follows from (3) and (8).

(10) We have an epimorphism from the completion of $\Omega_{S_1/\mathbb{C}}$ to a free S_1-module E_1 of rank $t-1$. In fact, let $E_1 = E/xE + Sh(dx)$, viewed as S_1-module.

(11) Lemma 5 follows from (9) and (10) by induction.

Remark. We apply Lemma 5 to $S = \hat{R}$ of Lemma 4. In view of Lemma 3, we can choose $x_1, \ldots, x_n \in M$ such that Ω^* is freely generated by $dx_1, \ldots, dx_n$ as an R-module. The proof of Lemma 4 shows that these generators induce a system of free generators for E in the proof of Lemma 5. The derivation ∂ (obtained by the first projection) has $\partial x_j = 0$ for all

$j > 1$. Thus $x_2, \ldots, x_n$ are in S_1. In this manner, the proof of Lemma 5 gives an isomorphism $S_0[[x_1, \ldots, x_n]] \rightrightarrows \hat{R}$. Since $R = \varinjlim \underline{O}_i$, there exists $i_0 \geq 0$ such that all the x_j come from $\underline{O}_i$ if $i \geq i_0$ where $\underline{O}_i = \underline{O}_{X_i, \xi_i}$. We then have natural homomorphisms

$$\mathbb{C}[[x]] \to \hat{\underline{O}}_i \to \hat{R} \to \mathbb{C}[[x]]$$

where $x = (x_1, \ldots, x_n)$ and the last map is obtained by taking modulo the maximal ideal of S_0. The composition is clearly an identity.

Let $J_i = \operatorname{Ker}(\hat{\underline{O}}_i \to \mathbb{C}[[x]])$. Then, since n is the dimension of X_i, the above result signifies that X_i has a non-singular analytic branch through ξ_i, which is defined by J_i. Futhermore, if Γ_i denotes this analytic branch of X_i, then $\sigma_i\colon X_{i+1} \longrightarrow X_i$ induces an isomorphism $\Gamma_{i+1} \rightrightarrows \Gamma_i$ for all $i \geq i_0$. We want to show that, for $i \gg i_0$, the analytic branches of X_i other than Γ_i move away from ξ_i, i.e., Γ_i is the only analytic branch of X_i at ξ_i. Here we have

Separation Lemma. Let X be an algebraic variety over $\mathbb{C}$, reduced and equidimensional. Assume that X has a non-singular analytic branch Γ through a given closed point $\xi \in X$. Let $\sigma\colon X' \longrightarrow X$ be the Nash blowing-up. Then there exists a unique analytic branch Γ' of X' such that σ induces an isomorphism $\tau\colon \Gamma' \rightrightarrows \Gamma$. Furthermore, let $\xi' \in X'$ be the unique point of Γ' corresponding to ξ. Let I be the intersection of the ideals (in $\hat{\underline{O}}_{X,\xi}$) of those irreducible analytic branches of X at ξ which are different from Γ. Similarly define I' in $\hat{\underline{O}}_{X',\xi'}$ for X' and Γ'. Then $\tau_*(I'\underline{O}_{\Gamma',\xi'})$ is strictly bigger than I.

Remark. The last assertion signifies that the intersection $\Gamma' \cap \Lambda'$ is strictly smaller than $\Gamma \cap \Lambda$ with respect to the isomorphism τ, where Λ, Λ' are the union of other irreducible analytic branches.

Corollary. The assumption being the same as in the lemma, consider a sequence

$$X = X_0 \underset{\sigma_0}{\leftarrow} X_1 \underset{\sigma_1}{\leftarrow} X_2 \underset{\sigma_2}{\leftarrow} \cdots$$
$$\xi = \xi_0 \leftarrow \xi_1 \leftarrow \xi_2 \leftarrow \cdots$$

where the σ_i have the properties as in the beginning of this paper. Assume that X_i has a non-singular analytic branch Γ_i, through ξ_i, such that σ_i

induces an isomorphism $\Gamma_{i+1} \rightrightarrows \Gamma_i$ for all i where $\Gamma = \Gamma_0$. Then there exists $i_1 \geq 0$ such that X_i is smooth at ξ_i so that Γ_i is the only analytic branch of X_i through ξ_i for all $i \geq i_1$.

Proof. The first assertion of the Separation Lemma is easy to see. In fact, if κ: $\tilde{X} \longrightarrow X$ is the normalization and $\tilde{\xi} \in \tilde{X}$ is the point where κ induces a formal isomorphism from $\tilde{X}$ to Γ, then $\kappa^*(\Omega_{X,\xi})/(\text{its torsion})$ is locally free at $\tilde{\xi}$. Therefore the rational map $\tilde{X} \rightarrow X'$ induces an analytic isomorphism from $\tilde{X}$ at $\tilde{\xi}$ to an analytic branch of X' through ξ'. This branch is Γ' of the lemma. Let $x = (x_1, \ldots, x_n)$ be a system of elements in the maximal ideal of $\underline{O}_{X,\xi}$ such that $(dx_1, \ldots, dx_n)$ induces a free base of $\sigma^*\Omega_{X,\xi}/(\text{its torsion})$ at ξ'. Then we have canonical isomorphisms $\mathbb{C}[[x]] \rightrightarrows \underline{\hat{O}}_{\Gamma,\xi} \rightrightarrows \underline{\hat{O}}_{\Gamma',\xi'}$. Clearly df with $f \in I$ vanishes on every irreducible analytic branch of X' at ξ', different from Γ'. Namely, if we write $df = \sum_i f_i dx_i$ with $f_i \in \underline{O}_{X',\xi'}$, then $f_i \in I'$ for all i. In other words, identifying $\underline{\hat{O}}_{\Gamma,\xi}$ and $\underline{\hat{O}}_{\Gamma',\xi'}$ with $\mathbb{C}[[x]]$ by the above isomorphisms, we have

$$I'\mathbb{C}[[x]] \supset \sum_{i=1}^{n} \frac{\partial}{\partial x_i} I\mathbb{C}[[x]]$$

which implies the last assertion of the lemma.

Let us next prove its corollary. Since each σ_i is the Nash blowing-up followed by some birational morphism, the strict inclusion of ideals at the end of the lemma holds for σ_i instead of the Nash blowing-up alone. The assertion of the corollary follows immediately because we cannot have an infinite sequence of strictly increasing ideals in $\mathbb{C}[[x]]$.

Back to the proof of the theorem, this corollary of the Separation Lemma implies that, if $i \gg i_0$, Γ_i is the sole analytic branch of X_i at ξ_i so that X_i is smooth at ξ_i and σ_i is isomorphic at ξ_i.

Application. Consider the case of $n = 2$ and a sequence $\{\sigma_i\}$, where σ_i is the Nash blowing-up followed by normalization. Let π: $\tilde{X} \longrightarrow X$ be the minimal resolution of the surface X, and define π_i: $\tilde{X}_i \longrightarrow X_i$ and $\tilde{\sigma}_i$: $\tilde{X}_{i+1} \longrightarrow \tilde{X}_i$ by taking $\tilde{X}_i$ to be the graph of the birational correspondence between $\tilde{X}_{i-1}$ and X_i, where $\tilde{X}_0 = \tilde{X}$. We claim that π_i is an isomorphism for all $i \gg 0$. If not, there exists an irreducible curve Γ_i in $\tilde{X}_i$ for all i, such that $\tilde{\sigma}_i(\Gamma_{i+1}) = \Gamma_i$ and $\pi_i(\Gamma_i)$ is a point, say $\xi_i \in X_i$. The Γ_i defines a divisorial valuation ring V of K such that $\xi_i \in X_i$ is its center for all i. By the theorem, X_i is smooth at ξ_i for $i \gg 0$. Then the rational map $X_i \to \tilde{X}$ is regular at ξ_i, which implies that π_i is locally isomorphic at ξ_i. This is a contradiction for the assumption of Γ_i. In conclusion, X_i dominates the minimal resolution $\tilde{X}$ of X for $i \gg 0$ so that X_i has only rational singularities.

References

[1] Gonzalez-Sprinberg, G., "Resolution de Nash des points doubles rationnels," Mimeographed Note, Centre de Math., Ecole Polytech., France, (October, 1980).

[2] Nobile, A., "Some properties of the Nash blowing-up," Pacific J. Math. *60*, pp. 297–305 (1975).

[3] Zariski, O., "Some open questions in the theory of singularities," Bull. Am. Math. Soc. 77 pp. 481–491 (1971).

Received July 27, 1982

Professor Heisuke Hironaka
Department of Mathematics
Harvard University
Cambridge, Massachusetts 02138

Arrangements of Lines and Algebraic Surfaces

F. Hirzebruch

To I. R. Shafarevich

Introduction

In recent years the Chern numbers c_1^2 and c_2 of algebraic surfaces have aroused special interest. For a minimal surface of general type they are positive and satisfy the inequality $c_1^2 \leq 3c_2$ (see Miyaoka [22] and Yau [31]) where the equality sign holds if and only if the universal cover of the surface is the unit ball $|z_1|^2 + |z_2|^2 < 1$ (see Yau [31] and [23] §2 for the difficult and [11] for the easy direction of this equivalence). It is interesting to know which positive rational numbers ≤ 3 occur as c_1^2/c_2 for a minimal surface of general type. For a long time (before 1955) it was believed that $c_1^2/c_2 \leq 2$, in other words that the signature of a surface of general type is nonpositive. It is interesting to find surfaces with $2 < c_1^2/c_2 \leq 3$. Some were found by Kodaira [19]. Other authors constructed more surfaces of this kind (Holzapfel [13], [14], Inoue [16], Livné [21], Mostow-Siu [25], Miyaoka [23]). Recently Mostow [24] and Deligne-Mostow [5] used a paper of E. Picard of 1885 to construct discrete groups of automorphisms of the unit ball leading to interesting surfaces. In fact, the author was stimulated by Mostow's lecture at the Arbeitstagung 1981 to study the surface Y_1 (see §3.2) which is related to one of the surfaces of Mostow.

For a surface of general type there is a unique minimal model in its birational equivalence class. The Chern numbers of the minimal model are therefore invariants of the birational equivalence class, i.e., they are invariants of the field of meromorphic functions on the surface. In the present paper we study surfaces whose function fields are Kummer extensions $K(a_1^{1/n}, a_2^{1/n}, \ldots, a_{k-1}^{1/n})$ of exponent n of the rational function field, i.e., of the function field $K = \mathbb{C}(z_1/z_0, z_2/z_0)$ of the complex projective plane ($a_i \in K$). We consider the following Kummer extensions: For k distinct lines $l_1 = 0, l_2 = 0, \ldots, l_k = 0$ in the plane (where l_i is a linear form in z_0, z_1, z_2) we take $a_1 = l_2/l_1, a_2 = l_3/l_1, \ldots, a_{k-1} = l_k/l_1$. Under some assumptions on the arrangement of lines these Kummer extensions ($n \geq 2$) determine surfaces of general type. For three special arrangements of lines we obtain a minimal surface of general type with $c_1^2/c_2 = 3$. Therefore the unit ball appears as an infinite branched cover of the projective plane (with points blown up), the branching locus consisting of lines and the exceptional curves coming from the blown-up points. We also get surfaces with c_1^2/c_2 near to 3, but not arbitrarily close to 3. We show for example that 5/2 is an accumulation point of accumulation points of the values c_1^2/c_2 for the minimal models of our Kummer extensions.

The topology of the complement of an arrangement of lines in the projective plane is very interesting ([1], [3], [26]), the investigation of the fundamental group of the complement very difficult. We need very little information on the complement (the Euler number is sufficient), because the unramified coverings of the complement occuring in this paper are always abelian. The real simplicial arrangements play an important role in the present paper and it should be mentioned that the complement in the complex projective plane of the complexification of such an arrangement is an Eilenberg-MacLane space [4]. There exists much work on real simplicial arrangements (see Grünbaum [7], [8]), but the classification of these arrangements does not seem to be known.

The inequality $c_1^2 \leq 3c_2$ gives theorems on arrangements of lines (see §3.1) which up to now could not be proved directly.

§1 Arrangements of Lines

We consider the complex projective plane $P_2(\mathbb{C})$ with homogeneous coordinates $z_0 : z_1 : z_2$. An arrangement of k lines is a set of k distinct lines in $P_2(\mathbb{C})$. They can be given by linear forms $l_1, \ldots, l_k$ in z_0, z_1, z_2. The union of the lines is the reducible curve

$$l_1 l_2 \ldots l_k = 0.$$

Let $t_r (r \geq 2)$ be the number of r-fold points of this curve, i.e., the number of points lying on exactly r lines of the arrangement. Then we have

$$\frac{k(k-1)}{2} = \sum t_r \frac{r(r-1)}{2}$$

For the whole paper *we shall always assume that $k \geq 3$ and $t_k = 0$.* (The arrangement should not be a "pencil", i.e., not all lines should pass through one point.)

1.1 Real arrangements

If all linear forms $l_1, \ldots, l_k$ have real coefficients, then $l_1 l_2 \ldots l_k = 0$ is an arrangement of lines in the real projective plane $P_2(\mathbb{R})$ which, of course, can also be regarded as an arrangement in $P_2(\mathbb{C})$. A real arrangement defines a cellular decomposition of $P_2(\mathbb{R})$, each cell is bounded by an r-gon. Let p_r be the number of cells bounded by r-gons. Then $p_2 = 0$ (because of our assumption that we do not have a pencil). Let f_0 be the number of vertices, f_1 the number of edges, f_2 the number of cells. The Euler-Poincaré characteristic of $P_2(\mathbb{R})$ equals 1, so

(1) $$f_0 - f_1 + f_2 = 1$$

It is easy to see that

(2) $$f_1 = \sum r \cdot t_r = \frac{1}{2} \sum r \cdot p_r$$

Since $f_0 = \Sigma t_r$ and $f_2 = \Sigma p_r$, we conclude from (1) that

(3) $$3 + \Sigma(r-3)t_r + \Sigma(r-3)p_r = 0$$

Since $p_2 = 0$, we have

$$t_2 \geq 3.$$

An arrangement is called simplicial if $p_r = 0$ for $r > 3$. By (3)

$$\text{simplicial} \Leftrightarrow 3 + \Sigma(r-3)t_r = 0.$$

For a simplicial arrangement

(4) $$f_1 = 3f_0 - 3, \quad f_2 = 2f_0 - 2.$$

If we dualize an arrangement of lines, we get an *arrangement of points* which are not collinear and t_r is the number of lines containing exactly r of the points, t_2 is the number of the so-called ordinary lines. In this setup, Sylvester claimed (1893) that $t_2 > 0$ for real arrangements. The above proof for $t_2 \geq 3$ is due to Melchior (1940). For the history of this problem see Grünbaum's book [8] which is our standard reference for real arrangements. It was proved recently by Sten Hansen [9] that

(5) $$t_2 \geq \left[\frac{k}{2}\right]$$

Grünbaum ([7],[8]) has tried to classify ***simplicial real arrangements.*** There is first the near-pencil $A_0(k)$ which consists of k lines of which $k-1$ pass through one point ($t_{k-1} = 1$, $t_2 = k-1$, $t_r = 0$ otherwise). Then he defines the regular arrangements $A_1(2m)$ for $m \geq 3$ and $A_1(4m+1)$ for $m \geq 2$. We quote from Grünbaum [8]: The regular arrangement $A_1(2m)$ consists of $2m$ lines, of which m are the lines determined by the edges of a regular m-gon in the Euclidean plane, while the other m are the lines of symmetry of that m-gon. The regular arrangement $A_1(4m+1)$ consists of $4m+1$ lines of which $4m$ form the arrangement $A_1(4m)$ while the last one is the line at infinity. Grünbaum ([7],[8]) conjectures that up to ***combinatorial equivalence*** there are besides the three series $A_0(k)$, $A_1(2m)$, $A_1(4m+1)$ exactly 91 simplicial real arrangements. He lists 90 of them in [7] and gives the extra one in [8]. He lists them (together with the three series) as $A_i(k)$ where k is the number of lines and i a counting index. Grünbaum also gives the values of the t_r. For $A_1(2m)$ we have (for $m > 3$)

$$t_2 = m, t_3 = \frac{m(m-1)}{2}, t_m = 1, t_r = 0 \quad \text{otherwise.}$$

For $A_1(4m+1)$ we have (for $m > 2$)

$$t_2 = 3m, \ t_3 = m(2m-2), \ t_4 = m, \ t_{2m} = 1, \ t_r = 0 \quad \text{otherwise.}$$

The arrangement $A_1(6)$ is projectively equivalent to any complete quadrilateral

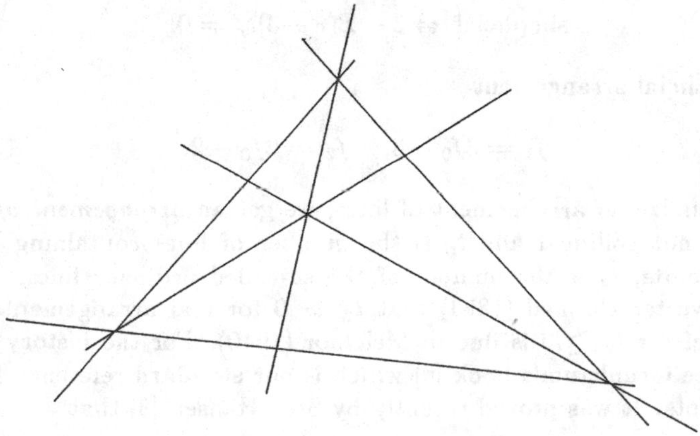

$$t_2 = 3, \ t_3 = 4, \ t_r = 0 \quad \text{otherwise.}$$

The arrangement $A_1(8)$ looks as follows

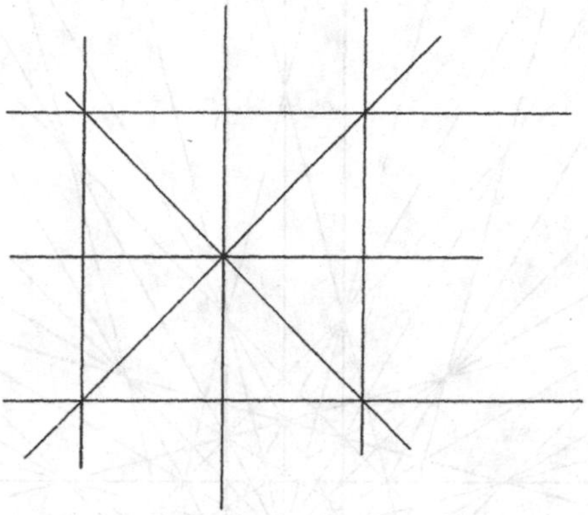

$$t_2 = 4,\ t_3 = 6,\ t_4 = 1,\ t_r = 0 \quad \text{otherwise.}$$

We get $A_1(9)$ if we add the infinite line.
For $A_1(9)$ we have

$$t_2 = 6,\ t_3 = 4,\ t_4 = 3,\ t_r = 0 \quad \text{otherwise.}$$

For $k = 19$ Grünbaum has the following list (the near-pencil is omitted)

	f_0	t_2	t_3	t_4	t_5	t_6
$A_1(19)$	49	21	18	6	0	4
$A_2(19)$	51	21	18	6	6	0
$A_3(19)$	49	24	12	6	6	1
$A_4(19)$	51	20	20	6	4	1
$A_5(19)$	51	20	20	6	4	1
$A_6(19)$	51	20	20	6	4	1
$A_7(19)$	52	21	15	15	0	1

$t_r = 0$ otherwise

The maximal k for Grünbaum's 91 exceptional arrangements is 37. Besides $A_0(37)$ and $A_1(37)$ belonging to a series he has $A_2(37)$ and $A_3(37)$ where $A_2(37)$ is reproduced here from [7]. (It contains the infinite line.)

$A_2(37)$

$t_2 = 72,\ t_3 = 72,\ t_4 = 12,\ t_5 = 24,\ t_{12} = 1$, all other $t_r = 0$, $f_0 = 181$.

(1.2) Arrangements defined by reflection groups

We first consider real arrangements. A finite reflection group G in the Euclidean space $\mathbb{R}^l$ has finitely many reflecting hyperplanes. Let $L(G)$ be the set of linear subspaces of $\mathbb{R}^l$ which are intersections of reflecting hyperplanes. If V is a 3-dimensional subspace belonging to $L(G)$, we take the set $A(V)$ of all 2-dimensional elements of $L(G)$ which are contained in V. If we pass from V to the projective plane defined by V, then $A(V)$ defines an arrangement of lines which is simplicial (see Orlik and Solomon [27], [28], and the literature given there). The paper [27] gives a complete list of the arrangements obtainable in this way, however the values of the t_r are not given. But Orlik and Solomon have communicated to me a list of the t_r of these arrangements. J. Tits had mentioned to me before that I should look at such arrangements. He gave me the values of the t_r for one arrangement coming from the Weyl group of E_8.

Let G be the Weyl group of E_8 acting as reflection group in $\mathbb{R}^8$. By the above method we obtain 8 arrangements of lines. There are two arrangements with 19 lines. They occur as $A_1(19)$ and $A_3(19)$ in Grünbaum's list.

We can take in particular the arrangements defined by reflection groups in $\mathbb{R}^3$, ($V = \mathbb{R}^3$, k = number of reflecting hyperplanes). We have the symmetry groups of the regular tetrahedron, the cube, the icosahedron of orders $24, 48, 120$ respectively (improper rotations included). For the tetrahedron and the cube we get the Weyl groups of A_3 and B_3 and the arrangements $A_1(6)$ and $A_1(9)$ mentioned in 1.1. For the icosahedron we get the arrangement $A_1(15)$ of [7] with $t_2 = 15$, $t_3 = 10$, $t_5 = 6$, $t_r = 0$ otherwise.

If we consider finite unitary reflection groups G in the hermitian vector space $\mathbb{C}^l$, we can consider $L(G)$ and $A(V)$ as before in the real case and obtain interesting arrangements of lines (see [28]). Orlik and Solomon have communicated to me a complete list with the values of the t_r.

The groups $G(m,p,n)$ (see the classification in Shephard and Todd [30]) give rise to the following arrangements [28] ($m \geq 2$)

name	k	t_2	t_3	t_m	t_{m+1}	t_{m+2}
$A_3^0(m)$	$3m$	0	m^2	3	0	0
$A_3^1(m)$	$3m+1$	m	m^2	1	2	0
$A_3^2(m)$	$3m+2$	$2m$	m^2	0	2	1
$A_3^3(m)$	$3m+3$	$3m$	m^2	0	0	3

All other $t_r = 0$. We do not say more about these arrangements here, because they occur in some other context in 1.3. For $m = 2, 3$ the above table has to be read in an obvious way, e.g., $A_3^0(3)$, $k = 9$, $t_2 = 0$, $t_3 = 12$, all other $t_r = 0$. For $m = 2$ we get the real arrangements $A_1(6)$, $A_1(7)$, $A_1(8)$, $A_1(9)$ of [7]. (Compare 1.1).

In the following table we list only the arrangements coming from irreducible unitary reflection groups acting on $\mathbb{C}^3$ (omitting the $G(m,p,3)$ and the real groups). The groups have "registration numbers" according to the Shephard-Todd classification [30].

S-T number	$\lvert G\rvert$	k	t_2	t_3	t_4	t_5	projective
24	336	21	0	28	21	0	G_{168}
25	648	12	12	0	9	0	G_{216}
26	1296	21	36	0	9	12	G_{216}
27	2160	45	0	120	45	36	G_{360}

The corresponding groups of projective transformations are the simple group of order 168 (studied by F. Klein), the automorphism group of the Hesse pencil (see 1.3) of order 216 and the Valentiner-Wiman group of order 360 (isomorphic to the alternating group of six letters, see [6], Volume 2, Anhang). The group G_{168} has 21 elements of order 2. The group G_{360} has 45 elements of order 2. The arrangements consist of the pointwise fixed lines of these 21 (or 45) involutions of $P_2(\mathbb{C})$.

(1.3) Arrangements defined by cubic curves

Cubic curves give us arrangements in the dual setup, where we look for arrangements of points and where t_r denotes the number of lines containing exactly r points. Let C be a smooth cubic curve in $P_2(\mathbb{C})$. The group structure is well-defined once we have chosen one of the 9 inflection points as neutral element. The sum $p_1+p_2+p_3$ of $p_1,p_2,p_3 \in C$ equals 0 if and only if $\{p_1,p_2,p_3\}$ is the intersection of C with a line. The inflection points are the points with $3p=0$. Let U be a finite subgroup of C of order k. Consider the arrangement consisting of the points of U. The k tangents to C in points of U are ordinary lines containing two points of the arrangement provided they are not tangents in inflection points. Therefore $t_2 = k-w$, where w is the number of inflection points contained in U. Since $t_r = 0$ for $r>3$ and $\frac{k(k-1)}{2} = t_2 + 3t_3$, we have

Lemma. *Let U be a subgroup of order k of a smooth cubic. For this arrangement of k points we have*

$$t_2 = k-w, \quad t_3 = \frac{k(k-3)}{6} + \frac{w}{3}, \quad t_r = 0 \quad \textit{otherwise},$$

where w is the number of inflection points contained in U.

Cyclic groups U of order k can be realized over the reals. Then $w = 1$ or $w = 3$ and $t_3 = [\frac{k(k-3)}{6}] + 1$. These arrangements are not simplicial, except for $k = 4$ and $k = 6$. Our basic reference for this construction and many historical remarks about the "orchard problem" is the paper [2] by Burr, Grünbaum and Sloane.

If we take as U the group of the 9 inflection points we get (dually) the arrangement $A_3^0(3)$, see 1.2. The 12 lines responsible for $t_3 = 12$ make up a line arrangement to be mentioned in a moment.

The pencil of cubics passing through the nine inflection points of the smooth cubic C is called Hesse pencil. The Hesse pencil does not depend (up to projective equivalence) on the given cubic. The group of projective automorphisms of $P_2(\mathbb{C})$ which carry the Hesse pencil of the cubic C to itself is the group G_{216} mentioned in 1.2. The Hesse pencil of C has 4 singular cubics consisting of three lines each. These 12 lines are an arrangement with $t_2 = 12$, $t_4 = 9$ ($t_r = 0$ otherwise), the nine quadrupel points being the inflection points. It belongs to the group with No. 25 in the Shephard-Todd classification (see 1.2). If we take these 12 lines together with 9 additional lines E_x to be described we get the arrangement belonging to the group carrying S-T number 26. For each inflection point x we have a line E_x which intersects each cubic C_t of the pencil in $\alpha'_t - x$, $\alpha''_t - x$, $\alpha'''_t - x$, where α'_t, α''_t, α'''_t are the points of order 2 of C_t (and where we use the addition on C_t). Through each vertex of a singular cubic of the pencil pass 3 of these 9 lines.

The construction of point arrangements can also be done with singular cubics. Let C be a singular cubic consisting of three lines C_1, C_2, C_3 defining a triangle with distinct vertices $A_1 = C_2 \cap C_3$, $A_2 = C_3 \cap C_1$, $A_3 = C_1 \cap C_2$. Then $C' = C - \{A_1, A_2, A_3\}$ can be given a group structure

$$C' \simeq \mathbb{C}^* \times \mathbb{Z}/3\mathbb{Z}$$

such that three points B_1, B_2, B_3 of C' (with $B_i \in C_i$) are collinear if and only if $B_1 \cdot B_2 \cdot B_3 = 1$ (Theorem of Menelaos). If in such a group structure we choose the $3m$ points of C' which as points of $\mathbb{C}^* \times \mathbb{Z}/3\mathbb{Z}$ have m^{th}-roots of unity as first component, we get the dual version of the arrangement $A_3^0(m)$ of 1.2. If we add one, two, or three of the points A_1, A_2, A_3 to this arrangement we obtain $A_3^1(m)$, $A_3^2(m)$, $A_3^3(m)$. In a similar way the arrangement $A_1(2m)$ of 1.1 can be obtained by choosing a singular cubic consisting of a conic and a line.

§2 Algebraic Surfaces Associated to Arrangements

For an arrangement $l_1 = 0, \ldots, l_k = 0 (k \geq 3, t_k = 0)$ of lines and a natural number $n \geq 2$ we consider the function field

$$\mathbb{C}(z_1/z_0, z_2/z_0)((l_2/l_1)^{1/n}, (l_3/l_1)^{1/n}, \ldots, (l_k/l_1)^{1/n})$$

which is an abelian extension of the function field $\mathbb{C}(z_1/z_0, z_2/z_0)$ of $P_2(\mathbb{C})$ of degree n^{k-1} and Galois group $(\mathbb{Z}/n\mathbb{Z})^{k-1}$. (Of course, it does not matter which line of the arrangement is given the subscript 1.) The function field determines an algebraic surface X with normal singularities which ramifies over the plane with the arrangement as locus of ramification:

$$\pi: X \longrightarrow P_2(\mathbb{C}).$$

If the point $p \in P_2(\mathbb{C})$ lies on r lines of the arrangement ($r \geq 0$), then $\pi^{-1}(p)$ consists of n^{k-1-r} points which are an orbit of the Galois group $(\mathbb{Z}/n\mathbb{Z})^{k-1}$ acting on X. We desingularize all singularities of X in the minimal way (see §2.1) and obtain a smooth algebraic surface Y depending on the arrangement and on n. The surfaces Y will be investigated in this section.

(2.1) Desingularization

If a point of the plane lies on r lines, we choose an affine coordinate system u, v centered at p with a line of the arrangement not passing through p as line at infinity. The local ring of a point of X lying over p is the extension of the regular local ring at p by $(\tilde{l}_1)^{1/n}, \ldots, (\tilde{l}_r)^{1/n}$, where $\tilde{l}_1 = 0, \ldots, \tilde{l}_r = 0$ are the equations of the lines passing through p written in terms of the u, v coordinate system. We have singular points over p if and only if $r \geq 3$. Such a singularity is resolved by blowing up p in a projective line E_p. The reader should observe the exceptional role of $r = 2$ in all forthcoming calculations.

A singular point q of X over p is blown up into a smooth curve C which covers the projective line E_p and is the "Riemann surface" over E_p with n^{r-1} sheets defined by the functions $(\tilde{l}_2/\tilde{l}_1)^{1/n}, \ldots, (\tilde{l}_r/\tilde{l}_1)^{1/n}$ where u, v are now regarded as homogeneous coordinates on E_p. The Euler number $e(C) = 2 - 2g(C)$ is easy to calculate (Hurwitz)

$$e(C) = n^{r-1}(2 - r) + r \cdot n^{r-2}. \tag{1}$$

We apply this process in all singular points q of X and obtain our smooth surface Y. We have natural maps

$$\rho: Y \longrightarrow X, \quad \sigma: Y \longrightarrow \hat{P}_2, \quad \tau: \hat{P}_2 \longrightarrow P_2(\mathbb{C})$$

where $\hat{P}_2$ is the projective plane with all points p with $r \geq 3$ blown up. Of course, $\pi\rho = \tau\sigma$. The map σ is of degree n^{k-1} and ramifies over the proper transforms in $\hat{P}_2$ of the lines of the arrangement and over the curves E_p, each ramification index equals n. Furthermore, $\sigma^* E_p$ is a divisor in Y consisting of n^{k-1-r} disjoint curves C each with multiplicity n. All these curves are mapped to each other under the Galois group. As far as selfintersection numbers go we have the equation

$$(\sigma^* E_p) \cdot (\sigma^* E_p) = n^{k-1}(E_p \cdot E_p) = -n^{k-1}$$

which implies for each curve C

$$C \cdot C = -n^{r-2}. \tag{2}$$

(2.2) Characteristic numbers

We have defined a smooth algebraic surface Y for each arrangement of lines and any natural number $n \geq 2$. We wish to calculate the Chern numbers c_2, c_1^2 of Y. The number c_2 is the Euler number of Y, whereas c_1^2 is the selfintersection number of a canonical divisor of Y.

For the calculation of c_2 we use the principle that Euler numbers behave like the cardinal numbers of sets. We denote Euler numbers by e. Let $L_1, \ldots, L_k$ be the lines of the arrangement, L their union, and $\operatorname{sing} L$ the set of singular points of L. We first calculate the Euler number of the surface X with the set $\operatorname{sing} X$ of its singular points removed. We have

$$\begin{aligned} &e(X - \operatorname{sing} X) \\ &\quad = n^{k-1} e(P_2(\mathbb{C}) - L) + n^{k-2} e(L - \operatorname{sing} L) + n^{k-3} t_2 \\ &\quad = n^{k-1}\left(3 - 2k + \sum (r-1) t_r\right) + n^{k-2}\left(2k - \sum r t_r\right) + n^{k-3} t_2. \end{aligned} \tag{3}$$

Because of (1) we get for the Euler number of Y

$$e(Y) = e(X - \operatorname{sing} X) + \sum_{r \geq 3} n^{k-1-r} t_r \left(n^{r-1}(2-r) + r n^{r-2}\right) \tag{4}$$

which gives the following

Proposition. *Let Y be the smooth algebraic surface associated to an arrangement of k lines covering the plane with degree n^{k-1}. Then*

$$\begin{aligned}(5)\qquad e(Y)/n^{k-3} &= c_2(Y)/n^{k-3} \\ &= n^2(3-2k+f_1-f_0)+2n(k-f_1+f_0)+f_1-t_2.\end{aligned}$$

Here we have used the abbreviations

$$f_1=\sum rt_r,\quad f_0=\sum t_r$$

which in the case real arrangements have a geometric meaning (1.1).

We now calculate the other characteristic number $c_1^2(Y)$. For $p\in\operatorname{sing}L$ let r_p be the number of lines passing through p. If $r_p\geq 3$, then we have the exceptional curve E_p in $\hat{P}_2$. For the divisor $L=L_1+L_2+\cdots+L_k$ we consider its improper transform (lift) τ^*L to $\hat{P}_2$ (see 2.1.), then

$$\tau^*L-\sum_{\substack{p\in\operatorname{sing}L\\ r_p\geq 3}} r_pE_p$$

is the proper transform of L. Consider as in 2.1. the map $\sigma\colon Y\longrightarrow\hat{P}_2$. The divisors $\sigma^*(\tau^*L-\sum r_pE_p)$ and $\sigma^*(E_p)$ of Y ($p\in\operatorname{sing}L$, $r_p\geq 3$) are divisible by n. If K is a canonical divisor of $P_2(\mathbb{C})$, then $\tau^*K+\sum E_p$ is a canonical divisor of $\hat{P}_2$. Applying the well-known facts on the behaviour of canonical divisors under ramified coverings we obtain

Lemma. *For a canonical divisor K on $P_2(\mathbb{C})$ consider the divisor (with rational coefficients)*

$$K_Y=\tau^*K+\sum E_p+\frac{n-1}{n}\Big(\sum E_p+\tau^*L-\sum r_pE_p\Big)$$

on $\hat{P}_2$, the sums being taken over all $p\in\operatorname{sing}L$ with $r_p\geq 3$. Then σ^*K_Y is a canonical divisor of Y.

In $\hat{P}_2$ we have

$$K_YK_Y=\left(-3+k\frac{n-1}{n}\right)^2-\sum_{r\geq 3}t_r\left(1+\frac{n-1}{n}(1-r)\right)^2$$

which yields for $c_1^2(Y) = n^{k-1}K_Y K_Y$ the following formula

$$(6) \quad c_1^2(Y)/n^{k-3} = ((k-3)n-k)^2 - \sum_{r\geq 2} t_r((r-2)n-r+1)^2 + t_2$$

Using $\frac{k(k-1)}{2} = \sum t_r \frac{r(r-1)}{2}$ and the abbreviations $f_1 = \sum rt_r$ and $f_0 = \sum t_r$ we get

$$(7) \quad \begin{aligned} c_1^2(Y)/n^{k-3} &= n^2(-5k+9+3f_1-4f_0) \\ &\quad + 4n(k-f_1+f_0) + f_1 - f_0 + k + t_2. \end{aligned}$$

(2.3) Rough classification I

It is not difficult to determine the canonical dimension of the surfaces Y and to carry out the rough classification in the sense of Enriques-Kodaira [20]. We first remark without proof that for an arrangement of k lines ($k \geq 4$) in general position (i.e., $t_r = 0$ for $r \geq 3$) the surface Y (covering $P_2(\mathbb{C})$ with degree n^{k-1}) is a complete intersection of $k-3$ Fermat surfaces of degree n in $P_{k-1}(\mathbb{C})$ where a Fermat surface is given by an equation $\sum_{i=0}^{k-1} a_i z_i^n$.

The Galois group of Y over $P_2(\mathbb{C})$ is the obvious automorphism group of such a complete intersection. (For $k = 3$, the surface Y is a projective plane.) The characteristic numbers of complete intersections are given by well-known formulas [10] which can be checked with (5) and (6). For $k \geq 6$ and $n \geq 2$ such a complete intersection is of general type (i.e., the canonical dimension equals 2), except for $k = 6$ and $n = 2$ where we have a $K3$-surface. We shall prove a similar statement for the surfaces Y belonging to arrangements which need not be in general position.

For the following discussion we always assume

$$(8) \qquad k \geq 6, \quad n \geq 2, \quad t_k = t_{k-1} = t_{k-2} = 0.$$

The curve E_p in $\hat{P}_2$ is defined only if $r_p \geq 3$. This is always understood in the following formulas. In $\hat{P}_2$ we have to consider the proper transforms L_j' of the lines L_j of the arrangement:

$$L_j' = \tau^* L_j - \sum_{p\in L_j} E_p.$$

We now calculate $K_Y E_p$ and $K_Y L'_j$ (see the lemma in 2.2 and observe that $\tau^* K$ and $\tau^* L_j$ have with all E_p the intersection number 0).

$$\begin{aligned} K_Y E_p &= -1 + \frac{n-1}{n}(r_p - 1) \geq 0 \\ K_Y E_p &= 0 \quad \Leftrightarrow \quad n = 2 \text{ and } r_p = 3. \end{aligned} \tag{9}$$

We have $K_Y L'_j = -3 + \frac{n-1}{n}k - \sum_{p \in L_j} K_Y E_p$.

Formula (9) implies

$$\sum_{p \in L_j} K_Y E_p = \frac{n-1}{n}(k-1) - \frac{n-1}{n} \#\{p \in L_j \mid r_p = 2\} - \#\{p \in L_j \mid r_p \geq 3\}.$$

Therefore

$$K_Y L'_j = -2 - \frac{1}{n} + \#\{p \in L_j \mid r_p \geq 2\} - \frac{1}{n} \#\{p \in L_j \mid r_p = 2\}. \tag{10}$$

In view of our assumption (8) it can be checked

$$\begin{aligned} K_Y L'_j &\geq 0 \quad \text{if} \quad \#\{p \in L_j \mid r_p \geq 2\} \geq 3 \\ K_Y L'_j &> 0 \quad \text{if} \quad \#\{p \in L_j \mid r_p \geq 2\} \geq 3 \text{ and } n \geq 3. \end{aligned} \tag{11}$$

If $\#\{p \in L_j \mid r_p \geq 2\} = 2$, then the arrangement looks as follows

(12)

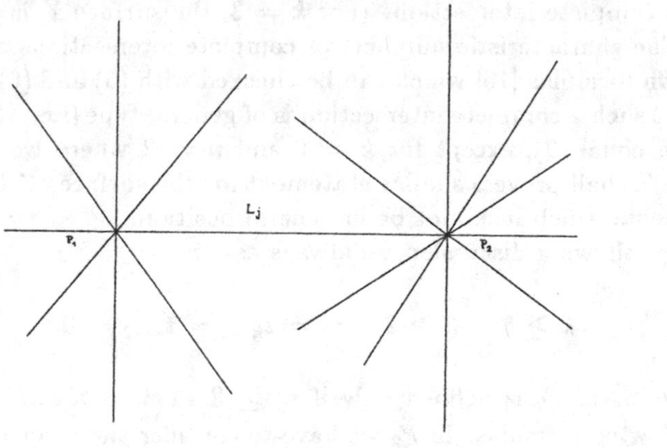

$$u + v - 1 = k, \quad u \geq 4, \quad v \geq 4$$

$$t_u = t_v = 1, \qquad t_2 = (u-1)(v-1), \quad t_r = 0 \quad \text{otherwise}.$$

We have $K_Y L'_j = -\frac{1}{n}$.

Theorem. *Assume the arrangement satisfies* (8) *and is not of type* (12). *Then the surface is minimal, i.e., does not contain exceptional curves. For an arrangement of type* (12) *the divisor* $\sigma^* L'_j$ *on* Y *consists of* n^{k-3} *disjoint exceptional curves (each with multiplicity* n*); blowing them down gives a minimal surface* Y_0 *which is a product of curves* C_1, C_2 *with Euler numbers*

$$e(C_1) = n^{u-1}(2-u) + un^{u-2}, \quad e(C_2) = n^{v-1}(2-v) + vn^{v-2}.$$

All the surfaces Y *belonging to an arrangement satisfying* (8) *are of general type for* $n \geq 3$. *For* $k = 6$ *and* $n = 2$ *the surface* Y *is a* $K3$*-surface and for* $k \geq 7$ *and* $n = 2$ *it is elliptic (of canonical dimension* ≥ 0*) or of general type.*

Proof. The divisor K_Y of the lemma in 2.2 can be written as

$$K_Y = \tau^* K + \left(2 - \frac{1}{n}\right) \sum E_p + \frac{n-1}{n} \sum L'_j.$$

We want to choose K such that K_Y is effective. Since $t_k = t_{k-1} = t_{k-2} = 0$, we can find six lines $L_1, L_2, \ldots, L_6$ in the arrangement such that not more than 3 of them pass through one point. For K on $P_2(\mathbb{C})$ we take $-\frac{1}{2}(L_1 + L_2 + \cdots + L_6)$ allowing half-integral coefficients (multiplying by 2 gives a double canonical divisor). Then K_Y is effective and positive, except for $k = 6$ and $n = 2$ where it vanishes (over the rationals). [In fact, in this case Y is a $K3$-surface, because the original surface X has only rational double points (of type A_1), their number being $4t_3$, and by Brieskorn's theory Y is a deformation of the surface which we get if we have an arrangement in general position, but then Y is a $K3$-surface.] The divisor $\sigma^* K_Y$ is an effective canonical divisor of Y (with rational coefficients). Therefore it contains all exceptional curves of Y. For an exceptional curve F on Y we have $\sigma^* K_Y \cdot F = -1$. Since K_Y is a linear combination of the E_p and L_j^1, the exceptional curve F is a component of some divisor $\sigma^* E_p$ or $\sigma^* L'_j$ (all components of such a divisor being transforms of each other under the Galois group). Therefore $\sigma^* K_Y \cdot F < 0$ is impossible because of (9) and (11) except for arrangements (12). All statements on an arrangement (12) can be checked easily, and we remark that $C_1 \times C_2$ is of general type, except if $n = 2$ and u or v equals 4. In the latter case C_1 or C_2 is elliptic. We still have to prove that Y is of general type if $n \geq 3$ (we can exclude arrangements (12)). We must show that $K_Y K_Y > 0$. But for $n \geq 3$ the divisor K_Y contains all curves E_p and L'_j with positive multiplicity, and $K_Y K_Y > 0$ follows from (9) and (11). The assertions for $n = 2$ follow because K_Y is effective for $k \geq 7$.

Remark. For $n = 2$ and $k \geq 7$ the surface Y is elliptic if and only if the arrangement is one of the following or of type (12) with u or v equal to 4.

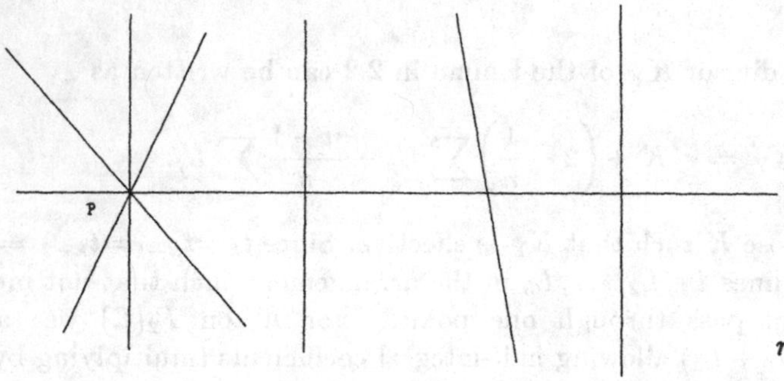

$r_p \geq 4$

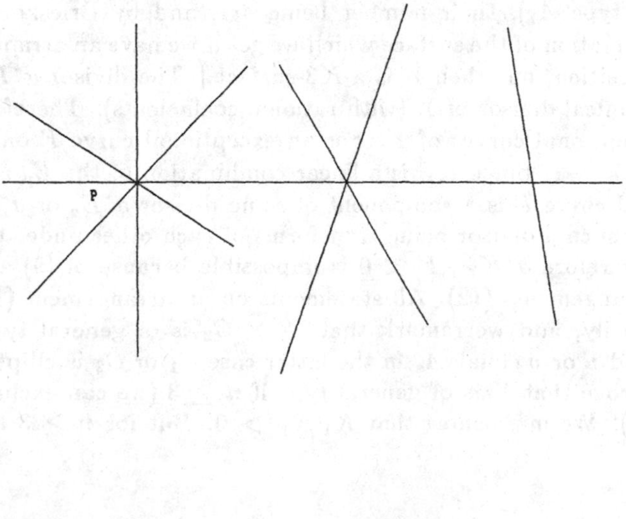

$r_p \geq 4$

(2.4) Rough classification II (for $t_{k-1} = 1$ or $t_{k-2} = 1$, and $k \geq 6$)

An arrangement

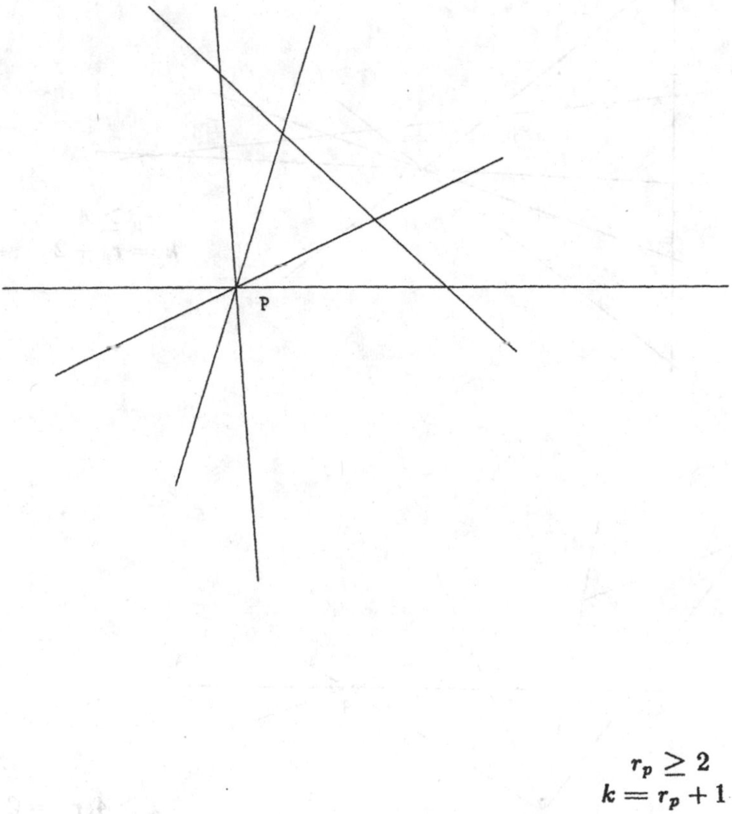

$$r_p \geq 2$$
$$k = r_p + 1$$

is called a near-pencil [8]. Then the surface Y is birationally equivalent to a ruled surface for all $n \geq 2$ and has canonical dimension $-\infty$.

The arrangements

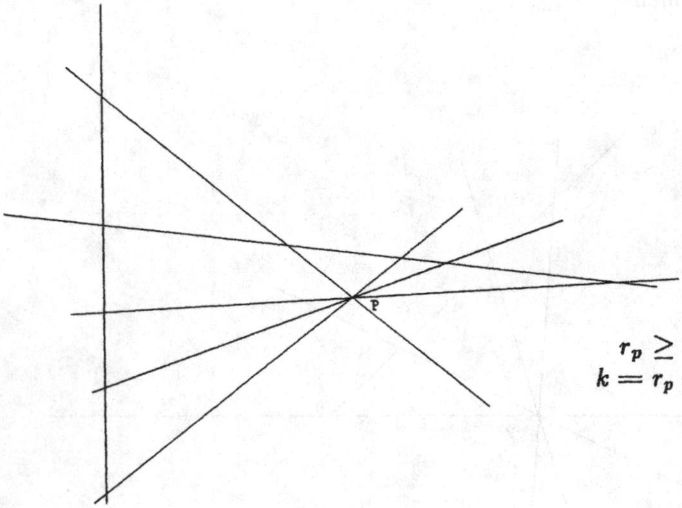

$r_p \geq 4$
$k = r_p + 2$

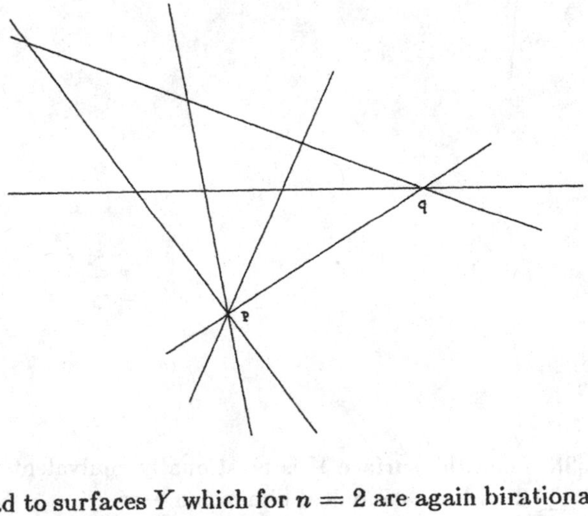

$r_p \geq 4, r_q = 3$
$k = r_p + 2$

lead to surfaces Y which for $n = 2$ are again birationally equivalent to ruled surfaces and have canonical dimensions $-\infty$. For $n = 3$ we get elliptic surfaces with canonical dimension ≥ 0 and for $n \geq 4$ surfaces of general type.

The theorem in 2.3 and the above remarks complete the rough classification of the surfaces Y for all arrangements with $k \geq 6$. In particular, we have

Theorem. *For an arrangement with $k \geq 6$ and $t_k = t_{k-1} = 0$, the surface Y has canonical dimension ≥ 0 for $n \geq 3$. For an arrangement with $k \geq 6$ and $t_k = t_{k-1} = t_{k-2} = 0$ the canonical dimension is ≥ 0 for $n \geq 2$. (In these cases no ruled surfaces occur.)*

We leave it to the reader to investigate the surfaces for $k = 4, 5$ ($t_k = t_{k-1} = 0$). The surface Y is rational, or of canonical dimension ≥ 0.

(2.5) The Miyaoka-Yau inequality

We consider the Chern numbers c_1^2 and c_2 for an algebraic surface S. If S is of general type and minimal, then

$$c_1^2(S) \leq 3c_2(S), \tag{13}$$

and equality holds if and only if the universal cover of S is biholomorphically equivalent to the unit ball $|z_1|^2 + |z_2|^2 < 1$ (compare the introduction). If (13) holds for a surface S, then it is also true for any surface obtained by blowing up points of S. Therefore (13) is true for all surfaces of general type. The Enriques-Kodaira classification gives also a list of minimal algebraic surfaces. The inequality (13) is wrong for the minimal algebraic surface S if and only if S is a ruled surface with a curve of genus $g \geq 2$ as base curve. In this case $c_1^2 = 4(1-g)$, $c_2 = 8(1-g)$.

Therefore, (13) *is true if S is not birationally equivalent to a ruled surface with base curve of genus $g \geq 2$.* In particular, (13) is true if the canonical dimension of S is ≥ 0 or if the surface is rational.

The results of 2.3 and 2.4 imply

Theorem. *For an arrangement of k lines with $t_k = t_{k-1} = t_{k-2} = 0$ the algebraic surface Y satisfies the Miyaoka-Yau inequality for all $n \geq 2$. If $t_k = t_{k-1} = 0$ the inequality is true for all $n \geq 3$.*

§3 The Miyaoka-Yau Inequality and Applications

For an algebraic surface S of general type the Euler number $c_2(S)$ is positive, the ratio $c_1^2(S)/c_2(S)$ is less or equal 3. The signature is positive if and only if $c_1^2(S)/-c_2(S)$ is greater than 2. We want to construct surfaces using arrangements of lines for which the ratio is 3 or near to 3. In any case the whole range between 2 and 3 is interesting. For the surfaces coming from arrangements of lines the number $3c_2(S)-c_1^2(S)$ is calculated in terms of the combinatorial data of the arrangement. Since $3c_2(S)-c_1^2(S) \geq 0$, this gives restrictions on the combinatorial data and results on arrangements which could not be proved directly up to now.

(3.1) The quadratic polynomial associated to an arrangement

Let Y be the smooth algebraic surface associated to an arrangement of k lines covering the plane with degree n^{k-1}. Then

$$(1)\quad \begin{aligned}&(3c_2(Y)-c_1^2(Y))/n^{k-3} = \\ &\qquad n^2(f_0-k)+n(-2f_1+2f_0+2k)+2f_1+f_0-k-4t_2.\end{aligned}$$

This follows from equations (5) and (7) in §2. If we replace in (1) the "variable" n by $x+1$ we obtain the following quadratic polynomial

$$(2)\quad F(x)=x^2(f_0-k)-2x(f_1-2f_0)+4(f_0-t_2).$$

The polynomial F is called the *quadratic polynomial of the arrangement.* The coefficients f_1-2f_0 and f_0-t_2 are positive except if the arrangement is in general position when both vanish. The coefficient f_0-k is positive if $t_k=t_{k-1}=0$. This follows from the theorem in 2.5. The inequality $f_0 \geq k$ (if $t_k=0$) is an old result on arrangements of lines valid in every projective plane [8].

The theorem in 2.5 implies

Theorem. *For an arrangement of k lines in the complex projective plane we consider the associated quadratic polynomial $F(x)$. If $t_k=t_{k-1}=t_{k-2}=0$, then $F(x)\geq 0$ for all integers x. If $t_k=t_{k-1}=0$, then $F(x)\geq 0$ for all integers $x\neq 1$; in particular $F(2)\geq 0$ which means*

$$(3)\quad t_2+t_3 \geq k+t_5+2t_6+3t_7+\cdots.$$

The inequality (3) shows for example that arrangements of k lines with $t_2 = t_3 = t_k = 0$ are impossible in the complex projective plane.

Conjecture. *The quadratic form*

$$(f_0 - k)u^2 - 2(f_1 - 2f_0)uv + 4(f_0 - t_2)v^2$$

is positive semidefinite if $t_k = t_{k-1} = t_{k-2} = 0$, *in other words we conjecture*

$$(f_1 - 2f_0)^2 \leq 4(f_0 - k)(f_0 - t_2). \tag{4}$$

The conjecture can be verified for all arrangements mentioned in §1. For a real simplicial arrangement we have $f_1 = 3f_0 - 3$ and

$$F(x) = x^2(f_0 - k) - 2x(f_0 - 3) + 4(f_0 - t_2). \tag{5}$$

The conjecture claims (if $t_k = t_{k-1} = t_{k-2} = 0$)

$$(f_0 - 3)^2 \leq 4(f_0 - k)(f_0 - t_2). \tag{6}$$

Of course, we are interested in cases when $F(x) = 0$ for some natural number x. Then we get a surface with $c_1^2 = 3c_2$. Up to now only the following arrangements could be found

I. The real simplicial arrangement $A_1(6)$ (see §1) with $k = 6$, $f_0 = 7$, $t_2 = 3$ and

$$F(x) = (x - 4)^2.$$

II. The arrangement of the 12 lines in the Hesse pencil (see 1.2, S-T number 25, and 1.3) with $k = 12$, $t_2 = 12$, $t_4 = 9$ and

$$F(x) = 9(x - 2)^2.$$

III. The arrangement $A_3^0(3)$ coming from the nine inflection points of a cubic by dualizing. In this case $k = 9, t_3 = 12$ and

$$F(x) = 3(x - 4)^2$$

Arrangement II is the only arrangement known with equality sign in (3).

(3.2) Three surfaces with $c_1^2 = 3c_2$

Using the arrangements I, II, III in 3.1 we get three surfaces Y_1, Y_2, Y_3 with $c_1^2 = 3c_2$, they are coverings of the plane of degrees 5^5, 3^{11}, 5^8 respectively. (Recall that $n = x + 1$.) The Euler numbers c_2 can be calculated by §2(5).

$$c_2(Y_1) = 15 \cdot 5^3,\ c_2(Y_2) = 48 \cdot 3^9,\ c_2(Y_3) = 111 \cdot 5^6.$$

Ishida [17] has calculated the irregularity of these surfaces. He obtained

$$q(Y_1) = 30,\ q(Y_2) = 154,\ q(Y_3) = 96.$$

Inoue [16] and Livné [21] constructed several surfaces of general type with $c_1^2 = 3c_2$ using elliptic modular surfaces. For convenience we copy the explanation of these surfaces from Ishida [18]: Let $E(n)$ be the elliptic modular surface of level $n > 3$. The elliptic fibration $E(n) \to X(n)$ has n^2 sections. The union D of all the sections is divisible in $\mathrm{Pic}((E(n))$ by n if n is odd and by $n/2$ if n is even. Hence for each divisor d of n or $n/2$, we can take a cyclic cover $S = S_d(n)_\Delta \to E(n)$ of order d which is ramified over D. It is not unique in general, but depends on the choice of the element $\Delta \in \mathrm{Pic}(E(n))$ with $d\Delta = D$. Inoue and Livné showed $c_1(S)^2 = 3c_2(S)$ for the cases

$$(n, d) = (7,7), (8,4), (9,3) \text{ and } (12,2).$$

In the case $(n, d) = (5,5)$, Δ is unique and the inverse images of the twenty-five sections of $E(n)$ in S are exceptional curves of the first kind. The nonsingular surface $S(5)$ obtained by contracting all these curves has Chern numbers $c_1^2 = 3c_2 = 225$.

The surface Y_1 admits an action of $A = (\mathbb{Z}/5\mathbb{Z})^5$. There are many subgroups A_2 of A of order 25 which operate freely on Y_1. Dividing Y_1 by such a subgroup gives a surface with $c_1^2 = 3c_2 = 225$, and A_2 can be chosen in such a way that $Y_1/A_2 = S(5)$ (Hirzebruch and Ishida [18], compare also [12]). Similarly Y_2 admits an action of $A = (\mathbb{Z}/3\mathbb{Z})^{11}$. Then A has a subgroup A_5 of order 3^5 which operates freely, the quotient being $S_3(9)_\Delta$ for the unique $\Delta \in \mathrm{Pic}(E(9))$ such that all automorphisms of $E(9)$ lift to $S_3(9)_\Delta$. This is a result of Ishida [18]. The surface $Y_2/A_5 = S_3(9)_\Delta$ has Chern numbers $c_1^2 = 3c_2 = 16 \cdot 3^6$.

(3.3) Behaviour of c_1^2/c_2 for $n \to \infty$

If S is minimal surface of general type, then both Chern numbers $c_1^2(S)$ and $c_2(S)$ are positive and we have

$$\frac{1}{5}(c_2(S) - 36) \leq c_1^2(S) \leq 3c_2(S),$$

the first inequality being due to M. Noether. Suppose we have a sequence S_n of minimal surfaces of general type with $\lim_{n\to\infty} c_2(S_n) = \infty$, then

$$\text{(7)} \qquad \frac{1}{5} \leq \underline{\lim}\frac{c_1^2(S_n)}{c_2(S_n)} \leq \overline{\lim}\frac{c_1^2(S_n)}{c_2(S_n)} \leq 3.$$

Problem. *Which numbers r in the interval $[\frac{1}{5}, 3]$ occur as*

$$r = \lim_{n\to\infty} \frac{c_1^2(S_n)}{c_2(S_n)}$$

for a sequence S_n of minimal surfaces of general type with $c_1^2(S_n)/c_2(S_n) \neq r$? Let $\mathfrak{R}$ be the set of these r, in other words $\mathfrak{R}$ is the set of accumulation points of the set of values c_1^2/c_2 of minimal surfaces of general type. Which numbers of the interval $[\frac{1}{5}, 3]$ are accumulation points of R?

For an arrangement of k lines with $k \geq 6$ and $t_k = t_{k-1} = t_{k-2} = 0$ which is not of type (12) (see §2), the surface Y_n associated to the arrangement (covering the plane with degree n^{k-1}) is minimal of general type for $n \geq 3$. Since $c_2(Y_n) > 0$, we have $f_1 - f_0 \geq 2k - 3$ (see the polynomial (5) in §2). Therefore, the linear coefficient of this polynomial is negative and hence $f_1 - f_0 > 2k - 3$. It also follows that the leading coefficient of the polynomial (7) in §2 is positive. We obtain

$$\text{(8)} \qquad \lim_{n\to\infty} \frac{c_1^2(Y_n)}{c_2(Y_n)} = \frac{5}{2} - \frac{3f_0 - f_1 - 3}{2(3 - 2k + f_1 - f_0)}$$

where $3 - 2k + f_1 - f_0 > 0$. The limit in (8) will be called the characteristic number γ of the arrangement.

Proposition. *The characteristic number γ of a real arrangement is less or equal $5/2$. We have $\gamma = 5/2$ if and only if the real arrangement is simplicial.*

Proof. By §1(3) we have $3f_0 - f_1 - 3 = \sum_{r\geq 3}(r-3)p_r \geq 0$.

Remark. In his paper [15] Iitaka has studied the characteristic number γ from some other point of view. He erroneously claimed $\gamma \leq 5/2$. But this can be wrong for arrangements not definable over the reals, for example $\gamma > \frac{5}{2}$ if $t_2 = 0$.

If the arrangement is in general position ($f_1 = 2f_0$), then $\gamma < 2$, but $\lim_{k\to\infty} \gamma = 2$. Therefore, the number 2 is an accumulation point (from below) of the set $\mathfrak{R}$. According to [19], $2 + \frac{1}{2n-1} \in \mathfrak{R}$, so 2 is also an accumulation point from above.

For the arrangements mentioned in the lemma of 1.3 we have

$$\gamma = \frac{5}{2} - (k-3-w)/2(k(k-6)/3 - w/3 + 3).$$

Therefore $5/2$ is an accumulation point (from below) of the set $\mathfrak{R}$.

For the arrangement $A_3^0(m)$ of 1.2 we get

$$\gamma = \frac{5}{2} + \frac{3m-6}{2(2m^2-3m)}.$$

Therefore $5/2$ is an accumulation point (from above) of the set $\mathfrak{R}$.

According to Holzapfel[13] the number 3 belongs to $\mathfrak{R}$. We could not prove this using the surfaces coming from line arrangements. However, it is possible to construct surfaces (for some arrangements and some n) with c_1^2/c_2 fairly near to 3.

(3.4) Surfaces with c_1^2/c_2 near to 3

According to §2(5), (7) c_1^2/c_2 is a quotient of quadratic polynomials in n. We give some examples. Let us first take the arrangements I, II, III of 3.1. We obtain

I. $3 - c_1^2/c_2 = (n-5)^2/(2n^2 - 10n + 15)$
For $n = 6$ we have $c_1^2/c_2 = 2\frac{26}{27}$

II. $3 - c_1^2/c_2 = 3(n-3)^2/(6n^2 - 18n + 16)$
For $n = 4$ we have $c_1^2/c_2 = 2\frac{37}{40}$

III. $3 - c_1^2/c_2 = 3(n-5)^2/(3n^2 - 10n + 12)$
For $n = 6$ we have $c_1^2/c_2 = 2\frac{59}{60}$

For the icosahedral arrangement (see 1.2, called $A_1(15)$ in Grünbaum's list) we obtain

$$3 - c_1^2/c_2 = \frac{16n^2 - 88n + 136}{32n^2 - 88n + 75}$$

For $n = 3$ this gives $c_1^2/c_2 = 2\frac{83}{99}$.

For the arrangement $A_1(19)$ with 19 lines arising from the Lie group E_8 (see 1.2) we have

$$3 - c_1^2/c_2 = \frac{30n^2 - 152n + 234}{60n^2 - 152n + 123}$$

For $n = 3$ we have $c_1^2/c_2 = 2\frac{53}{69}$.

For the arrangement with S–T number 24 belonging to the simple group of order 168 we have

$$3 - c_1^2/c_2 = \frac{7(n^2 - 7n + 13)}{20n^2 - 49n + 42}$$

For $n = 4$ we have $c_1^2/c_2 = 2\frac{159}{166}$.

The reader may try some more of the arrangements considered in §1. The resulting surfaces should be investigated more thoroughly, e.g., the irregularities should be calculated. The results of this paper can be generalized in many ways, e.g., one can consider conics instead of lines.

(3.5) Remark

It might be that the set $\mathfrak{R}$ introduced in 3.3 is the whole interval $[\frac{1}{5}, 3]$. In fact, it follows from Persson's results [29] that $[\frac{1}{5}, 2] \subset \mathfrak{R}$. But not much seems to be known for the interval $[2, 3]$. It also occurs in [21] that 5/2 is an accumulation point of $\mathfrak{R}$.

References

[1] Brieskorn, E.: Sur les groupes de tresses (d'après V.I. Arnold), Séminaire Bourbaki, 24e année, 1971/72, Lecture Notes 317, p. 21-44, Springer 1973.

[2] Burr, St., Grünbaum, B., and Sloane, N. J. A.: The orchard problem, Geometriae Dedicata 2, p. 397–424 (1974).

[3] Cartier, P.: Les arrangement d'hyperplans: Un chapitre de géométrie combinatoire, Séminaire Bourbaki, 33e année, 1980/81, Lecture Notes 901, p. 1–22, Springer 1981.

[4] Deligne, P.: Les immeubles des groupes de tresses généralisés, Inv. Math. 17, p. 273–302 (1972).

[5] Deligne, P., and Mostow, G.D.: On a construction of Picard.

[6] Fricke, R. und Klein, F.: Vorlesungen über die Theorie der automorphen Funktionen, zwei Bände, Nachdruck der 1. Auflage von 1897, Teubner Stuttgart 1965.

[7] Grünbaum, B.: Arrangements of hyperplanes, in Proc. Second Louisiana Conference on Combinatorics and Graph Theory, p. 41–106, Baton Rouge 1971.

[8] Grünbaum, B.: Arrangements and spreads, Regional Conference Series in Mathematics, Number 10, Amer. Math. Soc. 1972.

[9] Hansen, St.: Contributions to the Sylvester-Gallai-Theory, Polyteknisk Forlag København 1981.

[10] Hirzebruch, F.: Der Satz von Riemann-Roch in Faisceau-theoretischer Formulierung, Proc. Intern. Congress Math. 1954, Vol. III, p. 457–473, North-Holland Publishing Company, Amsterdam 1956.

[11] Hirzebruch, F.: Automorphe Formen und der Satz von Riemann-Roch, Symp. Intern. Top. Alg. 1956, p. 129–144, Universidad de México 1958.

[12] Hirzebruch, F.: Some examples of algebraic surfaces, Proc. 21st Summer Research Institute Australian Math. Soc. 1981, p. 55–71, Contemporary Mathematics Vol. 9, Amer. Math. Soc. 1982.

[13] Holzapfel, R.-P.: A class of minimal surfaces in the unknown region of surface geography, Math. Nachr. 98, p. 211–232 (1980).

[14] Holzapfel, R.-P.: Invariants of arithmetic ball quotient surfaces, Math. Nachr. 103, p. 117–153 (1981).

[15] Iitaka, S.: Geometry on complements of lines in P^2, Tokyo J. Math. 1, p. 1–19 (1978).

[16] Inoue, M.: Some surfaces of general type with positive indices, preprint 1981.

[17] Ishida, M.-N.: The irregularities of Hirzebruch's examples of surfaces of general type with $c_1^2 = 3c_2$, Math. Ann. (to appear).

[18] Ishida, M.-N.: Hirzebruch's examples of surfaces of general type with $c_1^2 = 3c_2$, Japan-France Seminar 1982.

[19] Kodaira, K.: A certain type of irregular algebraic surfaces, J. Analyse Math. 19, p. 207–215 (1967).

[20] Kodaira, K.: On the structure of complex analytic surface IV, Amer. J. of Math. 90, p. 1048–1066 (1967).

[21] Livné, R. A.: On certain covers of the universal elliptic curve, Ph. D. thesis, Harvard University 1981.

[22] Miyaoka, Y.: On the Chern numbers of surfaces of general type, Inv. Math. 42, p. 225–237 (1977).

[23] Miyaoka, Y.: On algebraic surfaces with positive index, preprint 1980, SFB Theor. Math. Bonn.

[24] Mostow, G.D.: Complex reflection groups and non-arithmetic monodromy, lecture at the Bonn Arbeitstagung June 1981, see also Proc. Nat. Acad. Sci. USA (to appear).

[25] Mostow, G.D., and Siu, Y.-T.: A compact Kähler surface of negative curvature not covered by the ball, Ann. of Math. 112, p. 321–360 (1980).

[26] Orlik, P., and Solomon, L.: Combinatorics and topology of complements of hyperplanes, Inv. Math. 56, p. 167–189 (1980).

[27] Orlik, P., and Solomon, L.: Coxeter arrangements, Amer. Math. Soc. Proc. Symp. Pure Math 40 (to appear).

[28] Orlik, P., and Solomon, L.: Arrangements defined by unitary reflection groups, Math. Ann. (to appear).

[29] Persson, U.: Chern invariants of surfaces of general type, Comp. Math. 43, p. 3–58 (1981).

[30] Shephard, G. C., and Todd, J. A.: Finite unitary reflection groups, Can. J. Math. 6, p. 274–304 (1954).

[31] Yau, S.-T.: Calabi's conjecture and some new results in algebraic geometry, Proc. Nat. Acad. Sci. USA 74, p. 1798–1799 (1977).

Remarks added in proof:

1) The conjecture in 3.1 is wrong as Dr. Werner Meyer pointed out:

Take arrangements A_1 and A_2 of k_1 and k_2 lines respectively which are in general position to each other. Consider the combined arrangement $A_1 \cup A_2$. Then the numerical characteristics t_r are additive except for t_2 which satisfies

$$t_2(A_1 \cup A_2) = t_2(A_1) + t_2(A_2) + k_1 k_2$$

Let F_1, F_2, F be the quadratic polynomials for $A_1, A_2, A_1 \cup A_2$. Then

$$F(x) = F_1(x) + F_2(x) + k_1 k_2 x^2$$

If we take for A_2 a pencil with k_2 large with respect to k_1, then $F(x)$ becomes indefinite. Maybe the conjecture remains true if one assumes that the arragement "does not contain large pencils".

2) The inequality (3) in 3.1 can be improved by using results of F. Sakai (Semi-Stable Curves on Algebraic Surfaces and Logarithmic Pluricanonical Maps, Math. Ann 254, p. 89–120 (1980)). We have

$$t_2 + \frac{3}{4} t_3 \geq k + t_5 + 2t_6 + 3t_7 + \cdots$$

This is sharp for $A_1(6), A_1(9), A_3^0(3), A_3^0(4)$ and the arrangements with S–T numbers 24 and 25.

3) B. Grünbaum has written to me that the arrangements $A_2(17)$ and $A_7(17)$ of his list are isomorphic.

Received November 22, 1982

Professor Dr. F. Hirzebruch
Max-Planck-Institut für Mathematik
Gottfried-Claren-Strasse 26
5300 Bonn 3
Federal Republic of Germany

Regular Functions on Certain Infinite-dimensional Groups

Victor G. Kac and Dale H. Peterson

To Igor Rostislavovich Shafarevich on his 60th birthday

§0. Introduction

In the paper [18], we began a detailed study of the "smallest" group G associated to a Kac-Moody algebra $\mathfrak{g}(A)$ and of the (in general infinite-dimensional) flag varieties $\mathbf{P}\mathcal{V}_\Lambda$ associated to G. In the present paper we introduce and study the algebra $\mathbf{F}[G]$ of "strongly regular" functions on G. We establish a Peter-Weyl-type decomposition of $\mathbf{F}[G]$ with respect to the natural action of $G \times G$ (Theorem 1) and prove that $\mathbf{F}[G]$ is a unique factorization domain (Theorem 3).

These considerations are intimately related to the study of the algebra $\mathbf{F}[\mathcal{V}_\Lambda]$ of polynomial functions on the variety $\mathcal{V}_\Lambda$ (Theorem 2) and the so-called Bruhat and Birkhoff decompositions of $\mathcal{V}_\Lambda$.

The group G is a (possibly infinite-dimentional) algebraic group in the sense of Shafarevich [20], and belongs to one of the following three classes (we assume A to be indecomposable):

1) Finite type groups. These are connected simply-connected split simple finite-dimensional algebraic groups. In this case almost all the results of the paper are well-known.

2) Affine type groups. Such a G is an $\mathbf{F}^*$-extension of the group of regular maps from $\mathbf{F}^*$ to a group of finite type, or a "twisted" analogue. The simplest flag variety may be regarded as the space of based polynomial loops on a compact Lie group (in the case $\mathbf{F} = \mathbf{C}$).

3) "Wild" type groups. No "concrete" realization of these groups or their flag varieties is known.

The study of the groups G and the varieties $\mathcal{V}_\Lambda$ in the affine case is of particular importance because of applications to topology [2], [8], analysis [1], [9], soliton equations [4], etc.

Throughout the paper the base field $\mathbf{F}$ is of characteristic zero.

We thank M. Hochster for numerous consultations in commutative algebra.

§1. Kac-Moody Algebras and Associated Groups. Integrable Representations

1A) A *symmetrizable generalized Cartan matrix* $A = (a_{ij})_{i,j \in I}$ indexed by a nonempty finite set I is a matrix of integers satisfying: $a_{ii} = 2$ for all i; $a_{ij} \leq 0$ if $i \neq j$; DA is symmetric for some nondegenerate diagonal matrix D. We fix such a matrix A, assumed for simplicity to be indecomposable.

Choose a triple $(\mathfrak{h}, \Pi, \Pi^\vee)$, unique up to isomorphism, where $\mathfrak{h}$ is a vector space over $\mathbf{F}$ of dimension $|I| + \operatorname{corank} A$, and $\Pi = \{\alpha_i\}_{i \in I} \subset \mathfrak{h}^*$, $\Pi^\vee = \{h_i\}_{i \in I} \subset \mathfrak{h}$ are linearly independent indexed sets satisfying $\alpha_j(h_i) = a_{ij}$.

The *Kac-Moody algebra* $\mathfrak{g} = \mathfrak{g}(A)$ is the Lie algebra over $\mathbf{F}$ generated by $\mathfrak{h}$ and symbols e_i and $f_i (i \in I)$ with defining relations:

$$(1.1)\qquad \begin{aligned} &[\mathfrak{h}, \mathfrak{h}] = (0); \quad [e_i, f_j] = \delta_{ij} h_i; \\ &[h, e_i] = \alpha_i(h) e_i, \quad [h, f_i] = -\alpha_i(h) f_i \quad (h \in \mathfrak{h}); \end{aligned}$$

$$(1.2)\qquad (ad\, e_i)^{1-a_{ij}}(e_j) = 0, \quad (ad\, f_i)^{1-a_{ij}}(f_j) = 0 \quad (i \neq j).$$

We have the canonical embedding $\mathfrak{h} \subset \mathfrak{g}$ and linearly independent *Chevalley generators* $e_i, f_i (i \in I)$ for the derived algebra $\mathfrak{g}'$ of $\mathfrak{g}$. The center $\mathfrak{c}$ of $\mathfrak{g}$ lies in $\mathfrak{h}' := \mathfrak{h} \cap \mathfrak{g}' = \sum \mathbf{F} h_i$. Every ideal of $\mathfrak{g}$ contains $\mathfrak{g}'$ or is contained in $\mathfrak{c}$ [7].

Define an involution ω of $\mathfrak{g}$ by requiring: $\omega(e_i) = -f_i$, $\omega(f_i) = -e_i$, $\omega(h) = -h (h \in \mathfrak{h})$. Let $\mathfrak{n}_+$ be the subalgebra of $\mathfrak{g}$ generated by the $e_i (i \in I)$, and put $\mathfrak{n}_- = \omega(\mathfrak{n}_+)$. We have the vector space decomposition $\mathfrak{g} = \mathfrak{n}_- \oplus \mathfrak{h} \oplus \mathfrak{n}_+$.

We have the *root space decomposition* $\mathfrak{g} = \bigoplus_{\alpha \in \mathfrak{h}^*} \mathfrak{g}_\alpha$, where $\mathfrak{g}_\alpha = \{x \in \mathfrak{g} \mid [h, x] = \alpha(h) x \text{ for all } h \in \mathfrak{h}\}$. Put $Q = \sum_{i \in I} \mathbf{Z} \alpha_i$, $Q_+ = \sum_{i \in I} \mathbf{Z}_+ \alpha_i$ (where $\mathbf{Z}_+ = \{0, 1, \dots\}$), and define a partial order on $\mathfrak{h}^*$ by: $\lambda \geq \mu$ if $\lambda - \mu \in Q_+$. A *root* (resp. *positive root*) is an element of $\Delta := \{\alpha \in \mathfrak{h}^* \mid \alpha \neq 0, \mathfrak{g}_\alpha \neq (0)\}$ (resp. $\Delta_+ := \Delta \cap Q_+$). We have: $\mathfrak{h} = \mathfrak{g}_0$, $\mathfrak{n}_\pm = \bigoplus_{\alpha \in \Delta_+} \mathfrak{g}_{\pm\alpha}$. For $\alpha = \sum k_i \alpha_i \in \Delta$, we write $ht\, \alpha = \sum k_i$.

Define *fundamental reflections* $r_i \in \operatorname{Aut}_{\mathbf{F}}(\mathfrak{h})$, $i \in I$, by $r_i(h) = h - \alpha_i(h) h_i$. They generate the *Weyl group* W, which is a Coxeter group on $\{r_i\}_{i \in I}$, with length function $w \mapsto l(w)$. W preserves the root system Δ. A *real root* is an element of $\Delta^{re} := \{w(\alpha) \mid w \in W, \alpha \in \Pi\}$. If $\alpha \in \Delta^{re}$, then $\dim \mathfrak{g}_\alpha = 1$. Put $\Delta^{re}_+ = \Delta^{re} \cap \Delta_+$. For $\alpha \in \Delta^{re}$, write $\alpha = w(\alpha_i)$ for some $w \in W$ and $i \in I$; then $r_\alpha := w r_i w^{-1}$ depends only on α.

We choose a nondegenerate $\mathfrak{g}$-invariant symmetric $\mathbf{F}$-bilinear form $(.|.)$ on $\mathfrak{g}$ such that $(h_i \mid h_i)$ is positive rational for all $i \in I$. $(.|.)$ is nondegenerate and W-invariant on $\mathfrak{h}$, and hence induces an isomorphism $\nu: \mathfrak{h} \longrightarrow \mathfrak{h}^*$ and a form $(.|.)$ on $\mathfrak{h}^*$ [10, Chapter II].

1B) Consider a $\mathfrak{g}'$-module V, or (V, π), where $\pi: \mathfrak{g}' \longrightarrow \mathrm{End}_{\mathbf{F}}(V)$. Let $V_{fin} = \{v \in V \mid \text{for every } \alpha \in \Delta^{re} \text{ there exists } N \text{ such that } \pi(\mathfrak{g}_\alpha)^N(v) = 0\}$ ($\alpha \in \pm\Pi$ suffices). V_{fin} is a $\mathfrak{g}'$-submodule of V; the $\mathfrak{g}'$-module V is called *integrable* if $V = V_{fin}$. $(\mathfrak{g}, ad)$ is an integrable $\mathfrak{g}'$-module.

Remark 1.1. We feel that the functor $V \mapsto V_{fin}$ from the category of all $\mathfrak{g}'$-modules to the category of integrable $\mathfrak{g}'$-modules is important.

Lemma 1.1. *Let (V, π) be a locally-finite $sl_2(\mathbf{F})$-module and let $\{e, f, h\}$ be the standard basis of $sl_2(\mathbf{F})$. Let $x \in \mathrm{End}_{\mathbf{F}}(V)$ satisfy*

(1.3) $$[\pi(f), x] = 0 \text{ and } [\pi(h), x] = ax, \quad \text{where } -a \in \mathbf{Z}_+.$$

Then $(ad\pi(e))^{1-a} x = 0$.

Proof. The $sl_2(\mathbf{F})$-module V decomposes into a direct sum of finite-dimensional submodules: $V = \bigoplus_i V_i$. Then x has a "block decomposition": $\sum_{i,j} x_{ij}$, where $x_{ij} \in \mathrm{Hom}_{\mathbf{F}}(V_i, V_j)$. (1.3) holds for each x_{ij}, and hence we have $(ad\,\pi(e))^{1-a} x_{ij} = 0$ for all i, j by the finite-dimensional representation theory of sl_2. Q.E.D.

One immediately deduces the following corollary, which allows us to "differentiate" integrable G-modules (i.e., modules such that the $U_\alpha (\alpha \in \Delta^{re})$ act locally unipotently).

Corollary 1.1. *Let $\tilde{\mathfrak{g}}'$ be the Lie algebra on generators e_i, f_i, h_i ($i \in I$) with defining relations (1.1), with $\mathfrak{h}$ replaced by $\sum \mathbf{F} h_i$. Let (V, π) be a $\tilde{\mathfrak{g}}'$-module such that all $\pi(e_i)$, $\pi(f_i)$ are locally nilpotent. Then $\pi(e_i)$ and $\pi(f_i)$ satisfy relations (1.2), so that we may regard (V, π) as an integrable $\mathfrak{g}'$-module.*

1C) We now recall the construction of the group G associated to the Lie algebra $\mathfrak{g}'$ [18].

Let G^* be the free product of the additive groups $\mathfrak{g}_\alpha$, $\alpha \in \Delta^{re}$, with canonical inclusions i_α: $\mathfrak{g}_\alpha \longrightarrow G^*$. For any integrable $\mathfrak{g}'$-module (V, π), define a homomorphism π^*: $G^* \longrightarrow \mathrm{Aut}_{\mathbf{F}}(V)$ by $\pi^*(i_\alpha(e)) = \exp \pi(e)$. Let N^* be the intersection of all $\mathrm{Ker}(\pi^*)$, put $G = G^*/N^*$, and let q: $G^* \longrightarrow G$ be the canonical homomorphism. For $e \in \mathfrak{g}_\alpha (\alpha \in \Delta^{re})$, put $\exp e = q(i_\alpha(e))$, so that $U_\alpha := \exp \mathfrak{g}_\alpha$ is an additive one-parameter subgroup of G. The $U_\alpha (\alpha \in \pm\Pi)$ generate G, and G is its own derived group. Denote by U_+ (resp. U_-) the subgroup of G generated by the U_α (resp. $U_{-\alpha}$), $\alpha \in \Delta_+^{re}$.

Example 1.1. a) Let A be the Cartan matrix of a split simple finite-dimensional Lie algebra $\mathfrak{g}$ over $\mathbf{F}$. Then the group G associated to $\mathfrak{g} \cong \mathfrak{g}'(A)$ is the group $\underline{G}(\mathbf{F})$ of $\mathbf{F}$-valued points of the connected simply-connected algebraic group $\underline{G}$ associated to $\mathfrak{g}$, and $U_+ a \underline{U}(\mathbf{F})$ for some maximal unipotent subgroup $\underline{U}$ of $\underline{G}$. These groups G are called *groups of finite type.*

b) Let $\mathfrak{g}$ be as in a), and let $\tilde{A}$ be the extended Cartan matrix of $\mathfrak{g}$. Then the group G associated to $\mathfrak{g}'(\tilde{A})$ is a central extension by $\mathbf{F}^*$ of $\underline{G}(\mathbf{F}[z, z^{-1}])$, and

$$U_+ \cong \{g \in \underline{G}(\mathbf{F}[z]) \mid g|_{z=0} \in \underline{U}(\mathbf{F})\}.$$

These groups G and their twisted analogues are called *groups of affine type.*

To any integrable $\mathfrak{g}'$-module (V, π) we associate the homomorphism (again denoted by) π: $G \longrightarrow \mathrm{Aut}_{\mathbf{F}}(V)$ satisfying $\pi(\exp e) = \exp \pi(e)$ for $e \in \mathfrak{g}_\alpha (\alpha \in \Delta^{re})$. The homomorphism associated to $(\mathfrak{g}, ad)$, denoted Ad, maps G into $\mathrm{Aut}_{\mathbf{F}}(\mathfrak{g})$. The kernel of Ad is the center C of G, and $Ad(G)$ acts faithfully on $\mathfrak{g}'/\mathfrak{c}$. We have $\pi(Ad(g)x) = \pi(g)\pi(x)\pi(g)^{-1}$ for any integrable $\mathfrak{g}'$-module (V, π) and all $g \in G$, $x \in \mathfrak{g}'$. It follows that if (V, π) is an integrable $\mathfrak{g}'$-module with $\mathrm{Ker}\, \pi \subset \mathfrak{c}$, then (on G) $\mathrm{Ker}\, \pi \subset C$.

For each $i \in I$ we have a unique homomorphism φ_i: $SL_2(\mathbf{F}) \longrightarrow G$ satisfying:

$$\varphi_i\begin{pmatrix} 1 & t \\ 0 & 1 \end{pmatrix} = \exp te_i, \quad \varphi_i\begin{pmatrix} 1 & 0 \\ t & 1 \end{pmatrix} = \exp tf_i \quad (t \in \mathbf{F}).$$

Let $G_i = \varphi_i(SL_2(\mathbf{F}))$, $H_i = \varphi_i(\{\mathrm{diag}(t, t^{-1}) \mid t \in \mathbf{F}^*\})$, and let N_i be the normalizer of H_i in G_i. Let H (resp. N) be the subgroup of G generated by the H_i (resp. N_i); H is an abelian normal subgroup of N. The φ_i are monomorphisms and H is the direct product of the H_i. We have an isomorphism φ: $W \longrightarrow N/H$ such that $\varphi(r_i)$ is the coset $N_i H \setminus H$. We identify W and N/H using φ; this gives sense to expressions such as wH

and wU_+w^{-1} occurring in the sequel. If $h \in \mathfrak{h}$, $w \in W$ and $n \in wH$, then $Ad(n)h = w(h)$. We put $B_+ = HU_+$, $B_- = HU_-$.

1D) Choose $\Lambda_i \in \mathfrak{h}^*(i \in I)$ satisfying $\Lambda_i(h_j) = \delta_{ij}(j \in I)$. Put P_+ (resp. P_{++})$= \{\sum_i k_i\Lambda_i \mid k_i \in \mathbf{Z}$ and $k_i \geq 0$ (resp. $> 0)\}$. Given $\Lambda \in P_+$ (or more generally, $\Lambda \in \mathfrak{h}^*$ such that all $\Lambda(h_i) \in \mathbf{Z}_+$), there exists an irreducible $\mathfrak{g}$-module $(L(\Lambda), \pi_\Lambda)$, unique up to isomorphism, containing a $v_\Lambda \neq 0$ satisfying: $\pi_\Lambda(\mathfrak{n}_+)v_\Lambda = (0)$; $\pi_\Lambda(h)v_\Lambda = \Lambda(h)v_\Lambda(h \in \mathfrak{h})$. $L(\Lambda)$ is an absolutely irreducible integrable $\mathfrak{g}'$-module, and we have $L(\Lambda) = \pi_\Lambda(U(\mathfrak{n}_-))v_\Lambda$, $\mathrm{End}_{\mathfrak{g}'}(L(\Lambda)) = \mathbf{F}I_{L(\Lambda)}$. The module $L(\Lambda)$ is called an *integrable module with highest weight* Λ [11]. Recall that $\bigoplus_{i\in I} L(\Lambda_i)$ is a faithful G-module [18].

We have the *weight space decomposition* $L(\Lambda) = \bigoplus_{\lambda\in\mathfrak{h}^*} L(\Lambda)_\lambda$, where $L(\Lambda)_\lambda = \{v \in L(\Lambda) \mid h(v) = \lambda(h)v$ for all $h \in \mathfrak{h}\}$. Elements of $P(\Lambda) := \{\lambda \in \mathfrak{h}^* \mid L(\Lambda)_\lambda \neq (0)\}$ are called *weights* of $L(\Lambda)$. We have $\mathbf{F}v_\Lambda = L(\Lambda)_\Lambda = \{v \in L(\Lambda) \mid \mathfrak{n}_+(v) = (0)\}$; elements of $\mathbf{F}^*v_\Lambda$ are called *highest weight vectors*. We have $P(\Lambda) \subset \Lambda - Q_+$, and $\dim L(\Lambda)_{w(\lambda)} = \dim L(\Lambda)_\lambda$ if $\lambda \in \mathfrak{h}^*$, $w \in W$; in particular, $\dim L(\Lambda)_{w(\Lambda)} = 1$.

Regarded as a $\mathfrak{g}$-module under $\pi^*_\Lambda := \pi_\Lambda \circ \omega$, $L(\Lambda)$ is denoted $L^*(\Lambda)$ and v_Λ is denoted v^*_Λ. There exists a unique $\mathfrak{g}$-invariant bilinear form on $L(\Lambda) \times L^*(\Lambda)$ satisfying $\langle v_\Lambda, v^*_\Lambda\rangle = 1$; it is nondegenerate. Using $\langle\ ,\ \rangle$ we regard $L^*(\Lambda)$ as a subspace of $L(\Lambda)^*$, the algebraic dual of $L(\Lambda)$.

Note that if a statement holds for $\mathfrak{n}_+, U_+$ or $L(\Lambda)$, then a similar statement holds for $\mathfrak{n}_-, U_-$ or $L^*(\Lambda)$ using ω. We keep this observation in mind in the sequel.

§2. A Peter-Weyl-Type Theorem

2A) For every real root α, we fix a non-zero element e_α of $\mathfrak{g}_\alpha$, and coordinatize U_α by putting $x_\alpha(t) = \exp te_\alpha(t \in \mathbf{F})$. Furthermore, for $\overline{\beta} = (\beta_1, \ldots, \beta_k) \in (\Delta^{re})^k$, define a map $x_{\overline{\beta}}\colon \mathbf{F}^k \longrightarrow G$ by

$$x_{\overline{\beta}}(t_1, \ldots, t_k) = x_{\beta_1}(t_1)\ldots x_{\beta_k}(t_k),$$

and denote by $U_{\overline{\beta}}$ the image of $x_{\overline{\beta}}$.

We call a function $f\colon G \longrightarrow \mathbf{F}$ *weakly regular* if $f \circ x_{\overline{\beta}}\colon \mathbf{F}^k \longrightarrow \mathbf{F}$ is a polynomial function for all $\overline{\beta} \in (\Delta^{re})^k$ and $k \in \mathbf{Z}_+$ ($\overline{\beta} \in (-\Pi \cup \Pi)^k$ suffices). Denote by $\mathbf{F}[G]_{w.r.}$ the algebra of all weakly regular functions.

Let V be a $\mathfrak{g}'$-module; given $v^* \in V^*$ and $v \in V_{fin}$, we get a weakly regular function $f_{v^*,v}$ on G, called a *matrix coefficient*: $f_{v^*,v}(g) = \langle g(v), v^* \rangle$. The matrix coefficients $\theta_\Lambda := f_{v_\Lambda^*, v_\Lambda}$ are especially important.

Lemma 2.1. a) *The ring* $\mathbf{F}[G]_{w.r.}$ *is an integral domain.*

b) *If* $f \in \mathbf{F}[G]_{w.r.}$ *is such that* $\bigcap_{k>0} f^k \mathbf{F}[G]_{w.r.} \neq (0)$, *then* $f \in \mathbf{F}^*$.

c) *Every unit of* $\mathbf{F}[G]_{w.r.}$ *lies in* $\mathbf{F}^*$.

d) *Any* $f \in \mathbf{F}[G]_{w.r}$ *is determined by its restriction to* $U_- H U_+$.

Proof. a) holds since any $g_1, g_2 \in G$ lie in some $U_{\overline{\beta}}$. b) and c) hold by considering pullbacks under $x_{\overline{\beta}}$. d) holds, using a), since for $\Lambda \in P_{++}$ the Birkhoff decomposition [18] gives:

$$U_- H U_+ = \{g \in G \mid \theta_\Lambda(g) \neq 0\}. \qquad \text{Q.E.D.}$$

2B) Put $\tilde{H} = \mathrm{Hom}(Q, \mathbf{F}^*)$, and define a homomorphism $Ad\colon \tilde{H} \longrightarrow \mathrm{Aut}_{\mathbf{F}}(\mathfrak{g})$ by $Ad(h)x = h(\alpha)x$ if $x \in \mathfrak{g}_\alpha$. Ad induces an action of $\tilde{H}$ on G, defining $\tilde{H} \ltimes G$, to which Ad extends in the obvious way. We extend the action of G on $L(\Lambda)$ to $\tilde{H} \times G$ by requiring $\tilde{H}$ to fix v_Λ.

Subgroups U'_+ of U_+ and U'_- of U_- are called *large* if there exist $g_1, \dots, g_m \in G$ such that $\bigcap_{j=1}^m g_j U_\pm g_j^{-1} \subset U'_\pm$.

Lemma 2.2. a) *A subgroup U' of U_+ is large if and only of it contains the stabilizer*[1] *in G of some finite-dimensional subspace of* $\bigoplus_{\Lambda \in P_+} L(\Lambda)$.

b) *Let U' be a large subgroup of U_+. Then:*

(i) *for every* $\alpha \in \Delta^{re}$, *the subgroup* $\bigcap_{u \in U_\alpha} uU'u^{-1}$ *of* U_+ *is large;*

(ii) *the subgroup* $\bigcap_{h \in \tilde{H}} hU'h^{-1}$ *of* U_+ *is large;*

(iii) *there exists* $\overline{\beta}$ *such that* $U_{\overline{\beta}} U' = U_+$.

Proof. a) holds since $U_+ = \{g \in G \mid g(v_{\Lambda_i}) = v_{\Lambda_i} \text{ for all } i \in I\}$ [18], and since $L(\Lambda)$, $\Lambda \in P_+$, is spanned by $G(v_\Lambda)$. b (i) and (ii) hold by a). To prove b (iii), we may assume that U' is the stabilizer in U_+ of a U_+-invariant finite-dimensional subspace V of $\bigoplus L(\Lambda)$. Let π_V be the restriction

---

[1]We define the *stabilizer* (resp. *normalizer*) in G of a subset M of a G-set to be the group $\{g \in G \mid g(v) = v \text{ (resp. } g(v) \in M) \text{ for all } v \in M\}$.

of the action of U_+ to V. U_+ acts as a finite-dimensional unipotent group $U := \pi_V(U_+)$ on V. The one-parameter subgroups $\pi_V(U_\alpha)$ $(\alpha \in \Delta_+^{re})$ generate U, hence $\pi_V(U_{\overline{\beta}}) = U$ for some $\overline{\beta} \in (\Delta_+^{re})^k$, $k \in \mathbf{Z}_+$. Hence $U_+ = U_{\overline{\beta}}U'$. Q.E.D.

2C) We call a weakly regular function f *strongly regular* if there exist large subgroups $U'_\pm$ of $U_\pm$ such that $f(u_-gu_+) = f(g)$ for all $g \in G$ and $u_\pm \in U'_\pm$. Note that the matrix coefficients $f_{v^*,v}$, where $v^* \in L^*(\Lambda)$ and $v \in L(\Lambda)$, are strongly regular functions. We denote by $\mathbf{F}[G]_{s.r.}$, or $\mathbf{F}[G]$ for short, the algebra of all strongly regular functions on G. $\mathbf{F}[G]$ is a $G \times G$-module under π_{reg}, where $(\pi_{reg}(g_1, g_2)f)(g) = f(g_1^{-1}gg_2)$.

Now we can prove the following analogue of the Peter-Weyl theorem.

Theorem 1. *The linear map ϕ: $\bigoplus_{\Lambda \in P_+} L^*(\Lambda) \otimes L(\Lambda) \longrightarrow \mathbf{F}[G]$ defined by $\phi(v^* \otimes v) = f_{v^*,v}$ is an isomorphism of $G \times G$-modules.*

Proof. From Lemmas 2.1d and 2.2b (i and iii) it follows that every $U_\alpha \times U_\beta (\alpha, \beta \in \Delta^{re})$ acts locally unipotently on $\mathbf{F}[G]$. Hence there exist unique locally nilpotent elements of $\operatorname{End}_{\mathbf{F}} \mathbf{F}[G]$, which we denote by $\pi(e_i, 0)$, $\pi(0, e_i)$, $\pi(f_i, 0)$, $\pi(0, f_i)$, such that $\pi_{reg}(\exp te_i, 1) = \exp t\,\pi(e_i, 0)$, etc. Then Corollary 1.1 shows that there exists a unique homomorphism π: $\mathfrak{g}' \times \mathfrak{g}' \longrightarrow \operatorname{End}_{\mathbf{F}} \mathbf{F}[G]$ with the given values on $(e_i, 0)$, etc. Using Lemmas 2.1d and 2.2b (ii and iii), there exists a Q-gradation $\mathbf{F}[G] = \bigoplus R_\alpha$, where $R_\alpha = \{f \in \mathbf{F}[G] \mid f(h^{-1}gh) = h(\alpha)f(g) \text{ for all } h \in \tilde{H}\}$.

Thus, $(\mathbf{F}[G], \pi)$ is an integrable $\mathfrak{g}' \times \mathfrak{g}'$-module and $(\mathbf{F}[G], \pi_{reg})$ is the associated $G \times G$-module. Using Lemma 2.2b (iii), $U_- \times U_+$ acts locally-finitely on $\mathbf{F}[G]$. Hence, $\mathfrak{n}_- \times \mathfrak{n}_+$ acts locally-finitely, and therefore, using the Q-gradation, locally-nilpotently. Using the complete reducibility theorem [14, Proposition 2.9]), we deduce that the $G \times G$-module $\mathbf{F}[G]$ is isomorphic to a direct sum of modules of the form $L^*(\lambda) \otimes L(\mu)$ $(\lambda, \mu \in P_+)$.

Now, regarding $v^* \otimes v$ as an operator on $L(\lambda)$, we have: $f_{v^*,v}(g) = tr(v^* \otimes v)\pi_\lambda(g)$, so that ϕ is a well-defined $G \times G$-module homomorphism. ϕ is injective since the $L^*(\lambda) \otimes L(\lambda)$ are irreducible and inequivalent $G \times G$-modules, and $\phi(v_\lambda^* \otimes v_\lambda) = \mathcal{O}_\lambda \neq 0$. On the other hand, let ψ: $L^*(\lambda) \otimes L(\mu) \longrightarrow \mathbf{F}[G]$ be a $G \times G$-module homomorphism. Considering the action of $B_- \times U_+$ on $\psi(v_\lambda^* \otimes v_\mu)$ and using Lemma 2.1d, we obtain that $\psi(v_\lambda^* \otimes v_\mu) \in \mathbf{F}\mathcal{O}_\lambda$. Hence, ϕ is surjective. Q.E.D.

2D) Later we will need the following corollary of the proof of Theorem 1.

Corollary 2.1. *Let $f_1, f_2 \in \mathbf{F}[G]$, $f_1 \neq 0$, and suppose that for each $k \in \mathbf{Z}_+$ and $\overline{\beta} \in (\Delta^{re})^k$ there exists a polynomial function $q_{\overline{\beta}}$: $\mathbf{F}^k \longrightarrow \mathbf{F}$ such that $q_{\overline{\beta}}(f_1 \circ x_{\overline{\beta}}) = f_2 \circ x_{\overline{\beta}}$. Then $f_1^{-1} f_2 \in \operatorname{Fract} \mathbf{F}[G]$ lies in $\mathbf{F}[G]$.*

Proof is that of Theorem 1, replacing $\mathbf{F}[G]$ by the subalgebra of $\operatorname{Fract} \mathbf{F}[G]$ consisting of all $f_1^{-1} f_2$, where f_1, f_2 satisfy the hypothesis of the corollary. Q.E.D.

For a subgroup P of G, let $\mathbf{F}[G]^P = \{f \in \mathbf{F}[G] \mid f(gp) = f(g)$ for all $g \in G$ and $p \in P\}$. This is a subalgebra of $\mathbf{F}[G]$, and G acts on it by left multiplication: $(g \cdot f)(g') = f(g^{-1}g')$. For $\Lambda \in P_+$, put:

$$S_\Lambda = \{f \in \mathbf{F}[G] \mid f(gb) = \theta_\Lambda(b) f(g) \text{ for all } g \in G \text{ and } b \in B_+\}.$$

This is a G-submodule of $\mathbf{F}[G]^{U_+}$.

Corollary 2.2. a) $\mathbf{F}[G]^{U_+} = \bigoplus_{\Lambda \in P_+} S_\Lambda$.

b) *The map $L^*(\Lambda) \to S_\Lambda$ defined by $v \mapsto f_{v,v_\Lambda}$ is a G-module isomorphism.*

c) $S_\Lambda S_M = S_{\Lambda+M}$.

Proof is immediate by Theorem 1 and the following facts: $L(\Lambda)^{U_+} = \mathbf{F} v_\Lambda$; $\theta_\Lambda \theta_M = \theta_{\Lambda+M}$; multiplication is G-equivariant. Q.E.D.

Remark 2.1. a) The algebra $\mathbf{F}[G]^{U_+}$ can be constructed without reference to the group G. Indeed, for $\Lambda, M \in P_+$ we have the Cartan product $L^*(\Lambda) \otimes L^*(M) \xrightarrow{\phi} L^*(\Lambda + M)$, defined by the properties that ϕ is a $\mathfrak{g}$-module homomorphism and $\phi(v_\Lambda^* \otimes v_M^*) = v_{\Lambda+M}^*$. Under the Cartan product, the space $\bigoplus_{\Lambda \in P_+} L^*(\Lambda)$ becomes an algebra, isomorphic to $\mathbf{F}[G]^{U_+}$ by Corollary 2.2.

b) Corollary 2.2 can be viewed as a Borel-Weil-type theorem. (A special case of this is considered in [13]). It should not be difficult, using the method of [5], to extend it to a Borel-Weil-Bott-type theorem.

Corollary 2.3. *Let $\Lambda \in P_+ \setminus \{0\}$ and let $B_\Lambda = \{b \in B_+ \mid \theta_\Lambda(b) = 1\}$. Then, provided that $\mathbf{F}$ is algebraically closed, one has:*

$$\mathbf{F}[G]^{B_\Lambda} \cong \bigoplus_{n \geq 0} L^*(n\Lambda),$$

where $L^*(n\Lambda)L^*(m\Lambda) = L^*((n+m)\Lambda)$ under the Cartan product.

Example 2.1. a) If G is of finite type, then $\mathbf{F}[G] = \mathbf{F}[G]_{w.r.}$ is the coordinate ring of the finite-dimensional affine variety G.

b) Let G be of affine type as in Example 1.1b. Then the only strongly regular functions f such that $f(cg) = f(g)$ for all $c \in \mathbf{F}^* \subset C$ and $g \in G$ are constants (by Theorem 1). On the other hand, given a rational N-dimensional representation π of $\underline{G}$, let $\pi(g) = (\sum_k a_{ij}^k(g)z^k)_{i,j=1}^N$ for $g \in \underline{G}(\mathbf{F}[z, z^{-1}])$; then the pullback of each function $g \mapsto a_{ij}^k(g)$ is a weakly regular function on G.

2E) We introduce the Zariski topology on G defined by strongly regular functions, i.e., a closed subset is the set of zeros of an ideal of $\mathbf{F}[G]$.

Note that the stabilizer or normalizer of a finite-dimensional subspace of $\bigoplus_{\Lambda \in P_+} L(\Lambda)$ or $\bigoplus_{\Lambda \in P_+} L^*(\Lambda)$ is a closed subgroup of G. It follows that $U_\pm$ and $B_\pm$ are closed subgroups and hence $H = B_+ \cap B_-$ is a closed subgroup. Similarly, the $G_iU_\pm$ are closed subgroups and hence the subgroups $G_i = (G_iU_+) \cap (G_iU_-)$ are closed. It is easy to show that $\varphi_i\colon SL_2(\mathbf{F}) \longrightarrow G_i$ is a Zariski homeomorphism. One can also show that H is homeomorphic to $(\mathbf{F}^*)^I$ and $U_+ \cap (wU_-w^{-1})$ to $\mathbf{F}^{l(w)}$.

For $\overline{\beta} \in (\Delta^{re})^k$, let $\mathbf{F}[G]_{w.r.}^{\overline{\beta}} = \{f \in \mathbf{F}[G]_{w.r.} \mid f \text{ vanishes on } U_{\overline{\beta}}\}$. Taking the $\mathbf{F}[G]_{w.r.}^{\overline{\beta}}$ for a basis of neighborhoods of 0 makes $\mathbf{F}[G]_{w.r.}$ into a Hausdorff complete topological ring.

Remark 2.2. We have the canonical inclusion $G \to \operatorname{Specm} \mathbf{F}[G]$ (= set of all closed ideals of codimension 1). Let G be of infinite type (i.e., $\dim \mathfrak{g} = \infty$). Then $\mathfrak{m} := \phi(\bigoplus_{\Lambda \in P_+ \setminus \{0\}} L^*(\Lambda) \otimes L(\Lambda))$ is a closed ideal of codimension 1 in $\mathbf{F}[G]$ (this follows from the well-known fact that $(Q_+|_{\mathfrak{h}'}) \cap (P_+|_{\mathfrak{h}'}) = \{0\}$ in the infinite type case). Since $\mathfrak{m}$ is $G \times G$-invariant, we deduce that $\mathfrak{m} \in (\operatorname{Specm} \mathbf{F}[G]) \setminus G$.

§3. The Varieties $\mathcal{V}_\Lambda$

3A) Given a decomposition of a vector space V into a direct sum of finite-dimensional subspaces, $V = \bigoplus_\alpha V_\alpha$, we denote by $\mathbf{F}[V]$ the symmetric algebra over $\bigoplus_\alpha V_\alpha^* \subset V^*$. We call elements of $\mathbf{F}[V]$ *strongly regular functions* on V. The algebra $\mathbf{F}[V]$ is a polynomial algebra on a basis of $\bigoplus V_\alpha^*$ (it may be viewed as the coordinate ring of $\prod_\alpha V_\alpha$). It is a subalgebra of the algebra $\mathbf{F}[V]_r$ of *regular functions*, i.e., $\mathbf{F}$-valued functions on V whose restriction to any finite-dimensional subspace is a polynomial function. Taking the ideals of finite-dimensional subspaces of V for a basis of neighborhoods of zero makes $\mathbf{F}[V]_r$ into a complete topological ring.

We introduce the Zariski topology on V defined by strongly regular functions. For a closed subset $\mathcal{V}$ of V (resp. the zero set $\mathcal{V}$ of an ideal of $\mathbf{F}[V]_r$), we denote by $\mathbf{F}[\mathcal{V}]$ (resp. $\mathbf{F}[\mathcal{V}]_r$) the restriction of $\mathbf{F}[V]$ (resp. $\mathbf{F}[V]_r$) to $\mathcal{V}$.

These definitions will be applied in this section to the vector spaces $L(\Lambda)$ and $L^*(\Lambda)$ with the weight space decompositions and $\mathfrak{g}$ with the root space decomposition. Here $\mathbf{F}[L(\Lambda)] = \operatorname{Sym} L^*(\Lambda)$ and $\mathbf{F}[L^*(\Lambda)] = \operatorname{Sym} L(\Lambda)$.[2]

Remark 3.1. It is easy to see that the canonical map $V \to \operatorname{Specm} \mathbf{F}[V]$ is a bijection (cf. Remark 2.2); more generally, we have a bijection $\mathcal{V} \to \operatorname{Specm} \mathbf{F}[\mathcal{V}]$ for every closed subset $\mathcal{V}$ of V.

3B) For each $\alpha \in \Delta \cup \{0\}$, choose dual bases $\{e_\alpha^{(i)}\}$ of $\mathfrak{g}_\alpha$ and $\{f_\alpha^{(i)}\}$ of $\mathfrak{g}_{-\alpha}$. Let $\Lambda \in P_+$. Denote by $\mathcal{V}_\Lambda$ the set of all $v \in L(\Lambda)$ which satisfy the following equation in $L(\Lambda) \otimes L(\Lambda)$:

$$(\Lambda \mid \Lambda) v \otimes v = \sum_{\alpha \in \Delta \cup \{0\}} \sum_i f_\alpha^{(i)}(v) \otimes e_\alpha^{(i)}(v). \tag{3.1}$$

Note that the sum on the right-hand side is finite. (3.1) is equivalent to a (possibly infinite) system of equations of the form $P = 0$, where $P \in \operatorname{Sym}^2 L^*(\Lambda)$. We call these polynomials P (and their analogues in $\operatorname{Sym}^2 L(\Lambda)$) *Plücker polynomials.*

We have shown in [18] that:

$$\mathcal{V}_\Lambda = G(\mathbf{F} v_\Lambda). \tag{3.2}$$

---

[2] Here and further on, $\operatorname{Sym} V = \bigoplus_{k \geq 0} \operatorname{Sym}^k V$ denotes the symmetric algebra over a vector space V.

By the complete reducibility theorem ([12], [14]), we have: $\mathrm{Sym}^k L(\Lambda) = L(k\Lambda) \oplus J_k$, $\mathrm{Sym}^k L^*(\Lambda) = L^*(k\Lambda) \oplus J_k^*$, where $L(k\Lambda)$ (resp. $L^*(k\Lambda)$) is the $U(\mathfrak{g})$-submodule generated by v_Λ^k (resp. $(v_\Lambda^*)^k$) and J_k (resp. J_k^*) is the (unique) complementary submodule. Set $J = \bigoplus_{k\geq 2} J_k$, $J^* = \bigoplus_{k\geq 2} J_k^*$. Note that the restriction map ϕ: $\mathbf{F}[L(\Lambda)] \longrightarrow \mathbf{F}[\mathcal{V}_\Lambda]$ is a G-module homomorphism by (3.2); it is surjective by definition. Note also that: $(v_\Lambda^*)^k(tv_\Lambda) = t^k$ and $J_k^*(tv_\Lambda) = 0$. Hence, J^* is the ideal of $G(\mathbf{F}v_\Lambda)$ in $\mathbf{F}[L(\Lambda)]$, so that by (3.2) and Remark 2.1a we have:

Lemma 3.1. *$\mathbf{F}[\mathcal{V}_\Lambda] \cong \mathbf{F}[L(\Lambda)]/J^*$ is isomorphic to $\bigoplus_{k\geq 0} L^*(k\Lambda)$ with the Cartan product.*

Theorem 2. a) *The ideals J and J^* are generated by the Plücker polynomials.*

b) *The algebra $\mathbf{F}[G]^{U_+}$ is the associative commutative $\mathbf{F}$-algebra with unity on generators $\bigoplus_i L^*(\Lambda_i)$ with defining relations*

$$(\Lambda_i \mid \Lambda_j)uv = \sum_{\alpha\in\Delta\cup\{0\}} \sum_s f_\alpha^{(s)}(u)\, e_\alpha^{(s)}(v),$$

where $i, j \in I$, $u \in L^*(\Lambda_i)$, $v \in L^*(\Lambda_j)$.

The proof of Theorem 2 uses the Casimir operator introduced in [11] (cf. [14]):

$$\Omega = 2\nu^{-1}(\rho) + \sum_s f_0^{(s)} e_0^{(s)} + 2 \sum_{\alpha\in\Delta_+} \sum_s f_\alpha^{(s)} e_\alpha^{(s)},$$

where $\rho = \sum_i \Lambda_i$. Recall that Ω acts on $L(\Lambda)$ as a scalar $c_\Lambda := (\Lambda + 2\rho \mid \Lambda)$ ([11], [12]).

Lemma 3.2. *Let $\Lambda, \Lambda' \in \mathfrak{h}^*$ satisfy $\Lambda(h_i), \Lambda'(h_i) \in \mathbf{Z}_+$ for all $i \in I$. If $\Lambda > \Lambda'$, <u>then</u> $c_\Lambda - c_{\Lambda'} > 0$.*

Proof. $c_\Lambda - c_{\Lambda'} = (\Lambda + \Lambda' + 2\rho | \Lambda - \Lambda') > 0$. Q.E.D.

Proof of Theorem 2. To prove a), it suffices to consider J. Using Lemma 3.2 and the complete reducibility theorem applied to $\mathrm{Sym}^k L(\Lambda)$ we have:

$$J_k = (\Omega - c_{k\Lambda})\ \mathrm{Sym}^k L(\Lambda). \tag{3.3}$$

It follows from (3.1) and (3.3) that J_2 is the space of Plücker polynomials. Furthermore, we have by an easy calculation in $\mathrm{Sym}^k L(\Lambda)$, $k \geq 2$:

$$(\Omega - c_{k\Lambda})v^k = \frac{1}{2}k(k-1)((\Omega - c_{2\Lambda})v^2)v^{k-2},$$

which shows that J_2 generates the ideal J.

The proof of b) is similar, using the identity

$$\Omega(xyz) = \Omega(xy)z + \Omega(yz)x + \Omega(zx)y - \Omega(x)yz - \Omega(y)zx - \Omega(z)xy.$$

Q.E.D.

Example 3.1. a) Let $G = SL_n(\mathbf{F})$; then $L(\Lambda_k)$ is the G-module $\Lambda^k\mathbf{F}^n$ and $\mathcal{V}_{\Lambda_k}$ is the set of all decomposable k-vectors, so that $\mathbf{P}\mathcal{V}_{\Lambda_k}$ is the Grassmann variety of k-dimensional subspaces of $\mathbf{F}^n$. The ideal of $\mathcal{V}_{\Lambda_k}$ is generated by the classical Plücker relations; this result is due to Plücker. Theorem 2 in the finite-dimensional case is due to Kostant. Our proof is essentially the same as Kostant's (presented in [16]).

b) Let G be a group of affine type and $L(\Lambda_0)$ its basic representation (see [12], [15]). Then the Plücker relations are equivalent to the hierarchy of Hirota bilinear equations studied in [4], the simplest case $A_1^{(1)}$ being equivalent to the celebrated KdV hierarchy. Theorem 2a shows that the ideal of equations satisfied by all polynomial solutions of these hierarchies is generated by Hirota bilinear equations.

c) Let K be a connected compact Lie group of type $X_n (= A_n, B_n, \ldots, E_8)$. Let σ be an automorphism of K of finite order m. Let k be the minimal positive integer such that σ^k is an inner automorphism of K and let K_0 be the fixed point set of σ. Let $S^1 = \{z \in \mathbf{C} \mid |z| = 1\}$ be the unit circle; denote by $\Omega_\sigma(K)$ the space of all σ-equivariant polynomial loops on K, i.e., polynomial maps $g\colon S^1 \longrightarrow K$ such that $\sigma(g(z)) = g(z \exp \frac{2\pi i}{m})$. Then K_0 operates by right multiplication on $\Omega_\sigma(K)$ and we may consider the space $\Omega_\sigma(K)/K_0$. Let A be the generalized Cartan matrix of type $X_n^{(k)}$ [12] and G the associated group. Then we have a homeomorphism of topological spaces $\Omega_\sigma(K)/K_0 \cong \mathbf{P}\mathcal{V}_\Lambda$ for a suitable Λ (cf. [18]). Note that $\Omega_1(K)/K$ is the space of based loops; in this case A is of type $X_n^{(1)}$ and we have: $\Omega_1(K)/K \cong \mathbf{P}\mathcal{V}_{\Lambda_0}$. This allows one to compute the homology of certain loop spaces (cf. [2] and Theorem 4e in §4).

3C) In this subsection we use some elements of the theory of Coxeter groups, which can be found e.g. in [3]. Let $\Lambda \in P_+$; consider the orbit

$W(\Lambda)$. Recall the definition of the *Bruhat order* $\succeq$ on the set $W(\Lambda)$ [18]. This is the partial order generated by: $r_\alpha(\lambda) \succeq \lambda$ if $\alpha \in \Delta^{re}$ and $r_\alpha(\lambda) \geq \lambda$.

If $\Lambda \in P_{++}$, we may identify W with the set $W(\Lambda)$ by $w \Leftrightarrow w(\Lambda)$. We write $w' \leq w$ if $w'(\Lambda) \succeq w(\Lambda)$. It is easy to see that this definition is independent of the choice of $\Lambda \in P_{++}$ and coincides with the usual definition of the Bruhat order on W as the partial order generated by:

$$r_{i_1} \dots r_{i_{s-1}} r_{i_{s+1}} \dots r_{i_k} < w \quad (1 \leq s \leq k),$$

where $w = r_{i_1} \dots r_{i_k}$ is a reduced expression; or, equivalently, generated by: $w_1 w_2 < w_1 r_i w_2$ if $w_1(\alpha_i) > 0$ and $w_2^{-1}(\alpha_i) > 0$.

Note that $w(\Lambda_i) < \Lambda_i$ if and only if w contains r_i in one (and hence every) reduced expression. Therefore, we have:

$$w(\Lambda_i) < \Lambda_i \quad \text{iff} \quad r_i \leq w. \tag{3.4}$$

The following lemma summarizes some of the results of [18, Theorem 1 and Corollaries 2 and 5].

Lemma 3.3. a) *For $\Lambda \in P_+$ and $\lambda \in W(\Lambda)$ let*

$$\mathcal{V}_\Lambda(\lambda)_\pm = U_\pm(L(\Lambda)_\lambda \setminus \{0\}).$$

Then[3]

(i) $\mathcal{V}_\Lambda \setminus \{0\} = \coprod_{\lambda \in W(\Lambda)} \mathcal{V}_\Lambda(\lambda)_\pm$;

(ii) $\overline{\mathcal{V}_\Lambda(\lambda)_+} \setminus \{0\} = \coprod_{\mu \succeq \lambda} \mathcal{V}_\Lambda(\mu)_+$;

(iii) $\overline{\mathcal{V}_\Lambda(\lambda)_-} \setminus \{0\} = \coprod_{\lambda \succeq \mu} \mathcal{V}_\Lambda(\mu)_-$.

b) *Given a subset X of I, let $W_X = \langle r_i \mid i \in X \rangle \subset W$ and $P_X = B_+ W_X B_+ \subset G$. Then:*

(i) $G = \coprod_{w \in W/W_X} B_+ w P_X$ (*Bruhat decomposition*);

---

[3]Here and further on, $\overline{M}$ denotes the Zariski closure of M unless otherwise specified.

(ii) $G = \coprod_{w \in W/W_X} B_- w P_X$ (*Birkhoff decomposition*);
(iii) $G = \bigcup_{w \in W} w B_- B_+$.

Remark 3.2. a) If G is of finite type, then Lemma 3.3b (i) and (ii) give equivalent decompositions of G. This decomposition is due to Gauss, Gelfand-Naimark, Bruhat and Harish-Chandra.

b) Let G be a group of affine type from Example 1.1b, so that $G/\mathbf{F}^* = \underline{G}(\mathbf{F}[z, z^{-1}])$. Let $\underline{H}$ be a Cartan subgroup of $\underline{G}$. Then $W = W_X \ltimes T$ is the affine Weyl group of $\underline{G}$, where W_X is the Weyl group of $\underline{G}$ and T is the group of "translations", which is isomorphic to $\underline{H}(\mathbf{F}[z, z^{-1}])/\underline{H}(\mathbf{F})$. Furthermore, $P_X/\mathbf{F}^* = G(\mathbf{F}[z])$ and $B_-/\mathbf{F}^* \subset \underline{G}(\mathbf{F}[z^{-1}])$. Then Lemma 3.3b (ii) gives:

$$\underline{G}(\mathbf{F}[z, z^{-1}]) = \underline{G}(\mathbf{F}[z^{-1}])\underline{H}(\mathbf{F}[z, z^{-1}])\,\underline{G}(\mathbf{F}[z]),$$

a result usually attributed to Grothendieck [8]. A special case of this is the decomposition:

$$SL_n(\mathbf{F}[z, z^{-1}]) = \coprod_{\substack{k_1 \leq \cdots \leq k_n \\ \sum k_i = 0.}} SL_n(\mathbf{F}[z^{-1}])\, \mathrm{diag}\,(z^{k_1}, \ldots, z^{k_n}) SL_n(\mathbf{F}[z]).$$

This is due to Dedekind-Weber and Birkhoff [1].

Lemma 3.4. *Let $\mathcal{T}$ be a G-biinvariant topology on G such that (i) Zariski-closed subsets are $\mathcal{T}$-closed, and (ii) G_i lies in the $\mathcal{T}$-closure of $U_{-\alpha_i} H_i U_{\alpha_i}$ for all $i \in I$. Then for all $w \in W$, we have:*

(*a*) $$\coprod_{w' \leq w} B_+ w' B_+ \text{ is the } \mathcal{T}\text{-closure of } B_+ w B_+$$

(*b*) $$\coprod_{w \leq w'} B_- w' B_+ \text{ is the } \mathcal{T}\text{-closure of } B_- w B_+.$$

Proof. Fix $\Lambda \in P_{++}$ and consider the map $\phi \colon G \longrightarrow \mathcal{V}_\Lambda$ defined by $g \mapsto g(v_\Lambda)$. The map ϕ is Zariski-continuous and [18]:

$$\phi^{-1}(U_\pm(L(\Lambda)_{w(\Lambda)})) = B_\pm w B_+.$$

Hence, by (i) and Lemma 3.3 a, the $\mathcal{T}$-closure of $B_{\pm}wB_+$ is contained in the union in question.

In order to prove the reverse inclusion in a), suppose that $w = w_1 r_i w_2$, where $w_1(\alpha_i) > 0$, $w_2^{-1}(\alpha_i) > 0$. Then we have:

$$B_+wB_+ = B_+w_1r_iw_2B_+ = B_+(w_1r_iU_{-\alpha_i}r_iw_1^{-1})w_1r_iw_2(w_2^{-1}U_{\alpha_i}w_2)B_+$$
$$= B_+w_1r_iU_{-\alpha_i}H_iU_{\alpha_i}w_2B_+.$$

Since $\mathcal{T}$ is biinvariant we get (here $\overline{M}$ denotes the $\mathcal{T}$-closure of M):

$$\overline{B_+wB_+} \supset B_+w_1r_i\overline{(U_{-\alpha_i}H_iU_{\alpha_i})}w_2B_+.$$

Since, by (ii), $N_i \subset \overline{U_{-\alpha_i}H_iU_{\alpha_i}}$, we deduce that $\overline{B_+wB_+} \supset B_+w_1w_2B_+$. Similarly, we have:

$$B_-w_1w_2B_+ = B_-(w_1U_{-\alpha_i}w_1^{-1})w_1w_2(w_2^{-1}U_{\alpha_i}w_2)B_+$$
$$= B_-w_1U_{-\alpha_i}H_iU_{\alpha_i}w_2B_+$$

and hence

$$\overline{B_-w_1w_2B_+} \supset B_-w_1\overline{U_{-\alpha_i}H_iU_{\alpha_i}}w_2B_+ \supset B_-w_1r_iw_2B_+ = B_-wB_+,$$

proving the reverse inclusion in b). Q.E.D.

Remark 3.3. The Zariski topology on G satisfies the hypothesis of Lemma 3.4.

Let $\Gamma_i = \{g \in G \mid \theta_{\Lambda_i}(g) = 0\}$. We deduce from (3.4) and Remark 3.3:

Corollary 3.1. a) $\Gamma_i = \overline{B_-r_iB_+}(i \in I)$.

b) $U_-HU_+ = G \setminus \bigcup_i \Gamma_i$ *is open in* G, *and therefore* $G = \bigcup_{w \in W} wU_-HU_+$ *is a covering of* G *by open sets.*

Remark 3.4. One can show that $w' \leq w$ if and only if

$$B_+wB_+ \cap B_-w'B_+ \neq \emptyset.$$

§4. The Structure of the Algebra $\mathbf{F}[G]$

4A) Recall that the Lie algebra $\mathfrak{g}$ carries the *principal gradation* $\mathfrak{g} = \bigoplus_{j\in\mathbf{Z}} \mathfrak{g}_j$ defined by $\deg e_i = -\deg f_i = 1$, $\deg \mathfrak{h} = 0$; let $\mathfrak{g}_{(j)} = \bigoplus_{i\geq j} \mathfrak{g}_i$ be the associated filtration. Let $U_j (j \geq 1)$ be the descending central series of U_+: $U_1 = U_+$, $U_{j+1} = (U_+, U_j)$ for $j \geq 1$.

Lemma 4.1. *For any $u \in U_j (j \geq 1)$ there exists a unique element $\phi_j(u)$ of $\mathfrak{g}_j$ such that*

(i) $Ad(u)x - x \equiv [\phi_j(u), x] \bmod \mathfrak{g}_{(j+k+1)}$ *for all* $x \in \mathfrak{g}_{(k)}$, $k \in \mathbf{Z}$; *moreover, we have:*

(ii) $\phi_1(\exp te_i) = te_i \quad (i \in I)$, $\phi_1(\exp te_\alpha) = 0$ *if* $\alpha \in \Delta_+^{re} \setminus \Pi$;

(iii) $\phi_j(uu') = \phi_j(u) + \phi_j(u')$;

(iv) $\phi_{j+j'}((u, u')) = [\phi_j(u), \phi_{j'}(u')]$;

(v) $\phi_j \quad (j \geq 1)$ *is surjective.*

Proof. (i) for $k = 0$ gives uniqueness of $\phi_j(u)$. It is easy to check that (i) implies (ii), (iii), (iv). We construct $\phi_j(u)$ satisfying (i) by induction on $j \geq 1$ using (ii), (iii), (iv). Finally, (v) follows from (ii), (iii) and (iv). Q.E.D.

Corollary 4.1. *Let $h \in \mathfrak{h}$ be such that $\alpha(h) \neq 0$ for all $\alpha \in \Delta$. Then for all $k \geq 1$, $Ad(U_+)$ acts transitively on $(h + \mathfrak{n}_+) \bmod \mathfrak{g}_{(k)}$.*

4B) Given two sets B_1 and B_2, we have the canonical inclusion

$$\mathbf{F}^{B_1} \otimes \mathbf{F}^{B_2} \to \mathbf{F}^{B_1 \times B_2}$$

given by $(\phi_1 \otimes \phi_2)(b_1, b_2) = \phi_1(b_2)\phi_2(b_2)$. Let P be a group, $\mathbf{F}[P]$ an algebra of $\mathbf{F}$-valued functions on P containing $\mathbf{F}$. We say that $\mathbf{F}[P]$ is *naturally a Hopf algebra* if for the multiplication map μ: $P \times P \longrightarrow P$, we have $\mu^*(\mathbf{F}[P]) \subset \mathbf{F}[P] \otimes \mathbf{F}[P]$, and for the inversion map i: $P \longrightarrow P$, we have $i^*(\mathbf{F}[P]) = \mathbf{F}[P]$.

For any subgroup U of U_+ or U_- considered in the sequel, we denote by $\mathbf{F}[U]$ the restriction of $\mathbf{F}[G]$ to U.

Lemma 4.2. *$\mathbf{F}[U_+]$ and $\mathbf{F}[U_-]$ are naturally Hopf algebras.*

Proof. We prove the lemma for U_+. By Theorem 1, every $f \in \mathbf{F}[U_+]$ is a linear combination of functions $f_{v^*,v}$ where $v \in L(\Lambda)$, $v^* \in L^*(\Lambda)(\Lambda \in P_+)$. Since U_+ acts locally unipotently on $L(\Lambda)$, and $\pi_\Lambda(u^{-1}) = \exp(-\log \pi_\Lambda(u))$, the lemma is clear. Q.E.D.

Remark 4.1. $\mathbf{F}[G]$ is naturally a Hopf algebra if and only if $\dim \mathfrak{g} < \infty$. If $\dim \mathfrak{g} = \infty$, then $\mu^*(\mathbf{F}[G]) \not\subset \mathbf{F}[G] \otimes \mathbf{F}[G]$ and $i^*(\mathbf{F}[G]) \not\subset \mathbf{F}[G]$. Note that $i^*(\mathbf{F}[G]) = \omega^*(\mathbf{F}[G])$.

4C) **Lemma 4.3.** a)*Let $h \in \mathfrak{h}$ be such that $\alpha(h) \neq 0$ for all $\alpha \in \Delta$. Then the map ψ: $U_+ \longrightarrow \mathfrak{n}_+$ defined by $\psi(u) = Ad(u)h - h$ induces an isomorphism $\psi^* : \mathbf{F}[\mathfrak{n}_+] \rightrightarrows \mathbf{F}[U_+]$.*

b) *Fix $\Lambda \in P_{++}$; $\mathbf{F}[U_+]$ is a polynomial algebra on generators $f_{v^*_\Lambda, x_i(v_\Lambda)}$ (restricted to U_+), where $\{x_i\}$ is a basis of $\mathfrak{n}_-$.*

Proof. b) follows from a) and the formula:

$$\langle a(v_\Lambda), v^*_\Lambda\rangle = \big(a \mid \nu^{-1}(\Lambda)\big) \quad \text{for } a \in \mathfrak{g}.$$

Indeed, by this formula, $\big(Ad(u)\nu^{-1}(\Lambda) \mid a\big) = f_{v^*_\Lambda, a(v_\Lambda)}(u^{-1})$, and we apply Lemma 4.2.

To prove a), fix $\lambda \in P_{++}$; by [18, Lemma 5b], the map ϕ: $\mathfrak{n}_+ \longrightarrow L^*(\lambda)$ defined by $\phi(n) = n(v^*_\lambda)$ is injective. Hence, we may identify $\mathfrak{n}_+$ with its image in $L^*(\lambda)$ and $\mathbf{F}[\mathfrak{n}_+]$ with the restriction of $\mathbf{F}[L^*(\lambda)]$ to $\phi(\mathfrak{n}_+)$. Take $v \in L(\lambda)$. By Lemma 4.2, we may write: $u^{-1}(v) = \sum_i f_i(u)v_i$ (finite sum) for $u \in U_+$, where $f_i \in \mathbf{F}[U_+]$ and $v_i \in L(\lambda)$. Hence, the function $u \mapsto \langle v, \big(Ad(u)h\big)v^*_\lambda\rangle = -\sum_i f_i(u)\langle u\big(h(v_i)\big), v^*_\lambda\rangle$ lies in $\mathbf{F}[U_+]$, showing that $\psi^*(\mathbf{F}[\mathfrak{n}_+]) \subset \mathbf{F}[U_+]$. ψ^* is injective by Corollary 4.1.

To show that ψ^* is surjective, choose a basis $e^{(j)}_\alpha$ of $\mathfrak{g}_\alpha$ for each $\alpha \in \Delta_+$, such that $e^{(1)}_{\alpha_i} = e_i$. Then we have: $Ad(u)h = h + \sum_{\alpha\in\Delta_+}\sum_j \varphi^{(j)}_\alpha(u)e^{(j)}_\alpha$, where $\varphi^{(j)}_\alpha \in B := \psi^*(\mathbf{F}[\mathfrak{n}_+])$. Choose $h' \in \mathfrak{h}$ such that $\alpha(h') \neq 0$ for all non-zero $\alpha \in Q_+$. Then from $[Ad(u)h, Ad(u)h'] = 0$ we deduce that $Ad(u)h' = h' + \sum_{\alpha\in\Delta_+}\sum_j \tilde\varphi^{(j)}_\alpha(u)e^{(j)}_\alpha$, where $\tilde\varphi^{(j)}_\alpha \in B$ by induction on $ht\,\alpha$. Using this, the equation $[Ad(u)h', Ad(u)f_i] = -\alpha_i(h')Ad(u)f_i$ gives:

$$Ad(u)f_i = f_i - \alpha_i(h')^{-1}\tilde\varphi^{(1)}_{\alpha_i}(u)h_i + \sum_{\alpha\in\Delta_+}\sum_j \varphi^{(j)}_{\alpha,i}(u)e^{(j)}_\alpha,$$

where $\varphi^{(j)}_{\alpha,i} \in B$, again by induction on $ht\,\alpha$.

Now, functions of the form $f_{v^*,v}$, where $v = f_{i_1} \ldots f_{i_k}(v_\mu)$, $\mu \in P_+$, $i_1, \ldots, i_k \in I$ and $v^* \in L^*(\mu)$, generate $\mathbf{F}[U_+]$. But

$$f_{v^*,v}(u) = \langle (Ad(u)f_{i_1}) \ldots (Ad(u)f_{i_k})v_\mu, v^* \rangle$$

so that $f_{v^*,v} \in B$ since the $\varphi_{\alpha,i}^{(j)} \in B$. Q.E.D.

Remark 4.2. The map $\psi\colon U_+ \longrightarrow \mathfrak{n}_+$ is injective; however, ψ is surjective only if $\dim \mathfrak{g} < \infty$.

4D) Put $S = \{\theta_\lambda | \lambda \in P_+\} \subset \mathbf{F}[G]$. This is a multiplicative set since $\theta_\lambda \theta_\mu = \theta_{\lambda+\mu}$. We put $\theta_i = \theta_{\Lambda_i}$ for short. Denote by $\mathbf{F}[H]$ the algebra of functions on H generated by S and S^{-1}. We have: $\mathbf{F}[H] \cong \mathbf{F}[\theta_i, \theta_i^{-1};\ \ i \in I]$, the coordinate ring of $(\mathbf{F}^*)^I$.

Lemma 4.4. *The map $\phi\colon U_- \times H \times U_+ \longrightarrow G$ defined by*

$$\phi(u_-, h, u_+) = u_- h u_+$$

induces an isomorphism $\phi^*\colon\ S^{-1}\mathbf{F}[G] \rightrightarrows \mathbf{F}[U_-] \otimes \mathbf{F}[H] \otimes \mathbf{F}[U_+]$. In particular, (by Lemma 4.3b), $S^{-1}\mathbf{F}[G]$ is a unique factorization domain.

Proof. Using Theorem 1, one can easily check that

$$\phi^*(S^{-1}\mathbf{F}[G]) \subset \mathbf{F}[U_-] \otimes \mathbf{F}[H] \otimes \mathbf{F}[U_+];$$

ϕ^* is injective by Lemma 2.1d.

To prove surjectivity of ϕ^* we use the formulas:

$$\phi^*(\theta_\lambda) = 1 \otimes \theta_\lambda|_H \otimes 1,$$
$$\phi^*(\theta_\lambda^{-1} f_{v_\lambda^*,v}) = 1 \otimes 1 \otimes f_{v_\lambda^*,v}|_{U_+},$$
$$\phi^*(\theta_\lambda^{-1} f_{v^*,v_\lambda}) = f_{v^*,v_\lambda}|_{U_-} \otimes 1 \otimes 1,$$

and apply Lemma 4.3b. Q.E.D.

Corollary 4.2. *Let* $\mathbf{F}$ *be algebraically closed. If* $\mathfrak{a}$ *is a finitely generated ideal of* $S^{-1}\mathbf{F}[G]$ *and* $f \in S^{-1}\mathbf{F}[G]$ *vanishes on the zero set of* $\mathfrak{a}$ *in* U_-HU_+, *then* $f \in \sqrt{\mathfrak{a}}$.

Proof. Recall the map $\psi: U_+ \longrightarrow \mathfrak{n}_+$ defined in Lemma 4.3; similarly, we define the map $\psi_-: U_- \longrightarrow \mathfrak{n}_-$. Define a map $\sigma: U_-HU_+ \longrightarrow \mathfrak{n}_- \times H \times \mathfrak{n}_+$ by $\sigma(u_-hu_+) = \big(\psi_-(u_-), h, \psi(u_+)\big)$. Then, by Lemmas 4.3 and 4.4, σ induces an isomorphism $\sigma^*: \mathbf{F}[\mathfrak{n}_-] \otimes \mathbf{F}[H] \otimes \mathbf{F}[\mathfrak{n}_+] \rightrightarrows S^{-1}\mathbf{F}[G]$. By Corollary 4.1, given $q_1, \ldots, q_s \in \mathbf{F}[\mathfrak{n}_-] \otimes \mathbf{F}[H] \otimes \mathbf{F}[\mathfrak{n}_+]$ and $x \in \mathfrak{n}_- \times H \times \mathfrak{n}_+$, there exists $x' \in \sigma(U_-HU_+)$ such that $q_i(x) = q_i(x')(1 \le i \le s)$. Now we apply Hilbert's Nullstellensatz to $(\sigma^*)^{-1}f$ and $(\sigma^*)^{-1}\mathfrak{a}$. Q.E.D.

Lemma 4.5. a) *Let* $w \in W$ *and let* $U_1 = U_+ \cap wU_-w^{-1}$, $U_2 = U_+ \cap wU_+w^{-1}$. *Then the map* $\psi: U_1 \times U_2 \longrightarrow U_+$ *defined by* $\psi(u_1, u_2) = u_1u_2$ *induces an isomorphism* $\psi^*: \mathbf{F}[U_+] \rightrightarrows \mathbf{F}[U_1] \otimes \mathbf{F}[U_2]$.

b) *Moreover, let* $\alpha \in \Pi$ *be such that* $w(\alpha) \in \Delta_+$ *and let* $U_3 = U_+ \cap (wr_\alpha)U_-(wr_\alpha)^{-1}$. *Then the map* $\phi: U_1 \times U_{w(\alpha)} \longrightarrow U_3$ *defined by* $\phi(u, u') = uu'$ *induces an isomorphism* $\phi^*: \mathbf{F}[U_3] \rightrightarrows \mathbf{F}[U_1] \otimes \mathbf{F}[U_{w(\alpha)}]$.

c) *Let* $\beta \in \Delta^{re}$. *Then* $\mathbf{F}[U_\beta]$ *is a polynomial algebra over* $\mathbf{F}$ *in one variable* x, *where* $x(\exp te_\beta) = t$.

Proof. By Lemma 4.2, $\psi^*(\mathbf{F}[U_+]) \subset \mathbf{F}[U_1] \otimes \mathbf{F}[U_2]$. ψ^* is injective since ψ is onto by [18, Corollary 5b]. To see that $\psi^*(\mathbf{F}[U_+]) \supset \mathbf{F}[U_1] \otimes \mathbf{F}[U_2]$, fix $\lambda \in P_{++}$ and choose $n \in wH$. Then, for $v \in L(\lambda)$ and $v^* \in L^*(\lambda)$, we have:

$$\psi^*(f_{v^*,n(v_\lambda)}) = f_{v^*,n(v_\lambda)}|_{U_1} \otimes 1,$$
$$\psi^*(f_{n(v_\lambda^*),v}) = 1 \otimes f_{n(v_\lambda^*),v}|_{U_2}.$$

But the $f_{v^*,n(v_\lambda)}|_{U_1}$ (resp. $f_{n(v_\lambda^*),v}|_{U_2}$) generate $\mathbf{F}[U_1]$ (resp. $\mathbf{F}[U_2]$), as seen by applying Lemmas 4.2 and 4.3b and conjugating by n. This proves a).

Since $U_{w(\alpha)} \subset U_2$ and $U_1U_{w(\alpha)} = U_3$, b) is clear from a) by restriction. c) for $\beta \in \Pi$ follows from the proof of a) in the case $w = r_\beta$; the general case then follows by conjugating by elements of N. Q.E.D.

4E) We proceed to prove the main result of this section:

Theorem 3. *The ring* $\mathbf{F}[G]$ *is a unique factorization domain (UFD).*

The proof is based on the following simple fact. (Its proof can be easily extracted from [17, p. 43].)

Lemma 4.6. *Let R be an integral domain and $p_1, \ldots, p_m$ prime elements of R (p is called prime if $p \neq 0$ and (p) is a prime ideal). Suppose that:*

(i) $\bigcap_{k=1}^{\infty}(p_i^k) = 0$ *for all* i;
(ii) $S^{-1}R$ *is a UFD, where S is the multiplicative system generated by* $p_1, \ldots, p_m$.

Then R is a UFD.

We apply this lemma to $R = \mathbf{F}[G]$ and $p_i = \theta_i (i \in I)$. Using Lemmas 2.1 a,b and 4.4, it suffices to show that the elements θ_i are prime.

For $f \in \mathbf{F}[G]$ and $n \in G$, we denote by ${}^n f$ the strongly regular function ${}^n f(g) = f(ng)$, $g \in G$. We will deduce that θ_i is prime from the following lemma.

Lemma 4.7. *For $i \in I$ and $n \in N$, ${}^n\theta_i$ is either a prime element or a unit in $S^{-1}\mathbf{F}[G]$.*

We may (and will) assume in the proof of Lemma 4.7 and the following deduction from it that θ_i is prime that $\mathbf{F}$ is algebraically closed.

Assume that Lemma 4.7 holds. Suppose that θ_i divides $f_1 f_2$, where $f_1, f_2 \in \mathbf{F}[G]$; we must show that θ_i divides one of f_1, f_2. By Corollary 3.1a, the set Γ_i of zeros of θ_i on G is the closure of $B_- r_i B_+ = r_i U_-^{\alpha_i} B_+$, where $U_-^{\alpha_i} = U_- \cap r_i U_- r_i^{-1}$ (cf. [18]). By Lemmas 4.4 and 4.5a, the restriction of $\mathbf{F}[G]$ to $U_-^{\alpha_i} B_+$ is an integral domain. Hence, one of the f_k, say f_1, vanishes on Γ_i. Lemma 4.7 and Corollary 4.2 now imply that $({}^n\theta_i)^{-1}({}^n f_1) \in S^{-1}\mathbf{F}[G]$ for all $n \in N$. Corollaries 3.1b and 2.1 now force $\theta_i^{-1} f_1 \in \mathbf{F}[G]$, proving Theorem 3. Q.E.D.

Proof of Lemma 4.7. We proceed by induction on $l(w)$, where $n \in wH$, $w \in W$. If $l(w) = 0$, i.e. $n \in H$, then ${}^n\theta_i \in \mathbf{F}^* S$ is a unit in $S^{-1}\mathbf{F}[G]$. Otherwise, choose $j \in I$ such that $l(r_j w) < l(w)$. Put $w' = r_j w$, choose $n_j \in r_j H$ and put $n' = n_j^{-1} n$. If $j \neq i$, then ${}^n\theta_i = \theta_i(n_j)({}^{n'}\theta_i) \in \mathbf{F}^*({}^{n'}\theta_i)$ is prime or a unit in $S^{-1}\mathbf{F}[G]$ by the inductive assumption. If $j = i$, put $U_0 = U_- \cap w^{-1}U_-w$, $U_1 = U_{-\alpha_i}$, $U_2 = U_- \cap (w'^{-1}U_+w')$, and define the map

$$\psi: (U_0 \times U_1 \times U_2) \times H \times U_+ \longrightarrow G$$

by $\psi(u_0, u_1, u_2, h, u_+) = u_0(n'^{-1}u_1n')u_2hu_+$. Then Lemmas 4.2, 4.4 and 4.5 a,b show that ψ induces an isomorphism

$$\psi^*: S^{-1}\mathbf{F}[G] \longrightarrow \mathbf{F}[U_0] \otimes \mathbf{F}[U_1] \otimes \mathbf{F}[U_2] \otimes \mathbf{F}[H] \otimes \mathbf{F}[U_+].$$

Put $f = \theta_i^{-1}({}^n\theta_i)$, $f' = \theta_i^{-1}({}^{n'}\theta_i)$, $x = {}^{n_i}\theta_i|_{U_1}$. Then x generates the polynomial algebra $\mathbf{F}[U_1]$ by Lemma 4.5c and we compute, using $n_iu_1^{-1}n_i^{-1}(v^*_{\Lambda_i}) = v^*_{\Lambda_i} + x(u_1)n_i(v^*_{\Lambda_i})$, that:

$$\psi^*(f') = 1 \otimes 1 \otimes f'|_{U_2} \otimes 1 \otimes 1. \tag{4.1}$$

$$\psi^*(f) = 1 \otimes x \otimes f'|_{U_2} \otimes 1 \otimes 1 + 1 \otimes 1 \otimes f|_{U_2} \otimes 1 \otimes 1. \tag{4.2}$$

Suppose that ${}^n\theta_i$ is not prime or a unit in $S^{-1}\mathbf{F}[G]$, so that $\psi^*(f)$ is not prime or a unit. Since $S^{-1}\mathbf{F}[G]$ is a UFD, (4.1) and (4.2) show that $\psi^*(f)$ and $\psi^*(f')$ have a nontrivial common factor. Hence, by the inductive assumption, $\psi^*(f')$ is prime. Now, the set $P := \left((w')^{-1}\overline{B_-r_iB_+}\right) \cap (U_-HU_+)$ is non-empty by Corollary 4.2, since $(w')^{-1}\overline{B_-r_iB_+}$ is the set of zeros of ${}^{n'}\theta_i$ on G and U_-HU_+ is open (see Corollary 3.1). But ${}^n\theta_i$ vanishes nowhere on P, since $nP \subset U_-HU_+$. Hence, $\psi^*(f')$ does not divide $\psi^*(f)$. This contradiction completes the proof of the lemma and of Theorem 3. Q.E.D.

Remark 4.3. It is easy to see that $f \in \mathbf{F}[G]$ is divisible by θ_i if it vanishes on Γ_i, even if $\mathbf{F}$ is not algebraically closed.

Corollary 4.3. a) *$\mathbf{F}[G]^{U_+}$ is a UFD.*

b) *$\mathbf{F}[G]$, $\mathbf{F}[G]^{U_+}$ and $\mathbf{F}[\mathcal{V}_\Lambda]$, $\Lambda \in P_+$, are integrally closed.*

c) *$\mathbf{F}[\mathcal{V}_\Lambda]$, $\Lambda \in P_+$, is a UFD if and only if $\Lambda = \Lambda_i$ for some $i \in I$ or $\Lambda = 0$.*

Proof. The group U_+ acts by automorphisms locally unipotently on the UFD $\mathbf{F}[G]$ with unit group $\mathbf{F}^*$; a) follows. Since a UFD is integrally closed, and since the ring of invariants of a group acting by automorphisms on an integrally closed domain is integrally closed, b) follows from Theorem 3, using Corollary 2.3 and Lemma 3.1. c) is proved using the P_+-gradation $\mathbf{F}[G]^{U_+} \cong \bigoplus_\Lambda L^*(\Lambda)$ (see Corollary 2.2) and Lemma 3.1. Q.E.D.

Remark 4.4. a) The fact that the coordinate ring of a connected simply-connected simple algebraic group is a UFD is well-known. The earliest reference that we know is Voskresenskii [22] (see also [19]).

b) It is not difficult to see that the results of [21] can be extended to our setup.

Remark 4.5. Assume that **F** is algebraically closed. Let M be a subset of G. A function f on M is called ***strongly regular*** if for every $x \in M$ there exist a neighborhood U of x and functions $f_1, f_2 \in \mathbf{F}[G]$, such that f_2 does not vanish on $M \cap U$, for which $f = f_1/f_2$ on $M \cap U$. Denote by $\mathbf{F}[M]$ the ring of strongly regular functions on M. This definition coincides with the original one when $M = G$, $U_+ \cap wU_\pm w^{-1} (w \in W)$ or H.

Remark 4.6. It is clear that $\mathbf{F}[G]|_H$ is spanned by the characters of $H \simeq (\mathbf{F}^*)^I$ which appear as weights of the G-modules $L(\Lambda)(\Lambda \in P_+)$. But the union of the sets of weights of all $L(\Lambda)(\Lambda \in P_+)$, restricted to $\mathfrak{h}'$, coincides with $\sum_i \mathbf{Z}\Lambda_i|_{\mathfrak{h}'}$ if and only if G is of finite type. It follows that $\mathbf{F}[G]|_H = \mathbf{F}[H]$ if and only if G is of finite type. This phenomenon is related to Remark 2.2.

4F) The inversion map $i: G \longrightarrow G$ clearly induces an automorphism

$$i^*: \mathbf{F}[G]_{w.r.} \longrightarrow \mathbf{F}[G]_{w.r.}.$$

However, $i^*(\mathbf{F}[G]) \neq \mathbf{F}[G]$ if G is of infinite type (cf. Remarks 4.1 and 4.6). Let $\mathbf{F}[G]_r$ denote the closure in $\mathbf{F}[G]_{w.r.}$ of the subalgebra generated by $\mathbf{F}[G]$ and $i^*(\mathbf{F}[G])$. Elements of $\mathbf{F}[G]_r$ are called ***regular***.

We now indicate how G may be viewed as an infinite-dimensional affine group in the sense of Shafarevich [20] with coordinate ring $\mathbf{F}[G]_r$. Let $\mathbf{A} = \bigoplus_i \mathbf{F}_i$ be a direct sum, possibly infinite, of copies of **F**. Then we can apply the terminology of subsection 3A. Let $\mathcal{V} \subset \mathbf{A}$ be the zero set of an ideal of $\mathbf{F}[\mathbf{A}]_r$; closed subvarieties of intersections of $\mathcal{V}$ with finite-dimensional subspaces of **A** are called ***finite subvarieties*** of $\mathcal{V}$. Such a subset $\mathcal{V}$ of **A** with a group structure is called an ***affine group of Shafarevich type*** if the multiplication map $\mu: \mathcal{V} \times \mathcal{V} \longrightarrow \mathcal{V}$ and the inversion map $i: \mathcal{V} \longrightarrow \mathcal{V}$ have the following property: for every finite subvariety M of $\mathcal{V}$ there exists a finite subvariety N of $\mathcal{V}$ such that $\mu(M \times M) \subset N$, $i(M) \subset N$ and the induced maps $\mu: M \times M \longrightarrow N$, $i: M \longrightarrow N$ are morphisms.

Now we construct an injection

$$\phi\colon G \to \mathbf{A} := \left(\bigoplus_{i\in I} L(\Lambda_i)\right) \oplus \left(\bigoplus_{i\in I} L^*(\Lambda_i)\right).$$

Let $v = \sum_i v_{\Lambda_i}$, $v^* = \sum_i v^*_{\Lambda_i}$, and define $\phi(g) = g(v+v^*)$; ϕ is injective by [18, Corollary 3a]. Furthermore, $\phi(G) \subset \mathbf{A}$ is defined by the following system of equations: $x = \sum x_i + \sum x_i^*$, where $x_i \in L(\Lambda_i)$, $x_i^* \in L^*(\Lambda_i)$, lies in $\phi(G)$ if and only if

$$x_i \otimes x_j \in L(\Lambda_i + \Lambda_j) \subset L(\Lambda_i) \otimes L(\Lambda_j),$$
$$x_i^* \otimes x_j^* \in L^*(\Lambda_i + \Lambda_j) \subset L^*(\Lambda_i) \otimes L^*(\Lambda_j),$$

and $\langle x_i, x_i^* \rangle = 1$ for all $i, j \in I$. This follows easily from [18, Theorem 1b], using the idea of the proof of Theorem 2.

Furthermore, one can show that ϕ induces an isomorphism

$$\phi^*\colon \mathbf{F}[\phi(G)]_r \rightrightarrows \mathbf{F}[G]_r$$

and that $\phi(G)$ is an affine group of Shafarevich type with Lie algebra $\mathfrak{g}'$. One can show that G operates morphically on $L(\Lambda)$ $(\Lambda \in P_+)$ and $\mathfrak{g}'$; in particular, the matrix coefficients of G on $L(\Lambda)$, $L^*(\Lambda)$ and $\mathfrak{g}'$ are regular.

4G) Let $\mathbf{F}$ be a non-discrete locally-compact topological field. We call a subset U of G *open* if $x_{\overline{\beta}}^{-1}(U) \subset \mathbf{F}^k$ is open for all $\overline{\beta} \in (\Delta^{re})^k$, $k \in \mathbf{Z}_+$. G is a Hausdorff σ-compact topological group (and hence is paracompact) in this topology. Similarly, we call a subset U of $L(\Lambda)$ *open* if $x^{-1}(U) \subset \mathbf{F}^k$ is open for all $x \in \mathrm{Hom}_{\mathbf{F}}(\mathbf{F}^k, L(\Lambda))$, $k \in \mathbf{Z}_+$. The following results will be proved in a subsequent paper. (See [18] for definitions.)

Theorem 4. *Let $\Lambda \in P_+$ and let $X = \{i \in I \mid \Lambda(h_i) = 0\}$. Then:*
a) *The multiplication map $U_- \times H \times U_+ \to U_- H U_+$ is a homeomorphism and $U_- H U_+$ is open in G.*

b) *The canonical map $G \to G/P_X$ is a fibration, and the map $gP_X \mapsto g(\mathbf{F}^* v_\Lambda)$ of G/P_X onto $\mathbf{P}\mathcal{V}_\Lambda$ is a homeomorphism.*

c) *If $\mathbf{F} = \mathbf{C}$, then G is a connected simply-connected topological group.*

d) *If $\mathbf{F} = \mathbf{C}$, then $H_+ \times U_+$ is contractible and the multiplication map $K \times H_+ \times U_+ \to G$ is a homeomorphism.*

e) *If* $\mathbf{F} = \mathbf{C}$, *then* G/P_X *is a CW-complex with cells* B_+wP_X/P_X, *where* $w \in W/W_X$, *of dimension* $2d_X(w)$, *where* $d_X(w)$ *is the length of the shortest element of* wW_X.

4H) *Open problems.*

a) Is it true that the rings $\mathbf{F}[\overline{\mathcal{V}_\Lambda(\lambda)_\pm}]$ are integrally closed? (This would imply that the closures of finite Schubert cells are normal (see [18, Remark (iii)]), as is known for finite type groups [6].)

b) Compute Specm $\mathbf{F}[G]$. (Recall that Specm $\mathbf{F}[G]$ is larger than G if G is of infinite type, by Remark 2.2).

c) Is it true that the sum of two closed ideals of $\mathbf{F}[G]$ (or $\mathbf{F}[G]_{w.r.}$) is closed? In particular, is it true that every finitely-generated ideal of $\mathbf{F}[G]$ is closed?

d) Let $\mathbf{F}$ be algebraically closed. Is it true that every proper finitely-generated ideal of $\mathbf{F}[G]$ vanishes at some point of G? (It is obviously true for principal ideals.)

e) Is it true that $\mathbf{F}[G]_{w.r.} = \mathbf{F}[G]_r$?

References

[1] Birkhoff G.D., A theorem on matrices of analytic functions, Math. Ann. 74(1913), 122–133.

[2] Bott R., An application of the Morse theory to the topology of Lie groups, Bull. Soc. Math. France 84(1956), 251–281.

[3] Bourbaki N., Groupes et Algebres de Lie (Hermann, Paris), Chap. 4,5 and 6, 1968.

[4] Date E., Jimbo M., Kashiwara M., Miwa T., Transformation groups for soliton equations, preprints (1981–82).

[5] Demazure M., A very simple proof of Bott's theorem, Inventiones Math. 33(1976), 271–272.

[6] Demazure M., Désingularisation des variétés de Schubert généralizees, Annales Sci. l'École Norm. Sup., 4e ser., 7(1974), 53–88.

[7] Gabber O., Kac V.G., On defining relations of certain infinite-dimensional Lie algebras, Bull. Amer. Math. Soc., New ser., 5(1981), 185–189.

[8] Grothendieck A., Sur la classification des fibres holomorphes sur la sphere de Riemann, Amer. J. Math. 79(1957), 121–138.

[9] Gohberg I., Feldman I.A., Convolution equations and projection methods for their solution, Transl. Math. Monographs, 41, Amer. Math. Soc., Providence 1974.

[10] Kac V.G., Simple irreducible graded Lie algebras of finite growth, Math. USSR-Izvestija 2(1968), 1271–1311.

[11] Kac V.G., Infinite-dimensional Lie algebras and Dedekind's η-function, J. Funct. Anal. Appl. 8(1974), 68–70.

[12] Kac V.G., Infinite-dimensional algebras, Dedekind's η-function, classical Möbius function and the very strange formula, Advances in Math. 30(1978), 85–136.

[13] Kac V.G., Peterson D.H., Spin and wedge representations of affine Lie algebras, Proc. Natl. Acad. Sci. USA 78(1981), 3308–3312.

[14] Kac V.G., Peterson D.H., Infinite-dimensional Lie algebras, theta functions and modular forms, Adv. Math., 50(1983).

[15] Kac V.G., Kazhdan D.A., Lepowsky J., Wilson R.L., Realization of the basic representations of the Euclidean Lie algebras, Adv. Math., 42(1981), 83–112.

[16] Lancaster G., Towber J., Representation-functions and flag-algebras for the classical groups I, J. of Algebra 59(1979), 16–38.

[17] Nagata M., Local rings, Interscience Publishers, 1962.

[18] Peterson D.H., Kac V.G., Infinite flag varieties and conjugacy theorems, Proc. Natl. Acad. Sci. USA, 80(1983), 1778–1782.

[19] Popov V.L., Picard groups of homogeneous spaces of linear algebraic groups and 1-dimensional homogeneous vector bundles, Math. USSR-Izvestija, 8(1974), 301–327.

[20] Shafarevich I.R., On certain infinite-dimensional groups II, Math. USSR-Izvestija, 18(1981), 185-194.

[21] Vinberg E.B., Popov V.L., On a class of quasihomogeneous affine varieties, Math. USSR-Izvestija 6(1972), 743–758.

[22] Voskresenskii V.E., Picard groups of linear algebraic groups, Studies in number theory of Saratov University, 3(1969), 7–16 (in Russian).

Received January 7, 1983

Partially supported by NSF grant no. MCS–8203739

Professor Victor G. Kac
Department of Mathematics
Massachusetts Institute of Technology
Cambridge, Massachusetts 02139

Professor Dale H. Peterson
Department of Mathematics
University of Michigan
Ann Arbor, Michigan 48109

Examples of Surfaces of General Type with Vector Fields

William E. Lang

To I.R. Shafarevich

The purpose of this paper is to introduce some new surfaces of general type, called generalized Raynaud surfaces, and to prove that in many cases these surfaces possess global vector fields, contradicting a guess of Rudakov-Shafarevich [3].

In a lecture at M.I.T. in October 1981, H. Kurke announced that he and P. Russell had found surfaces of general type with vector fields. These surfaces were of the form Y^D, where Y is a ruled surface, and D is a p-closed vector field with divisorial singularities. While all details were not given, the calculations seemed rather involved. The structure of the resulting surface, however, was quite simple. Inspired by Kurke's talk (and by conversations with M. Artin), I tried to generalize the elementary construction of Raynaud surfaces in characteristic three studied in [1] to higher characteristic, and finished the construction given here in November 1981. These surfaces are also of the form Y^D, and I suspect that they are deformations of the Kurke-Russell examples; however, both the construction of the surfaces and the method used to prove that some of the surfaces have vector fields are quite different from those of Kurke and Russell, and I hope more transparent.

1. Construction of Generalized Raynaud Surfaces

Let p be a prime number, and let n and d be positive integers, such that if $p \neq 2$, and d is odd, n is also odd. Let k be an algebraically closed field of characteristic p.

Definition. A *generalized Tango curve* over k of type (p, n, d) is a triple (C, L, dt), where C is a smooth curve over k, L is a line bundle on C of degree d, and dt is a nowhere vanishing section of $K_C \otimes L^{\otimes p(1-mp)}$ which is locally exact.

It is easy to see that a generalized Tango curve is the same as a curve C together wih an exact rational differential df such that

$$\operatorname{div}(df) = p(np-1)D,$$

where D is a divisor on C. Then $\mathcal{L} = \mathcal{O}_X(D)$. If we let $u_{\alpha\beta}^{-1}$ be the transition functions for $\mathcal{L}$ on an affine open cover $\{U_\alpha\}$ of C, then

$$dt \in H^0\Big(C, K_C \otimes \mathcal{L}^{\otimes p(1-np)}\Big)$$

gives rise to functions $t_\alpha \in H^0(C, U_\alpha)$ satisfying $dt_\alpha = u_{\alpha\beta}^{p(np-1)} dt_\beta$. Then on $U_\alpha \cap U_\beta, t_\alpha = u_{\alpha\beta}^{p(np-1)} t_\beta - r_{\alpha\beta}^p$. (see [1] for a more detailed discussion in the case $(p, n, d) = (3, 1, d)$, which generalizes immediately to the current situation.) Note that $K_C \simeq \mathcal{L}^{\otimes p(np-1)}$, so if g is the genus of C, $2g - 2 = p(np-1)d$.

Given a Tango curve C of type (p, n, d), we will construct a complete surface X together with a map $f\colon X \longrightarrow C$, imitating the special case $p = 3$, $n = 1$, done in [1]. (The original Raynaud surfaces constructed in [2] do not fit into this framework, unless $p = 3$.)

Let $f^{-1}(U_\alpha)$ be the closed subscheme of $\tilde{\mathbb{P}}^2 \times U_\alpha$ defined by

$$y_\alpha^{np-1} z_\alpha = x_\alpha^p + t_\alpha z_\alpha^{np}.$$

Here $\tilde{\mathbb{P}}^2$ is weighted projective space, where x_α has weight n, and y_α and z_α have weight 1. We patch these surfaces together over $U_\alpha \cap U_\beta$ by

$$\begin{aligned} x_\alpha &= u_{\alpha\beta}^{np-1} x_\beta + r_{\alpha\beta}^n z_\beta^n \\ y_\alpha &= u_{\alpha\beta}^p y_\beta \\ z_\alpha &= z_\beta. \end{aligned}$$

We notice that $f^{-1}(U_\alpha)$ is covered by two affine pieces, which are the subschemes of $A^2 \times U_\alpha$ defined by the following equations:

1) $\quad Y_\alpha^{np-1} = X_\alpha^p + t_\alpha,\ Y_\alpha = y_\alpha / z_\alpha,\ X_\alpha = x_\alpha / z_\alpha^n$

2) $\quad Z_\alpha = X_\alpha^p + t_\alpha Z_\alpha^{np},\ Z_\alpha = z_\alpha / y_\alpha,\ X_\alpha = x_\alpha / y_\alpha^n.$

We refer to these affine pieces as the $X_\alpha Y_\alpha$ chart and the $X_\alpha Z_\alpha$ chart, respectively. We will call the union of the $X_\alpha Y_\alpha$ charts the XY chart.

We notice that, if $n = 1$, the $X_\alpha Z_\alpha$ chart of our surface is a Russell-type unipotent group scheme over U_α [4], and that the patching data preserves the group structure. Thus, it is reasonable to hope to find vector fields if the normal bundle to the zero-section has sections. We will see that this is the case, although this point of view is not used in the proof.

We will call the surface X constructed above a *generalized Raynaud surface* of type (p, n, d).

2. A Lower Bound on the Number of Vector Fields

From the above description of X, we see that X is smooth and that all fibers of f are irreducible rational curves with one singular point. The curve of singularities D lives in the XY-chart and is locally defined by $Y_\alpha = 0$. Putting $t =$ constant, we find that the singularity of a fiber is isomorphic to the singularity $Y^{np-1} = X^p$. This is resolved by n blowups. The 0th, 1st, ..., $(n-2)$nd infinitely near points have multiplicity p, while the last has multiplicity $p-1$. Therefore the arithmetic genus of a fiber F satisfies

$$g(F) = \tfrac{1}{2}(n-1)p(p-1) + \tfrac{1}{2}(p-1)(p-2).$$

(The proof must be slightly modified when $p = 2$, but the formula is still correct.) Notice that $g(F) > 0$ unless $n = 1$, $p = 2$. We will discard this case, in which the surface constructed is ruled. Since $g(C) > 1$, we see that our surface is minimal. Furthermore, X is of general type if $g(F) > 1$, which is true unless $p = 2$, $n = 2$, or $p = 3$, $n = 1$. In the exceptional cases, the surface is quasi-elliptic.

We have a section S of f lying in the XZ-chart, locally defined by $X_\alpha = Z_\alpha = 0$. S is isomorphic to C, therefore the genus g of S satisfies

$$2g(S) - 2 = p(np-1)d.$$

Lemma 1. *The normal bundle of S is isomorphic to $\mathcal{L}$ (using the isomorphism between S and C). Therefore $S^2 = d$.*

Proof. One checks easily that X_α is a local equation for S near S, while Z_α vanishes to order p along S. We thus get, after restricting to S, $dX_\alpha = u_{\alpha\beta}^{-1} dX_\beta$, from which the result follows easily. (The reader concerned about the sign may consult [1].)

Notational Remark. In the next two lemmas, we show that K_X and $\mathcal{O}_X(D)$ are linear combinations of $f^*\mathcal{L}$ and $\mathcal{O}_X(S)$ in $\mathrm{Pic}(X)$. We find it convenient to show first that these bundles are Z-linear combinations of $f^*\mathcal{L}$ and $\mathcal{O}_X(S)$, and then to compute the coefficients using the genus formula and obvious intersection properties. These computations can be carried out in the Neron-Severi group of divisors modulo numerical equivalence, and we shall do so. Therefore, throughout the rest of this section, an expression of the form $D = adF + bS$ should be understood to mean

$$\mathcal{O}_X(D) \simeq \mathcal{O}_X(bS) \otimes f^*\mathcal{L}^{\otimes a}$$

(using the fact that since $\mathcal{L}$ has degree d, the divisor of $f^*\mathcal{L}$ is numerically equivalent to dF).

Lemma 2. *The canonical divisor of X is given by*

$$K_X = (np + p - 1)dF + \big(n(p^2 - p) - 2p\big)S$$

(using the conventions of the notational remark).

Proof. It is easy to see that $dX_\alpha \wedge dY_\alpha$ is a 2-form on $f^{-1}(U_\alpha)$ with no zeros or poles away from S, and we have $dX_\alpha \wedge dY_\alpha = u_{\alpha\beta}^{np-1+p} dX_\beta \wedge dY_\beta$. (Note that since $Y_\alpha^{np-2} dY_\alpha = -dt_\alpha$, terms such as $du_{\alpha\beta} \wedge dY_\alpha$ are 0.) Hence K_X is in the subgroup of $\mathrm{Pic}(X)$ generated by $f^*\mathcal{L}$ and $\mathcal{O}_X(S)$. We now write $K_X = aF + bS$ and compute the coefficients. Since $F^2 = 0$, $S^2 = d$, and $F.S = 1$, we find that $b = K_X.F$. By the genus formula, $K_X.F = (n-1)p(p-1) + p^2 - 3p = n(p^2 - p) - 2p$. Again by the genus formula, $K.S + S^2 = p(np-1)d$, so $K_X.S = d(np^2 - p - 1)$. Now $K_X.S = a + db$, so, solving for a, we find $a = d(np + p + 1)$. This completes the proof.

Now consider the curve of singularities D, which lies in the XY-chart and which is locally defined by $Y_\alpha = 0$. Now Y_α is a rational function on $f^{-1}(U_\alpha)$ which vanishes to order 1 along D, has poles along S, and has no zeros or poles elsewhere. Furthermore, $Y_\alpha = u_{\alpha\beta}^p Y_\beta$, so D is in the subgroup of $\mathrm{Pic}(X)$ generated by $f^*\mathcal{L}$ and $\mathcal{O}_x(S)$. Since $D.S = 0$, $D.F = p$, we compute the coefficients as in the proof of Lemma 2, and obtain

Lemma 3. $D = -pdF + pS$.

Lemma 4. *We have an exact sequence*

$$0 \to f^*\Omega^1_C \otimes \mathcal{O}_X\big((np-2)D\big) \to \Omega^1_X \to M \to 0,$$

where M is a line bundle.

Proof. This is clear away from D, since then f is a smooth morphism. Near D. X_α and Y_α form a local coordinate system, and we have

$$dt_\alpha = -Y_\alpha^{np-2}dY_\alpha.$$

The result is now clear. (Compare [1], Lemma 4.1.)

Since $\Omega^1_C \simeq L^{\otimes p(np-1)}$, we find that (using the conventions of the notational remark) Ω^1_X contains a subbundle $\mathcal{O}_X(E)$ with invertible quotient, where $E = p(np-1)dF + (np-2)(-pdF + pS) = pdF + p(np-2)S$.

Let Θ_X be the tangent bundle of X. Since $\Theta_X \simeq \Omega^1_X \otimes K_X^{-1}$, we find that $\mathcal{O}_X(G) \subset \Theta_X$ where $\mathcal{O}_X(G) = \mathcal{O}_X(E) \otimes K_X^{-1}$. We compute G using Lemma 2, and find

$$G = d(1-np)F + npS = dF + nD.$$

Therefore we get the following theorem.

Theorem 1. *If X is a generalized Raynaud surface over a Tango curve $(C, \mathcal{L}, dt_\alpha)$, then $h^0(\Theta_X) \geq h^0(\mathcal{L})$.*

Proof. We have an exact sequence

$$0 \to \mathcal{O}_X(G) \to \Theta_X \to \mathcal{O}_X(-G-K) \to 0,$$

then since X is not ruled, and G is effective (if G is not effective, then $h^0(\mathcal{L}) = 0$, and there is nothing to prove), we find that

$$H^0(X, \Theta_X) = H^0(X, \mathcal{O}_X(G)) \geq H^0(C, \mathcal{L}).$$

Theorem 2. *If X is a generalized Raynaud surface of type (p, n, d), then*

$$K_X^2 = d(-n^2p^2 - 4np^2 - 2np^3 + 2np + 4p + n^2p^4)$$
$$e(X) = 2pd(1-np)$$
$$\chi(\mathcal{O}_X) = (1/12)d(-n^2p^2 - 6np^2 - 2np^3 + 2np + 6p + n^2p^4).$$

Proof. Left to the reader.

3. Examples

To construct generalized Raynaud surfaces of type (p, n, d), we need only construct generalized Tango curves of type (p, n, d). I expect that these exist for all possible values. At the moment, however, I have only a few simple examples.

Consider the curve C with affine equation $y^2 = x^w - 1$, where w is an odd integer. If p does not divide w, and $p \neq 2$, then C is a hyperelliptic curve over a field of characteristic p of genus g, where $2g - 2 = w - 3$. The differential dx/y has divisor $(w-3)P$, where P is the point at infinity. Thus, if dx/y is exact, and $w - 3 = p(np - 1)d$, then C ic a Tango curve of type (p, n, d), with $\mathcal{L} \simeq \mathcal{O}_C(dP)$, hence $h^0(\mathcal{L}) > 0$, and we obtain a generalized Raynaud surface with a global vector field.

To check exactness of dx/y, we need only show that $\mathcal{C}(dx/y) = 0$, where $\mathcal{C}$ is the Cartier operator. For this, we notice that

$$y\mathcal{C}(dx/y) = \mathcal{C}(y^p(dx/y)) = \mathcal{C}(y^{p-1}dx) = \mathcal{C}((x^w - 1)^{1/2(p-1)}dx).$$

One now checks easily that if we put $w = p(p-1) + 3$, then $\mathcal{C}(dx/y) = 0$ if and only if $p = 2 \pmod 3$. Thus, we get a generalized Raynaud surface of type $(p, 1, 1)$ with at least one vector field for all odd $p \equiv 2 \pmod 3$.

To get a Raynaud surface with arbitrarily many vector fields, put $p = 5$, $w = 20d + 3$. Then we get Raynaud surfaces of type $(5, 1, d)$, and the number of vector fields is greater than or equal to $[d/2]$.

References

[1] Lang, W. E., Quasi-elliptic surfaces in characteristic three. Ann. Sci. École Norm. Sup. (4) *12*, 473–500 (1979).

[2] Raynaud, M., Contre-example de "vanishing de Kodaira" sur une surface lisse en car. $p > 0$. In: C. P. Ramanujan: A Tribute. Berlin, Heidelberg, New York: Springer, 273–278 (1978).

[3] Rudakov, A., Shafarevich, I., Inseparable morphisms of algebraic surfaces. Math. USSR-Izv. *10*, 1205–1237 (1976).

[4] Russell P., Forms of the affine line and its additive group, Pacific J. Math. *32*, 527–539 (1970).

[5] Tango, H., On the behavior of extensions of vector bundles under the Frobenius map. Nagoya Math. J. *48*, 73–89 (1972).

Received May 17, 1982
Supported in part by an N.S.F. grant

Professor W. Lang
Department of Mathematics
University of Minnesota
Minneapolis, Minnesota 55455

Flag Superspaces and Supersymmetric Yang-Mills Equations

Yu. I. Manin

To I. R. Shafarevich

1. Introduction

1. The self-dual Yang-Mills and Einstein equations have a simple geometric meaning, since they imply the vanishing of a part of the curvature tensor of a connection. This connection, the physicists' gauge potential, is given either on an external vector bundle (the Yang-Mills case) or on the spinorial bundle (the Einstein case) over space-time. After a suitable base change, the relevant part of the curvature becomes the total curvature of the lifted connection along the leaves of a foliation. At least locally (with respect to the initial base manifold), this foliation is a fibration and the self-dual field in question can be represented by the vector bundle of horizontal sections along the leaves on the base space of the foliation (Yang-Mills) or by the base space itself (Einstein). This representation is called the Penrose transform. The idea is closely related to the classical Radon transform. One of Penrose's discoveries was the possibility of using the rigidity of the holomorphic geometry to effectively construct the solutions of the differential equations by geometric means. A mathematician may profitably consult M. F. Atiyah [1] and the references cited therein.

The content of this article is a mathematical introduction to the supersymmetric extension of the Penrose twistorial approach. Supergeometry in general means "geometry with both commuting and anticommuting coordinates". It is now being vigorously developed by physicists working in supergravity and unified field theories (see e.g. the recent report by P. van Nieuwenhuizen [2]). Supertwistors were introduced by A. Ferber [3] and E. Witten [4]. For a mathematical introduction to differential supergeometry see D. A. Leites [5] and B. Kostant [6]. We shall deal mainly with analytic superspaces, briefly reviewed in §2 below.

To understand better the motivation of our work it is helpful to look at the classical situation in some detail.

2. We fix a four-dimensional complex vector space T (the Penrose twistor space). Let $M = G(2;T)$ be the grassmannian of 2-planes in T, S the tautological sheaf on it, $\tilde{S} = (O_M \otimes T/S)^*$, $\mathcal{T}M$ the tangent sheaf, $\Omega^1 M = (\mathcal{T}M)^*$ (everthing is complex analytic). There is a canonical isomorphism $\Omega^1 M = S \otimes \tilde{S}$. Hence

$$\Omega^2 M = S^2(S) \otimes \wedge^2(\tilde{S}) \oplus \wedge^2(S) \otimes S^2(\tilde{S}) = \Omega^2_+ M \oplus \Omega^2_- M,$$

and

$$S^2(\Omega^1 M) = S^2(S) \otimes S^2(\tilde{S}) \oplus \wedge^2(S) \otimes \wedge^2(\tilde{S}).$$

Thus the "spinorial decomposition" defines simultaneously the holomorphic conformal metric $g\colon S^2(\mathcal{T}M) \longrightarrow \wedge^2(S^*) \otimes \wedge^2(\tilde{S}^*)$ on M and the corresponding decomposition of 2-forms into $\pm$-parts, usually introduced by means of the Hodge star operator.

Suppose now a locally free sheaf $\mathcal{E}$ with connection $\nabla\colon \mathcal{E} \longrightarrow \mathcal{E} \otimes \Omega^1 U$ is given over a domain $U \subset M$. The curvature $F(\nabla)$ is a section of $\operatorname{End}\mathcal{E} \otimes \Omega^2 U$. Put $F(\nabla) = F_+(\nabla) + F_-(\nabla)$, $F_\pm(\nabla) \in \Gamma(U, \operatorname{End}\mathcal{E} \otimes \Omega^2_\pm U)$. The equation $F_-(\nabla) = 0$ is called the self-dual Yang-Mills equation. The equation $\tilde{\nabla}(F_\pm(\nabla)) = 0$ is called the Yang-Mills equation (in vacuo). From the Bianchi identity $\tilde{\nabla}(F(\nabla)) = 0$ it follows that the self dual fields automatically satisfy the complete Yang-Mills system. (We denote by $\tilde{\nabla}$ the covariant differential induced by ∇ on $\operatorname{End}\mathcal{E}$ and extended to $\operatorname{End}\mathcal{E} \otimes \Omega^{\cdot} U$ in a standard way.)

Consider now the projection $\pi_2\colon F \longrightarrow M$, where F is the space $F(2,3;T)$ of the $(2,3)$-flags in T, and also the projection $\pi_1\colon F \longrightarrow L = G(3;T) = P(T^*)$. Lift $\mathcal{E}$ to $F(U) = \pi_2^{-1}(U)$ and restrict the induced connection on the fibers of π_1: $\nabla_{F/L}\colon \pi_2^*(\mathcal{E}) \longrightarrow \pi_2^*(\mathcal{E}) \otimes \Omega^1 F/L$. The self-duality condition $F_-(\nabla) = 0$ is equivalent to $\nabla^2_{F/L} = 0$. The fibers of π_1 project onto the α-null planes in M. Hence, if on the intersections of the null planes with U the connection ∇ has trivial monodromy, we get a locally free sheaf of relative horizontal sections $\mathcal{E}_L = \pi_{1*}(\operatorname{Ker} \nabla_{F/L})$ on $L(U)$. It is geometrically trivial on all lines $\pi_1 \pi_2^{-1}(x) = L(x) \subset U$, $x \in U$, and contains complete information about the initial self-dual field. All instantons were classified in this way [1]; recently essential progress in the classification of the Yang-Mills-Higgs monopoles was made.

3. One can try to represent the total Yang-Mills equations as the integrability conditions in a similar vein. An ingenious way to do this was suggested by E. Witten [4] and J. Isenberg, Ph. B. Yasskin and P. S. Green

[7]. This time we start with the double fibration $\pi_2\colon F \longrightarrow M$, $\pi_1\colon F \longrightarrow L$, where $F = F(1,2,3;T)$ (complete flags),

$$L = F(1,3;T) \subset G(1;T) \times G(3;T) = P(T) \times P(T^*).$$

Since the fibers of π_1 are now one-dimensional null-lines, the integrability condition for $\nabla_{F/L}$ holds trivially, and hence we can construct $\mathcal{E}_L = \pi_{1*}$ (Ker $\nabla_{F/L}$) starting from any $(\mathcal{E}, \nabla)$ (locally), and then reconstruct $(\mathcal{E}, \nabla)$ from $\mathcal{E}_L$. To get nontrivial equations one should require that the extension of $\mathcal{E}_L$ from L (or rather $L(U)$) to the third infinitesimal neighbourhood $L^{(3)}$ of L in $H = P(T) \times P(T^*)$ be possible. As was shown in [4] and [7], this condition is exactly equivalent to the Yang-Mills equations for $(\mathcal{E}, \nabla)$. In the papers by G. M. Henkin and the author [8]–[10], it was demonstrated that extension problems for various cohomology classes of $\mathcal{E}_L$ are equivalent to a number of physically interesting equations, such as the Dirac equation in the Yang-Mills background.

Although the extension property does not look like an integrability condition, it can be interpreted as such with some strain, if one endows with auxiliary nilpotents not only L, but F and M as well. In [4], E. Witten suggested a much more significant extension by anticommuting nilpotents. One great technical advantage of this approach is that one can still deal with the (super) smooth spaces although they are nonreduced from the naive (commutative) viewpoint. In this way one gets the so called supersymmetric Yang-Mills equations as integrability conditions along the "fat" null-lines bearing spinorial anticommuting nilpotents along with the usual "bosonic" coordinate.

4. In this paper we try to present the beautiful idea of Witten in the general context of the geometry of homogeneous superspaces. Having summarily explained the relevant background in §2, we proceed to describe the whole class of superhomogeneous double fibrations, which after reducing nilpotents, turn into the triple (M, F, L) discussed in the previous section, and to investigate the integrability conditions along null-lines on them. The main results are stated in §3 and proved in §§ 4–6.

2. Analytic Superspaces; Structure of Supergrassmannians

1. The reader is invited to consult [5] and [6] about the elements of superalgebra and differential supergeometry. Our brief exposition here

mainly fixes notation. Recall that all additive groups in a superalgebra are $\mathbf{Z}_2$-graded. A ring $A = A_0 \bigoplus A_1$ is supercommutative if $ab = (-1)^{\tilde{a}\tilde{b}}ba$ for all homogeneous elements $a, b \in \tilde{A}$ of degrees $\tilde{a}, \tilde{b}$ correspondingly. An A-module T is a $\mathbf{Z}_2$-graded A-bimodule such that its left and right multiplications are compatible with respect to the same sign rule: $at = (-1)^{\tilde{t}\tilde{a}}ta$, $a \in A$, $t \in T$. A-modules form a tensor category in the terminology of Deligne [11], p. 105, with the identity object $(A, 1)$ and the commutativity constraint $\psi_{S,T}\colon S \otimes T \longrightarrow T \otimes S$, $s \otimes t \to (-1)^{\tilde{s}\tilde{t}} t \otimes s$. This category has an internal $\underline{\mathrm{Hom}}$: in fact, $\underline{\mathrm{Hom}}(S, T)$ can be realized as the set of all right A-module homomorphisms. We put $T^* = \underline{\mathrm{Hom}}(T, A)$. An important role is played by the parity-change functor $\Pi : (\Pi T) = T_1$, $(\Pi T)_1 = T_0$; right multiplication by A coincides on T and ΠT. In the following, all tensor algebra operations refer to the tensor structure described above.

An A-module is called free of rank $d_0|d_1$ if it is isomorphic to $A^{d_0} \bigoplus (\Pi A)^{d_1}$. We sometimes write d_0 instead of $d_0|0$. A submodule $S \subset T$ is called direct if it can be split off as a direct summand. A morphism $f\colon S \longrightarrow T$ (even or odd) is called direct if $\operatorname{Ker} f$ and $\operatorname{Im} f$ are direct in S and T. A bilinear form $b\colon T \times T \longrightarrow A$ is called direct if the corresponding morphism $\beta\colon T \longrightarrow T^*$ is direct. It is called nondegenerate if β is an isomorphism. Similar conventions are applied to an odd bilinear form $b\colon T \times T \longrightarrow \Pi A$. Bilinearity conditions of course include the omnipresent sign conventions: $b(at, t') = (-1)^{\tilde{b}\tilde{a}} ab(t, t')$, etc.

Consider a bilinear form $b\colon T \times T \longrightarrow A$ (or ΠA). Let

$$b^{\tau}(t, t') = (-1)^{\tilde{t}\tilde{t}'} b(t', t) \quad \text{and} \quad b^{\Pi}(t, t') = (-1)^{\tilde{t}} b(t, t').$$

Then b^{τ} is also bilinear on T and b^{Π} is bilinear on ΠT. The form b is called symmetric if $b^{\tau} = b$ and alternate if $b^{\tau} = -b$. Instead of "even symmetric, even alternate, odd symmetric, odd alternate" we shall speak of forms "of the type $OSp, SpO, \Pi O, \Pi Sp$" respectively. The map $b \mapsto b^{\Pi}$ changes $SpO, \Pi O$ to $OSp, \Pi Sp$ respectively.

Let T be a projective A-module T with a nondegenerate form $\beta\colon T \longrightarrow T^*$ (or ΠT^*). For a direct submodule $S \subset T$, let $S^{\perp} = \{f \in T^* | S \subset \operatorname{Ker} f\}$ and $S_b^{\perp} = \beta^{-1}(S)$ (or $\beta^{-1}(\Pi S)$). This is a direct submodule of T. Its rank is equal to the corank of S if b is even, and to the Π-inverted corank of S if b is odd. If $S \subset S_b^{\perp}$, S is called isotropic. The form b is called split if its matrix in an appropriate base takes a standard form: see [5].

2. In this paper we work with superschemes, analytic superspaces (over C), coherent sheaves and differential operators on them. A superscheme $(M, \mathcal{O})$ is a topological space with a sheaf of supercommutative rings $\mathcal{O} = \mathcal{O}_0 \oplus \mathcal{O}_1$ such that $(M, \mathcal{O}_0)$ is a scheme and $\mathcal{O}_1$ is $\mathcal{O}_0$-quasicoherent. The definition of an analytic superspace is similar. A superscheme of finite type over C has a canonical analytic structure: $\mathcal{O}_{0,an}$ is defined by $\mathcal{O}_0$ in the standard way and $\mathcal{O}_{1,an} = \mathcal{O}_{0,an} \otimes_{\mathcal{O}_0} \mathcal{O}_1$. We shall call superspace an object of one of these two categories, usually the analytic one.

A (left) superderivation X of a supercommutative ring A (of degree $\tilde{X}$) is an additive map $X: A \longrightarrow A$ with the Leibnitz rule

$$X(ab) = (Xa)b + (-1)^{\tilde{X}\tilde{a}} a(Xb).$$

Superderivations form a Lie superalgebra with respect to the supercommutator $[X, Y] = XY - (-1)^{\tilde{x}\tilde{y}} YX$. Relative superderivations and superderivations with values in an A-module are defined in an obvious way. For a superspace $(M, \mathcal{O}_M)$ we put $\mathcal{T}M =$ the sheaf of superderivations of $\mathcal{O}_M$ (over C), $\Omega^1 M = (\mathcal{T}M)^* = \underline{\underline{\mathrm{Hom}}}(\mathcal{T}M, \mathcal{O}_M)$ (double underlining means the sheaf of internal Hom), $\Omega^i = \wedge^i(\Omega^1 M)$. For a morphism of superspaces $\varphi: M \longrightarrow N$ put

$$\mathcal{T}M/N = \mathrm{Ker}(\varphi: \mathcal{T}M \longrightarrow \varphi^*(\mathcal{T}N)),$$
$$\Omega^1 M/N = \underline{\underline{\mathrm{Hom}}}(\mathcal{T}M/N, \mathcal{O}_M),$$
$$\Omega^i(M/N) = \wedge^i(\Omega^1 M/N).$$

The deRham complex is defined in the usual way.

3. Let $(M, \mathcal{O})$ be a superspace. It is intrinsically endowed with a sheaf of ideals $J = \mathcal{O}_1 + \mathcal{O}_1^2$. Using its powers one constructs various canonical filtrations on sheaves, cohomology etc. Work with the associated Z-graded objects and spectral sequences is the basic technique in supergeometry. In the smooth category it usually takes form of choosing a system of local coordinates x^i (even) and ξ^j (odd) and then expanding everything into Grassmann monomials in ξ^j. Physicists sometimes call this "componentwise analysis".

We shall use the following notation:

$$\mathrm{Gr}\,\mathcal{O} = \bigoplus_{i \geq 0} \mathrm{Gr}_i\,\mathcal{O} = \bigoplus_{i \geq 0} J^i/J^{i+1},$$

$$\mathrm{Gr}\,\mathcal{E} = \bigoplus_{i \geq 0} \mathrm{Gr}_i\,\mathcal{E} = \bigoplus_{i \geq 0} \mathcal{E}J^i/J^{i+1}$$

for a quasicoherent $\mathcal{O}$-module $\mathcal{E}$. Furthermore we put $\mathrm{Gr}\, M = (M, \mathrm{Gr}\, \mathcal{O})$, $M_{rd} = (M, \mathrm{Gr}_0\, \mathcal{O})$. The space M_{rd} need not be reduced in general; it is a canonical closed subspace of M.

We will call M smooth if the following conditions are fulfilled:

a. M_{rd} is reduced and smooth in the usual sense;

b. $\mathrm{Gr}\, \mathcal{O} = S(\mathrm{Gr}_1\, \mathcal{O})$ (symmetric algebra over $\mathrm{Gr}_0\, \mathcal{O}$) and $\mathrm{Gr}_1\, \mathcal{O}$ is locally free on M_{rd};

c. Locally, M is isomorphic to $\mathrm{Gr} M$; i.e., $\mathcal{O}_M$ locally is a Grassmann algebra over the conventional analytic functions.

Many of the elementary properties of smooth manifolds carry over to smooth superspaces. In particular, the tangent sheaf is locally free, the Frobenius integrability condition is verbally the same, etc. For an algebraic geometer it is this combination of smoothness and nilpotents which makes the supergeometry particularly rewarding.

The following simple result often helps to analyze superspaces componentwise.

4. **Proposition.** *Let M be a smooth superspace, $\Omega^1 = \Omega^1 M$, $\mathcal{E}$ a locally free sheaf on M. Then one has the following ismorphisms:*

a. $\mathrm{Gr}_i \mathcal{E} = \mathrm{Gr}_0 \mathcal{E} \otimes S^i(\mathrm{Gr}_1 \mathcal{O})$ *(over* $\mathrm{Gr}_0 \mathcal{O}$*)*;

b. $\mathrm{Gr}_1 \mathcal{O} = (\mathrm{Gr}_0 \Omega^1)_1$ *(here the external index 1 refers to the* $\mathbf{Z}_2$*-gradation and the isomorphism is induced by the differential* $d\colon J \longrightarrow \Omega^1$*).*

The easy proof is omitted.

A smooth irreducible space M has a well-defined (super) dimension equal to the rank of $\mathcal{T}M$.

We now finish with the generalities and introduce grassmannians and flag superspaces.

5. Let M be a superspace, $\mathcal{T}$ a locally free sheaf on it of finite rank. A flag of length κ in $\mathcal{T}$ is a sequence of locally direct subsheaves

$$\mathcal{T}_1 \subset \mathcal{T}_2 \subset \cdots \subset \mathcal{T}_{\kappa+1} = \mathcal{T}$$

(this means that all inclusions $\mathcal{T}_i \subset \mathcal{T}_j$ are locally direct). The type of the flag is the sequence $(rk\mathcal{T}_1, \ldots, rk\mathcal{T}_\kappa)$. If $\varphi\colon N \longrightarrow M$ is a morphism, φ^* induces a map of flags in $\mathcal{T}$ into flags in $\varphi^*(\mathcal{T}) = \mathcal{T}_N$ of the same type.

We shall also consider sheaves $\mathcal{T}$, endowed with one of the following structures:

a. a Π-symmetry, i.e., an isomorphism $p\colon \mathcal{T} \longrightarrow \Pi\mathcal{T}$, $p^2 = id$; sometimes it is convenient to view p as an odd involution of $\mathcal{T}$;

b. a nondegenerate form b of the types $OSp, SpO, \Pi Sp, \Pi O$.

The lifted sheaves $\mathcal{T}_N$ inherit such a structure. Its existence makes it possible to define Π-symmetric and isotropic flags in $\mathcal{T}$ and $\mathcal{T}_N$.

A pair $(M, \mathcal{T})$, a structure on $\mathcal{T}$, and a type being fixed, we can consider the corresponding flag functor. We shall use the following notation for these functors $(\text{Superspaces}/M)^0 \to \text{Sets}$:

$F_M(\text{type}; \mathcal{T})\colon (N,\varphi) \mapsto \{\text{flags of given type in } \varphi^*(\mathcal{T})\}$;
$F\Pi_M(\text{type}; \mathcal{T})\colon (N,\varphi) \mapsto \{\Pi\text{-symmetric flags of given type in } \varphi^*(\mathcal{T})\}$;
$FI_M(\text{type}; \mathcal{T}, \text{form type})\colon (N,\varphi) \mapsto \{\text{isotropic flags of given type in } \varphi^*(\mathcal{T})\}$

Flag functors of length one are grassmannians; we usually denote them by $G_M, G\Pi_M, GI_M$ respectively.

All flag functors are represented by superspaces of relatively finite type over M. The natural inclusions $F\Pi_M, FI_M \subset F_M$ are represented by closed embeddings. The natural projections onto the subflags of smaller type are represented by relatively smooth morphisms (whatever this means...). Grassmannians, Π-symmetric grassmannians and isotropic grassmannians for split forms are covered by open superspaces which are relative superaffine spaces $\operatorname{Spec} A[x_1, \ldots, x_n; \xi_1, \ldots, \xi_m]$. The proof will be published elsewhere, and anyway can safely be left to the reader.

Let F_M be one of the flag spaces over M, $\pi\colon F_M \longrightarrow M$ the structural morphism. We will usually denote the tautological flag on F_M by $S_F^{d_1} \subset S_F^{d_2} \subset \cdots \subset \pi^*(\mathcal{T}) = \mathcal{T}_F^d$, superscripts referring to ranks. Of importance for us will be also the orthogonal flag on F_M whose components are denoted

$$\tilde{S}_F^{d-d_\kappa} \subset \tilde{S}_F^{d-d_{\kappa-1}} \subset \cdots \subset \tilde{S}_F^{d-d_1} \subset \pi^*(\mathcal{T}^*), \quad \tilde{S}_F^{d-d_i} = (S_F^{d_i})^\perp.$$

6. Let $\pi\colon G \longrightarrow M$ be one of the grassmannians of subsheaves of $\mathcal{T}$, and let $S \subset \pi^*(\mathcal{T})$ be the tautological sheaf on G. Consider a local vertical vector field X on G (i.e., a section of $\mathcal{T}G/M$) and define a natural action of X on G (lifted sections are horizontal). The Leibnitz formula shows

that the map $\overline{X}: S \longrightarrow \pi^*(\mathcal{T})/S$, $\overline{X}s = Xs \bmod S$ is linear in s. Moreover, the map $X \mapsto \overline{X}$ is linear in X. Therefore we obtain a morphism of $\mathcal{O}_G$-modules $\mathcal{T}G/M \to \underline{\mathrm{Hom}}(S, \pi^*(\mathcal{T})/S) = S^* \otimes \tilde{S}^*$. For $G = G\Pi$ or GI the image of this morphism is not all of $S^* \otimes \tilde{S}^*$. In fact, consider $G\Pi_M(d; \mathcal{T})$ as a closed subspace of $G_M(d; \mathcal{T})$ invariant with respect to the involution induced by p, which we also denote p. This involution acts on the tangent sheaf, and $\mathcal{T}G\Pi$ is the invariant part of $\mathcal{T}G$ restricted onto $G\Pi$. It is easy to relate this action to the action of p on S and $\tilde{S}$, but we shall not need the precise formula. Furthermore, for isotropic grassmannians the form b_G on $\pi^*(\mathcal{T})$ makes it possible to construct a map λ: $\tilde{S}^* \longrightarrow S^*$ (for even b) or $\Pi\tilde{S}^* \to S^*$ (for odd b). Below, we shall identify the image of $\mathcal{T}G/M$ in $S^* \otimes S^*$ or in $\Pi(S^* \otimes S^*)$ (with respect to $id \otimes \lambda$). Namely, we shall essentially use the following theorem which we state here without proof.

7. Theorem. *The relative tangent sheaves on grassmannians of various types are described by the following isomorphisms and exact sequences:*

$$\mathcal{T}G_M(d;\mathcal{T})/M = S^* \otimes \tilde{S}^*,$$
$$\mathcal{T}G\Pi_M(d;\mathcal{T})/M = \underline{\mathrm{Hom}}^p(S,\tilde{S}^*) = (S^* \otimes \tilde{S}^*)^p,$$

$$0 \to S^* \otimes (S^{\perp}_{\pi^*(b)}/S) \to \mathcal{T}GI_M(d;\mathcal{T},b)/M \to \mathcal{R} \to 0,$$

where $\mathcal{R} = \Lambda^2(S^*)$, $S^2(S^*)$, $\Pi S^2(S^*)$, $\Pi\Lambda^2(S^*)$ *for the form b of type OSp, SpO, ΠSp, ΠO respectively.*

We can combine theorem 2.7 and proposition 2.4 to analyze grassmannians componentwise. As an example we describe the associated graded sheaf of a simple grassmannian. Take a $\mathbf{Z}_2$-graded linear $\mathbf{C}$-space $T = \mathbf{C}^{d_0+c_0|d_1+c_1} = T_0 \oplus T_1$ (considered as a sheaf over $\operatorname{Spec}\mathbf{C}$) and put $G = G(d_0|d_1;T)$, $G_1 = G(0|d_1;T_1)$, $G_0 = G(d_0|0;T_0)$. The superspaces G_0, G_1 are essentially conventional grassmannians.

8. Proposition. *One can define the following canonical isomorphisms:*

a. $\mathrm{Gr}\, d = G_0 \times G_1$.
b. $\mathrm{Gr}\, \mathcal{O}_G = S(S_{G_0} \boxtimes \tilde{S}_{G_1} \otimes \tilde{S}_{G_0} \boxtimes S_{G_1})$.
c. $\mathrm{Gr}\, S_G = (S_{G_0} \boxplus S_{G_1}) \otimes \mathrm{Gr}\, \mathcal{O}_G$ (over $\mathrm{Gr}_0\, \mathcal{O}_G$).

Proof. The sheaf $S_{G_0} \boxplus S_{G_1}$ on $G_0 \times G_1$ is induced by a morphism

$\rho: G_0 \times G_1 \longrightarrow G$, i.e., $S_{G_0} \boxplus S_{G_1} = \rho^*(S_G)$. It is easy to convince oneself (say, by computing in coordinates) that ρ is an isomorphism of $G_0 \times G_1$ with $G_{rd} = (G, O_G/J)$, and hence $\mathrm{Gr}_0 S_G$ identifies naturally with $S_{G_0} \boxplus S_{G_1}$. Now $\Omega^1 G = S_G^{d_0|d_1} \otimes \tilde{S}_G^{c_0|c_1}$ in view of theorem 2.7. Hence we can write

$$(\mathrm{Gr}_0 \Omega^1 G)_1 = S_{G_0}^{d_0|0} \boxtimes \tilde{S}_{G_1}^{0|c_1} \oplus \tilde{S}_{G_0}^{c_0|0} \boxtimes S_{G_1}^{0|d_1}.$$

It remains to apply proposition 4.

We shall see later that $G(d;T)$ is not in general isomorphic to $\mathrm{Gr}\, G(d;T)$, contrary to the situation in differential geometry, where smooth superspaces are always isomorphic (noncanonically) to their associated graded objects.

3. Principal Results

1. We shall call an admissible triple a diagram of analytic superspaces $L \overset{\pi_1}{\leftarrow} F \overset{\pi_2}{\rightarrow} M$ with the following properties:

a. The reduced diagram $L_{rd} \leftarrow F_{rd} \rightarrow M_{rd}$ is isomorphic to the diagram $F(1,3;T) \leftarrow F(1,2,3;T) \rightarrow G(2;T)$, $T \simeq \mathbf{C}^4$ of 1.3.

b. The superspaces L, F, M belong to the class of flag superspaces described in 2.5, over $\mathbf{C}$, and the morphisms π_1, π_2 are the standard "projections onto a subflag".

Our first result is a list of all admissible triples. To shorten the statement we only give the structure of M explicitly. It will be clear how to supply the missing L's and F's from the proof in §4.

We note that one could consider a more natural definition, of admissible triples as diagrams $G/P_L \leftarrow G/P_F \rightarrow G/P_M$ with the property *a.*, where G is an analytic supergroup over $\mathbf{C}$ and P_L, $P_M \supset P_F$ its parabolic subgroups. It is possible that supergroups of Cartan type would supply new items in the list. The flag realization of the corresponding homogeneous superspaces will be discussed elsewhere.

2. **Theorem** *The following list contains all flag spaces M (over $\mathbf{C}$)*

with $M_{rd} = G(2|0; T^{4|0})$. *The dimensions of T and M are shown explicitly.*

a. $F(2|0, 2|N; T^{4|N})$, $4|4N$.

b. $F(2|0, 4|0; T^{4|N})$ *and* $F(0|N, 2|N; T^{N|4})$, $4|4N$.

c. $G(2|0; T^{4|N})$ *and* $G(2|N; T^{4|N})$, $4|2N$.

Besides, there are cases $\Pi a.$, $\Pi b.$, $\Pi c.$, *which are deduced from a., b., c. by the change* $T \to T' = \Pi T$, *e.g.,* $\Pi a.\ F(0|2, N|2; T^{N|4})$.

d. $G\Pi(2|2; T^{4|4})$, $4|4$.

e. $GI(1|0; T^{6|2N}, OSp)$ *and* $GI(0|1; T^{2N|6}, SpO)$, $4|2N$.

f. $GI(2|2; T^{4|4}, \Pi Sp)$ *and* $GI(2|2; T^{4|4}, \Pi O)$, $4|4$.

g. $GI(2|0; T^{4|4}, \Pi Sp)$, $4|7$; *and* $GI(0|2; T^{4|4}, \Pi O)$, $4|5$.

h. $FI(2|0, 2|2; T^{4|4}, \Pi Sp)$ and $FI(0|2, 2|2; T^{4|4}, \Pi O)$, $4|8$.

i. $FI(2|0, 4|0; T^{4|4}, \Pi Sp)$; *and* $FI(0|2, 0|4; T^{4|4}, \Pi O)$, $4|6$.

3. There are isomorphisms between some superspaces in this list, e.g., a non-degenerate even form $\beta\colon T \longrightarrow T^*$ permutes subcases in *b.* and *c.*. We shall not try to describe all isomorphisms here. Of more importance for us are some physically motivated restrictions on interesting models of the Minkowski space including anticommuting coordinates, such as the possibility to extend the standard real structure or the spin $\frac{1}{2}$ of the fermionic coordinates. We also expect nontrivial integrability conditions along the null-lines, i.e. we want the dimension of fibres of π_1 to be $1|a$, $a > 0$. Motivated by these considerations we chose for the further investigation the following three types of admissible triples (the notation conforms with V. Kac's list of the simple Lie superalgebras):

$$\begin{aligned} A_N : L^{5|2N} &= F(1|0, 3|N; T^{4|N}) \overset{\pi_1}{\leftarrow} \\ F^{6|4N} &= F(1|0, 2|0, 2|N, 3|N; T^{4|N}) \overset{\pi_2}{\rightarrow} \\ M^{4|4N} &= F(2|0, 2|N; T^{4|N}). \end{aligned}$$

$$
\begin{aligned}
P: L^{5|5} &= GI(1|1; T^{4|4}, \Pi S_p) \overset{\pi_1}{\leftarrow} \\
F^{6|6} &= FI(1|1, 2|2; T^{4|4}, \Pi S_p) \overset{\pi_2}{\rightarrow} \\
M^{4|4} &= GI(2|2; T^{4|4}, \Pi S_p). \\
Q: L^{5|5} &= F\Pi(1|1, 3|3; T^{4|4}) \overset{\pi_1}{\leftarrow} \\
F^{6|6} &= F\Pi(1|1, 2|2, 3|3; T^{4|4}) \overset{\pi_2}{\rightarrow} \\
M^{4|4} &= G\Pi(2|2; T^{4|4}).
\end{aligned}
$$

Of these the cases A_N(and their curved versions) correspond to the most important supersymmetry and supergravity models of physics; cases P and Q are in many respects exotic, but are at least mathematically interesting.

In what follows, $U \subset M$ means a connected open superspace $F(U) = \pi_2^{-1}(U)$, $L(U) = \pi_1(F(U))$. Let $\mathcal{E}$ be a locally free sheaf on U. A (super)connection on $\mathcal{E}$ is given by an even covariant differential

$$\nabla : \mathcal{E} \to \mathcal{E} \otimes \Omega^1 U$$

with the usual properties. A morphism $(\mathcal{E}, \nabla) \to (\mathcal{E}', \nabla')$ is a morphism of sheaves commuting with ∇, ∇'. Similarly, let $\mathcal{E}_F$ be a locally free sheaf on $F(U)$. A relative connection along the fibers of π_1 is an even covariant differential $\nabla_{F/L}$: $\mathcal{E}_F \longrightarrow \mathcal{E}_F \otimes \Omega^1 F/L$. The usual formalism extends readily to superspaces. In particular, the curvatures are represented by morphisms ∇^2: $\mathcal{E} \longrightarrow \mathcal{E} \otimes \Omega^2 U$, $\nabla^2_{F/L}$: $\mathcal{E} \longrightarrow \mathcal{E} \otimes \Omega^2 F/L$ or by the sections $F(\nabla)\Gamma(U, \underline{\mathrm{End}}\,\mathcal{E} \otimes \Omega^2 U)$, $f(\nabla_{F/L}) \in \Gamma(F(U), \underline{\mathrm{End}}\,\mathcal{E}_F \otimes \Omega^2 F/L)$.

4. Theorem. *Consider the following categories:*

a. *The category of pairs* $(\mathcal{E}, \nabla)$ *on* U.

b. *The category of pairs* $(\mathcal{E}_F, \nabla_{F/L})$ *on* $F(U)$, *such that the natural map* $\pi_2^* \pi_{2*} \mathcal{E}_F \to \mathcal{E}_F$ *is an isomorphism.*

Then there is a natural lifting functor a. $\to$ *b. which is an equivalence of categories in the cases* A_N *and* Q.

5. We shall say that $(\mathcal{E}, \nabla)$ is integrable along null-lines if $F(\nabla_{F/L}) = 0$, where $\nabla_{F/L}$ is the lifted connection, restricted to the fibers of π_1. The componentwise analysis of this integrability condition will be given elsewhere. Here we state two theorems, of which the first one is a supersymmetric

version of the Penrose-Ward transform, and the second of which shows the relevance of the integrability condition to the usual Yang-Mills fields.

6. Theorem. *For the diagrams A_N, Q the following categories are equivalent:*

a. *The category of pairs $(\mathcal{E}, \nabla)$ on U, satisfying the ingrability condition and having trivial monodromy along the null-lines.*

b. *The category of locally free sheaves $\mathcal{E}_L$ on $L(U)$ having trivial restrictions on all closed superspaces $\pi_1\pi_2^{-1}(x)$, $x \in U$.*

(If i: $Y \longrightarrow L$ is a closed embedding, $\mathcal{E}_L$ a sheaf on L, we call $i^*(\mathcal{E}_L)$ the restriction of $\mathcal{E}_L$ on Y; the restriction is trivial, if $i^*(\mathcal{E}_L) \simeq \mathcal{O}_Y^{p|q}$).

7. Theorem. a. *The category of holomorphic solutions of the vacuum Yang-Mills equations $\tilde{\nabla}(F_{\pm}(\nabla)) = 0$ in a connected domain of*

$$U \subset G(2|0; T^{4|0}),$$

with trivial monodromy along the null-lines, has a natural embedding into the category of pairs $(\mathcal{E}, \nabla)$ with integrability condition, defined on the superspace extension U, corresponding to the case A_3 (Witten).

b. *For the cases P, Q a similar statement is true for all the trivial monodromy Yang-Mills fields on U.*

4. Proofs I

1. In this section we prove theorems 3.2 and 3.7, and indicate how to enumerate all admissible triples. The principle of classification is quite simple. Suppose F is a flag space with $F_{rd} = G(2|0; T^{4|0})$. Let $F \to G$ be the projection of F onto the grassmannian of the last component of the flag. Then either $G_{rd} = G(2|0; T^{4|0})$, or $G_{rd} = \operatorname{Spec} \mathbf{C}$. The reduced grassmannians are calculated with the help of 2.7–2.8, which imposes strong restrictions on the dimensions of T and on the last component. It is then

easy to sort out the remaining parameters. Subsections 2–5 below give some details for all types of flag spaces.

2. First of all,

$$G(d_0|d_1;T^{d_0+d_1|c_0+c_1})_{rd} = G(d_0|0;T^{d_0+c_0|0}) \times G(0|d_1;T^{0|d_1+c_1}),$$

in view of Prop. 2.8. Hence $G_{rd} = \operatorname{Spec}\mathbf{C}$ iff either $c_0 = d_1 = 0$, or $c_1 = d_0 = 0$ (excluding the trivial cases $c_0 = c_1 = 0$ and $d_0 = d_1 = 0$). These cases are permuted by the Π-transform, so that we need consider only $G(d|0;T^{d|c})$. If the initial flag has a component of dimension $d_1|0, d_1 < d$, then necessarily $d_1 = 2, d = 4$, since $F(d_1|0,d|0;T^{d|c})_{rd} = G(d_1|0;T^{d|0})$ and additional components would enlarge the even dimension.

Otherwise the reduced grassmannian of the last component can be only $G(2|N;T^{4|N}), G(2|0;T^{4|N})$, and the Π-transformed versions of these. Without changing G_{rd} we can add only components of dimensions $2|0$ and $2|N$ correspondingly. This exhausts cases a. – b. and Πa. – Πb. in theorem 3.2. The reader can supply F's and L's using A_N as a pattern.

3. Turning to the Π-symmetric flags, we first observe that if T admits a Π-symmetry p, then $rk\,T = r|r$, and all pairs (T,p) of the same rank are isomorphic. Furthermore, $G\Pi(d|d;T^{c+d|c+d})_{rd} = G(d|0;T^{c+d|0})$. Hence if $F\Pi_{rd} = G(2|0;T^{4|0})$, then $F\Pi = G\Pi(2|2;T^{4|4})$. The only admissible triple of Π-symmetric flags is Q.

4. For the OSp-isotropic flags we have

$$GI(r|s;T^{m|n},OSp)_{rd} = GI(r|0;T^{m|0},O) \times GI(s|0;T^{n|0},Sp) \quad \text{(exercise)}.$$

Hence the reduced grassmannian of the last components cannot be $\operatorname{Spec}\mathbf{C}$. As $G(2|0;T^{4|0})$ is not a direct product, we have either $s = 0$, or $r = 0$. In the first case the equation for the even part of dimension

$$r(m-2r) + \frac{r(r-1)}{2} = 4$$

has the unique solution $r = 1$, $m = 6$. In the second case the equation

$$s(n-2s) + \frac{s(s+1)}{2} = 4$$

also has a solution $s = 1, n = 5$, but an OSp-form on the space $T^{m|5}$ is necessarily degenerate.

To observe that $GI(1|0;T^{6|2N},OSp)_{rd} = G(2|0;T^{4|0})$, one realizes the last grassmannian as the Plücker quadric of decomposable bivectors in $P(\wedge^2 T^{4|0})$. The corresponding admissible triples are

$$GI(2|0;T^{6|2N},OSp) \overset{\pi_1}{\leftarrow} FI(1|0,2|0;T^{6|2N},OSp) \overset{\pi_2}{\rightarrow} GI(1|0;T^{6|2N},OSp).$$

Unfortunately, the fibers of π_1 are $1|0$-dimensional and do not give rise to nontrivial integrability conditions.

5. Finally, consider ΠSp- and ΠO-isotropic flags. Here again $rk\, T = m|m$; moreover, odd and even parts of T are dual. Hence $GI(r|s;T^{m|m}, \Pi Sp \text{ or } \Pi O)_{rd}$ can be realized as the relative grassmannian $G_H(s|0;\tilde{S}^{m-r|0})$ over the base grassmannian $H = G(r|0;T^{m|0})$. Hence we can get Spec C only for $GI(m|0;T^{m|m})$ or $GI(0|m;T^{m|m})$. This leads to case i of theorem 3.2. The remaining cases correspond to the reduced grassmannian of the last components.

6. We proceed now to the proof of theorem 3.7. Denote by $L^{(m)}$ the m-th infinitesimal neighbourhood of $L = F(1|0,3|0;T^{4|0}) \subset P(T) \times P(T^*)$, as in 1.3. This is a conventional analytic space with nilpotents. On the other hand, let $L(A_N), L(P), L(Q)$ be the corresponding flag spaces of our triples. Below, we shall show the existence of natural surjective maps

$$L(A_3) \to L^{(3)}, \quad L(P) \to L^{(2)}, \quad L(Q) \to L^{(2)},$$

identical on the reduced spaces. (We identify $T^{4|0}$ with $(T^{4|3})_0$ or $(T^{4|4})_0$ respectively). It remains then to apply the following results of [6], [7], [9]. To each Yang-Mills field on U with trivial monodromy along null-lines, there corresponds a canonically defined locally free sheaf $\mathcal{E}_L^{(2)}$ on $L^{(2)}(U)$ which can be extended onto $L^{(3)}(U)$ iff the initial field fulfills the complete Yang-Mills equations. Lifting $\mathcal{E}_L^{(2)}$ from $L^{(2)}$ onto $L(P), L(Q)$ or $\mathcal{E}_L^{(3)}$ from $L^{(3)}$ to $L(A_3)$, one gets the corresponding functors by means of the inverse superspace Penrose-Ward transform implied by theorem 3.6.

a. Case A_N. Let x_i, ξ_j ($i = 1,\ldots,4; j = 1,\ldots,N$) be a base of $(T^{4|N})^*$, and x^i, ξ^j the dual base of $T^{4|N}$. One can interpret these elements as sections of the sheaves $(S^{1|0})^*$ on $G(1|0;T^{4|N})$ and $G(1|0;(T^{4|N})^*)$ respectively. The subspace $L(A_N) \subset G(1|0;T^{4|N}) \times G(1|0;(T^{4|N})^*)$ is given by the incidence relation

$$\sum_{i=1}^{4} x_i \otimes x^i + \sum_{j=1}^{N} \xi_j \otimes \xi^j = 0$$

(the left part is a section of the sheaf $(S^{1|0} \boxtimes S^{1|0})^*$). On the other hand, $L^{(N)}$ is defined by the equations

$$\xi_j = \xi^j = 0, \quad \left(\sum_{i=1}^{4} x_i \otimes x^i\right)^{N+1} = 0.$$

Since

$$\left(\sum_{j=1}^{N} \xi_j \otimes \xi^j\right)^{N+1} = 0,$$

the identification of the reduced subspaces $L(A_N)_{rd} = L^{(N)}_{red}$ naturally extends to a morphism $L(A_N) \to L^{(N)}$. To convince oneself that it is a surjective map, one passes to the standard affine covering and uses

$$\left(\sum_{j=1}^{N} \xi_j \otimes \xi^j\right)^{N} \neq 0.$$

b. Case P. Choose a base in $T^{4|4}$ and define the ΠSp-form by the Gram matrix $\begin{pmatrix} 0 & E \\ E & 0 \end{pmatrix}$. The grassmannian $G = G(1|1; T^{4|4})$ can be then described in the following way. Its "homogeneous coordinates" are the elements of the matrix

$$Z = \begin{pmatrix} x_1 & x_2 & x_3 & x_4 & \xi_1 & \xi_2 & \xi_3 & \xi_4 \\ \eta_1 & \eta_2 & \eta_3 & \eta_4 & y_1 & y_2 & y_3 & y_4 \end{pmatrix}.$$

More precisely, G is covered by open subspaces G_{ij}, $i, j = 1, \ldots, 4$. Putting

$$Z_{ij} = \begin{pmatrix} x_i & \xi_j \\ \eta_i & y_j \end{pmatrix}$$

one can define G_{ij} as the (super)spectrum of the ring generated by the elements of the matrix $Z_{ij}^{-1} Z$. The sheaf $S_G^{1|1}$ is given on G_{ij} together with a trivialization: it is generated by two lines of the same matrix. The transition functions, both for G and $S_G^{1|1}$, are evident. The subspace $L(P) \subset G$ is defined by the isotropy equations

$$\sum_{i=1}^{4} x_i y_i + \sum_{i=1}^{4} \xi_i \eta_i = 0, \quad \sum_{i=1}^{4} y_i \eta_i = 0.$$

Unlike the case A_3, however, here the transition rules and the second equation imply $(\sum_{i=1}^4 x_i y_i)^3 = 0$, and only $(\sum_{i=1}^4 x_i y_i)^2 \neq 0$. In fact, consider the example of G_{11}, which is the spectrum of the ring generated by the entries of the matrix

$$\begin{pmatrix} 1 & X_2 & X_3 & X_4 & 0 & \Xi_2 & \Xi_3 & \Xi_4 \\ 0 & H_2 & H_3 & H_4 & 1 & Y_2 & Y_3 & Y_4 \end{pmatrix}.$$

The equations of $L(P)$ on G_{11} are

$$1 + \sum_{i=2}^{4} X_i Y_i + \sum_{i=2}^{4} \Xi_i H_i = 0, \quad \sum_{i=2}^{4} Y_i H_i = 0.$$

It follows from the first equation that at least one of the coordinates Y_i is invertible at each point. The second equation then shows that the corresponding H_i can be expressed linearly in terms of the remaining ones. Hence

$$H_2 H_3 H_4 = 0 \quad \text{and} \quad \left(\sum_{i=2}^{4} \Xi_i H_i\right)^3 = 0.$$

c. Case Q. Here again we consider the closed embedding

$$L(Q) \subset G\Pi(1|1; T^{4|4}) \times G\Pi(1|1; T^{4|4})^*)$$

as an "incidence quadric". The base in $T^{4|4}$ is taken in the form $(e_i, p(e_i))$, where p is the Π-symmetry. The grassmannian $G\Pi(1|1; T^{4|4})$ is covered by open subspaces with the coordinates.

$$\begin{pmatrix} x_i & \xi_i \\ -\xi_i & x_i \end{pmatrix}^{-1} \begin{pmatrix} x_1 & x_2 & x_3 & x_4 & \xi_1 & \xi_2 & \xi_3 & \xi_4 \\ -\xi_1 & -\xi_2 & -\xi_3 & -\xi_4 & x_1 & x_2 & x_3 & x_4 \end{pmatrix}.$$

We denote the corresponding coordinates on the other grassmannian by superscripts. The incidence equations are $\sum_{i=1}^4 x_i x^i + \sum_{j=1}^4 \xi_j \xi^j = 0$, $\sum_{i=1}^4 x_i \xi^i - \sum_{j=1}^4 x^j \xi_j = 0$. Computing to the affine coordinates we find as before that $(\sum_{i=1}^4 x_i x^i)^3 = 0$, $(\sum_{i=1}^4 x_i x^i)^2 \neq 0$. This finishes the proof.

5. Auxiliary Calculations

1. We continue to consider admissible triples of the types A_N, P, Q. In this section the structure of some sheaves and their cohomology on the components of these triples is made explicit.

2. *The structure of F/M.* We recall that on a flag space, say
$$M = F(2|0, 2|N; T^{4|N}),$$
we denote by $S_M^{2|0}, S_M^{2|N}$ the components of the tautological flag and by $\tilde{S}_M^{2|0}, \tilde{S}_M^{2|N}$ the components of the orthogonal flag. The corresponding flags lie in $\mathcal{T}^{4|N} = \mathcal{O}_M \otimes T^{4|N}$, or in $(\mathcal{T}^{4|N})^*$. In this notation we can visualize $\pi_2: F \longrightarrow M$ as the structural morphism of a relative grassmannian or as a fibered product of such morphisms:
$$A_N: F = G_M(1|0; S_M^{2|0}) \underset{M}{\times} G_M(1|0; \tilde{S}_M^{2|0}) \to M,$$
$$P: F = G_M(1|1; S_M^{2|2}) \to M,$$
$$Q: F = G\Pi_M(1|1; S_M^{2|2}) \underset{M}{\times} G\Pi_M(1|1; \tilde{S}_M^{2|2}) \to M.$$
The second factor in A_N is obtained in this form if we consider the orthogonal flag $\tilde{S}^{1|0} \subset \tilde{S}^{2|0} \subset (\mathcal{T}^{4|N})^*$ in place of the flag
$$S^{2|N} \subset S^{3|N} \subset \mathcal{T}^{4|N}.$$
A similar trick applies to Q.

It follows that locally F over M is a direct product. The fiber is the two-dimensional quadric $\mathbf{P}^1 \times \mathbf{P}^1$ in the case A_N, the "P-quadric" $G(1|1; S^{2|2})$ and the "Q-quadric" $G\Pi(1|1; S^{2|2}) \times G\Pi(1|1; \tilde{S}^{2|2})$ in the remaining cases. By Proposition 2.8,

$$\operatorname{Gr} G(1|1; S^{2|2}) =$$
$$(\mathbf{P}^1 \times \mathbf{P}^1; \mathcal{O}_{\mathbf{P}^1 \times \mathbf{P}^1} \oplus \Pi[\Omega_l^1 P^1(1,-1) \oplus \Omega_r^1 P^1(-1,1)] \oplus \Omega^2(P^1 \times P^1)),$$
where Ω_l^1, Ω_r^1 denote the sheaves of 1-forms lifted from the first and second factors respectively. One can prove similarly that
$$G\Pi(1|1; S^{2|2}) = (P^1; \mathcal{O}_{P^1} \oplus \Pi\Omega^1 P^1).$$
This shows, in particular, that P- and Q-quadrics are not isomorphic, although they have the common dimension $2|2$. Applying a base change theorem for superspaces, obvious in our situation, we get:

3. Proposition a. $\pi_{2*}\mathcal{O}_F = \mathcal{O}_M$,

b. $R^1\pi_{2*}\mathcal{O}_F = 0$ *in the cases* A_N, P; $R^1\pi_{2*}\mathcal{O}_F$ *is locally free of rank* $0|2$ *in the case* Q.

c. $R^2\pi_{2*}\mathcal{O}_F = 0$ *in the case* A_N; $R^2\pi_{2*}\mathcal{O}_F$ *is locally free of rank* $1|0$ *in the cases* P, Q.

We note two characteristic cohomological peculiarities of superspaces. First, from the (topological) properness of G it does not follow automatically that $\Gamma(G, \mathcal{O}_G) = \mathbf{C}$. Trivial counterexamples are supplied by $(M, \wedge(\mathcal{E}))$, where M is projective and $\mathcal{E}$ has nontrivial sections. In general, the middle term in the standard decomposition $M \to (N, \pi_*\mathcal{O}_M) \to (N, \mathcal{O}_N)$ of a smooth proper morphism π: $M \longrightarrow N$ of smooth superspaces can be nontrivial and nonsmooth. Second, the example of P- and Q-quadrics shows that an equivariant invertible sheaf on a proper homogeneous space (the structure sheaf in this case) can have several nonvanishing cohomology groups. The Borel-Weil-Bott theorem for superspaces must describe a more complicated situation than the classical one.

4. *The structure of* $\Omega^1 M$. In the cases P, Q the tangent sheaf $\mathcal{T}M$ can be calculated with the help of theorem 2.7. To treat the case A_N we introduce the left and right spaces

$$M_l = G(2|N; T^{4|N}) = G(2|0; (T^{4|N})^*), \quad M_r = G(2|0; T^{4|N}).$$

We put then

$$\Omega^1_l M = \Omega^1 M / M_r,$$
$$\Omega^1_r M = \Omega^1 M / M_l,$$
$$\Omega^1_0 M = \mathrm{Ker}(res\colon \Omega^1 M \longrightarrow \Omega^1_l M \oplus \Omega^1_r M).$$

Clearly, M is a relative grassmannian over both M_l and M_r:

$$M = G_{M_l}\left(2|0; S^{2|N}_{M_l}\right) = G_{M_r}\left(0|N; \mathcal{T}^{4|N}/S^{2|0}_{M_r}\right).$$

Hence Ω^1_r and Ω^1_l are calculable by 2.7. We obtain the exact locally split sequence

$$A_N\colon 0 \to \Omega^1_0 M \to \Omega^1 M \to \Omega^1_l M \oplus \Omega^1_r M \to 0,$$
$$(\text{ranks } 4|0, 4|4N, 0|2N + 0|2N),$$

$$\Omega_l^1 M = \underline{\underline{\mathrm{Hom}}}\left(\mathcal{T}_M^{4|N}/S_M^{2|N}, S_M^{2|N}/S_M^{2|0}\right) = S_M^{2|N}/S_M^{2|0} \otimes \tilde{S}_M^{2|0},$$
$$\Omega_r^1 M = \underline{\underline{\mathrm{Hom}}}\left(S_M^{2|N}/S_M^{2|0}, S_M^{2|0}\right) = S_M^{2|0} \otimes \left(S_M^{2|N}/S_M^{2|0}\right)^*, \tag{1}$$
$$\Omega_0^1 M = S_M^{2|0} \otimes \tilde{S}_M^{2|0}.$$

The last isomorphism is defined with the help of the dual sequence

$$0 \to \mathcal{T}M/M_r \oplus \mathcal{T}M/M_l \to \mathcal{T}M \to \mathcal{T}_0 M \to 0,$$

where $\mathcal{T}_0 M = (\Omega_0^1 M)^*$. The supercommutator induces a bilinear form $\mathcal{T}M/M_r \times \mathcal{T}M/M_l \to \mathcal{T}_0 M$, which can be identified with the composition map

$$\underline{\underline{\mathrm{Hom}}}\left(S_M^{2|0}, S_M^{2|N}/S_M^{2|0}\right) \times \underline{\underline{\mathrm{Hom}}}\left(S_M^{2|N}/S_M^{2|0}, \mathcal{T}_M^{4|N}/S_M^{2|N}\right)$$
$$\to \underline{\underline{\mathrm{Hom}}}\left(S_M^{2|0}, \left(\tilde{S}_M^{2|0}\right)^*\right).$$

To this we add the information about P- and Q-cases:

$$P\colon \Omega^1 M = \Pi S^2(S_M^{2|2}), \tag{2}$$

$$Q\colon \Omega^1 M = \underline{\underline{\mathrm{Hom}}}^p\left(\mathcal{T}_M^{4|4}/S_M^{2|2}, S_M^{2|2}\right) = \left(S_M^{2|2} \otimes \tilde{S}_M^{2|2}\right)^p. \tag{3}$$

5. *The structure of* $\Omega^1 F/L$. One readily sees, that in the A_N-case, F over L is a relative flag space of the following type:

$$F = F_L(1|0, 1|N; S_L^{3|N}/S_L^{1|0}).$$

As earlier, we put $F_l = G_L(1|N; S_L^{3|N}/S_L^{1|0})$, $F_r = G_L(1|0; S_L^{3|N}/S_L^{1|0})$, and $\Omega_l^1 F/L = \Omega^1 F/F_r$, $\Omega_l^1 F/L = \Omega^1 F/F_l$. The following display is to be compared with (1):

$$A_N\colon 0 \to \Omega_0^1 F/L \to \Omega^1 F/L \to \Omega_l^1 F/L \oplus \Omega_r^1 F/L \to 0$$
$$(\text{ranks } 1|0, 1|2N, 0|N + 0|N)$$

$$\Omega_l^1 F/L = \underline{\underline{\mathrm{Hom}}}\left(S_F^{3|N}/S_F^{2|N}, S_F^{2|N}/S_F^{2|0}\right) = S_F^{2|N}/S_F^{2|0} \otimes \tilde{S}_F^{2|0}/\tilde{S}_F^{1|0},$$
$$\Omega_2^1 F/L = \underline{\underline{\mathrm{Hom}}}\left(S_F^{2|N}/S_F^{2|0}, S_F^{2|0}/S_F^{1|0}\right) = S_F^{2|0}/S_F^{1|0} \otimes \left(S_F^{2|N}/S_F^{2|0}\right)^*,$$
$$\Omega_0^1 F/L = S_F^{2|0}/S_F^{1|0} \otimes \tilde{S}_F^{2|0}/\tilde{S}_F^{1|0}. \tag{1'}$$

In the case P, clearly $F/L = GI_L(1|1;(S_L^{1|1})^{\perp}/S_L^{1|1},\Pi Sp)$. Moreover, the tautological bundle on F corresponding to this representation of F as a relative grassmannian equals $S_F^{2|2}/S_F^{1|1}$, where S_F are the components of the tautological flag in the representation $F\Pi(1|1,2|2;T^{4|4})$. Hence

$$P: \Omega^1 F/L = \Pi S^2\left(S_F^{2|2}/S_F^{1|1}\right). \tag{2'}$$

Finally, in the Q-case we have $F/L = G\Pi_L(1|1;S_L^{3|3}/S_L^{1|1})$, the corresponding tautological bundle being $S_F^{2|2}/S_F^{1|1}$, and we get

$$\begin{aligned} Q: \Omega^1 F/L &= \underline{\mathrm{Hom}}^p\left(S_F^{3|3}/S_F^{2|2}, S_F^{2|2}/S_F^{1|1}\right) \\ &= \left(S_F^{2|2}/S_F^{1|1} \otimes \tilde{S}_F^{2|2}/\tilde{S}_F^{1|1}\right)^p. \end{aligned} \tag{3'}$$

Since all these identifications are natural, one can guess and check that the restriction map $\pi_2^*\Omega^1 M \overset{res}{\to} \Omega^1 F/L$ is induced by the identifications $\pi_2^* S_M^{a|b} = S_F^{a|b}$ (and similarly for $\tilde{S}$) in all cases. This makes it possible to understand the structure of the sheaf $\mathcal{N}$ which is defined as the kernel of res:

$$0 \to \mathcal{N} \to \pi_2^*\Omega^1 M \to \Omega^1 F/L \to 0. \tag{4}$$

Namely, comparing (1)–(3) with (1′)–(3′), we can construct the following exact sequences:

$$A_N: 0 \to \mathcal{N}_0 \to \mathcal{N} \to \mathcal{N}_l \oplus \mathcal{N}_r \to 0, \tag{5}$$

$$0 \to S_F^{1|0} \otimes \tilde{S}_F^{1|0} \to S_F^{2|0} \otimes \tilde{S}_F^{1|0} \oplus S_F^{1|0} \otimes \tilde{S}_F^{2|0} \to \mathcal{N}_0 \to 0, \tag{6}$$

$$N_l = S_F^{2|N}/S_F^{2|0} \otimes S_F^{1|0}, \tag{7}$$

$$N_r = S_F^{1|0} \otimes (S_F^{2|N}/S_F^{2|0})^*; \tag{8}$$

$$P: \quad 0 \to S^2(S_F^{1|1}) \to S_F^{1|1} \cdot S_F^{2|2} \to \Pi\mathcal{N} \to 0. \tag{9}$$

(The middle sheaf here is the image of $S_F^{1|1} \otimes S_F^{2|2}$ in $S^2(S_F^{2|2})$).

$$Q: 0 \to \left(S_F^{1|1} \otimes \tilde{S}_F^{1|1}\right)^p \to \left(S_F^{2|2} \otimes \tilde{S}_F^{1|1} \oplus S_F^{1|1} \otimes \tilde{S}_F^{2|2}\right)^p \to \mathcal{N} \to 0. \tag{10}$$

6. Proposition. a. *$R^i\pi_{2*}\mathcal{N} = 0$ for all $i \geq 0$ in the case A_N.*
b. *$\pi_{2*}\mathcal{N} = 0$ in the case Q.*

Proof. a. Recall that in the case A_N, the space F/M is a relative quadric, the sheaves $S_F^{1|0}$ and $\tilde{S}_F^{1|0}$ are locally (over M) $\mathcal{O}(-1,0)$ and $\mathcal{O}(0,-1)$ respectively, and $S_F^{2|0}$, $S_F^{2|N}$, $\tilde{S}_F^{2|0}$ are lifted from M. Furthermore,

$$H^i(\mathbf{P}^1 \times \mathbf{P}^1, \mathcal{O}(a,b)) = 0$$

for all i, if $a = -1$ or $b = -1$. From this and (5)–(8) it follows that $R^i\pi_{2*}\mathcal{N} = 0$.

b. We shall show that

$$\pi_{2*}\left(S_F^{1|1}\right) = 0, \quad \pi_{2*}\left(\tilde{S}_F^{1|1}\right) = 0, \quad R^1\pi_{2*}\left(S_F^{1|1} \otimes \tilde{S}_F^{1|1}\right) = 0.$$

Using (10), proposition 5.3, and the fact that $S_F^{2|2}$ and $\tilde{S}_F^{2|2}$ are lifted from M, we obtain *b*.

We shall check the vanishing of the relevant direct images fiberwise. Since the Q-quadric is a direct product (5.2), we can essentially calculate on

$$G = G\Pi\left(1|1; S^{2|2}\right) = \left(\mathbf{P}^1 = G\left(1|0 : S^{2|0}\right), \mathcal{O}_{P^1} \oplus \Pi\Omega^1\mathbf{P}^1\right).$$

It is easy to show that

$$\operatorname{Gr} S_G^{1|1} = [\mathcal{O}_{P^1}(-1) \oplus \Pi\mathcal{O}_{P^1}(-1)] \otimes_{\mathcal{O}_{P^1}} [\mathcal{O}_{P^1} \oplus \Pi\Omega^1 P^1]. \qquad (11)$$

From this and from the exact sequence

$$0 \to \operatorname{Gr}_1 S_G^{1|1} \to S_G^{1|1} \to \operatorname{Gr}_0 S_G^{1|1} \to 0$$

it follows that $H^0(S_G^{1|1}) = 0$, $H^1(S_G^{1|1}) \simeq S^{2|2}$. Therefore $\pi_{2*}(S_F^{1|1}) = 0$, $\pi_{2*}(\tilde{S}_F^{1|1}) = 0$. For $S_G^{1|1} \otimes \tilde{S}_G^{1|1}$ the calculation is similar. First, putting $H = \mathbf{P}^1 \times \mathbf{P}^1$, we find from (11) (omitting evident subscripts and superscripts):

$$\operatorname{Gr}(S \otimes \tilde{S}) = [\mathcal{O}_H(-1,-1)^2 \oplus \Pi\mathcal{O}_H(-1,-1)^2] \otimes_{\mathcal{O}_H} [\mathcal{O}_H \oplus \Omega^2 H \oplus \Pi(\Omega_l^1 P^1 \oplus \Omega_r^1 P^1)].$$

Hence $H^0(\mathrm{Gr}_2(\mathcal{S}\otimes\tilde{\mathcal{S}})) = H^1(\mathrm{Gr}_2(\mathcal{S}\otimes\tilde{\mathcal{S}})) = 0$ and $H^0(\mathcal{S}\otimes\tilde{\mathcal{S}}) = H^0(\mathcal{F})$, where $\mathcal{F} = (\mathcal{S}\otimes\tilde{\mathcal{S}})/\mathrm{Gr}_2(\mathcal{S}\otimes\tilde{\mathcal{S}})$. Furthermore,

$$0 \to \mathrm{Gr}_1(\mathcal{S}\otimes\tilde{\mathcal{S}}) \to \mathcal{F} \to \mathrm{Gr}_0(\mathcal{S}\otimes\tilde{\mathcal{S}}) \to 0$$

is exact and all cohomology of Gr_1 vanishes. Thus finally

$$H^0(\mathcal{F}) = H^0(\mathrm{Gr}_0(\mathcal{S}\otimes\tilde{\mathcal{S}})) = 0.$$

We note that a similar calculation in the case P shows that $\pi_{2*}\mathcal{N} \neq 0$.

6. Proofs II

1. In this section we prove the theorems 3.4 and 3.6. The notations of §3 are retained. We describe the relevant inverse and direct image functors, and rely upon calculations of §5 to establish their properties.

2. *The lift from U to $F(U)$.* Starting from $(\mathcal{E},\nabla)$ we construct $\mathcal{E}_F = \pi_2^*(\mathcal{E})$ and $\nabla_{F/L}$: $\mathcal{E}_F \xrightarrow{\pi_2^*(\nabla)} \mathcal{E}_F\otimes\Omega^1F(U) \overset{res}{\to} \mathcal{E}_F\otimes\Omega^1F(U)/L(U)$. It follows from proposition 5.3 that $\pi_{2*}(\mathcal{E}_F) = \mathcal{E}$. This functor is obviously compatible with the tensor structure of both categories, and with reductions of the structure group compatible with the connection.

3. *The descent from $F(U)$ to U.* Starting from $(\mathcal{E}_F, \nabla_{F/L})$ on $F(U)$, we construct $\mathcal{E} = \pi_{2*}(\mathcal{E}_F)$, ∂: $\pi_{2*}(\nabla_{F/L})$: $\mathcal{E} \to \pi_{2\cdot}(\mathcal{E}_F\otimes\Omega^1F/L)$. Generally we can only guarantee that ∂ is a differential operator of order ≤ 1. We must check that under the conditions of the theorem 3.4. it is essentially a covariant differential. First of all, since $\mathcal{E}_F \to \pi_2^*\mathcal{E}$ is an isomorphism, we have $\pi_{2*}(\mathcal{E}_F\otimes\Omega^1F/L) = \mathcal{E}\otimes\pi_{2*}(\Omega^1F/L)$. From proposition 5.6 and (4), §5, it follows that $\pi_{2*}(\Omega^1F/L) = \Omega^1M$ in the case A_N, the isomorphism being induced by the descent of the restriction map

$$\pi_2^*(\Omega^1M) \to \Omega^1F/L.$$

In the case Q however, we get only an injection $\Omega^1M \subset \pi_{2*}(\Omega^1F/L)$. We must show that the image of ∂ lies in $\mathcal{E}\otimes\Omega^1U$, and that ∂ verifies the Leibnitz rule $\partial(sf) = (\partial s)f + s\otimes df$ for a section of s of $\mathcal{E}$ and a function f on U. In fact, we have $\partial(sf) - \partial s \otimes f = sA(f)$, where $A(f)\colon \mathcal{E} \longrightarrow \mathcal{E} \otimes \pi_{2*}(\Omega^1 F/L)$ is a linear operator. Looking at the latter as a section of $\operatorname{End} \mathcal{E} \otimes \pi_{2*}(\Omega^1 F/L)$ one easily obtains the functional equation $fA(g)+A(f)g = A(fg)$. Therefore the map $A\colon \mathcal{O}_U \longrightarrow \operatorname{End} \mathcal{E} \otimes \pi_{2*}(\Omega^1 F/L)$ is a composition $\mathcal{O}_U \xrightarrow{d} \Omega^1 U \xrightarrow{b} \operatorname{End} \mathcal{E} \otimes \pi_{2*}(\Omega^1 F/L)$, where b is an $\mathcal{O}$-linear map. Using the functorality in $\mathcal{E}$ and the direct calculation for $\mathcal{E} = \mathcal{O}$, one sees that $b(w) = id_{\mathcal{E}} \otimes j(w)$, where $j\colon \Omega^1 M \longrightarrow \pi_{2*}(\Omega^1 F/L)$ is the injection constructed above. In particular, $A(f) = id_{\mathcal{E}} \otimes df$, so that ∂ is a covariant differential.

To prove that the constructed functors are quasi-inverses of each other, it is convenient to interpret the morphisms $(\mathcal{E}_1, \nabla_1) \to (\mathcal{E}_2, \nabla_2)$ as the horizontal sections of $\mathcal{E}_1^* \otimes \mathcal{E}_2$ relative to the induced connection $\nabla_1^* \otimes 1 + 1 \otimes \nabla_2$, and similarly on F. Then the main point is that $\operatorname{Ker} \nabla = \pi_{2*} \operatorname{Ker} \nabla_{F/L}$. We omit the details.

4. *The descent from $F(U)$ to $L(U)$*. Starting from $(\mathcal{E}_F, \nabla_{F/L})$ on $F(U)$, we construct the sheaf $\pi_{1*}(\operatorname{Ker} \nabla_{F/L})$. The vanishing of the relative curvature tensor $\nabla^2_{F/L}$ and the triviality of monodromy of $\nabla_{F/L}$ along the fibers of π_1 means that the restriction of $\mathcal{E}_F$ to each fiber is trivialized by the subsheaf of $\nabla_{F/L}$-horizontal sections. It follows that $\pi_{1*}(\operatorname{Ker} \nabla_{F/L}) = \mathcal{E}_L$ is a locally free sheaf of the same rank as $\mathcal{E}_F$. Moreover, if $\mathcal{E}_F = \pi_2^*(\mathcal{E})$, then $\mathcal{E}_F$ is trivial along the fibers of π_2 embedded into L. Since $\pi_1^* \mathcal{E}_L = \mathcal{E}_F$, the sheaf $\mathcal{E}_L$ is also trivial along the π_1-images of these fibers.

5. *The lift from $L(U)$ to $F(U)$*. Starting with any locally free sheaf $\mathcal{E}_L$ on L, one can construct a couple $(\mathcal{E}_F = \pi_1^* \mathcal{E}_L, \nabla_{F/L})$, $\nabla_{F/L}$ being the unique connection for which all the sections of $\pi_1^{-1}(\mathcal{E}_L)$ are horizontal. If moreover $\mathcal{E}_L$ is trivial on the π_1-images of the fibers of π_2, then the map $\mathcal{E}_F \to \pi_2^* \pi_{2*} \mathcal{E}_F$ is an isomorphism, since the implied trivialization is canonical in view of Proposition 3.5.

References

[1] Atiyah, M.F., Geometry of Yang-Mills fields, Lezioni Fermiani, Pisa, 1979.

[2] van Nieuwenhuizen, P., Supergravity, Phys. Reports, 68, (1981), 189–398.

[3] Ferber, A., Supertwistors and conformal supersymmetry, Nuclear Physics, B 132 (1978), 55–64.

[4] Witten, E., An interpretation of classical Yang-Mills theory, Phys. Letters, 77B (1978), 394–400.

[5] Leites, D.A., Introduction to the theory of supermanifolds, Uspekhi 53:1 (1980), (in Russian).

[6] Kostant, B., Graded manifolds, graded Lie theory, and prequantization, Springer Lecture Notes in Math. 570 (1977), 177–306.

[7] Isenberg, J., Yasskin Ph. B., Green, P.S., Non-selfdual gauge fields, Phys. Letters, 78B, (1978), 464–468.

[8] Manin, Yu. I., Gauge fields and holomorphic geometry, in Sovremennye Problemy Matematiki, 17 (1981), Moscow, VINITI (in Russian).

[9] Henkin, G.M., Manin, Yu. I., Twistor description of classical Yang-Mills-Dirac fields, Phys. Letters 95B (1980), 405–408.

[10] Henkin, G.M., Manin Yu. I., On the cohomology of twistor flag spaces, Compositio Math. 44, (1981), 103–111.

[11] Deligne, P., Milne, J.S., Ogus, A., Shin, K., Hodge cycles, motives and Shimura varieties, Springer Lecture Notes in Math. 900 (1982).

Received July 9, 1982

Professor Yurii Ivanovic Manin
Steklov Institute of Mathematics
ul. Vavilova, 42
Moscow 117966 GSP-1, USSR

Algebraic Surfaces and the Arithmetic of Braids, I

B. Moishezon

Dedicated to I.R. Shafarevich

Introduction

The present work is an informal continuation of [2]. Our terminology here is slightly different from that in [2]. We chose to work in affine spaces assuming always that algebraic varieties, which we consider, are in general positions to hyperplanes at ∞ and to centers of projections.

For an oriented Euclidian plane L with a finite subset K, $\#(K) = n$, we denote the corresponding braid group B_n (see [2], p. 111) by $B[L,K]$. Let $K = \{A_1, \dots, A_n\}$, $\{X_1, \dots, X_{n-1}\}$ be a system of simple paths such that $\forall i = 1, \dots, n-1$, $\partial X_i = \{a_i, a_{i+1}\}$, $X_i \cap K = \partial X_i$ and

$$\forall i,j = 1,\dots,n-1 \quad X_i \cap X_j = \begin{cases} 0 & \text{if } |i-j| \geq 2 \\ a_{i+1} & \text{if } j = i+1 \end{cases}.$$

We associate with each X_i an element of $B[L,K]$ represented by a positive half-twist corresponding to X_i. It is convenient to denote this element of $B[L,K]$ again by X_i. It is well known that such set of braids $\{X_1, \dots, X_{n-1}\}$ is a system of generators of $B[L,K]$. We call it a "good ordered system of generators of $B[L,K]$." All relations between $X_1, \dots, X_{n-1}$ follow from:

$$X_iX_j = X_jX_i \quad \text{if } |i-j| \geq 2 \qquad (i,j \in 1,2,\dots,n-1)$$

and

$$X_iX_{i+1}X_i = X_{i+1}X_iX_{i+1}, \qquad i \in 1,2,\dots,n-2.$$

We can assume that there exists an open neighborhood of ∞ in L, say U, such that $U \cap K = \emptyset$ and $\forall \alpha \in B[L,K] \exists$ a representing α homeomorphism α' with $\alpha' \mid U = Id_U$. Let $a \in U$ and $Y_1, \dots, Y_n$ be a system of simple paths connecting a with $a_1, \dots, a_n$ (that is, $\forall i = 1,2,\dots,n, \gamma_i \cap K = a_i$

and $\forall i, j = 1, 2, \ldots, n, \gamma_i \cap \gamma_j = a$). We assume that the order $\gamma_1, \ldots, \gamma_n$ is such that in a small disk d_a centered at a segments $\gamma_i \cap d_a, i = 1, \ldots, n$, follow each other (as radii) in the positive direction. Choosing small positively oriented circles around the points $a_1, \ldots, a_n$, we associate with $\gamma_1, \ldots, \gamma_n$ a system $\tilde{\gamma}_1, \ldots, \tilde{\gamma}_n$ of generators of $\pi_1(L - K, a)$, which we call a "good ordered system of generators." $B[L, K]$ acts (faithfully) as an automorphism group on $\pi_1(L - K, a)$. We shall say that $\{X_1, \ldots, X_{n-1}\}$ and $\{\tilde{\gamma}_1, \ldots, \tilde{\gamma}_n\}$ are related if $\forall i = 1, 2, \ldots, n-1$ $X_i(\tilde{\gamma}_{i+1}) = \tilde{\gamma}_i$ (see Figure 1). As usual we denote by $\Delta^2 = \Delta^2(B[L, K])$ the positive generator of the center of $B[L, K]$.

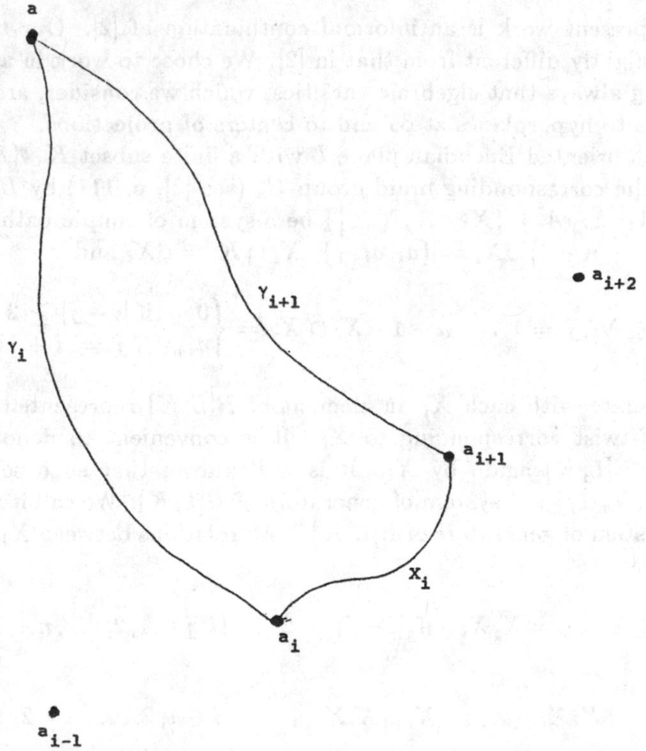

Figure 1

In $B[L,K]$ there is a (normal) semi-group $B_n^+ = B^+[L,K]$ which is generated by all $QX_1Q^{-1}, Q \in B[L,K]$. The definition of $B^+[L,K]$ depends only on the orientation of L (and not of the choice of X_1).

If $\sigma_1, \dots, \sigma_r, \sigma \in B^+[L,K]$ and $\sigma = \sigma_1, \dots, \sigma_r$, we say that $\prod_{i=1}^r \sigma_i$ is a positive factorization of σ or simply a factorization of σ.

A study of factorization properties of elements of the semigroup $B^+[L,K]$ we call "arithmetic of $B[L,K]$" or more generally "arithmetic of braids."

Because $B^+[L,K]$ is not commutative we must introduce an equivalence relation for factorizations. We say that $\sigma = \prod_{i=1}^{r'} \sigma'_i$ is obtained from $\sigma = \prod_{i=1}^r \sigma_i$ by an elementary transformation if $r' = r$ and for some $j \in (1, \dots, r-1)$ we have $\sigma'_j = \sigma_j \sigma_{j+1} \sigma_j^{-1}$, $\sigma'_{j+1} = \sigma_j$ and when $i < j$ or $i > j+1, \sigma'_i = \sigma_i$.

Definition. We say that two factorizations of $\sigma \in B^+[L,K]$ are *equivalent* if one of them could be obtained from another by a finite sequence of elementary transformations and their inverses.

If $\prod_{i=1}^r \sigma_i$ is equivalent to $\prod_{i=1}^{r'} \sigma'_i$ we shall write: $\prod_{i=1}^r \sigma_i \equiv \prod_{i=1}^{r'} \sigma'_i$. (Clearly then, $r = r'$).

Now let E be a plane affine algebraic curve of degree n, $E \subset \mathbb{C}^2, (x,y)$ be affine coordinates in $\mathbb{C}^2$, $\Pi\colon \mathbb{C}^2 \longrightarrow \mathbb{C}^1$ be defined by $\Pi((x,y)) = x$, $\pi = \Pi|_E$. Let $s(E) = \{u \in \mathbb{C}^1 \mid \#\pi^{-1}(u) < n\}$, $a \in \mathbb{C}^1 - s(E)$. There exists a well-defined homomorphism: $\varphi_{E,\Pi}\colon \pi_1(\mathbb{C}^1 - s(E), a) \longrightarrow B[\Pi^{-1}(a), \pi^{-1}(a)]$. We call $\varphi_{E,\Pi}$ "the braid monodromie corresponding to $E \subset \mathbb{C}^2$, Π and a." For an explicit description of $\varphi_{E,\Pi}$ we choose a good ordered system of generators of $\pi_1(\mathbb{C}^1 - s(E), a)$, say $\Gamma_1, \dots, \Gamma_\nu$. Then $\varphi_{E,\Pi}$ is defined by the set $\{\varphi_{E,\Pi}(\Gamma_i), i = 1, 2, \dots, \nu\}$. An important (and simple) fact is that

$$\Delta^2 = \Delta^2(B[\Pi^{-1}(a), \pi^{-1}(a)]) = \Pi_{i=1}^{\nu} \varphi_{E,\Pi}(\Gamma_i). \tag{1}$$

Replacing E by a generic nonsingular curve E' which is very close to E we get a splitting of the points of $s(E)$ into the (stable) branch points of $\Pi|_{E'}\colon E' \longrightarrow \mathbb{C}^1$. That actually means that

$$\forall i = 1, 2, \dots, \nu \quad \varphi_{E,\Pi}(\gamma_i) \in B^+[\Pi^{-1}(a), \pi^{-1}(a)].$$

Thus $\Delta^2 \in B^+[\Pi^{-1}(a), \pi^{-1}(a)]$ and (1) is a factorization of Δ^2 in

$$B^+[\Pi^{-1}(a), \pi^{-1}(a)].$$

It is easy to see that factorizations of Δ^2, which are equivalent to (1), correspond to expressions of $\varphi_{E,\Pi}$ for some other choices of $\{\Gamma_1,\dots,\Gamma_\nu\}$. Thus a class of equivalent factorizations of Δ^2 corresponds to $\varphi_{E,\Pi}$. (This class is unique up to inner automorphisms of $B[\Pi^{-1}(a),\pi^{-1}(a)]$.)

There is an empirical evidence that very often we can find for a singular curve a factorization of Δ^2 (corresponding to braid monodromie) which is remarkably short and natural. We call such factorizations "normal forms of braid monodromies."

Existence of "normal forms" gives a strong indication that an essential part of the (yet to be discovered) theory of singular plane curves is "written" in the language of (positive) factorizations of Δ^2. In other words, this theory is closely related to the arithmetic of braids.

Consider now algebraic surfaces in $\mathbb{C}^3$. In that case we have to introduce two braid monodromies, one of which we call "horizontal braid monodromie, " shortly "h.b.m.", and the second we call "vertical braid monodromie," or "v.b.m."

So let V be an algebraic surface in $\mathbb{C}^3$ of degree n, (x,y,z) be affine coordinates in $\mathbb{C}^3$, $\Pi_1\colon \mathbb{C}^3\longrightarrow\mathbb{C}^2$, $\Pi_2\colon \mathbb{C}^2\longrightarrow\mathbb{C}^1$ projections defined by $\Pi_1((x,y,z))=(x,y)$, $\Pi_2((x,y))=x$, $\pi_1=\Pi_1|_V$. Let

$$
\begin{aligned}
C &= \{v\in\mathbb{C}^2 \mid \#(\pi_1^{-1}(v))<n\},\\
\pi_2 &= \Pi_2|_{C^2},\\
s(C) &= \{u\in\mathbb{C}^1 \mid \#(\pi_2^{-1}(u))<\deg C\},\ a\in\mathbb{C}^1-s(C),\\
\tilde{\mathbb{C}}^1 &= \Pi_2^{-1}(a),\\
\tilde{\mathbb{C}}^2 &= \Pi_1^{-1}(\tilde{\mathbb{C}}^1),\\
\tilde{\Pi}_1 &= \Pi_1|_{\tilde{\mathbb{C}}^2}\ \tilde{\mathbb{C}}^2\longrightarrow\tilde{\mathbb{C}}^1,\\
E &= V\cap\tilde{\mathbb{C}}^2,\\
\tilde{\pi}_1 &= \pi_1|_E,\\
s(E) &= C\cap\tilde{\mathbb{C}}^1,\ b\in\tilde{\mathbb{C}}^1-s(E).
\end{aligned}
$$

Now we can consider two braid monodromies:

$$
\begin{aligned}
\varphi_{E,\tilde{\pi}_1}&\colon \pi_1(\tilde{\mathbb{C}}^1-s(E),b)\longrightarrow B[\tilde{\Pi}_1^{-1}(b),\tilde{\pi}_1^{-1}(b)],\\
\varphi_{C,\pi_2}&\colon \pi_1(\mathbb{C}^1-s(C),a)\longrightarrow B[\Pi_2^{-1}(a),\pi_2^{-1}(a)].
\end{aligned}
$$

We call $\varphi_{E,\tilde{\pi}_1}$ the vertical braid monodromie (v.b.m.) and φ_{C,π_2} the horizontal braid monodromie (h.b.m.) corresponding to V (or more precisely) to $\{V,\Pi_1,\Pi_2,a,b,\}$.

Choosing good ordered systems of generators $\{\delta_1, \ldots, \delta_{\deg C}\}$ for $\pi_1(\tilde{\mathbb{C}}^1 - s(E), b)$ and $\{\Gamma_1, \ldots, \Gamma_\nu\}$ for $\pi_1(\mathbb{C}^1 - s(C), a)$ we get two expressions:

$$\text{(2)} \qquad \Delta^2(B[\tilde{\Pi}_1^{-1}(b), \tilde{\pi}_1^{-1}(b)]) = \prod_{i=1}^{\deg C} \varphi_{E,\tilde{\Pi}_1}(\delta_i),$$

$$\text{(2')} \qquad \Delta^2(B[\Pi_2^{-1}(a), \pi_2^{-1}(a)]) = \prod_{j=1}^{\nu} \varphi_{C,\Pi_2}(\Gamma_j).$$

We say that (2) and (2′) are related if the factorization $\prod_{j=1}^{\nu} \varphi_{C,\Pi_2}(\Gamma_j)$ is explicitly expressed in a good ordered system of generators $\{X_1, \ldots, X_{degC-1}\}$ of $B[\Pi_2^{-1}(a), \pi_2^{-1}(a)]$, related to $\{\delta_1, \ldots, \delta_{\deg C}\}$. If (2) and (2′) are related and if they give some "normal forms" for v.b.m. and h.b.m., we say that we have related normal forms of v.b.m. and h.b.m. of V.

We call a surface in $\mathbb{C}^3$ defined by $a_1x^n + a_2y^n + a_3z^n = 1$ (all a_1, a_2, a_3 not zero) a "Fermat surface." We say that X_n is a generic Fermat surface in $\mathbb{C}^3$ if X_n is a generic surface in $\mathbb{C}^3$ which is very close to a Fermat surface. In [2] we described some related normal forms of v.b.m. and h.b.m. for generic Fermat surfaces (see [2], Thm. 1. §3). Considering a family $\{X(t), t \in [0,1]\}$ of algebraic surfaces in $\mathbb{C}^3$ with $X(0) = V$, $X(1) = X_n$ (a generic Fermat surface), all $X(t)$, $t \neq 0$, are nonsingular, we can obtain (2) and (2′) (v.b.m. and h.b.m. of V) by some transformations ("degenerations") applied to related normal forms of v.b.m. and h.b.m. of X_n. For $t = \epsilon$, $0 < \epsilon \ll 1$, we can reconstruct v.b.m. and h.b.m. of $X(\epsilon)$ applying to (2) and (2′) some simple "rules of regeneration." We will obtain expressions of v.b.m. and h.b.m. of $X(\epsilon)$ which is natural to call "ready to collapse" expressions. Replacing V by $X(\epsilon)$ and (2), (2′) by the corresponding "ready to collapse" expressions, we reduce the question of obtaining (2), (2′) from normal forms of h.b.m. and v.b.m. of X_n to the question of transformations of these normal forms to "ready to collapse" forms. Using this approach we can always be in the realm of nonsingular surfaces where related expressions of v.b.m. and h.b.m. are very closely connected to some finite presentations of B_n. Thus the problem of degeration of X_n to V becomes essentially a problem of transformation of a "universal" finite presentation of B_n to some "ready to collapse" finite presentations (see also [2], §4).

Classification of algebraic surfaces of general type depends very much on studying pluricanonical embeddings of such surfaces. Using generic projections to 3-space we can reduce the problem to studying "pluricanonical" surfaces in $\mathbb{C}^3$ having only ordinary singularities (and possibly also rational double points). Let V (considered above) be such a surface. Then we can assume that $E(= V \cap \tilde{\mathbb{C}}^2)$ is a nodal curve of the main stream[1] (that is it could be degenerated (in $\tilde{\mathbb{C}}^2$) in a union of n lines in general position). In that case we know normal forms for braid monodromies (see [2], §1). Thus transformation of vertical braid monodromies to "ready to collapse" normal forms is a relatively easy problem. Transformation of v.b.m. is actually a transformation of systems of generators of π_1 of the "base" of v.b.m.

To choose generators for π_1 of the "base" of v.b.m. means to choose generators for the braid group of the "fiber" of h.b.m. A hope is that "ready to collapse" normal forms for h.b.m. will be algebraically prescribed by a simplification process applied to 'universal" normal forms of generic Fermat surfaces rewritten in new generators of (the "fiber") braid groups.

As in the case of singular plane curves, we expect existance of a reach braid-arithmetic language carrying essential parts of the theory of algebraic surfaces. It is important already to translate to a braid-arithmetic language known descriptions of different classes of algebraic surfaces.[2]

Usually our explicit data are such that we can degenerate given surfaces to unions of minimal rational surfaces. Projecting everything to $\mathbb{CP}^3$ (or $\mathbb{C}^3$) we see that actually we have to study "big streams of degenerations" starting at nonsingular surfaces in $\mathbb{CP}^3$ and ending at unions of minimal rational surfaces. (This, of course, is a natural generalization of (Severi's)

---

[1]Let V be an affine part of a projection to $\mathbb{CP}^3$ of a surface $Y \subset \mathbb{CP}^N$ and the embedding $Y \subset \mathbb{CP}^N$ corresponds to $|2mK_Y|$. Assume that $m > 1$ is such that $|mK_Y|$ is an infinite linear system without base points. Then $\exists$ two nonsingular curves $S_1, S_2, \in |mK_Y|$, such that S_1 and S_2 intersect transversally. For $i = 1, 2$ we have $2(\text{genus}(S_i)) - 2 = (K_Y + mK_Y).mK_Y < 2mK_Y.mK_Y = 2mK_Y.S_i = \deg S_i$. From these inequalities it follows easily that generic projections, say S'_1, S'_2 of S_1, S_2 to $\mathbb{CP}^2$ could be degenerated to unions of n lines. Then the same property is true for $S'_1 \cup S'_2$, which we can consider as a degeneration of a generic hyperplane section of $\overline{V}$ (projective closure of V).

[2]An example of how we expect to use "braid arithmetic" for classification questions is the following: Big series of algebraic surfaces of general type were constructed in [3]. Constructions of [3] are such that for most underlying topological 4-manifolds obtained there) we feel that moduli spaces (of complex structures) have more than one connected component (for some subseries we actually can prove it). K. Chakiris suggested to distinguish such components by fundamental groups of complements in $\mathbb{CP}^2$ of branch curves of pluricanonical models. These fundamental groups are defined by the corresponding braid monodromies (see [2], §1 and §4).

main stream of degenerations of nodal curves in $\mathbb{CP}^2$ (see [2], §1 and §4)).

So an understanding of the braid-arithmetic material related to such situations is of great interest.

In the present article we begin such a program by considering degenerations of nonsingular hypersurfaces in $\mathbb{C}^3$ to union of a nonsingular hypersurface and planes in general positions. (The corresponding "big stream" starts at a nonsingular hypersurface of degree n and ends at unions of n planes in general position). Algebraically the problem is how to transform related normal forms of v.b.m. and h.b.m. of X_n (generic Fermat surface), obtained in [2], to "ready to collapse" forms, corresponding to splittings of planes from nonsingular surfaces.

To avoid technical details (which will increase too much the length of our article) we will consider here only the case of splitting just one plane from a generic Fermat surface X_n, more precisely degenerations $X_n \to X_{n-1} \cup W$ where W is a plane. This case (combined with the Theorem 2 of §3) plays the main role in the inductive description of the general case.

§1 Some Formulae from the "Arithmetic of Braids"

Denote (as above) by L an oriented Euclidean plane, and let K be a finite subset of L. For a simple path γ in L with $\partial\gamma \subset K$ we will usually denote by γ also the element of $B[L, K]$ represented by a positive half-twist corresponding to γ.

1. Let $K_1 = \{a_1, \ldots, a_\nu\}$ be a subset of K, α be a simple closed (connected) curve in L such that $\alpha \cap K = K_1$ and the bounded domain U_α with $\partial U_\alpha = \alpha$ has no points of K in it. Assume that the given (by indexes) ordering $\{a_1, \ldots, a_\nu\}$ is consistent with the positive orientation of α. It is convenient here to consider indexes in a_i's as residue classes (mod ν). Denote by α_i the part of α between a_i and a_{i+1}. Let $A(K_1, \alpha) = \prod_{i=\nu-1}^{1} \alpha_i (\in B[L, K])$.

It is easy to check that $\alpha_\nu = \alpha_0 = (\prod_{i=\nu-1}^{2} \alpha_i)\alpha_i(\prod_{i=\nu-1}^{2} \alpha_i)^{-1}$, and more generally $\big(A(K_1, \alpha)\big)^{-1}\alpha_i\big(A(K_1, \alpha)\big) = \alpha_{i+1}$. It implies that $\prod_{i=\nu-1}^{1} \alpha_i \approxeq \prod_{i=\nu-1}^{1} \alpha_{i+j}$. We call such transition from $\prod_{i=\nu-1}^{\alpha} \alpha_i$ to $\prod_{i=\nu-1}^{1} \alpha_{i+j}$ a j-cyclic shift (in the factorization $\prod_{i=\nu-1}^{1} \alpha_i$ of $A(K_1, \alpha)$). For $i, j \in (1, 2, \ldots, \nu)$, $i < j-1$, we denote by β_{ij} a simple curve in $L - U_\alpha$ connecting a_i and a_j and lying very close to $U_{r=i}^{j-1}\alpha_r$. It is clear that $\beta_{ij} = (\prod_{r=i}^{j-2} \alpha_r)\alpha_{j-1}(\prod_{r=i}^{j-2} \alpha_r)^{-1}$ (see Figure 2).

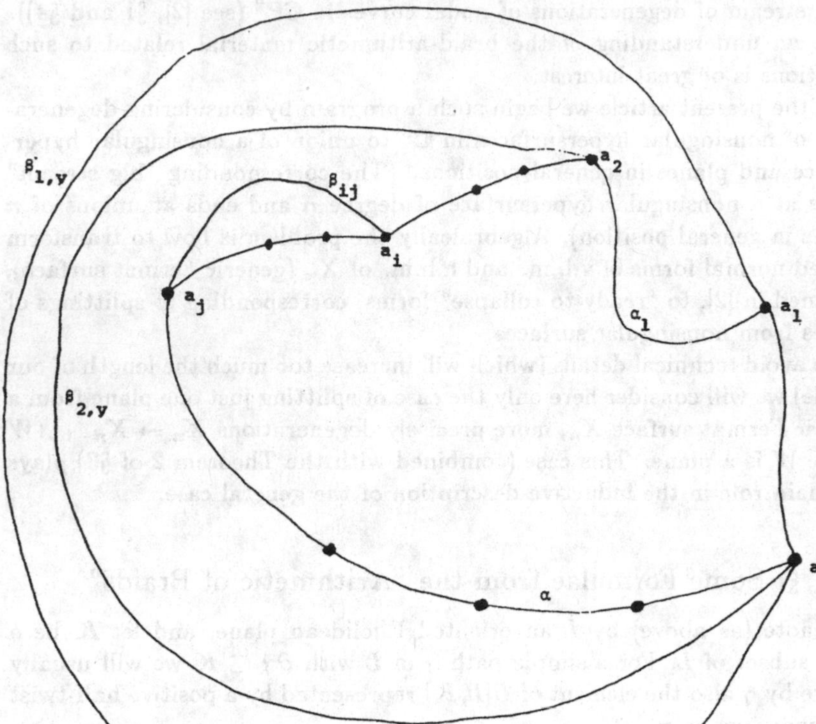

Figure 2

Denote by $A = A(K_1, \alpha)$.

Lemma 1. (i) *We have an equality:*

$$A^{\nu+1} = \alpha_{\nu-1}^3 \prod_{j=\nu-2}^{1} \left(\left(\prod_{l=\nu}^{j+2} \beta_{jl}^2 \right) \cdot \alpha_j^3 \right). \tag{3}$$

(ii) *$\forall k$ denote by $\beta_{i+k,j+k} = A^{-k}\beta_{ij}A^k$. Then*

$$\alpha_{\nu-1}^3 \cdot \prod_{j=\nu-2}^{1} \left(\left(\prod_{l=\nu}^{j+2} \beta_{jl}^2 \right) \cdot \alpha_j^3 \right) \cong \alpha_{\nu+k-1}^3 \cdot \prod_{j=\nu-2}^{1} \left(\prod_{l=\nu}^{j+2} \beta_{j+k,l+k}^2 \right) \cdot \alpha_{j+k}^3$$

(In other words a cyclic shift (in $\prod_{j=\nu-1}^{1} \alpha_i$) transforms the factorization of $A^{\nu+1}$ given by the right side of (3) in an equivalent factorization).

Proof. We present here only a proof for (i) ((ii) is easier). We use induction by ν. Denoting by $A' = \prod_{j=\nu-1}^{2} \alpha_j$ we assume that we know already that

$$(A')^\nu = \alpha_{\nu-1}^3 \prod_{j=\nu-2}^{2} \left(\left(\prod_{l=\nu}^{j+2} \beta_{jl}^2 \right) \cdot \alpha_j^3 \right).$$

An inductive formula for Δ^2 (see [2], §1, p.121) gives us:

$$A^\nu = (A')^{\nu-1} \cdot \left(\prod_{l=\nu}^{3} \beta_{1l}^2 \right) \alpha_1^2.$$

Because A' commutes with $(\prod_{l=\nu}^{3} \beta_{1l}^2)\alpha_1^2$ we can write:

$$\begin{aligned} A^{\nu+1} &= A^\nu \cdot A = (A')^{\nu-1} \cdot \left(\prod_{l=\nu}^{3} \beta_{1l}^2 \right) \alpha_1^2 \cdot A'\alpha_1 \\ &= (A')^\nu \cdot \left(\prod_{l=\nu}^{3} \beta_{1l}^2 \right) \alpha_1^2 \cdot \alpha_1 = \alpha_{\nu-1}^3 \prod_{j=\nu-2}^{1} \left(\left(\prod_{l=\nu}^{j+2} \beta_{jl}^2 \right) \cdot \alpha_j^3 \right). \end{aligned}$$

Q.E.D.

Let us consider now $A^{\nu-1}$. Denoting as above by $A' = \prod_{j=\nu-1}^{2} \alpha_j$ and using $A^{-1}\alpha_i A = \alpha_{i+1}$ we get

$$\begin{aligned} A^{\nu-1} &= A'\alpha_1 A^{\nu-2} = A' A^{\nu-2} \alpha_{\nu-1} = A' \cdot A'\alpha_1 A^{\nu-3} \alpha_{\nu-1} \\ &= (A')^2 A^{\nu-3} \alpha_{\nu-2}\alpha_{\nu-1} = \cdots = (A')^k A^{\nu-k-1} \alpha_{\nu-k} \cdots \alpha_{\nu-1} \\ &= (A')^k \cdot A'\alpha_1 A^{\nu-k-2} \alpha_{\nu-k} \ldots \alpha_{\nu-1} \\ &= (A')^{k+1} A^{\nu-k-2} \alpha_{\nu-k-1}\alpha_{\nu-k} \cdots \alpha_{\nu-1} \\ &= \cdots = (A')^{\nu-1} \alpha_1\alpha_2 \ldots \alpha_{\nu-1}. \end{aligned}$$

Denote by $\overline{A} = \prod_{j=2}^{\nu-1} \alpha_j$ and by $B(\alpha_2, \ldots, \alpha_{\nu-1})$ the subgroup of $B[L, K]$ generated by $\alpha_2, \ldots, \alpha_{\nu-1}$. It is clear that $B(\alpha_2, \ldots, \alpha_{\nu-1})$ is isomorphic to $B_{\nu-1}$ and that $(A')^{\nu-1} = \Delta^2(B(\alpha_2, \ldots, \alpha_{\nu-1})) = (\overline{A})^{\nu-1}$. Now we get

$$(4) \qquad A^{\nu-1} = \overline{A}^{\nu-1} \cdot \alpha_1 \alpha_2 \ldots \alpha_{\nu-1} = \overline{A}^{\nu-1} \cdot \alpha_1 \overline{A} = \overline{A}^{\nu} (\overline{A}^{-1} \alpha_1 \overline{A}).$$

Denote by $\overline{\beta}_{jl}$ a simple path in $\overline{U}_\alpha$ (closure of U_α) such that

$$\partial \overline{\beta}_{jl} = \{a_i, a_j\} = \overline{\beta}_{jl} \cap \alpha.$$

Let $\overline{\alpha} = (\cup_{j=2}^{\nu-1} \alpha_j) \cup \beta_{2,\nu}$ (see Figure 2). $\overline{K}_1 = \{a_\nu, \ldots, a_2\}$. Applying Lemma 1 (i) to $\overline{A} = A(\overline{K}_1, \overline{\alpha})$ we have:

$$(3)_{\nu-1} \qquad (\overline{A})^\nu = \alpha_2^3 \cdot \prod_{j=4}^{\nu} \left(\left(\prod_{l=2}^{j-2} \overline{\beta}_{jl}^2 \right) \cdot \alpha_{j-1}^3 \right).$$

Denote by $\overline{D} = (\overline{A})^\nu$ and by $\mathring{\overline{D}}$ the factorization of $(\overline{A})^\nu$ (product of cubes and squares) given by the right side of $(3)_{\nu-1}$.

It is easy to check that $\overline{A}^{-1} \alpha_1 \overline{A} = \beta_{1,\nu}$. From (4) and $(3)_{\nu-1}$ we get:

$$(4') \qquad A^{\nu-1} = \mathring{\overline{D}} \circ \beta_{1\nu}.$$

Now denote by $A'(-1) = \prod_{j=\nu-2}^{1} \alpha_j$ and by $\overline{A}(-1) = \prod_{j=1}^{\nu-2} \alpha_j$. Arguing as above, we can write:

$$\begin{aligned} A^{\nu-1} &= A^{\nu-2} \cdot \alpha_{\nu-1} \cdot A'(-1) = \alpha_1 \cdot A^{\nu-2} \cdot A'(-1) \\ &= \alpha_1 A^{\nu-3} \cdot \alpha_1 \cdot A'(-1) \cdot A'(-1) = \cdots \\ (4'') \qquad &= \alpha_1 \alpha_2 \ldots \alpha_{\nu-1} (A'(-1))^{\nu-1} \\ &= \overline{A}(-1) \alpha_{\nu-1} (A'(-1))^{\nu-1} \\ &= [\overline{A}(-1) \alpha_{\nu-1} (\overline{A}(-1))^{-1}] \cdot (\overline{A}(-1))^\nu. \end{aligned}$$

Clearly, $\overline{A}(-1) \alpha_{\nu-1} (\overline{A}(-1))^{-1} = \beta_{1,\nu}$.

Let $\overline{D}(-1) = (\overline{A}(-1))^\nu$ and $\mathring{\overline{D}}(-1)$ be the factorization

$$\alpha_1^3 \cdot \prod_{j=3}^{\nu-1} \left(\left(\prod_{l=1}^{j-2} \overline{\beta}_{jl}^2 \right) \cdot \alpha_{j-1}^3 \right) \quad \text{of } \overline{D}(-1)$$

provided by Lemma 1(i). (4)″ now has the form:

$$(4''') \qquad A^{\nu-1} = \beta_{1,\nu} \cdot \overset{\circ}{\overline{D}}(-1).$$

Clearly, $\forall i = 2, \ldots, \nu-2$, α_i commutes with $\beta_{1\nu}$ and $\beta_{1\nu}\alpha_1\beta_{1\nu}^{-1} = \beta_{2\nu}$ (see Figure 2). Thus the sequence

$$\{\beta_{1\nu}\alpha_i\beta_{1\nu}^{-1}, i = \nu-2, \ldots, 1\} = \{\alpha_{\nu-2}, \ldots, \alpha_2, \beta_{2\nu}\}$$

is obtained from $\{\alpha_{\nu-1}, \ldots, \alpha_2\}$ by a "(−1)-cyclic shift" on the curve $\overline{\alpha}$. From Lemma 1 (ii) it follows that the factorization $\beta_{1\nu}(\overset{\circ}{\overline{D}}(-1))\beta_{1\nu}^{-1}$ [3] is equivalent to $\overset{\circ}{\overline{D}}$. From (4‴),

$$\beta_{1\nu} \cdot \overset{\circ}{\overline{D}}(-1) \approx [\beta_{1\nu}(\overset{\circ}{\overline{D}}(-1))\beta_{1\nu}^{-1}] \cdot \beta_{1\nu} \approx \overset{\circ}{\overline{D}} \cdot \beta_{1\nu}$$

and from (4′) we obtain

Lemma 1′. $A^{\nu-1} = \beta_{1\nu} \cdot \overset{\circ}{\overline{D}}(-1) \approx \overset{\circ}{\overline{D}} \cdot \beta_{1\nu}$.

Now let $\hat{D} = A^{\nu-1}$ and $\hat{D}^{\circ}$ be the factorization of $\hat{D}$ equal to $\overset{\circ}{\overline{D}} \cdot \beta_{1\nu} (\approx \beta_{1\nu}\overset{\circ}{\overline{D}}(-1))$. Denote by $\hat{D}^{\circ}(j)$ the factorization of $\hat{D}$ obtained from $\hat{D}^{\circ}$ by the j-cyclic shift in A (that is, $\hat{D}^{\circ}(j) = A^{-j}(\hat{D}^{\circ})A^{j}$).

Lemma 1″. $\forall_j\, \hat{D}^{\circ}(j) \approx \hat{D}^{\circ}$.

Proof. It is enough to consider the case $j = -1$. Clearly, $A(\overset{\circ}{\overline{D}})A^{-1} = \overset{\circ}{\overline{D}}(-1)$. So we have

$$\begin{aligned}\hat{D}^{\circ}(-1) &\approx (A(\overset{\circ}{\overline{D}})A^{-1}) \cdot A\beta_{1\nu}A^{-1} \approx \overset{\circ}{\overline{D}}(-1) \cdot A\beta_{1\nu}A^{-1} \\ &\approx [\overline{D}(-1)A\beta_{1\nu}A^{-1}(\overline{D}(-1))^{-1}] \cdot \overset{\circ}{\overline{D}}(-1).\end{aligned}$$

Because $A\hat{D}A^{-1} = \hat{D} = \beta_{1\nu}\overline{D}(-1)$ we have $\overline{D}(-1)A\beta_{1\nu}A^{-1}(\overline{D}(-1))^{-1} = \beta_{1\nu}$.

---

[3]If $\prod_{i=1}^{r} \sigma_i$ is a factorization of σ we always denote by $\tau(\prod_{i=1}^{r} \sigma_i)\tau^{-1}$ the factorization of $\tau\sigma\tau^{-1}$ equal to $\prod_{i=1}^{r}(\tau\sigma_i\tau^{-1})$.

Finally (using Lemma 1′) we get

$$\hat{D}^{\circ}(-1) \approxeq \beta_{1\nu}\mathring{\tilde{D}}(-1) \approxeq \hat{D}^{\circ}.$$

Q.E.D.

The next lemma is very useful for transformations of factorizations:

Lemma 1‴. *Let $g_{ij} \in B^{+}[L,K]$, $i = 1,\ldots,l$, $j = 1,\ldots,m$, be such that any two g_{ij}, $g_{i'j'}$ with $i > i'$, $j < j'$ commute. Then*

$$\prod_{i=1}^{l}\prod_{j=1}^{m} g_{ij} \approxeq \prod_{j=1}^{m}\prod_{i=1}^{l} g_{ij}.$$

Proof. Let us use induction by l. We can write:

$$\prod_{i=1}^{l}\prod_{j=1}^{m} g_{ij} \approxeq \left(\prod_{i=1}^{l-1}\prod_{j=1}^{m} g_{ij}\right)\cdot\prod_{j=1}^{m} g_{lj} \approxeq \left(\prod_{j'=1}^{m}\prod_{i=1}^{l-1} g_{ij'}\right)\cdot\prod_{j=1}^{m} g_{lj}.$$

It is clear that each $g_{lj}(1 \leq j \leq m)$ commutes with $G_{j'} = \prod_{i=1}^{l-1} g_{ij'}$ when $j' > j$. Thus we have $(\prod_{j'=1}^{m}(\prod_{i=1}^{l-1} g_{ij'}))\cdot\prod_{j=1}^{m} g_{lj} \approxeq \prod_{j=1}^{m-1}[(\prod_{i=1}^{l-1} g_{ij})\cdot(\prod_{j'=j+1}^{m} G_{j'})(g_{lj})(\prod_{j'=j+1}^{m} G_{j'})^{-1})]\cdot(\prod_{i=1}^{l-1} g_{im})\cdot g_{lm} \approxeq \prod_{j=1}^{m}\prod_{i=1}^{l} g_{ij}$.

Q.E.D.

Let

$$b_{jl} = \begin{cases} \alpha_j^3 & \text{if } j - l = 1 \\ \alpha_l^3 & \text{if } l - j = 1 \\ \beta_{jl}^2 & \text{if } |j - l| \geq 2, \end{cases} \qquad \overline{D}(j) = A^{-j}\overline{D}A^j$$

and $\mathring{\overline{D}}(j) = A^{-j}(\mathring{\overline{D}})A^j$ (a factorization of $\overline{D}(j)$). Now we can write $(3)_{\nu-1}$ in the form:

$$(3')_{\nu-1} \qquad \mathring{\overline{D}} = \prod_{j=3}^{\nu}\prod_{l=2}^{j-1} b_{jl}.$$

Let $\tilde{\beta}_{\nu,\nu-1} = A\beta_{1\nu}A^{-1}$. From Lemma 1′ and 1″ it follows that

$$\begin{aligned}\hat{D}^{\circ} &\approxeq \hat{D}^{\circ}(-2) \approxeq A(\hat{D}^{\circ}(-1))A^{-1} \approxeq A\beta_{1\nu}A^{-1}\cdot(A(\mathring{\tilde{D}}(-1))A^{-1}) \\ &\approxeq \tilde{\beta}_{\nu,\nu-1}\cdot\mathring{\tilde{D}}(-2).\end{aligned}$$

Because $\hat{D}\tilde{\beta}_{\nu,\nu-1}\hat{D}^{-1} = A^{\nu-1}\tilde{\beta}_{\nu,\nu-1}A^{-(\nu-1)} = A^{-1}\tilde{\beta}_{\nu,\nu-1}A = \beta_{1\nu}$ we get (using also $\hat{D}^{\circ} = \beta_{1\nu}\mathring{\tilde{D}}(-1)$ (Lemma 1')) that

$$\begin{aligned}\hat{D}^{\circ}\cdot\hat{D}^{\circ} &\approx \hat{D}^{\circ}\cdot\beta_{\nu,\nu-1}\mathring{\tilde{D}}(-2) \approx (\hat{D}\tilde{\beta}_{\nu,\nu-1}\hat{D}^{-1})\cdot\hat{D}^{\circ}\cdot\mathring{\tilde{D}}(-2)\\ &\approx \beta_{1\nu}\cdot\beta_{1\nu}\cdot\mathring{\tilde{D}}(-1)\cdot\mathring{\tilde{D}}(-2),\end{aligned} \tag{5}$$

From $(3')_{\nu-1}$ it follows that

$$\mathring{\tilde{D}}(-1) \approx \prod_{j=2}^{\nu-1}\prod_{l=1}^{j-1} b_{jl},\quad \mathring{\tilde{D}}(-2) \approx \prod_{j=1}^{\nu-2}\prod_{l=0}^{j-1} b_{jl}$$

where by definition $b_{j0} = b_{j\nu}$.

It is easy to see that for $l < l' < j' < j$ the corresponding b_{jl} and $b_{j'l'}$ commute which gives us (by Lemma 1''') $\mathring{\tilde{D}}(-1) \approx \prod_{j=2}^{\nu-1}\prod_{l=1}^{j-1} b_{jl} \approx \prod_{l=1}^{\nu-2}\prod_{j=l+1}^{\nu-1} b_{jl}$. From (5) we get

$$\begin{aligned}\hat{D}^{\circ}\cdot\hat{D}^{\circ} &\approx \beta_{1\nu}\cdot\beta_{1\nu}\cdot\left(\prod_{l=1}^{\nu-2}\prod_{j=l+1}^{\nu-1} b_{jl}\right)\cdot\left(\prod_{j'=1}^{\nu-2}\prod_{l'=0}^{j-1} b_{j'l'}\right)\\ &\approx \beta_{1\nu}\cdot\beta_{1\nu}\cdot\left(\prod_{l=1}^{\nu-2}\prod_{j=l+1}^{\nu-1} b_{jl}\right)\cdot\left(\prod_{l'=1}^{\nu-2}\prod_{j'=0}^{l'-1} b_{j'l'}\right).\end{aligned} \tag{5'}$$

Because for $1 \le l' < l \le \nu-2$, $j \in (l+1,\dots,\nu-1)$, $j' \in (0, l'-1)$, the corresponding $b_{j'l'}$ and b_{lj} commute, we can transform (5′) as follows:

$$\begin{aligned}\hat{D}^{\circ}\cdot\hat{D}^{\circ} &\approx \beta_{1\nu}\cdot\beta_{1\nu}\cdot\prod_{l=1}^{\nu-2}\left(\left(\prod_{j=l+1}^{\nu-1} b_{jl}\right)\cdot\left(\prod_{j'=0}^{l-1} b_{j'l}\right)\right)\\ &\approx \beta_{1\nu}\cdot\beta_{1\nu}\cdot\prod_{l=1}^{\nu-2}\left(\left(\prod_{j=l+1}^{\nu} b_{jl}\right)\cdot\left(\prod_{j=1}^{l-1} b_{jl}\right)\right)\\ &\approx \beta_{1\nu}\cdot\beta_{1\nu}\cdot\prod_{l=1}^{\nu-2} B_l,\end{aligned} \tag{5''}$$

where for $l \ge 2$, $B_l \overset{\text{def}}{=} (\prod_{j=l+1}^{\nu} b_{jl})\cdot(\prod_{j=l}^{l-1} b_{jl})$ and $B_1 \overset{\text{def}}{=} \prod_{j=l+l}^{\nu} b_{jl}$.

Proposition 1. *Denote by* $A_0 = \prod_{i=2}^{\nu-2} \alpha_i$, $\overline{D}_0 = A_0^{\nu-1}$ *and by* $\mathring{\overline{D}}$ *the factorization* $\prod_{j=3}^{\nu-1}\prod_{l=2}^{j-1} b_{jl}$ *of* $\overline{D}_0$ *given by Lemma 1 (i) (applied to the curve* $\left((\cup_{i=\nu-2}^{2}\alpha_i)\cup\beta_{2,\nu-1}\right)$ *containing the set* $\{a_{\nu-1},\ldots,a_2\}$*). Let* $\hat{\beta}_{\nu j} = \beta_{1\nu}(\overline{\beta}_{1j})\beta_{l\nu}^{-1}$*. Then*

$$(\hat{D}^{\circ})^{\nu} \approx \left[\alpha_1^3\cdot(\beta_{1\nu}\alpha_1\beta_{1\nu}^{-1})^3\cdot(\beta_{1\nu}^2\alpha_1\beta_{1\nu}^{-2})^3\cdot\left(\prod_{l=3}^{\nu-1}(\overline{\beta}_{1l}^{2}\cdot\hat{\beta}_{\nu l}^{2})\right)\cdot\mathring{\overline{D}}_0\right]^{n-2}$$

$$\cdot\underbrace{[\beta_{1\nu}\cdot\ldots\beta_{1\nu}]}_{\nu\ times}$$

Proof. Using the formula (5″) above and also the fact that $B_{l-1} = A(B_l)A^{-1} = \hat{D}^{-1}(B_l)\hat{D}$ we can write

$$(\hat{D}^{\circ})^{\nu} \approx (\hat{D}^{\circ}\cdot\hat{D}^{\circ})\cdot(\hat{D}^{\circ})^{n-2} \approx \beta_{1\nu}\cdot\beta_{1\nu}\cdot\left(\prod_{l=1}^{\nu-2}B_l\right)\cdot(\hat{D}^{\circ})^{\nu-2}$$

(5‴)

$$\approx \beta_{1\nu}\cdot\beta_{1\nu}\cdot\prod_{l=1}^{\nu-2}\left[((\hat{D})^{-l-1}B_l(\hat{D})^{l-1})\cdot\hat{D}^{\circ}\right]$$

$$\approx \beta_{1\nu}\cdot\beta_{1\nu}\cdot[B_1\cdot\hat{D}^{\circ}]^{\nu-2}.$$

Let $\hat{b}_{jl} = \overline{A}^{-1}b_{jl}\overline{A}$. Because for $l < l' < j' < j, b_{jl}$ and $b_{j'l'}$ commute, we have also that the corresponding $\hat{b}_{jl}$ and $\hat{b}_{j'l'}$ commute. Using a cyclic 1-shift on the curve $[(\cup_{i=\nu-1}^{2}\alpha_i)\cup\beta_{2\nu}]$, Lemma 1 (ii) and Lemma 1‴ we get from $(3')_{\nu-1}$:

$$\mathring{\overline{D}} \approx \prod_{j=3}^{\nu}\prod_{l=2}^{j-1}b_{jl} \approx \prod_{j=3}^{\nu}\prod_{l=2}^{j-1}\hat{b}_{jl} \approx \prod_{l=2}^{\nu-1}\prod_{j=l+1}^{\nu}\hat{b}_{jl}$$

$$\approx \left(\prod_{j=3}^{\nu}\hat{b}_{j2}\right)\cdot\prod_{l=3}^{\nu-1}\prod_{j=l+1}^{\nu}\hat{b}_{jl} \approx \left(\prod_{j=3}^{\nu}\hat{b}_{j2}\right)\cdot\prod_{j=4}^{\nu}\prod_{l=3}^{j-1}\hat{b}_{jl}.$$

It is possible to see (from Figure 3) that $\hat{b}_{32} = \beta_{2\nu}^3$, $\forall_j = 4,\ldots,\nu$ $\hat{b}_{j2} = \hat{\beta}_{\nu,j-1}^{2}$ and $\forall i,l,\quad j\in(4,\ldots,\nu), l\in(3,\ldots,j-1)$, $\hat{b}_{jl} = b_{j-1,l-1}$.

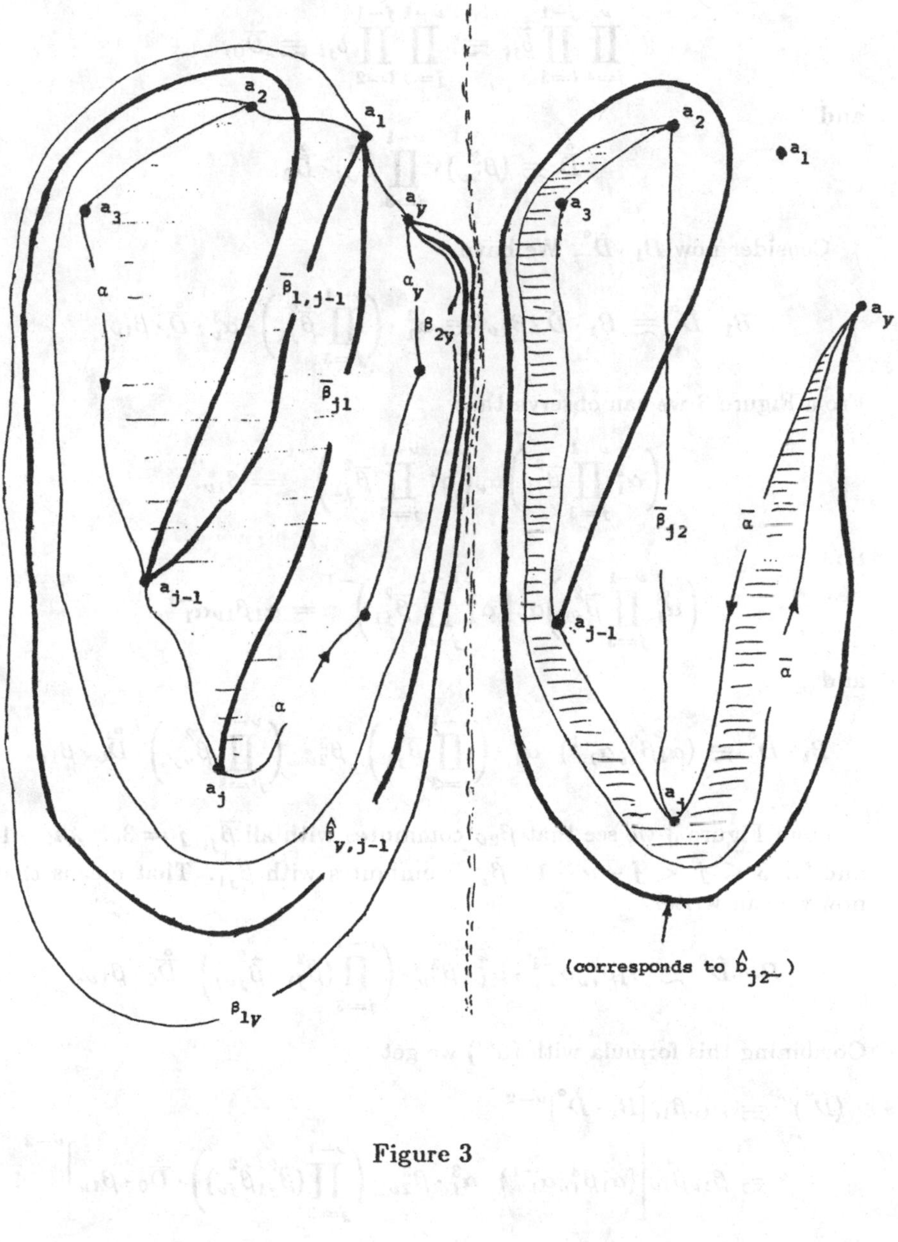
a2
a1
a3
ay
α
β̄1,j-1
αy
β2y
β̄j1
aj-1
α
aj
β̂y,j-1
β1y
a2
a1
a3
ay
β̄j2
ᾱ
aj-1
ᾱ
aj
(corresponds to b̂j2~)

Figure 3

Thus

$$\prod_{j=4}^{\nu}\prod_{l=3}^{j-1}\hat{b}_{jl} \approx \prod_{j=3}^{\nu-1}\prod_{l=2}^{j-1} b_{jl} \approx \mathring{D}_0,$$

and

$$\mathring{D} \approx (\beta_{2\nu}^3)\cdot\prod_{j=3}^{\nu-1}\hat{\beta}_{\nu j}^{3}\cdot\mathring{D}_0.$$

Consider now $B_1\cdot\hat{D}^{\circ}$. We have

$$B_1\cdot\hat{D}^{\circ} \approx B_1\cdot\mathring{D}\cdot\beta_{1\nu} \approx \alpha_1^3\cdot\left(\prod_{j=3}^{\nu-1}\overline{\beta}_{j1}^{2}\right)\cdot\alpha_\nu^3\cdot\mathring{D}\cdot\beta_{1\nu}.$$

From Figure 3 we can observe that

$$\left(\alpha_1^2\prod_{j=3}^{\nu-1}\overline{\beta}_{j1}^{2}\right)\alpha_\nu\left(\alpha_1^2\prod_{j=3}^{\nu-1}\overline{\beta}_{j1}^{2}\right)^{-1} = \beta_{1\nu}.$$

So

$$\left(\alpha_1^3\prod_{j=3}^{\nu-1}\overline{\beta}_{j1}^{2}\right)\alpha_\nu\left(\alpha_1^3\prod_{j=3}^{\nu-1}\overline{\beta}_{j1}^{2}\right)^{-1} = \alpha_1\beta_{1\nu}\alpha_1^{-1}.$$

and

$$B_1\cdot\hat{D}^{\circ} \approx (\alpha_1\beta_{1\nu}^3\alpha_1^{-1})\cdot\alpha_1^3\cdot\left(\prod_{j=3}^{\nu-1}\overline{\beta}_{j1}^{2}\right)\cdot\beta_{2\nu}^3\cdot\left(\prod_{j'=3}^{\nu-1}\hat{\beta}_{\nu j'}^{2}\right)\cdot\mathring{D}_0\cdot\beta_{1\nu}.$$

From Figure 3 we see that $\beta_{2\nu}$ commutes with all $\overline{\beta}_{jl}, j = 3,\ldots,\nu-1$, and for $3 \leq j' < j-\nu-1$ $\hat{\beta}_{\nu j'}$ commutes with $\overline{\beta}_{j1}$. That means that now we can write

$$B_1\cdot\hat{D}^{\circ} \approx \alpha_1\beta_{1\nu}^3\alpha_1^{-1}\cdot\alpha_1^3\cdot\beta_{2\nu}^3\cdot\left(\prod_{j=3}^{\nu-1}(\overline{\beta}_{j1}^{2}\cdot\hat{\beta}_{j\nu}^{2})\right)\cdot\mathring{D}_0\cdot\beta_{1\nu}.$$

Combining this formula with (5‴) we get

$$\begin{aligned}(\hat{D}^{\circ})^\nu &\approx \beta_{1\nu}\beta_{1\nu}[B_1\cdot\hat{D}^{\circ}]^{\nu-2}\\ &\approx \beta_{1\nu}\beta_{1\nu}\left[(\alpha_1\beta_{1\nu}^3\alpha_1^{-1})\cdot\alpha_1^3\cdot\beta_{2\nu}^3\cdot\left(\prod_{j=3}^{\nu-1}(\overline{\beta}_{j1}^{2}\hat{\beta}_{j\nu}^{2})\right)\cdot\mathring{D}_0\cdot\beta_{1\nu}\right]^{\nu-2}.\end{aligned}$$

It is easy to check that $\beta_{1\nu}$ commutes with $\overline{D}_0$ and with $\overline{\beta}_{j1}^2\hat{\beta}_{j\nu}$, $j=3,\ldots,\nu-1$. A simple direct verification shows also that $\beta_{1\nu}$ commutes with $[\alpha_1\beta_{1\nu}^3\alpha_1^{-1}\cdot\alpha_1^3\cdot\beta_{2\nu}^3]$ (see Lemma 1^{iv} below). Thus $\beta_{1\nu}$ commutes with $[\alpha_1\beta_{1\nu}^3\alpha_1^{-1}\cdot\alpha_1^3\cdot\beta_{2\nu}^3\cdot(\prod_{j=3}^{\nu-1}(\overline{\beta}_{j1}^2\hat{\beta}_{j\nu}^2))\cdot\overline{D}_0]$ and we can write:

$$(5'''')\quad (\hat{D}^\circ)^\nu \approx \left[(\alpha_1\beta_{1\nu}^3\alpha_1^{-1})\cdot\alpha_1^3\cdot\beta_{2\nu}^3\cdot\left(\prod_{j=3}^{\nu-1}(\overline{\beta}_{j1}^2\cdot\hat{\beta}_{j\nu}^2)\right)\cdot\mathring{D}_0\right]^{\nu-2}$$

$$\cdot\underbrace{[\beta_{1\nu}\cdot\ldots\cdot\beta_{1\nu}]}_{\nu\text{ times}}$$

Because $\alpha_1\beta_{1\nu}\alpha_1=\beta_{1\nu}\alpha_1\beta_{1\nu}$ we have $\alpha_1\beta_{1\nu}\alpha_1^{-1}=\beta_{1\nu}^{-1}\alpha_1\beta_{1\nu}$. From Figure 3 we see that $\beta_{2\nu}=\beta_{1\nu}\alpha_1\beta_{1\nu}^{-1}$. So

$$\begin{aligned}\alpha_1\beta_1^3\alpha_1^{-1}\cdot\alpha_1^3\cdot\beta_{2\nu}^3 &\approx \beta_{1\nu}^{-1}\alpha_1^3\beta_{1\nu}\cdot\alpha_1^3\cdot\beta_{1\nu}\alpha_1^3\beta_{1\nu}^{-1}\\ &\approx \beta_{1\nu}^{-1}[\alpha_1^3\cdot\beta_{1\nu}\alpha_1^3\beta_{1\nu}^{-1}\cdot\beta_{1\nu}^2\alpha_1^3\beta_{1\nu}^{-2}]\beta_{1\nu}.\end{aligned}$$

To get from (5'''') the statement of the Proposition 1 it is enough to prove the following (actually important)

Lemma 1''''. *Let X,Y be two braids such that $XYX=YXY$. Then $\forall$ integer k*

$$(X^{-1}YX)^3\cdot Y^3\cdot(XYX^{-1})^3\approx X^k[(X^{-1}YX)^3\cdot Y^3\cdot(XYX^{-1})^3]X^{-k}.$$

Proof of Lemma 1''''. It is enough to consider the case $k=1$. In that case we have $Y^3\cdot XY^3X^{-1}\cdot X^2Y^3X^{-2}\approx[(Y^3\cdot XY^3X^{-1})(X^2Y^3X^{-2})(Y^3XY^3X^{-1})^{-1}]\cdot Y^3\cdot XY^3X^{-1}$. It is easy to check (using $XYX=YXY$) that $(Y^3XY^3X^{-1})(X^2YX^{-2})(Y^3XY^3X^{-1})^{-1}=X^{-1}YX$. Thus

$$Y^3\cdot XY^3X^{-1}\cdot X^2Y^3X^{-2}\approx X^{-1}Y^3X\cdot Y^3\cdot XY^3X^{-1}.$$

Q.E.D.

Using Lemma 1'''' with $X=\beta_{1\nu}$, $Y=\alpha_1$ we replace in (5'''') the expression $\alpha_1\beta_{1\nu}^3\alpha_1^{-1}\cdot\alpha_1^3\cdot\beta_{2\nu}^3$ by $\alpha_1^3\cdot(\beta_{1\nu}\alpha_1\beta_{1\nu}^{-1})^3\cdot(\beta_{1\nu}^2\alpha_1\beta_{1\nu}^{-2})^3$. This finishes the proof of the Proposition 1. Q.E.D.

2. Now let us consider a second subset $K_2 \subset K$ with a simple closed curve $\alpha' \supset K_2$ (defined as $\alpha(\supset K_1)$ above). Assume that $K_1 \cap K_2 = \emptyset$, $\alpha \cap \alpha' = \emptyset$ and $\#(K_2) = \#(K_1) = \nu$. Let $K_2 = \{a'_1, \ldots, a'_\nu\}$ with the order in this sequence to be consistent with the positive orientation of α'. Let β_ν be a simple closed curve in L connecting α_ν with α'_ν and such that $\beta_\nu \cap K = \{\alpha_\nu, \alpha'_\nu\} = \beta_\nu \cap (\alpha \cup \alpha')$. Denote by

$$A_1 = A(K_1, \alpha),$$
$$A_2 = A(K_2, \alpha'),$$
$$\beta_i = (A_1A_2)^{\nu-i}(\beta_\nu)(A_1A_2)^{-(\nu-i)}$$

and by $P(K_1, K_2, \alpha, \alpha', \beta_\nu)$ the (product)-expression $\prod_{i=\nu}^{1} \beta_i$.

Lemma 2. *Let $P = P(K_1, K_2, \alpha, \alpha', \beta)$. Then $\forall_k$ the factorizations $P = \prod_{i=\nu}^{1} \beta_i$ and $(A_1A_2)^k P (A_1A_2)^{-k} = \prod_{i=\nu}^{1} [(A_1A_2)^k \beta_i (A_1A_2)^{-k}]$ are equivalent. (In other words, P is invariant with respect to simultaneous cyclic shifts in α and α').*

Proof. It is enough to consider the case $k = 1$. For $i = 1, \ldots, \nu$, let $\tilde{\beta}_i = (A_1A_2)\beta_i(A_1A_2)^{-1}$. Clearly, $\forall i = 1, \ldots, \nu - 1$, $\beta_i = \tilde{\beta}_{i+1}$ and

$$\prod_{i=\nu}^{1} \tilde{\beta}_i \approx \left(\prod_{i=\nu}^{2} \tilde{\beta}_i\right) \cdot \tilde{\beta}_i \approx \left(\prod_{i=\nu-1}^{1} \beta_i\right) \cdot \tilde{\beta}_1$$
$$\approx \left[\left(\prod_{i=\nu-1}^{1} \beta_i\right)(\tilde{\beta}_1)\left(\prod_{i=\nu-1}^{1} \beta_i\right)^{-1}\right] \cdot \prod_{i=\nu-1}^{1} \beta_i.$$

A direct verification shows that

$$\left(\prod_{i=\nu-1}^{1} \beta_i\right)\tilde{\beta}_1\left(\prod_{i=\nu-1}^{1} \beta_i\right)^{-1} = \beta_\nu.$$

Q.E.D.

Definition. Let $g_1, \ldots, g_m \in B^+[L, K]$, $g = g_1 \cdot \ldots \cdot g_m$. Using a sequence of equivalences

$$\begin{aligned}\prod_{i=1}^{m} g_i &\approx \left(\prod_{i=2}^{m}(g_1(g_i)g_1^{-1})\right)\cdot g_1\\ &\approx \left(\prod_{i=3}^{m}(g_1g_2(g_i)g_2^{-1}g_1^{-1})\right)\cdot g_1g_2g_1^{-1}\cdot g_1 \approx \cdots\\ &\approx \prod_{r=m}^{1}(g_i\cdots g_{r-1}(g_r)g_{r-1}^{-1}\cdots g_1^{-1})\end{aligned}$$

we get from the factorization $\prod_{i=1}^{m} g_i$ of g a new factorization

$$\prod_{r=m}^{1}(g_i\cdots g_{r-1}(g_r)g_{r-1}^{-1}\cdots g_1^{-1}).$$

We say that this last factorization is obtained from $\prod_{i=1}^{m} g_i$ by Δ-*transformation* (it corresponds to the so-called element Δ of B_m (see [1])). The inverse transformation we call Δ^{-1}-*transformation.*

Now denote by

$$\begin{aligned}A_1^* &= \prod_{i=1}^{\nu-1}\alpha_i,\\ A_2^* &= \prod_{i=1}^{\nu-1}\alpha_i',\\ \beta_j^* &= (A_1^*A_2^*)^{-(\nu-j)}(\beta_\nu)(A_1^*A_2^*)^{\nu-j}\end{aligned}$$

and by $P^*(K_1, K_2, \alpha, \alpha', \beta_\nu)$ the (product)-expression $\prod_{j=1}^{\nu}\beta_j^*$.

Lemma 2′. *Let* $P^* = P^*(K_1, K_2, \alpha, \alpha', \beta_\nu)$. *Then* $P^* \approx P$.

Proof. A direct verification shows that applying Δ-transformation to P we get P^*. Q.E.D.

3. Now let $K = \cup_{r=1}^{m} K^{(r)}$, $\forall r_1, r_2, r_1 \neq r_2$, $K^{(r_1)} \cap K^{(r_2)} = \emptyset$, $\alpha^{(r)}$, $r = 1, \ldots, m$, be mutually disjoint simple closed connected curves,

$K \cap \alpha^{(r)} = K^{(r)}$, such that the interior of each $\alpha^{(r)}$ has no points of K in it. Let $K^{(r)} = \{a_1^{(r)}, \ldots, a_\nu^{(r)}\}$ (ν the same for all $r = 1, 2, \ldots, m$), β be a simple closed connected curve in L containing $\{a_\nu^{(1)}, \ldots, a_\nu^{(m)}\}$ and such that the order in the sequence $\{a_\nu^{(1)}, \ldots, a_\nu^{(m)}\}$ is consistent with the positive orientation of β_ν. Assume also that $\beta_\nu \cap (\cup_{r=1}^m \alpha^{(r)}) = \{a_\nu^{(1)}, \ldots, a_\nu^{(m)}\}$.

Denote by $\beta_\nu^{(r)}$ the part of β_ν between $a_\nu^{(r)}$ and $a_\nu^{(r+1)}$ and by $P_r = P(K^{(r)}, K^{(r+1)}, \alpha^{(r)}, \alpha^{(r-1)}, \beta_\nu^{(r)})$(see p. 216). Let $D^{(r)}, r = 1, 2, \ldots, m$, be a system of mutually disjoint open disks in L such that $\forall r = 1, \ldots, m$ $D^{(r)} \supset (\alpha^{(r)} \cup (\text{interior of } \alpha^{(r)})$. Let $\overline{a}(r) \in D^{(r)}, \overline{K} = \{\overline{a}^{(1)}, \ldots, \overline{a}^{(m)}\}$ and denote by G a subgroup of $B[L, K]$ defined as follows:

$$G = \{g \in B[L, K] \mid g \text{ contains a homeomorphism } \tilde{g} \text{ with } \tilde{g}(D^{(r)}) \subseteq D^{(r')}, r, r' = 1, 2, \ldots, m\}.$$

There is an evident homomorphism $\varphi \colon G \longrightarrow B[L, \overline{K}]$ (φ corresponds to contractions of $D^{(r)}$'s to $\overline{a}^{(r)}$'s).

Lemma 3. *Let $p_r = \varphi(P_r)$. Then $\{p_r, r = 1, 2, \ldots, m\}$ is a good ordered system of generators of $B[L, \overline{K}]$ and there exists a monomorphism $\psi \colon B[L, \overline{K}] \longrightarrow G$ with $\varphi \circ \psi = Id_{B[L,\overline{K}]}$ and $\psi(p_r) = P_r$, $r = 1, 2, \ldots, m$.*

Proof. A direct verification shows that $P_r P_{r+1} P_r = P_{r+1} P_r P_{r+1}$, $r = 1, 2, \ldots, m-1$, and $\forall r, r' \in (1, 2, \ldots, m), r - r' \geq 2$, $P_r P_{r'} = P_{r'} P_r$. Lemma 3 immediately follows from these facts. Q.E.D.

Let $A^{(r)} = A(K^{(r)}, \alpha^{(r)})$, $\beta_i^{(r)} = (A^{(r)} A^{(r+1)})^{\nu - i} \beta_\nu^{(r)} (A^{(r)} A^{(r+1)})^{-(\nu - i)}$, $r = 1, \ldots, m-1$. In these notations $P_r = \prod_{i=\nu}^{1} \beta_i^{(r)}$.

Lemma 3′.

$$P_1 \ldots P_{m-1} \approx \prod_{i=\nu}^{1} \prod_{r=m-1}^{1} \beta_i^{(r)}.$$

Proof. From Figure 4 we see that $\beta_i^{(r)}$, $\beta_{i'}^{(r')}$ with $i' < i$, $r' > r$ commute.

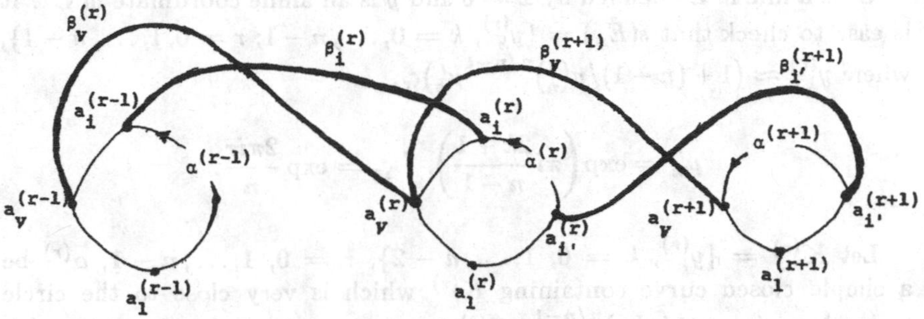

Figure 4

Now from Lemma 1‴ we get

$$\prod_{r=m-1}^{1}\prod_{i=\nu}^{1}\beta_i^{(r)} \approx \prod_{i=\nu}^{1}\prod_{r=m-1}^{1}\beta_i^{(r)}.$$

Q.E.D.

§2 Generic Fermat Surfaces

We shall use notations from the introduction. Consider a Fermat surface F_n defined by $z^n - y^n + x^n = -1$ and let X_n be a generic Fermat surface close to F_n. It is convenient to take X_n very close to an "intermediate" surface V_n, defined by

$$z^n + \epsilon(1+\rho x^{n-1})z - y^n + x^n = -1, \tag{6}$$

where $0 < \rho \ll \epsilon \ll 1$.

Take $a = \{x = 0\} \in \mathbb{C}^1$. Then $\tilde{\mathbb{C}}^2$ is defined by $x = 0$ in $\mathbb{C}^3$, (y, z) are affine coordinates in $\tilde{\mathbb{C}}^2$ and $E_n = V_n \cap \tilde{\mathbb{C}}^2$ is defined (in $\tilde{\mathbb{C}}^2$) by

$$z^n + \epsilon z - y^n = -1. \tag{7}$$

$\tilde{\mathbb{C}}^1$ is a line in $\tilde{\mathbb{C}}^2$ defined by $x = 0$ and y is an affine coordinate in $\tilde{\mathbb{C}}^1$. It is easy to check that $s(E_n) = \{y_k^{(r)},\ k = 0, \ldots, n-1;\ r = 0, 1, \ldots, n-1\}$, where $y_k^{(r)} \approx (1 + (n-1)/n(\frac{\varepsilon}{n})^{n/n-1}\mu_k')\varsigma_r$,

$$\mu_k' = \exp\left(\pi i \frac{2k+1}{n-1}\right), \quad \varsigma_r = \exp\frac{2\pi i r}{n}.$$

Let $K^{(r)} = \{y_k^{(r)},\ k = 0,\ 1, \ldots, n-2\}$, $r = 0,\ 1, \ldots, n-1$, $\alpha^{(r)}$ be a simple closed curve containing $K^{(r)}$ which is very close to the circle $|y - \varsigma_r| = (n-1)/n(\frac{\varepsilon}{n})^{n/n-1}$, $D^{(r)}$ be the open disk bounded by $\alpha^{(r)}$, $\sigma^{(r)}$ be the shortest curve in $\tilde{\mathbb{C}}^1 - \cup_{r=0}^{n-1} D^{(r)}$ connecting $y_0^{(r)}$ with $y_0^{(r+1)}$, $\sigma^{(r)} \cap s(E_n) = \{y_0^{(r)}, y_0^{(r+1)}\}$, $\alpha_k^{(r)}$ be the part of $\alpha^{(r)}$ between $y_k^{(r)}$ and $y_{k+1}^{(r)}$ (we consider indexes k here as residue classes (mod n)),

$$A_{n-1}^{(r)} = \prod_{j=n-3}^{0} \alpha_j^{(r)} (\in B[\tilde{\mathbb{C}}^1, s(E_n)]).$$

Introduce a good ordered system of generators of $B[\tilde{\mathbb{C}}^1, s(E_n)]$ as follows.

$$X_{kr} = \begin{cases} \alpha_k^{(r)} & \text{if } k = 0, 1, \ldots, n-3; \\ & \quad r = 0, 1, \ldots, n-1 \\ (A_{n-1}^{(r)})^{-(n-2)} \sigma^{(r)} (A_{n-1}^{(r)})^{n-2} & \text{if } k = n-2, \\ & \quad r = 0, 1, \ldots, n-2. \end{cases}$$

Define $X_{kr} < X_{k'r'}$ iff either $r > r'$ or $r = r'$, $k > k'$. Let $b = (y = M) \in \tilde{\mathbb{C}}^1$, $M \gg 1$, $\overline{\delta}_{oo}$ be the shortest path in $\tilde{\mathbb{C}}^1 - \cup_{r=0}^{n-1} D^{(r)}$ connecting b with $y_0^{(0)}$. For $k = 0, 1, \ldots, n-2$; $r = 0, 1, \ldots, n-1$ define

$$\overline{\delta}_{kr} = (\overline{\delta}_{oo}) \prod_{X_{k'r'} > X_{kr}} X_{k'r'} \quad \text{(the products in the descent order).}^4$$

Let $\{\delta_{kr}, k = 0, \ldots, n-2; r = 0, 1, \ldots, n-1\}$ be the system of generators of $\pi_1(\tilde{\mathbb{C}}^1 - s(E_n), b)$ corresponding to $\{\overline{\delta}_{kr}\}$. Define $\delta_{kr} < \delta_{k'r'}$, iff either

---

[4]We use evident action of $B[\tilde{\mathbb{C}}^1, s(E_n)]$ on paths connecting b with points of $s(E_n)$.

$r > r'$ or $r = r'$, $k > k'$. Evidently $\{\delta_{kr}\}$ is a good ordered system of generators of $\pi_1(\tilde{\mathbb{C}}^1 - s(E_n), b)$ related to $\{X_{kr}\}$.

For $k,l \in (0,1,\dots,n-2)$, $k < l$, $r \in (0,1,\dots,n-1)$ denote by $Z_{kl}^{(r)} = (\prod_{k'=k}^{l-2} X_{k'r})X_{l-1,r}(\prod_{k'=k}^{l-2} X_{k'r})^{-1}$.

It follows from Lemma 1 that

$$(3') \qquad (A_{n-1}^{(r)})^n = X_{n-3,r}^3 \cdot \prod_{j=n-4}^{0} \left(\prod_{l=n-2}^{j+2} (\tilde{Z}_{jl}^{(r)})^2 \cdot X_{jr}^3 \right).$$

Denote by $D_{n-1}^{(r)} = (A_{n-1}^{(r)})^n$ and the expression (product of cubes and squares) on the right part of (3′) by $\mathring{D}_{n-1}^{(r)}$. Let $\beta_{n-2}^{(r)} = (A_{n-1}^{(r+1)})(X_{n-2,r})(A_{n+1}^{(r+1)})^{-1}$,

$$\beta_i^{(r)} = (A_{n-1}^{(r)} A_{n-1}^{(r+1)})^{n-2-i} (\beta_{n-2}^{(r)}) (A_{n-1}^{(r)} A_{n-1}^{(r+1)})^{-(n-2-i)}$$

and $P_{n-1}^{(r)}$ be the (product) expression $\prod_{i=n-2}^{0} \beta_i^{(r)}$.

Consider now $\tilde{\Pi}_1^{-1}(b)$ and $\tilde{\pi}_1^{-1}(b)$. Clearly z is an affine coordinate in $\tilde{\Pi}_1^{-1}(b)$ and $\tilde{\pi}_1^{-1}(b) = \{z_j, j = 0,1,\dots,n-1\}$ where $z_j \approx M\varsigma_j$.

Denote by K_j the braid from $B[\tilde{\Pi}_1^{-1}(b), \tilde{\pi}_1^{-1}(b)]$ corresponding to $\langle z_j, z_{j+1}\rangle$ (segment of a straight line in $\tilde{\Pi}_1^{-1}(b)$ connecting z_j with z_{j+1}). Consider indexes in K_j's as residue classes (mod n). Using the proof of Theorem 1 of [2] we can easily deduce the following fact:

Theorem 1. *Related normal forms of v.b.m. and h.b.m. of X_n (generic surface which is very close to V_n (see (6)) are given by:*

$$(7)_v \qquad \varphi_{E_n}, \tilde{\pi}_1(\delta_{kr}) = K_{k-r} \quad (v.b.m.)$$

and

$$(7)_h \qquad \Delta^2(B[\Pi_2^{-1}(a), \pi_2^{-1}(a)] = \prod_{r=0}^{n-1} (\mathring{D}_{n-1}^{(r)})^{n-1} \cdot \left(\prod_{r'=n-2}^{0} P_{n-1}^{(r)} \right)^n$$

(normal form of h.b.m. of X_n related to $(7)_v$).

§3 Addition of Generic Planes

Let C' be a plane curve, $C' = C_1 \cup C_2, C_1, C_2$ algebraic curves intersecting transversally, $K = \Pi_2^{-1}(a) \cap C'$, $K^{(i)} = \Pi_2^{-1}(a) \cap C_i\, i = 1,2$, $\#K^{(i)} = \nu_i, \overset{\circ}{\Delta}{}_i^2$ be a factorization of $\Delta^2(B[\tilde{\mathbb{C}}^1, K^{(i)}])$ corresponding to the braid monodromie of C_i (more precisely of $C_i, \prod_2, a$). Let U_1, U_2 be two disjoint open (topological) disks in $\tilde{\mathbb{C}}^1 = \Pi_2^{-1}(a)$ with $U_i \supset K^{(i)}$, $b_i \in \partial U_i$, $i = 1,2$, $b \notin U_1 \cup U_2$, e_1 and e_2 be two simple paths from b to b_1, b_2 respectively, $e_1 \cap e_2 = b$, $e_i \cap (U_1 \cup U_2) = b_i$, $i = 1,2, \{\delta_1^{(i)}, \ldots, \delta_{\nu_i}^{(i)}\}$ be a good ordered system of paths in $\overline{U}_i$ (closure of U_i) connecting b_i with all the elements of $K^{(i)}, Y_{kl}(k = 1,2,\ldots,\nu_1;\ l = 1,2,\ldots,\nu_2)$ be an element of $B[\tilde{\mathbb{C}}^1, K]$ represented by a positive half-twist corresponding to the simple curve $\delta_k^{(1)} \cup e_1 \cup e_2 \cup \delta_l^{(2)}$. Identify $B[\tilde{\mathbb{C}}^1, K^{(i)}] = B[U_i, K^{(i)}]$ with a subgroup of $B[\tilde{\mathbb{C}}^1, K]$ corresponding to the embedding $U_i \subset \tilde{\mathbb{C}}^1$. Now we can consider $\overset{\circ}{\Delta}{}_1^2$ and $\overset{\circ}{\Delta}{}_2^2$ as certain (product) expressions in $B[\tilde{\mathbb{C}}^1, K]$. It is not difficult to prove that the braid monodromie of $C'(C', \Pi_2, a)$ could be expressed as the following factorization of $\Delta^2(B[\tilde{\mathbb{C}}^1, K])$:

$$\Delta^2 = \Delta^2(B[\tilde{\mathbb{C}}^1, K]) = \overset{\circ}{\Delta}{}_1^2 \cdot \left(\prod_{k=1}^{\nu_1} \prod_{l=1}^{\nu_2} Y_{kl}^2 \right) \cdot \overset{\circ}{\Delta}{}_2^2. \tag{8}$$

Consider now an algebraic surface V in $\mathbb{C}^3$ which has only ordinary singularities and let C and E be defined as in the introduction. Let W be a generic plane in $\mathbb{C}^3$, $D = \prod_1(V \cap W)$, $L = \tilde{\mathbb{C}}^2 \cap W$, $V' = V \cup W$, $C' = C \cup D$, $E' = E \cup L$. We want to describe how braid monodromies of V, namely $\varphi_{E,\tilde{\pi}_1}, \varphi_{C,\pi_2}$, are changed when we go from V to V'.

Because E and L are in general position, the formula (8) gives an immediate guidance for the vertical braid monodromie of V'. But C and D have many tangency points and $C \cup D$ could have triple points (coming from $C \cap D$), so for $C' = C \cup D$ the formula (8) has to be essentially modified. Clearly E is a curve with only nodes, so the corresponing $\varphi_{E,\tilde{\pi}_1}(\delta_i)$, $i = 1,2,\ldots,\deg C$ (we use notations from p. 202) have the form

$$\varphi_{E,\tilde{\pi}_1}(\delta_i) = Q_i K_1^{\rho_i} Q_i^{-1}, \quad \rho_i = 1 \text{ or } 2,$$

K_1 is a "good generator" of $B[\tilde{\Pi}_1^{-1}(b), \tilde{\Pi}_1^{-1}(b) \cap E]$, and v.b.m. of V could be expressed as follows:

$$(9)\qquad \Delta^2(B[\tilde{\Pi}_1^{-1}(b), \tilde{\Pi}_1^{-1}(b)\cap E]) = \prod_{i=1}^{m} Q_i K_1^{\rho_i} Q_i^{-1}, m = \deg C.$$

Let $\overset{\circ}{\Delta}{}^2_C$ (resp. $\overset{\circ}{\Delta}{}^2_D$) be a factorization of $\Delta^2(B[\tilde{\mathbb{C}}_1, \tilde{\mathbb{C}}_1 \cap C])$ (resp. of $\Delta^2(B[\tilde{\mathbb{C}}_1, \tilde{\mathbb{C}}_1 \cap D])$ corresponding to braid monodromie of C (resp. of D). If C and D would have only transversal intersections the braid monodromie of $C \cup D$ would be given by a formula similar to (8), namely as follows:

$$(8')\qquad \Delta^2 = \overset{\circ}{\Delta}{}^2_C \cdot \left(\prod_{k=1}^{m}\prod_{l=1}^{n} Y_{kl}^2\right) \cdot \overset{\circ}{\Delta}{}^2_D,$$

where Y_{kl} is defined ss above (for $C_1 \cup C_2$). We shall describe how to transform (8′) to a formula really giving an expression of the braid monodromie of $C' = C \cup D$.

First of all, $D \subset \mathbb{C}^2$ and $E \subset \tilde{\mathbb{C}}^2$ could be considered as having the same braid monodromies (they both are generic hyperplane sections of V). So we take an open disk $U_2 \subset \tilde{\mathbb{C}}^1$ with $U_2 \supset (\tilde{\mathbb{C}}^1 \cap D)$, $U_2 \cap (\tilde{\mathbb{C}}^1 \cap C) = \emptyset$ and identify $B[\tilde{\Pi}_1^{-1}(b), \tilde{\Pi}_1^{-1}(b) \cap E]$ with a subgroup G_D of $B[\tilde{\mathbb{C}}^1, \tilde{\mathbb{C}}^1 \cap (C \cup D)]$, corresponding to $B[U_2, \tilde{\mathbb{C}}^1 \cap D]$ and to embedding $U_2 \subset \tilde{\mathbb{C}}^1$. Now we can write

$$\overset{\circ}{\Delta}{}^2_D \approx \prod_{i=1}^{m} Q_i K_1^{\rho_i} Q_i^{-1}, \qquad \rho_i = 1 \text{ or } 2.$$

Clearly for each $k = 1, 2, \ldots, m$ $\prod_{l=1}^{n} Y_{kl}^2$ commutes with any $g \in G_D$. That means that we can rewrite (8′) as follows:

$$(8'')\qquad \begin{aligned} \Delta^2 &= \overset{\circ}{\Delta}{}^2_C \cdot \left(\prod_{k=1}^{m}\prod_{l=1}^{n} Y_{kl}^2\right) \cdot \overset{\circ}{\Delta}{}^2_D \\ &\approx \overset{\circ}{\Delta}{}^2_C \cdot \prod_{k=1}^{m}\left(\left(\prod_{l=1}^{n} Y_{kl}^2\right) \cdot Q_k K_1^{\rho_k} Q_k^{-1}\right). \end{aligned}$$

Let $D \cap \tilde{\mathbb{C}}^1 = \{a'_1, \ldots, a'_n\}$ and v_k be a simple path in U_2 connecting two elements of $D \cap \tilde{\mathbb{C}}^1$, say $c'_{k'}, c''_{k'}$, and corresponding to $Q_k K_1 Q_k^{-1}$. It is easy to rewrite each (product) expression $\prod_{l=1}^{n} Y_{kl}^2$ in a form $\prod_{l=1}^{n} \hat{Y}_{kl}^2$

($\approx \prod_{l=1}^{n} Y_{kl}^2$) where $\hat{Y}_{kl}$ corresponds to a path $\hat{\delta}_{kl}$ and $\hat{\delta}_{k1} \cap v_k = c'_k$, $\hat{\delta}_{k2} \cap v_k = c''_k$, $\forall l = 3, \ldots, n$, $\hat{\delta}_{kl} \cap v_k = 0$, $v_k^{-1} \hat{Y}_{k2} v_k = \hat{Y}_{k1}$. Consider now two possibilities:

1) $\rho_k = 1$. In this case we write:

$$\left(\prod_{l=1}^{n} Y_{kl}^2\right) \cdot v_k \approx \left(\prod_{l=1}^{n} \hat{Y}_{kl}^2\right) \cdot v_k \approx \hat{Y}_{k1}^2 \hat{Y}_{k2}^2 v_k \cdot \prod_{l=3}^{n} \hat{Y}_{kl}^2$$

$$\approx \hat{Y}_{k1}^2 \cdot v_k \cdot (v_k^{-1} \hat{Y}_{k2}^2 v_k) \cdot \prod_{l=3}^{n} \hat{Y}_{kl}^2 \approx \hat{Y}_{k1}^2 \cdot v_k \cdot Y_{k1}^2 \cdot \prod_{l=3}^{n} \hat{Y}_{kl}^2$$

$$\approx \hat{Y}_{k1}^2 \cdot Y_{k1}^2 \cdot (\hat{Y}_{k1}^{-2} v_k \hat{Y}_{k1}^2) \cdot \prod_{l=3}^{n} \hat{Y}_{kl}^2.$$

Replacing $\hat{Y}_{k1}^2 \cdot \hat{Y}_{k1}^2$ by $\hat{Y}_{k1}^4$ we get a factor corresponding to a tangency point of C and D.

(2) $\rho_k = 2$. Now we write:

$$\left(\prod_{l=1}^{n} Y_{kl}^2\right) \cdot v_k^2 \approx \hat{Y}_{k1}^2 Y_{k2}^2 v_k^2 \cdot \prod_{l=3}^{n} \hat{Y}_{kl}^2.$$

Considering $(Y_{k1}^2 Y_{k2}^2 v_k^2)$ as one factor we get a factor corresponding to a triple point of $C \cup D$ (coming from $C \cap D$).

After these purely algebraic remarks we present (without a proof) the following.

Theorem 2. *Let V, C, E, W, D, L, C', E' be as above, $K^{(1)} = C \cap \tilde{\mathbb{C}}^1$, $K^{(2)} = D \cap \tilde{\mathbb{C}}^1$, $K = K^{(1)} \cup K^{(2)}$, U_1 and U_2 be two mutually disjoint open (topological) disks in $\tilde{\mathbb{C}}^1$ with $U_i \supset K^{(i)}, i = 1, 2$. Take $b \notin U_1 \cup U_2$, $b_i \subset \partial U_i$, and let e_1, e_2 be two simple paths from b to b_1 and b_2 respectively, $e_1 \cap e_2 = b$, $e_i \cap (\overline{U}_1 \cup \overline{U}_2) = b_i$, $i = 1, 2$. Let $K^{(1)} = \{a_1, \ldots, a_m\}$, $\{e_{11}, \ldots, e_{1m}\}$ be a good ordered system of paths in $\overline{U}_1$ connecting b_1 with $a_1, \ldots, a_m$ respectively, $\overline{\delta}_k = e_1 \cup e_{1k}$, $k = 1, 2, \ldots, m$, $\{\delta_1, \ldots, \delta_m\}$ be a good ordered system of generators of $\pi_1(\tilde{\mathbb{C}}^1 - K^{(1)}, b)$ corresponding to $\{\overline{\delta}_1, \ldots, \overline{\delta}_m\}$, $\varphi_{E,\tilde{\Pi}_1}(\delta_k) = w_k^{\rho_k}$, $k = 1$ or 2, where w_k denotes also a simple path connecting two elements of*

$E \cap \tilde{\Pi}_1^{-1}(b)$, $d = W \cap \tilde{\Pi}_1^{-1}(b)$, $\{L_1, \ldots, L_n\}$ be a good ordered system of paths in $\tilde{\Pi}_1^{-1}(b)$ connecting d with the points of $\tilde{\Pi}_1^{-1}(b) \cap E$.

Then $\exists$ an (orientation preserving) homeomorphism ψ: $\tilde{\Pi}_1^{-1}(b) \longrightarrow U_2$ with $\psi(E \cap \tilde{\Pi}_1^{-1}(b)) = K^{(2)}$, $\psi(d) = b_2$, such that v.b.m. and h.b.m. could be described as follows: v.b.m. $\varphi_{E',\tilde{\Pi}_1}(\delta_i) = \varphi_{E,\tilde{\Pi}_1}(\delta_i)$, $i = 1, 2, \ldots, m$, (where $B[\tilde{\Pi}_1^{-1}(b), E \cap \tilde{\Pi}_1^{-1}(b)]$ is identified with a subgroup of $B[\tilde{\Pi}_1^{-1}(b), E' \cap \tilde{\Pi}_1^{-1}(b)]$ by using a domain U with $U \supset (E \cap \tilde{\Pi}_1^{-1}(b))$, $d \notin U$) and

$$\varphi_{E',\tilde{\Pi}_1}(\delta'_r) = L_r^2, \quad r = 1, 2, \ldots, n$$

where $\{\delta'_1, \ldots, \delta'_r\}$ is a good ordered system of generators of $\pi_1(\tilde{\mathbb{C}}^1 - K^{(2)}, b)$ corresponding to a system of paths $\{e_2 \cup \psi(L_r),\ r = 1, 2, \ldots, n\}$, $\pi_1(\tilde{\mathbb{C}}^1 - K^{(1)}, b)$ and $\pi_1(\tilde{\mathbb{C}}^1 - K^{(2)}, b)$ are identified with subgroups of $\pi_1(\tilde{\mathbb{C}}^1 - K, b)$ by using systems $\{\overline{\delta}_1, \ldots, \overline{\delta}_m\}$ and $\{e_2 \cup \psi(L_r),\ r = 1, 2, \ldots, n\}$ so that $\{\delta_1, \ldots, \delta_m, \delta'_1, \ldots, \delta'_n\}$ is a good ordered system of generators of $\pi_1(\tilde{\mathbb{C}}^1 - K, b)$; h.b.m. is given by

$$\Delta^2(B[\tilde{\mathbb{C}}^1, K]) = \mathring{\Delta}_C^2 \cdot \prod_{k=1}^{n} \left[T_k \cdot \left(\prod_{l=3}^{n} \hat{Y}_{kl}^2 \right) \right], \tag{10}$$

where

a) *$\mathring{\Delta}_C^2$ is a factorization of $\Delta^2(B[\tilde{\mathbb{C}}_1, K^{(1)}])$ corresponding to the braid monodromie of C, $B[\tilde{\mathbb{C}}_1, K^{(1)}] = B[U_1, K^{(1)}]$ is identified with a subgroup of $B[\tilde{\mathbb{C}}_1, K]$ by using embedding $U_1 \subset \tilde{\mathbb{C}}^1$;*

b) *Each $\hat{Y}_{kl}$ corresponds to the path $\delta_k \cup e_2 \cup e_{2kl}$, $k = 1, \ldots, m$, $l = 1, \ldots, n$, where $\forall k = 1, \ldots, m$ $\{e_{2kl}, l = 1, \ldots, n\}$ is a good ordered system of paths in U_2 connecting b_2 with the points of $K^{(2)}$ and such that $\forall l = 3, \ldots, n$, $e_{2kl} \cap \psi(w_k) = 0$, for $l = 1, 2$ $e_{2kl} \cap \psi(w_k) =$ (the end of e_{2kl}), $(e_{2k2})\psi_*(w_k) = e_{2k1}$ (ψ_*: $B[\tilde{\Pi}_1^{-1}(b), E \cap \tilde{\Pi}_1^{-1}(b)] \longrightarrow B[U_2, K^{(2)}]$ is the canonical isomorphism corresponding to ψ);*

c) $$T_k = \begin{cases} \hat{Y}_{k1}^4 \cdot (\hat{Y}_{k1}^{-2} \psi_*(w_k) \hat{Y}_{k1}^2) & \text{for } \rho_k = 1 \\ (\hat{Y}_{k1}^2 \hat{Y}_{k2}^2 (\psi_*(w_k))^2) & \text{for } \rho_k = 2 \end{cases}$$

(for $\rho_k = 1$ we consider T_k as having two factors: $\hat{Y}^4_{k1}$ and $\hat{Y}^{-2}_{k1}\psi_*(w_k)\hat{Y}^2_{k1}$ and for $\rho_k = 2T_k$ is considered as one factor).

The idea of the proof of Theorem 2 is to take W very close to $\tilde{\mathbb{C}}^2$ $(= \Pi_1^{-1}\Pi_2^{-1}(a))$ and to use projections for constructing ψ.

Now we want to apply Theorem 2 to the case when $V = X_{n-1}$ a generic Fermat surface which is very close to V_{n-1} defined by

$$(6)^\sim \qquad z^{n-1} + \epsilon(1+\rho x^{n-2})z - y^{n-1} + x^{n-1} = -1,$$

$$0 < \rho \ll \epsilon \ll 1$$

(see §2, p. 219) and

$$W = \{z - c = 0\}, \quad 1 \ll c \notin \mathbb{R}.$$

Let $E_{n-1} = V_{n-1} \cap \tilde{\mathbb{C}}^2$. As in §2 we can write

$$s(E_{n-1}) = \{\tilde{y}_m^{(k)},\ m = 0,1,\dots,n-3;\ k = 0,1,\dots,n-2\},$$

$$\tilde{y}_m^{(k)} \approx \left(1 + \frac{n-2}{n-1}\left(\frac{\epsilon}{n-1}\right)^{n/n-1}\eta'_m\right)\mu_k$$

where

$$\eta'_m = \exp\left(\pi i\frac{2m+1}{n-2}\right), \quad \mu_k = \exp\frac{2\pi i k}{n-1}.$$

Introduce $A^{(k)}_{n-2}$, $D^{(k)}_{n-2}$, $\mathring{D}^{(k)}_{n-2}$, $P^{(k')}_{n-2}$, $k = 0,1,\dots,n-2$, $k' = 0,1,\dots,n-3$ by the same way as $A^{(r)}_{n-1}$, $D^{(r)}_{n-1}$, $\mathring{D}^{(r)}_{n-1}$, $P^{(r')}_{n-1}$, $r = 0,1,\dots,n-1$, $r' = 0,1,\dots,n-2$, where introduced in §2 (we simply go everywhere from n to $n-1$).

Let $D = \prod_1(V_{n-1} \cap W)$. Clearly $D \cap \tilde{\mathbb{C}}_1 = \{\tilde{y}^{(i)}, i = 0,1,\dots,n-2\}$ which $\tilde{y}^{(i)} \approx c\mu_i$. Let $\tilde{K} = s(E_{n-1}) \cup (D \cap \tilde{\mathbb{C}}_1)$. Denote by $\delta_{ok,i}$ $(k, i = 0,1,\dots,n-2)$ the path in $\tilde{\mathbb{C}}^1$ consisting of the following three parts: (1) segment of a straight line from $\tilde{y}^{(i)} \approx c\mu_i$ to $2\mu_i$; (2) for $i \neq k$ an arc on the circle $|y| = 2$ from $2\mu_i$ to $2\mu_k$ where we go counter-clockwise if $k > i$ and clockwise if $k < i$; (3) segment of a straight line connecting $2\mu_k$ with $\tilde{y}_0^{(k)}$.

Let $Y_{ok,i}$ be the braid in $B[\tilde{\mathbb{C}}^1, \tilde{K}]$ corresponding (as a half twist) to $\delta_{ok,i}$ and $\forall m = 0, 1, \ldots, n-3$, $Y_{mk,i} = (A^{(k)}_{n-2})_T^{-m} Y_{ok,i} (A^{(k)}_{n-2})^m$. Now take $M > 0$ with $1 \ll M \ll c$, and $b = \{y = M\} (\in \tilde{\mathbb{C}}^1)$. Clearly

$$\Pi_1^{-1}(b) \cap V_{n-1} = \{\tilde{z}_i, i = 0, 1, \ldots, n-2\}$$

with $\tilde{z}_i \approx M\mu_i$ and $\Pi_1^{-1}(b) \cap W = C$. (We consider z as an affine coordinate in $\Pi_1^{-1}(b)$). Denote by $\tilde{K}_i$ (resp. $\tilde{K}_{o,o}$) the braid in $B[\Pi_1^{-1}(b), \Pi_1^{-1}(b) \cap (V_{n-1} \cup W)]$ corresponding to the segment of a straight line connecting $\tilde{z}_i$ with $\tilde{z}_{i+1}$ (resp. c with $\tilde{z}_0$) and by $\tilde{K}_{oi}$ the braid

$$\Big(\prod_{i'=n-3}^{0} \tilde{K}_{i'} \Big)^i \tilde{K}_{o,o} \Big(\prod_{i'=n-3}^{0} \tilde{K}_{i'} \Big)^{-i}.$$

Introduce a good ordered system of paths $\{\bar{\bar{\delta}}_{mk}, m = 0, \ldots, n-3; k = 0, 1, \ldots, n-2\}$ by the same way as $\{\bar{\bar{\delta}}_{kr}, k = 0, 1, \ldots, n-2; r = 0, 1, \ldots, n-2\}$ was introduced in §2. ($\bar{\bar{\delta}}_{mk}$ connects b with $\tilde{y}^{(k)}_m$ and $\bar{\bar{\delta}}_{mk} < \bar{\bar{\delta}}_{m'k'}$ iff either $k > k'$ or $k = k', m > m'$). Let $\{\tilde{\delta}_{mk}, m = 0, \ldots, n-3;\ k = 0, 1, \ldots, n-2\}$ be generators of $\pi_1(\tilde{\mathbb{C}}^1 - s(E_{n-1}), b)$ corresponding to $\{\bar{\bar{\delta}}_{mk}\}$. From Theorem 1 (§2) we get that vertical and horizontal braid monodromies of X_{n-1} are given by:

$$(7)_v^{\sim}\ v.b.m.: \qquad \varphi_{E_{n-1}, \tilde{\Pi}_1}(\tilde{\delta}_{mk}) = \tilde{K}_{m-k}, \text{ and}$$

$$(7)_h^{\sim}\ h.b.m.: \Delta^2\Big(B\Big[\Pi_2^{-1}(a), s(E_{n-1})\Big]\Big) = \prod_{k=0}^{n-2} (\mathring{D}^{(k)}_{n-2})^{n-2} \cdot \Big(\prod_{k'=n-3}^{0} p^{(k')}_{n-2} \Big)^n.$$

Now let $\delta'_{mk} = (\delta'_{mk})(A^{(k)}_{n-2})^k$ (braid acting on a path). From $(7)_v^{\sim}$ we get

$$(11) \qquad \varphi_{E_{n-1}, \tilde{\Pi}_1}(\delta'_{mk}) = \tilde{K}_m.$$

Denoting by δ'_o an element from $\pi_1(\tilde{\mathbb{C}}^1 - \tilde{K}, b)$ defined by the shortest path from b to $\tilde{y}^{(0)}$, by L_i the braid in $B[\tilde{\mathbb{C}}^1, \tilde{K}]$ corresponding to $\langle \tilde{y}^{(i)}, \tilde{y}^{(i+1)} \rangle$, by $\delta'_i = (\delta'_{i-1})L_{i-1}^{-1}$, $i = 1, \ldots, n-2$, by

$$Z_{mk,i} = (A_{n-2}^{(k)})^k Y_{n-3-m,k;i} (A_{n-2}^{(k)})^k$$

and by $E' = \tilde{\mathbb{C}}^2 \cap (V_{n-1} \cup W)$, we can easily deduce from Theorem 2 that v.b.m. and h.b.m. of $X_{n-1} \cup W$ are given by:

$$(12)_v v.b.m. : \quad \begin{array}{l} \varphi_{E',\tilde{\Pi}_1}(\delta'_{mk}) = \tilde{K}_m, \quad m = 0, \ldots, n-3; \quad k = 0, \ldots, n-2; \\ \varphi_{E',\tilde{\Pi}_1}(\delta'_i) = \tilde{K}^2_{oi}, \quad i = 0, 1, \ldots, n-2. \end{array}$$

$$(12)_h h.b.m. : \qquad \Delta^2(B[\tilde{\mathbb{C}}_1, \tilde{K}]) = \mathring{\Delta}^2_{C_{n-1}} \prod_{k=n-2}^{0} \mathring{R}_k$$

where

$$\mathring{\Delta}^2_{C_{n-1}} = \prod_{k=0}^{n-2} (\mathring{D}^{(k)}_{n-2})^{n-2} \cdot \Big(\prod_{k'=n-3}^{0} P^{(k')}_{n-2} \Big)^{n-1}$$

(see $(7)_{\widetilde{h}}$),

$$(13) \qquad R_k = \prod_{m=0}^{n-3} \Big(\prod_{i=0}^{n-2} Z^2_{mk,i} \Big) L_m$$

and $\mathring{R}_k$ denotes the (product) expression in the right side part of the following formula

$$(14)\; R_k = \prod_{m=0}^{n-3} \left[\Big(\prod_{i=0}^{m-1} Z^2_{mk,i} \Big) \cdot Z^4_{mk,m} \cdot (Z^{-2}_{mk,m}(L_m) Z^2_{mk,m}) \cdot \prod_{i=m+2}^{n-2} Z^2_{mk,i} \right].$$

($\prod$'s which don't make sense we always replace by the identity element.)

Our aim now is to obtain another form for $(12)_h$ (which is closer in some sense to $(7)_h$). The first step is a transformation for (14) (e.g., for $\mathring{R}_k$). We observe from Figure 5a that $\forall \tilde{m} < m$, $\prod_{i=m+1}^{n-2} Z^2_{\tilde{m}k,i}$ commutes with $(\prod_{i=0}^{m-1} Z^2_{mk,i})\cdot Z^4_{mk,m}$, $\prod_{i=m+r}^{n-2} Z^2_{mk,i}$ commutes with each $Z_{mk,m+r'}$, $0 \leq r' < r$, $\forall m' > m$, $\prod_{i=0}^{m} Z^2_{m'k,i}$ commutes with $\prod_{i=m+1}^{n-2} Z^2_{mk,i}$, $\prod_{i=0}^{m-r} Z^2_{m'k,i}$ commutes with each $Z_{mk,m-\bar{r}}$, $0 \leq \bar{r} < r$ and $\prod_{i=m+1}^{n-2} Z^2_{\tilde{m}k,i}$ commutes with $\prod_{i=0}^{m} Z^2_{m'k,i} (\tilde{m} < m < m')$.

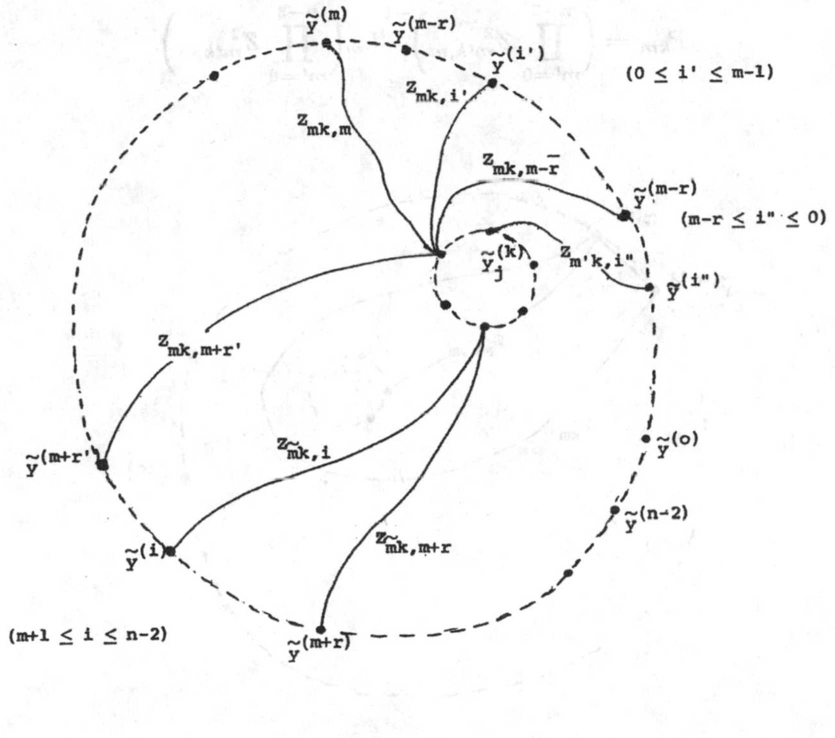

Figure 5a

Denote by $P_{km} = (\prod_{\tilde{m}=0}^{m-1} \prod_{i=m+1}^{n-2} Z^2_{\tilde{m}k,i})(\prod_{m'=m+1}^{n-3} \prod_{i=0}^{m} Z^2_{m'k,i})^{-1} \cdot [Z^{-2}_{mk,m} L_m Z^2_{mk,m}](\prod_{m'=m+1}^{n-3} \prod_{i=0}^{m} Z^2_{m'k,i})(\prod_{\tilde{m}=0}^{m-1} \prod_{i=m+1}^{n-2} Z^2_{\tilde{m}k,i})^{-1}$.

Using the observations made above about commutations it is not difficult to show that actually:

$$\mathring{R}_k \cong \prod_{i=0}^{n-3} \left[Z^4_{ik,i} \cdot \prod_{m=i+1}^{n-3} Z^2_{mk,i} \cdot P_{ki} \cdot \prod_{m=0}^{i-1} Z^2_{mk,i+1} \right]. \tag{15}$$

From Figure 5b we observe that

$$P_{km} = \left(\prod_{m'=0}^{n-3} Z^2_{m'k,m} \right)^{-1} L_m \left(\prod_{m'=0}^{n-3} Z^2_{m'k,m} \right).$$

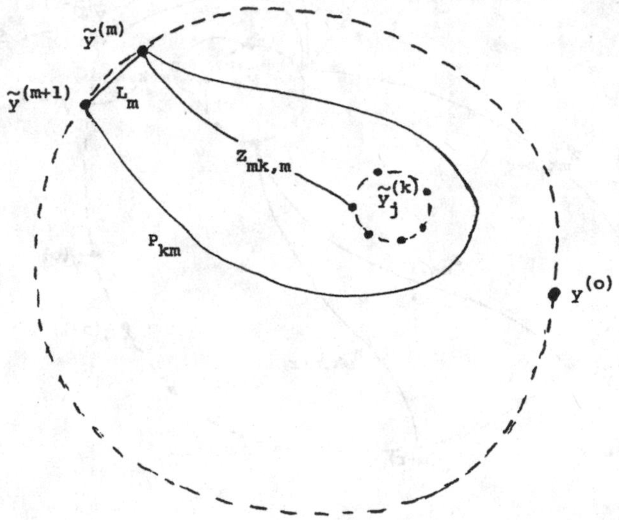

Figure 5b

From Figure 5c we see that $\forall i \leq k-1$

$$(16)\qquad \left(\prod_{i'=i}^{k-1} P_{ki'}\right)^{-1} Z_{mk,i}\left(\prod_{i'=i}^{k-1} P_{ki'}\right) = (A_{n-2}^{(k)})^{-(k-i)(n-2)} Z_{mk,k} (A_{n-2}^{(k)})^{(k-i)(n-2)},$$

and when $i \geq k$

$$(16')\qquad \left(\prod_{i'=k}^{i-1} P_{ki'}\right) Z_{mk,i}\left(\prod_{i'=k}^{i-1} P_{ki'}\right)^{-1} = (A_{n-2}^{(k)})^{-(k-i)(n-2)} Z_{mk,k} (A_{n-2}^{(k)})^{(k-i)(n-2)}$$

Denote by $\overline{Z}_{mk,k}^{(i)} = (A_{n-2}^{(k)})^{-(k-i)(n-2)} Z_{mk,k} (A_{n-2}^{(k)})^{(k-i)(n-2)}$.

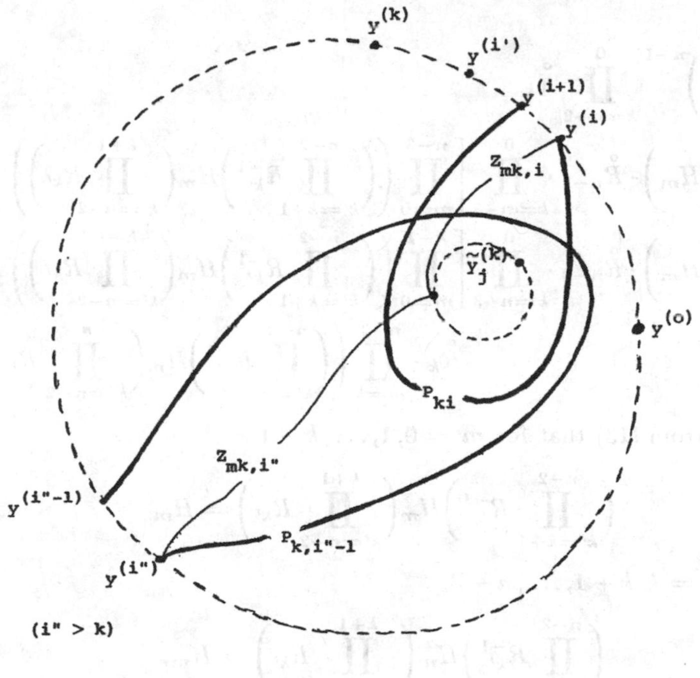

Figure 5c

Let us move now in (15) all P_{ki}'s with $i \leq k-1$ to the left end and all P_{ki}'s with $i \geq k$ to the right end. Using (16), (16′) we get from (15):

$$(17) \qquad \mathring{R}_k \approxeq \prod_{i=0}^{k-1} P_{ki} \cdot W_k \cdot \prod_{i=k}^{n-3} P_{ki},$$

where

$$W_k = \prod_{i=0}^{n-3}\left[(\overline{Z}^{(i)}_{ik,k})^4 \cdot \prod_{m=i+1}^{n-3} (\overline{Z}^{(i)}_{mk,k})^2 \cdot \prod_{m=0}^{i-1} (\overline{Z}^{(i+1)}_{mk,k})^2\right].$$

For $m = 0, 1, \ldots, n-3$ denote by $H_m = P^{(m)}_{n-2}$. From Lemma 3 it follows that $(\prod_{m=n-3}^{0} H_m)^{n-1} \approxeq (\prod_{m=0}^{n-3} H_m)^{n-1}$.

Now from $(12)_h$ we get:

$$(12') \qquad \Delta^2(B[\tilde{\mathbb{C}}_1, \tilde{K}] = \prod_{k=0}^{n-2} (\mathring{D}^{(k)}_{n-2})^{n-2} \cdot \left(\prod_{m=0}^{n-3} H_m\right)^{n-1} \prod_{k=n-2}^{0} \mathring{R}_k.$$

We have:

$$\left(\prod_{m=0}^{n-3} H_m\right)^{n-1} \prod_{k=n-2}^{0} \mathring{R}_k$$

$$\approxeq \left(\prod_{m=0}^{n-3} H_m\right) \cdot \mathring{R}_{n-2} \cdot \prod_{k=n-3}^{0}\left[\prod_{m=0}^{n-3}\left(\left(\prod_{k'=k+1}^{n-2} R_{k'}^{-1}\right) H_m \left(\prod_{k'=n-2}^{k+1} R_{k'}\right)\right) \cdot \mathring{R}_k\right]$$

$$\approxeq \left(\prod_{m=0}^{n-3} H_m\right) \cdot \mathring{R}_{n-2} \cdot \prod_{k=n-3}^{0}\left[\prod_{m=0}^{k-1}\left(\left(\prod_{k'=k+1}^{n-2} R_{k'}^{-1}\right) H_m \left(\prod_{k'=n-2}^{k+1} R_{k'}\right)\right)\right.$$

$$\left. \cdot \mathring{R}_k \cdot \prod_{m=k}^{n-3}\left(\left(\prod_{k'=k}^{n-2} R_{k'}^{-1}\right) H_m \left(\prod_{k'=n-2}^{k} R_{k'}\right)\right)\right].$$

It follows from (13) that for $m = 0, 1, \ldots, k-1$

$$\left(\prod_{k'=k+1}^{n-2} R_{k'}^{-1}\right) H_m \left(\prod_{k'=n-2}^{k+1} R_{k'}\right) = H_m$$

and for $m = k, k+1, \ldots, n-3$

$$\left(\prod_{k'=k}^{n-2} R_{k'}^{-1}\right) H_m \left(\prod_{k'=n-2}^{k+1} R_{k'}\right) = H_m.$$

Thus

$$\begin{aligned}(18)\quad \left(\prod_{m=0}^{n-3} H_m\right)^{n-1} \prod_{k=n-2}^{0} \mathring{R}_k &\approx \prod_{k=n-2}^{0} \left[\left(\prod_{m=0}^{k-1} H_m\right)\cdot \mathring{R}_k \cdot \left(\prod_{m=k}^{n-3} H_m\right)\right] \\ &\approx \prod_{k=n-2}^{0} \left[\left(\prod_{m=0}^{k-1} H_m\right)\cdot \left(\prod_{i=0}^{k-1} P_{ki}\right)\cdot W_k \cdot \prod_{i=k}^{n-3} P_{ki} \cdot \prod_{m=k}^{n-3} H_m\right].\end{aligned}$$

It is easy to see from Figure 6a that for $i = 0, \ldots, k-2$

$$\left(\prod_{m=i+1}^{k-1} H_m\right) P_{ki} \left(\prod_{m=i+1}^{k-1} H_m\right)^{-1} = P_{i+1,i}$$

and for $i = k, \ldots, n-3$

$$\left(\prod_{m=k}^{i} H_m\right)^{-1} P_{ki} \left(\prod_{m=k}^{i} H_m\right) = P_{i+1,i}.$$

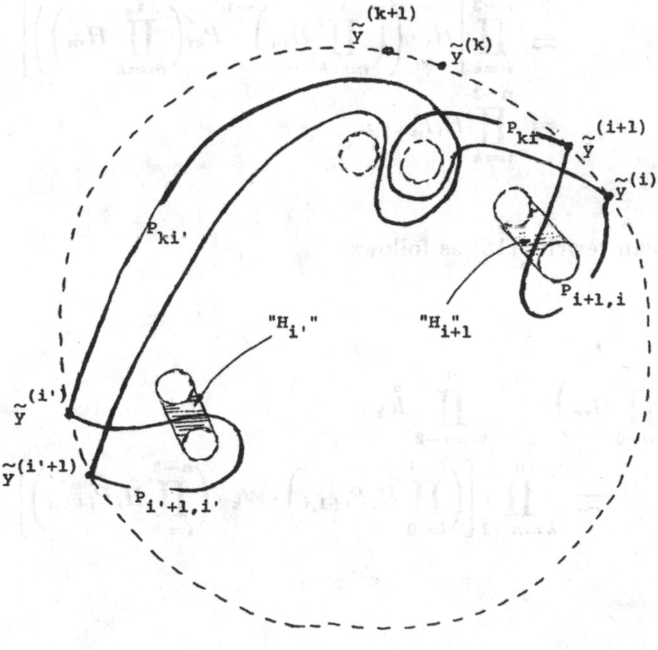

Figure 6a

From these equalities we get

$$\begin{aligned}&\left(\prod_{m=0}^{k-1} H_m\right)\cdot\left(\prod_{i=0}^{k-1} P_{ki}\right)\\ &\approx\left[\prod_{i=0}^{k-2}\left(H_i\cdot\left(\prod_{m=i+1}^{k-1} H_m\right)P_{ki}\left(\prod_{m=i+1}^{k-1} H_m\right)^{-1}\right)\right]\cdot H_{k-1}\cdot P_{k,k-1}\\ &\approx\prod_{i=0}^{k-1} H_iP_{i+1,i}\end{aligned}$$

and

$$\begin{aligned}&\left(\prod_{i=k}^{n-3} P_{ki}\right)\cdot\left(\prod_{m=k}^{n-3} H_m\right)\\ &\qquad\approx\prod_{i=k}^{n-3}\left[H_i\cdot\left(\left(\prod_{m=k}^{i} H_m\right)^{-1}P_{ki}\left(\prod_{m=k}^{i} H_m\right)\right)\right]\\ &\qquad\approx\prod_{i=k}^{n-3} H_iP_{i+1,i}.\end{aligned}$$

Now we can rewrite (18) as follows:

$$\begin{aligned}&\left(\prod_{m=0}^{n-3} H_m\right)^{n-1}\cdot\prod_{k=n-2}^{0}\mathring{R}_k\\ \text{(19)}\qquad &\approx\prod_{k=n-2}^{0}\left[\left(\prod_{i=0}^{k-1} H_iP_{i+1,i}\right)\cdot W_k\cdot\left(\prod_{i=k}^{n-3} H_iP_{i+1,i}\right)\right]\end{aligned}$$

From Figure 6b it follows that

$$(H_k P_{k+1,k}) W_k (H_k P_{k+1,k})^{-1} = W_{k+1},$$

$$\left(\prod_{i=0}^{n-3} H_i P_{i+1,i}\right) W_k \left(\prod_{i=0}^{n-3} H_i P_{i+1,i}\right)^{-1} = W_{k+1}$$

and

$$(H_k P_{k+1,k}) W_{k+1} (H_k P_{k+1,k})^{-1} = W_k.$$

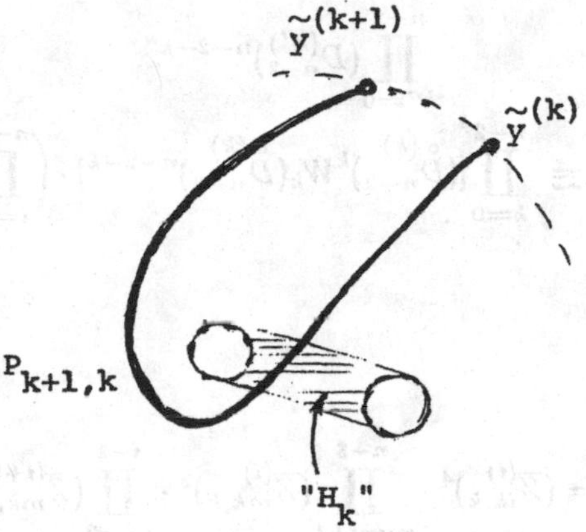

Figure 6b

Thus

$$\begin{aligned}
&\prod_{k=n-2}^{0}\left[\left(\prod_{i=0}^{k-1} H_i P_{i+1,i}\right)\cdot W_k \cdot \prod_{i=k}^{n-3} H_i P_{i+1,i}\right] \\
&\quad \approx \prod_{k=n-2}^{0}\left[W_0 \cdot \left(\prod_{i=0}^{n-3} H_i P_{i+1,i}\right)\right] \\
&\quad \approx \left(\prod_{k=n-2}^{0} W_{n-2-k}\right)\cdot\left(\prod_{i=0}^{n-3} H_i P_{i+1,i}\right)^{n-1} \\
&\quad \approx \left(\prod_{k=0}^{n-2} W_k\right)\cdot\left(\prod_{i=0}^{n-3} H_i P_{i+1,i}\right)^{n-1}.
\end{aligned}$$

Each $D^{(k)}_{n-2}$ commutes with $(\prod_{i=0}^{n-3} H_i P_{i+1,i})^{n-1}$ and with $W_{k'}$, $k' \neq k$. Now from (19) and (12′) we get

$$
\begin{aligned}
\Delta^2(B[\tilde{\mathbb{C}}_1, \tilde{K}]) &= \left(\prod_{k=0}^{n-2} (\overset{\circ}{D}{}^{(k)}_{n-2})^{n-2}\right) \cdot \prod_{k'=0}^{n-2} W_{k'} \cdot \left(\prod_{i=0}^{n-3} H_i P_{i+1,i}\right)^{n-1} \\
&\approx \prod_{k=0}^{n-2} (\overset{\circ}{D}{}^{(k)}_{n-2})^{k} \cdot \prod_{k'=0}^{n-2} W_{k'} \cdot \left(\prod_{i=0}^{n-3} H_i P_{i+1,i}\right)^{n-1} \\
&\qquad \prod_{k''=0}^{n-2} (D^{(k'')}_{n-2})^{n-2-k''} \\
&\approx \prod_{k=0}^{n-2} [(\overset{\circ}{D}{}^{(k)}_{n-2})^{k} W_k (\overset{\circ}{D}{}^{(k)}_{n-2})^{n-2-k}] \cdot \left(\prod_{i=0}^{n-3} H_i P_{i+1,i}\right)^{n-1}.
\end{aligned}
\tag{20}
$$

Denote by

$$
W^{(i)}_k = (\overline{Z}^{(i)}_{ik,k})^4 \cdot \prod_{m=i+1}^{n-3} (\overline{Z}^{(i)}_{mk,k})^2 \cdot \prod_{m=0}^{i-1} (\overline{Z}^{(i+1)}_{mk,k})^2.
$$

From (17) we have: $W_k = \prod_{i=0}^{n-3} W^{(i)}_k$. Hence

$$
\begin{aligned}
(\overset{\circ}{D}{}^{(k)}_{n-2})^{k} W_k (\overset{\circ}{D}{}^{(k)}_{n-2})^{n-2-k} &\approx (\overset{\circ}{D}{}^{(k)}_{n-2})^{k} \cdot \prod_{i=0}^{n-3} W^{(i)}_k \cdot (\overset{\circ}{D}{}^{(k)}_{n-2})^{n-2-k} \\
&\approx \prod_{i=0}^{k-1} \left[((D^{(k)}_{n-2})^{k-i} W^{(i)}_k (D^{(k)}_{n-2})^{-(k-i)}) \overset{\circ}{D}{}^{(k)}_{n-2}\right] W^{(k)}_k \cdot \overset{\circ}{D}{}^{(k)}_{n-2} \\
&\quad \prod_{i=k+1}^{n-3} \left[((D^{(k)}_{n-2})^{-(i-k)} W^{(i)}_k (D^{(k)}_{n-2})^{i-k}) \cdot \overset{\circ}{D}{}^{(k)}_{n-2}\right].
\end{aligned}
$$

We have for $m \geq i$

$$
\begin{aligned}
(D^{(k)}_{n-2})^{k-i}(\overline{Z}^{(i)}_{mk,k})(D^{(k)}_{n-2})^{-(k-i)}
&= (A^{(k)}_{n-2})^{(n-1)(k-i)}(A^{(k)}_{n-2})^{-(k-i)(n-2)} Z_{mk,k} \\
&\quad \cdot (A^{(k)}_{n-2})^{(k-i)(n-2)}(A^{(k)}_{n-2})^{-(n-1)(k-i)} \\
&= (A^{(k)}_{n-2})^{k-i} Z_{mk,k}(A^{(k)}_{n-2})^{k-i} \\
&= (A^{(k)}_{n-2})^{k-i}(A^{(k)}_{n-2})^{-k} Y_{n-3-m,k;k}(A^{(k)}_{n-2})^{-(k-i)} \cdot (A^{(k)}_{n-2})^{k} \\
&= (A^{(k)}_{n-2})^{-i} Y_{n-3-m,k;k}(A^{(k)}_{n-2})^{i} = Y_{n-3-m+i,k;k}
\end{aligned}
$$

and for $m \leq i-1$

$$
\begin{aligned}
(D^{(k)}_{n-2})^{k-i}\overline{Z}^{(i+1)}_{mk,k}(D^{(k)}_{n-2})^{-(k-i)}
&= (A^{(k)}_{n-2})^{(n-1)(k-i)}(A^{(k)}_{n-2})^{-(k-i-1)(n-2)}(Z_{mk,k}) \\
&\quad (A^{(k)}_{n-2})^{(k-i-1)(n-2)}(A^{(k)}_{n-2})^{(k-i-1)(n-2)} \\
&= (A^{(k)}_{n-2})^{n-2+k-i} Z_{mk,k}(A^{(k)}_{n-2})^{-(n-2+k-i)} \\
&= (A^{(k)}_{n-2})^{n-2-i} Y_{n-3-m,k;k}(A^{(k)}_{n-2})^{-(n-2-i)} \\
&= Y_{i-1-m,k;k}.
\end{aligned}
$$

Thus

$$
\begin{aligned}
\left[(\overset{\circ}{D}{}^{(k)}_{n-2})^{k-i}\, W^{(i)}_{k}(\overset{\circ}{D}{}^{(k)}_{n-2})^{-(k-i)}\right]
&\approx Y^4_{n-3,k;k} \cdot \prod_{m=i+1}^{n-3} Y^2_{n-3-m+i,k;k} \cdot \prod_{m=0}^{i-1} Y^2_{i-1-m,k;k} \\
&\approx Y^4_{n-3,k;k} \cdot \prod_{m=n-4}^{0} Y^2_{m,k;k}
\end{aligned}
$$

and the horizontal braid monodromie for $X_{n-1} \cup W$ is given by:

$$\Delta^2(B[\tilde{\mathbb{C}}^1, \tilde{K}]) = \left(\prod_{k=1}^{n-2} Y^4_{n-3,k;k} \cdot \prod_{m=n-4}^{0} Y^2_{m,k;k} \cdot \overset{\circ}{D}{}^{(k)}_{n-2} \right)^{n-2} \cdot \left(\prod_{i=0}^{n-3} H_i P_{i+1,i} \right)^{n-1} . \tag{21}$$

§4 Regeneration of $X_{n-1} \cup W$

Consider a nonsingular surface $X(\epsilon_1)$ which is very close to $X_{n-1} \cup W$. Let

$$S = \{u \in \mathbb{C}^2 \mid \#\Pi_1^{-1}(u) \cap X(\epsilon_1) < n\},$$
$$K_S = \tilde{\mathbb{C}}_1 \cap S,$$
$$\hat{E} = X(\epsilon_1) \cap \tilde{\mathbb{C}}_2.$$

To avoid the introducing of too many new notations we can assume that

$$K_S = \{\tilde{y}^{(k)}_m, \quad m = 0, 1, \ldots, n-3; \quad k = 0, 1, \ldots, n-2\} \cup \{\tilde{y}^{(i)}_1, \tilde{y}^{(i)}_2, \quad i = 0, 1, \ldots, n-2\}$$

(where the pair $\tilde{y}^{(i)}_1, \tilde{y}^{(i)}_2$ is the result of a splitting of former $\tilde{y}^{(i)}$), and $\hat{E} \cap \tilde{\Pi}_1^{-1}(b) = \{\tilde{z}_i, i = 0, 1, \ldots, n-2\} \cup \{z = c\} = E' \cap \Pi_1^{-1}(b)$. Let $\alpha_i, i \in (0, 1, \ldots, n-2)$, be the braid from $B[\tilde{\mathbb{C}}_1, K_S]$ corresponding to the segment of a straight line connecting $y^{(i)}_1$ with $y^{(i)}_2$. Each $\delta'_i (i \in 0, 1, \ldots, n-2)$ splits into a pair $\{\delta'_{i(1)}, \delta'_{i(2)}\}$ with $\delta'_{i(2)} = (\delta'_{i(1)})\alpha_i^{-1}$. Similarly, each former braid $Y_{m,k;k}$ splits into a pair $\{Y^{(1)}_{mk}, Y^{(2)}_{mk}\}$ with $Y^{(2)}_{mk} = \alpha_k Y^{(1)}_{mk} \alpha_k^{-1}$. Each $P_{i+1,i}$ produces now four braids $\{P_{i+1,i}(\tau, \tau'), \tau = 1, 2; \tau' = 1, 2\}$, where $P_{i+1,i}(\tau, \tau')$ corresponds to a path $P_{i+1,i}(\tau, \tau')$ connecting $y^{(i)}_\tau$ with $y^{(i+1)}_{\tau'}$ and

$$P_{i+1,i}(\tau, 2) = \alpha_{i+1} P_{i+1,i}(\tau, 1) \alpha_{i+1}^{-1},$$
$$P_{i+1,i}(2, \tau') = \alpha_i P_{i+1,i}(1, \tau') \alpha_i^{-1}.$$

It is convenient to assume also that $\forall i = 0, 1, \ldots, n-4$,

$$P_{i+1,i}(1,1) \cap P_{i+2,i+1}(1,1) = y_1^{(i+1)}$$

and $\forall i = 0, 1, \ldots, n-2$,

$$P_{i+1,i}(1,1) \cap Y_{mi}^{(1)} = y_1^{(i)},$$
$$P_{i+1,i}(1,1) \cap Y_{m,i+1}^{(1)} = y_1^{(i+1)}$$

($Y_{mi}^{(1)}$ denotes also a path corresponding to the (half-twist) braid $Y_{mi}^{(1)}$).

Clearly, all these splittings are defined up to conjugations by powers of elements $\alpha_i, i = 0, 1, \ldots, n-2$.

Let us describe what replacements we have to make in (21) when $X_{n-1} \cup W$ is regenerating to $X(\epsilon_1)$.

1. Replacement of $Y_{n-3;k,k}^4$. Geometrically this factor corresponds to a point of tangency of the branch curve of X_{n-1} with the curve $\Pi_1(X_{n-1} \cap W)$. Be direct computations it is possible to show that each of these tangency points produces three cusps of the branch curve of $X(\epsilon_1)$ and that the factor $Y_{n-3;k,k}^4$ has to be replaced by

$$(Y_{n-3}^{(1)})^3 \cdot (Y_{n-3}^{(2)})^3 \cdot (\alpha_k Y_{n-3,k}^{(2)} \alpha_k^{-1})^3.$$

The fact that this replacement is independent of the choice of splitting $\{Y_{n-3,k}^{(1)}, Y_{n-3,k}^{(2)}\}$ immediately follows from Lemma 1^{iv}.

2. Replacement of $Y_{mk,k}^2$, $m = n-4, \ldots, 1, 0$. Here we have to replace $Y_{mk,k}^2$ by $(Y_{mk}^{(1)})^2 \cdot (Y_{mk}^{(2)})^2$. Because

$$\begin{aligned}(Y_{mk}^{(1)})^2 \cdot (Y_{mk}^{(2)})^2 &\equiv (Y_{mk}^{(1)})^2 (\alpha Y_{mk}^{(1)} \alpha^{-1})^2 \\ &\equiv (\alpha Y_{mk}^{(1)} \alpha^{-1})^2 \cdot [(\alpha Y^{(1)} \alpha^{-1})^{-2} (Y_{mk}^{(1)})^2 (\alpha Y_{mk}^{(1)} \alpha^{-1})^2] \\ &\equiv (\alpha Y_{mk}^{(1)} \alpha^{-1})^2 \cdot \alpha^2 Y_{mk}^{(1)} \alpha^{-2},\end{aligned}$$

we see that our replacement does not depend on splitting.

3. Replacement of $(\prod_{i=0}^{n-3} H_i P_{i+1,i})^{n-1}$. We replace $(\prod_{i=0}^{n-3} H_i P_{i+1,i})^{n-1}$ by

$$\left[\prod_{i=0}^{n-3}(H_i \cdot P_{i+1,i}(1,1) \cdot P_{i+1,i}(2,2))\right]^{n-1}\left[\prod_{i'=0}^{n-2}\alpha_{i'}\right]^{n}.$$

Denote by $G = \left(\prod_{i=0}^{n-3}(H_i \cdot P_{i+1,i}(1,1)P_{i+1,i}(2,2))\right)^{n-1}$. Clearly G commutes with each α_i'. To show that the replacement is independent of a splitting satisfying the assumptions above it is enough to do the following:

$$\begin{aligned}
&\left(\prod_{i=0}^{n-3}(H_i\, P_{i+1,i}(1,1)P_{i+1,i}(2,2))\right)^{n-1} \cdot \alpha_{i'} \\
&\quad\equiv \alpha_{i'} \cdot \left[\prod_{i=0}^{n-3}\Big((\alpha_{i'}^{-1} H_i \alpha_{i'}) \cdot (\alpha_{i'}^{-1} P_{i+1,i}(1,1)\alpha_{i'}) \right. \\
&\qquad\qquad \left. \cdot (\alpha_{i'}^{-1} P_{i+1,i}(1,1)\alpha_{i'})\Big)\right]^{n-1} \\
&\quad\equiv \alpha_{i'} \cdot \left[\prod_{i=0}^{n-3}\Big(H_i \cdot (\alpha_{i'}^{-1} P_{i+1,i}(1,1)\alpha_{i'}) \right. \\
&\qquad\qquad \left. \cdot (\alpha_{i'}^{-1} P_{i+1,i}(1,1)\alpha_{i'})\Big)\right]^{n-1} \\
&\quad\equiv \left[\prod_{i=0}^{n-3}\Big(H_i \cdot (\alpha_{i'}^{-1} P_{i+1,i}(1,1)\alpha_{i'}) \right. \\
&\qquad\qquad \left. \cdot (\alpha_{i'}^{-1} P_{i+1,i}(1,1)\alpha_{i'})\Big)\right]^{n-1} \cdot \underbrace{G^{-1}\alpha_{i'} G}_{\alpha_{i'}}
\end{aligned}$$

Using 1., 2., 3., we get from (21) the following ("ready to collapse", see Introduction) formula describing the horizontal braid monodromie for $X(\epsilon_1)$:

$$\Delta^2(B[\tilde{\mathbb{C}}^1, K_S]) = \left[\prod_{k=0}^{n-2}\left((Y^{(1)}_{n-3,k})^3 (Y^{(2)}_{n-3,k})^3 (\alpha_k Y^{(2)}_{n-3,k}\alpha_k^{-1})^3\right) \cdot \prod_{m=n-4}^{0} \left((Y^{(1)}_{mk})^2 \cdot (Y^{(2)}_{mk})^2\right) \cdot \overset{\circ}{D}{}^{(k)}_{n-2}\right]^{n-2} \cdot \left[\prod_{i=0}^{n-3} (H_i, P_{i+1,i}(1,1) \cdot P_{i+1,i}(2,2))\right]^{n-1} \cdot \left(\prod_{i'=0}^{n-2} \alpha_{i'}\right)^n. \tag{22}$$

A related form for the vertical braid monodromie of $X(\epsilon_1)$ could be easily obtained from that of $X_{n-1} \cup W$ given by $(12)v$ and it is the following:

$$\begin{aligned} &\varphi_{\hat{E},\tilde{\Pi}_1}(\delta'_{mk}) = \tilde{K}_m, \quad m = 0, \ldots, n-3; \quad k = 0, \ldots, n-2, \\ &\varphi_{\hat{E},\tilde{\Pi}_1}(\delta'_{i(1)}) = \varphi_{\hat{E},\tilde{\Pi}_1}(\delta'_{i(2)}) = K_{oi}. \end{aligned} \tag{23}$$

§5 Transformations of Vertical Braid Monodromies

In the next sections we will describe how the formulae $(7)_v, (7)_h$ are transformed into (22), (23) when the surface X_n approaches $X_{n-1} \cup W$. In more explicit terms we have to analyze what happens when the equation (6) (for V_n) is transformed to an equation for $V_{n-1} \cup W$:

$$(z^{n-1} + \epsilon(1 + \rho x^{n-2})z - y^{n-1} + x^{n-1} + 1)(z - c) = 0 \tag{24}$$

(see p. 226).

It is convenient to separate transformations of equations in the following seven steps:

Step 1. From $z^n + \epsilon(1+\rho x^{n-1})z + x^n - y^n + 1 = 0$

$$\text{to} \quad z^n + \epsilon_0(\rho - \beta y^{n-1})z + x^n - y^n + 1 = 0,$$

where $\lambda \ll \beta \ll \epsilon_0 \ll 1$, $\epsilon_0\lambda = \epsilon$, using family:

$$z^n + \epsilon(\lambda - ty^{n-1} + sx^{n-1})z + x^n - y^n + 1 = 0,$$

$t, s \in \mathbb{R}$, t is changing from 0 to β, s is changing from ρ to 0.

Step 2. From $z^n + \epsilon_0(\lambda - \beta y^{n-1})z + x^n - y^n + 1 = 0$

$$\text{to} \quad z^n + \epsilon(1 - y^{n-1})z + x^n + 1 = 0,$$

using the family $z^n + \epsilon_0(\lambda - \beta y^{n-1})z + x^n - ty^n + 1 = 0$, $t \in \mathbb{R}$, t is changing from 1 to 0, and in the end replacing y by $y/\beta^{1/n-1}$.

Step 3. From $z^n + \epsilon(1 - y^{n-1})z + x^n + 1 = 0$

$$\text{to} \quad z^n + (1 - y^{n-1})z + \alpha_0(x^n + 1) = 0,$$

$\alpha_0 \ll 1$, using the family

$$z^n + t(1 - y^{n-1})z + s(x^n + 1) = 0, \quad t, s \in \mathbb{R},$$

t is changing from ϵ to 1, s is changing from 1 to α_0.

Step 4. From $z^n + (1 - y^{n-1})z + \alpha_0(x^n + 1) = 0$

$$\text{to} \quad z^n + (1 - y^{n-1})z - \alpha_0(x^n + 1) = 0,$$

using the family

$$z^n + (1 - y^{n-1})z + \alpha_0 e^{-i\varphi}(x^n + 1) = 0, \quad 0 \le \varphi \le \pi.$$

Step 5. From $z^n + (1 - y^{n-1})z - \alpha_0(x^n + 1) = 0$

to $(z^{n-1} + 1 - y^{n-1})(z - c_0) - \alpha_0(x^n + 1) = 0, \quad \alpha_0 \ll c_0 \ll 1,$

using the family

$$(z^{n-1} + 1 - y^{n-1})(z - t) - \alpha_0(x^n + 1) = 0,$$

$t \in \mathbb{R}$, t is changing from 0 to c_0.

Step 6. From $(z^{n-1} + 1 - y^{n-1})(z - c_0) - \alpha_0(x^n + 1) = 0$

to $(z^{n-1} + \epsilon(1 - \rho x^{n-2})z + 1 - y^{n-1})(z - c_0) - \alpha_1(x^n + 1) = 0,$

$\alpha_1 \ll \rho \ll \epsilon \ll \alpha_0$, using the family

$$(z^{n-1} + t(1 - \rho x^{n-2})z + 1 - y^{n-1})(z - c_0) - s(x^n + 1) = 0,$$

$t, s, \in \mathbb{R}$, t is changing from 0 to ϵ, s is changing from α_0 to α_1.

Step 7. From $(z^{n-1} + \epsilon(1 - \rho x^{n-2})z + 1 - y^{n-1})(z - c_0) - \alpha_1(x^n + 1) = 0$

to $(z^{n-1} + \epsilon(1 - \rho x^{n-2})z + 1 + x^{n-1} - y^{n-1})(z - c) - \alpha_1(x^n + 1) = 0,$

using the family

$$(z^{n-1} + \epsilon(1 - \rho x^{n-2})z + 1 + tx^{n-1} - y^{n-1})(z - s) - \alpha_1(x^n + 1) = 0,$$

$t, s, \in \mathbb{R}$, t is changing from 0 to 1, s is changing from c_0 to c.

For step $i, i \in \{1,2,3,4,5,6,7\}$ denote by $V[i]$ the surface obtained in this step, by $V[0] = V_{n-1}$ and $X[i]$ a generic surface close to $V[i]$.

Let $C[i] = \{u \in \mathbb{C}^2 \mid \#(\prod^{-1}(u) \cap X[i]) < n\}$, $E[i] = X[i] \cap \tilde{\mathbb{C}}^2$, $N[i] = \tilde{\mathbb{C}}^1 \cap C[i]$. Explicitly the sets $N[i]$ are:

$$N[0] = \{y_{kj} \approx (1 + \tfrac{1}{n} q_n^{1/n-1} \epsilon^{n/n-1} \mu'_k)\zeta_j,$$

$$k = 0, \ldots, n-2;\ j = 0, \ldots, n-1\}$$

$$\mu'_k = \exp\left(\pi i \frac{2k+1}{n-1}\right), \quad \zeta_j = \exp\frac{2\pi i j}{n}, \quad q_n = \frac{(n-1)^{n-1}}{n^n};$$

$$N[1] = \{y_{qj}[1] \approx (1 - \tfrac{1}{n} q_n^{1/n-1} \epsilon^{n/n-1} \mu_{-q})\zeta_j,$$

$$q_n = -\left[\frac{n-1}{2}\right], \left[\frac{n-1}{2}\right] + 1, \ldots, \left[\frac{n}{2}\right] - 1; \quad j = 0, 1, \ldots, n-1\}$$

$$\mu_k = \exp\frac{2\pi i k}{n-1};$$

$$N[2] = \{y_{qj}[2] \approx \frac{1}{q_n^{1/n(n-1)} \epsilon^{1/(n-1)}} \exp \pi i\left(\frac{2g}{n(n-1)} + \frac{2j}{n}\right),$$

$$q_n = -\left[\frac{n-1}{2}\right], -\left[\frac{n-1}{2}\right] + 1, \ldots, \left[\frac{n}{2}\right] - 1; \quad j = 0, 1, \ldots, n-1\};$$

$$N[3] = \{y_{jk}[3] \approx \left(1 + \frac{1}{n-1} \alpha_0^{n-1/n} \zeta_j\right)\mu_k,$$

$$j = 0, 1, \ldots, n-1, k = 0, 1, \ldots, n-2\},$$

$$N[4] = \{y_{jk}[4] \approx \left(1 - \frac{1}{n-1} \alpha_0^{n-1/n} \zeta'_j\right)\mu_k$$

$$j = 0, 1, \ldots, n-1, k = 0, 1, \ldots, n-2\},$$

$$\zeta'_j = \exp \pi_i\left(\frac{2j+1}{n}\right);$$

$$N[5] = \{y_{lk}[5] \approx \left(1 + \frac{\alpha_0}{n(n-1)c_0} + \frac{(n-2)\alpha_0^{n-1/n-2}}{n(n-1)^{n-1/n-2}c_0^{2(n-1)/(n-2)}}\eta'_l\right)\mu_k,$$

$$l = 0, 1, \ldots, n-3; \; k = 0, 1, \ldots, n-2,$$

and

$$\overline{y}_{rk}[5] \approx \left[(1 + c_0^{n-1})^{1/n-1} + 2\left(\frac{\alpha_0}{n-1}\right)^{1/2} \frac{(-1)^r}{c_0^{n/2}(1 + c_0^{n-1})} i\right]\mu_k$$

$$\{r = 0, 1; \quad k = 0, 1, \ldots, n-2\}$$

$$\eta'_l = \exp \pi i \frac{2l+1}{n-2}, \quad \text{(we can assume that } \alpha_0 \ll c_0^n);$$

$N[6]$ is practically the same as $N[5]$; $N[7]$ is like $N[5]$, where c_0 is replaced by $c\,(\gg 1)$.

Direct analysis gives the following picture for transitions:

$$N[i] \to N[i+1]:$$

Step 1. Denote by s_j a "circle" containing $\{y_{kj}[0], k = 0, 1, \ldots, n-2\}$ (see Figure 7). Then in Step 1 $N[1]$ is obtained from $N[0]$ be rotating each s_j on the angle $\pi + \pi/n - 1 - 2\pi j/n - 1$ (e.g. the rotation is negative if $j > n/2$).

Step 2. Keep the notations s_j for a "circle" containing

$$\left\{y_{qj}[1], \quad q = -\left[\frac{n-1}{2}\right], \ldots, \left[\frac{n}{2}\right] - 1\right\}.$$

Then in Step 2 each s_j is cut between $y_{-[n-1/2],j}[1]$ and $y_{[n/2]-1,j}[1]$ and unbended (see Figure 7) to become an arc $\hat{s}_j$ of a big "circle" $\hat{s}$ containing all the elements of $N[2]$.

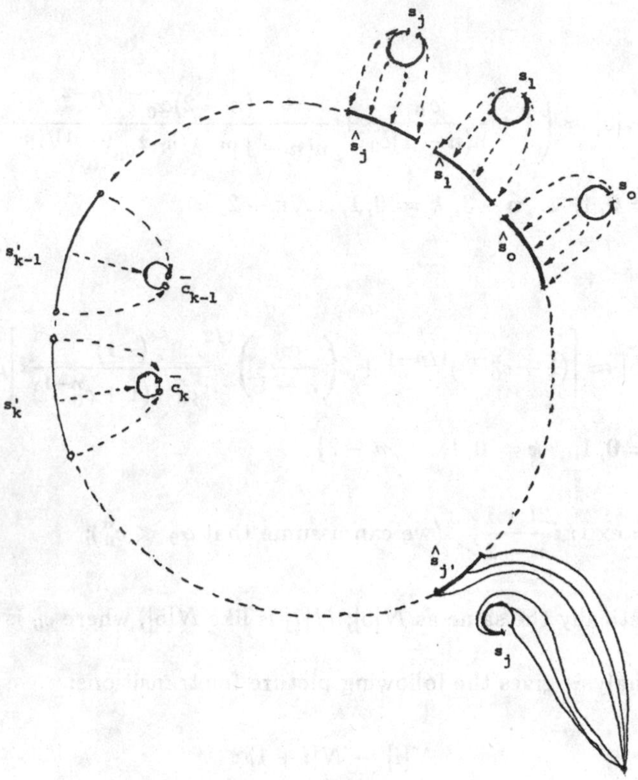

Figure 7

Step 3. Denote by s'_k, $k = 0, 1, \ldots, n-2$, successive arcs of $\hat{s}$ such that

$$s'_k \supset \{y_{-[n-1/2]+k,k}[2], \ldots, y_{[n/2]-1,k}[2], \\ y_{-[n-1/2],k+1}[2], \ldots, y_{-[n-1/2]+k,k+1}.\}$$

Then in Step 3 $\hat{s}$ is cut in the end-points of the arcs s'_k, $k = 0, 1, \ldots, n-2$, and then each s'_k is bended inside (see Figure 7) to become a "circle" $\bar{c}_k$ containing $\{y_{jk}[3], j = 0, 1, \ldots, n-1\}$.

Step 4. $N[4]$ is obtained from $N[3]$ by rotating each "circle" $\bar{c}_k$ negatively on the angle $\frac{n-1}{n}\pi$.

Step 5. Keep the notation $\bar{c}_k$ for a "circle" containing $\{y_{jk}[4], j=0,1,\dots,n-1\}$. Then in Step 5 each $\bar{c}_k$ is cut between the points $y_{0k}[4]$ and $y_{n-1,k}[4]$ to become a "horse shoe"; after this "horse shoe" is unbended all around and images of $y_{0k}[4]$ and $y_{n-1,k}[4]$ travel away along the radius $(\arg y = \frac{2\pi k}{n-1})$ to become $\bar{y}_{0k}[5]$ and $\bar{y}_{1k}[5]$. The images of the remaining points $\{y_{jk}[4], j=1,\dots,n-1\}$ become the points $y_{lk}[5], l=0,1,\dots,n-3$, of $N[5]$ (see Figure 8).

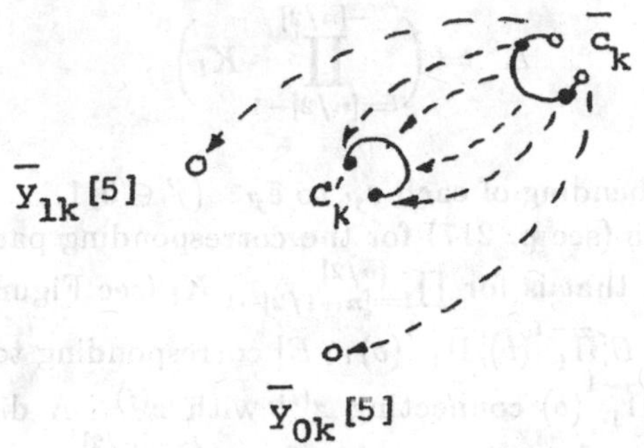

Figure 8

Step 6. There are no geometrically significant changes when $N[5]$ becomes $N[6]$.

Step 7. Denote by $y_{lk}[6]$ (resp. $\bar{y}_{rk}[6]$) a point obtained from $y_{lk}[5]$ (resp. $\bar{y}_{rk}[5]$) in Step 6 and by c'_k a "circle" containing $\{y_{lk}[6], l=0,1,\dots, n-3\}$. Then in Step 7 each pair $\{\bar{y}_{0k}[6], \bar{y}_{1k}[6]\}$ $(k=0,1,\dots,n-2)$ travels along the radius $(\arg y = \frac{2\pi k}{n-1})$ far away from c'_k.

Now the sequence of transformations of vertical braid monodromies for $X[i]$'s, $i=1,\dots,7$, can be described as follows (we relate everything to $B[\tilde{\Pi}_1^{-1}(b), \tilde{\Pi}_1^{-1}(b)\cap E]$: We start with the following expression of the vertical braid monodromie for $X[0]=X_n$ given by $(7)_v$.

$$(24)_0 \qquad \Delta^2 = \prod_{j=n-1}^{0} \prod_{k=n-2}^{0} K_{k-j},$$

where:

$$\Delta^2 = \Delta^2\left(B\tilde{\Pi}_1^{-1}(b),\ \tilde{\Pi}_1^{-1}(b) \cap E\right).$$

Here each $\prod_{k=n-2}^{0} K_{k-j}$ corresponds to s_j. In Step 1 this expression is transformed by a cyclic shift of $[\frac{n}{2}] - j$ steps, that is for $j > [\frac{n}{2}]$ we do $[\frac{n}{2}] - j$ positive steps and for $j < [\frac{n}{2}]$ we do $j - [\frac{n}{2}]$ negative steps. Thus $\prod_{k=n-2}^{0} K_{k-j}$ becomes $\prod_{l=[n-1/2]-1}^{-[n/2]} K_l$ and from $(24)_0$ we get

$$(24)_1 \qquad \Delta^2 = \left(\prod_{l=[n/2]-1}^{-[n/2]} K_l\right)^n.$$

In Step 2 the unbending of each $s_{j'}$ to $\hat{s}_{j'}$ $(j' \in 0, 1, \ldots, n-1)$ produces Δ^{-1}-transformation (see p. 217) for the corresponding part of the vertical braid monodromie, that is for $\prod_{l=[n-1/2]-1}^{-[n/2]} K_l$ (see Figure 7). Denote by L_{ij} the braid from $B[\tilde{\Pi}_1^{-1}(b), \tilde{\Pi}_1^{-1}(b) \cap E]$ corresponding to the segment of a straight line in $\tilde{\Pi}_1^{-1}(b)$ connecting $z^{(i)}$ with $z^{(j)}$. A direct verification shows that Δ^{-1}-transformation makes from $\prod_{l=[n-1/2]-1}^{-[n/2]} K_l$ the expression: $\prod_{-[n/2]+1}^{[n-1/2]} L_{q}, -[\frac{n}{2}]$.

Thus in Step 2 $(24)_1$ becomes

$$(24)_2 \qquad \Delta^2 = \left(\prod_{q=-[n/2]+1}^{[n-1/2]} L_{q,-[\frac{n}{2}]}\right)^n.$$

Step 3. Clearly to each s'_k (and to $\bar{c}_k$) corresponds the following part of $(24)_2$

$$\prod_{q=[n-1/2]-k}^{[n-1/2]} L_{q,-[\frac{n}{2}]} \cdot \prod_{q=-[n/2]+1}^{[n-1/2]-k} L_{q,-[\frac{n}{2}]}.$$

Thus we can express the vertical braid monodromie in Step 3 as follows:

$$(24)_3 \qquad \Delta^2 = \prod_{k=n-2}^{0} \left[\prod_{q=\frac{n-1}{2}-k}^{[n-1/2]} L_{q,-[\frac{n}{2}]} \cdot \prod_{q=-[n/2]+1}^{[n-1/2]-k} L_{q,-[\frac{n}{2}]}\cdot\right]$$

Step 4. Here we have to do "negative cyclic shift" of $[\frac{n-1}{2}]$ steps in each part of $(24)_3$ corresponding to $\bar{c}_k$ (and written in brackets in $[24]_3$).

A simple verification shows that as a result we will get

$$\left[\left(\prod_{\substack{q=-k\\ q\neq -[n/2]}}^{0} L_{q,-[\frac{n}{2}]}\right)\cdot \prod_{\substack{q=1\\ q\neq -[n/2]}}^{n-k} L_{q,-[\frac{n}{2}]}\cdot\right]$$

The vertical braid monodromie expression becomes:

$$(24)_4 \qquad \Delta^2 = \prod_{k=n-2}^{0}\left[\left(\prod_{\substack{q=-k\\ q\neq -[n/2]}}^{0} L_{q,-[\frac{n}{2}]}\right)\cdot \prod_{\substack{q=1\\ q\neq -[n/2]}}^{n-k} L_{q,-[\frac{n}{2}]}\cdot\right]$$

Step 5. Denote by

$$\Lambda_k = \prod_{\substack{q=-k\\ q\neq -[n/2]}}^{0} L_{q,-[\frac{n}{2}]} \prod_{\substack{q=1\\ q\neq -[n/2]}}^{n-k} L_{q,-[\frac{n}{2}]}.$$

First we move the first factor of Λ_k to the right end. We get

$$\Lambda_k \cong \Lambda'_k L_{n-k,-[\frac{n}{2}]} \cdot L_{n-k,-[\frac{n}{2}]}$$

where

$$\Lambda'_k = \prod_{\substack{q=-k\\ q\neq -[n/2]}}^{0} (L_{n-k,-[\frac{n}{2}]})(L_{q,-[\frac{n}{2}]})(L^{-1}_{n-k,-[\frac{n}{2}]})$$
$$\prod_{\substack{q=1\\ q\neq -[n/2]}}^{n-k} (L_{n-k,-[\frac{n}{2}]})(L_{q,-[\frac{n}{2}]})(L^{-1}_{n-k,-[\frac{n}{2}]}).$$

Λ'_k corresponds to $\{y_{lk}[5], l = 0, 1, \dots, n-3\}$. Because this set was obtained by an unbending process, we can see from Figure 8 that Δ-transformation (see p. 215) must be applied to Λ'_k. As a result we get

$$\Lambda'_k \cong \prod_{p=n-3}^{0} K'_{p-k},$$

where K'_l is defined as follows: for $l = 0, \ldots, [\frac{n-1}{2}] - 1$, $K'_l = K_l$; for $l = [\frac{n-1}{2}]$, $K'_{[n-1/2]} = K_{[n-1/2]} K_{[n+1/2]} K^{-1}_{[n-1/2]}$ and for $l = [\frac{n}{2}]+1, \ldots, n-2$, $K'_l = K_{l+1}$. For any other integer l we find $l' \in (0, 1, \ldots, n-2)$ with $l' \equiv l \,(\mathrm{mod}(n-1))$ and define $K'_l = K'_{l'}$.

Now from $(24)_4$ we obtain:

$$(24)_5 \qquad \Delta^2 = \prod_{k=n-2}^{0} \left[\left(\prod_{p=n-3}^{0} K'_{p-k} \right) \cdot L_{n-k,-[\frac{n}{2}]} L_{n-k,-[\frac{n}{2}]} \right].$$

Final step. We move all (pairs)-products $L_{n-k,-[\frac{n}{2}]} \cdot L_{n-k,-[\frac{n}{2}]}$ in $(24)_5$ to the left end of $(24)_5$.

Denote by $\overline{K}_0 = L^{-2}_{0,-[\frac{n}{2}]} K'_0 L^2_{0,-[\frac{n}{2}]}$, and $\forall l \not\equiv 0 \,(\mathrm{mod}(n-1))$ let $\overline{K}_l = K'_l$.

It is easy to check that $\forall k \in (0, 1, \ldots, n-2)$, $p \in (0, 1, \ldots, n-3)$

$$\left(\prod_{k'=k}^{0} L^2_{n-k',-[\frac{n}{2}]} \right)^{-1} K'_{p-k} \left(\prod_{k'=k}^{0} L^2_{n-k',-[\frac{n}{2}]} \right) = \overline{K}_{p-k}.$$

Thus finally we get the following expression for the vertical braid monodromie:

$$(25) \qquad \Delta^2 = \prod_{k=n-2}^{0} (L_{n-k,-[\frac{n}{2}]} \cdot L_{n-k,-[\frac{n}{2}]}) \cdot \prod_{k=n-2}^{0} \prod_{p=n-3}^{0} \overline{K}_{p-k}.$$

This expression is evidently "ready to collapse" to an expression of v.b.m. of $X_{n-1} \cup W$ (comp. formulae given on p. 228).

§6 Transformation of Normal Forms for Horizontal Braid Monodromies

To the sequence of formulae $(24)_i$, $i = 1, \ldots, 5$, (25) corresponds a sequence of related normal forms for the horizontal braid monodromies.

Step 1. Using the description of Step 1 on p. 245, formula $(7)_h$ and Lemma 2, it is easy to deduce that a normal form of h.b.m. of $X[1]$ (related to $(24)_1$) could be written as follows:

$$(26)\qquad \Delta^2 = \prod_{r=0}^{n-1} (\overset{\circ}{D}{}^{(r)}_{n-1})^{n-1} \cdot \left(\prod_{r=n-2}^{0} [(A^{(r)}_{n-1})^{-1} P^{(r)}_{n-1} (A^{(r)}_{n-1})] \right)^n ,$$

where

$$\Delta^2 = \Delta^2 \left(B[\Pi_2^{-1}(a), \Pi_2^{-1}(a) \cap C[1] \right).$$

Step 2. First we introduce some notations. $\forall u_1, u_2 \in \mathbb{C}^1$ denote by $\langle u_1, u_2 \rangle$ the segment of a straight line connecting u_1 with u_2 and use the same notation $\langle u_1, u_2 \rangle$ for the corresponding half twist and the braid. Let

$$\overline{y}_{qj} = y_{qj}[2],$$

$$q = -\left[\frac{n-1}{2}\right], \dots, \left[\frac{n}{2}\right] - 1;$$

$$j = 0, 1, \dots, n-1, \overline{Z}_{jq,j'q'} = \langle \overline{y}_{qj}, \overline{y}_{q'j'} \rangle,$$

$$\overline{A}^{(j)} = \prod_{q=-[n-1/2]}^{[n/2]-2} \overline{Z}_{jq,j,q+i}, \quad \overline{D}^{(j)} = (\overline{A}^{(j)})^n,$$

$$\overline{C}^j = \prod_{q=-[n-1/2]}^{[n/2]-1} [(\overline{A}^{(j-1)})^{-1} \overline{Z}_{j-1,q;j,q}) (\overline{A}^{(j-1)})], \quad \overline{C} = \prod_{j=n-1}^{1} \overline{C}_j.$$

It follows from Lemma 1 that

$$(27)\ \overline{D}^{(j)} = \overline{Z}^3_{j;-[\frac{n-1}{2}];j,-[\frac{n-1}{2}]+1} \cdot \prod_{q=-[n-1/2]+2}^{[n/2]-1} \left(\prod_{l=-[n-1/2]}^{q-2} Z^2_{j,q;j,l} \right) \cdot \overline{Z}^3_{jq,j,q+1}$$

Denote by $\overset{\circ}{\overline{D}}{}^{(j)}$ the (product of cubes and squares) expression on the right part of (27). From (26) we get a normal form of h.b.m. of $X[2]$:

$$(28)\qquad \Delta^2(B[\Pi_2^{-1}(a), \Pi_2^{-1}(a) \cap C[2]) = \left(\prod_{j=0}^{n-1} \overset{\circ}{\overline{D}}{}^{(j)} \right)^{n-1} \cdot \overline{C}^n.$$

Now we are going to transform (28) so that in Step 3 the new expression will give a normal form of h.b.m. of $X[3]$ (related to $(24)_3$).

We start with $\overline{C} = \prod_{j=n-1}^{1} \overline{C}_j$. Using Lemma 2 we can write

$$\overline{C}_j \approx \prod_{q=-[n-1/2]}^{[n/2]-1} \Big[(\overline{A}^{(j)}\overline{A}^{(j-1)})^{-(n-1-j)} (\overline{A}^{(j-1)})^{-1} (\overline{Z}_{j,q;\,j-1,q}) \overline{A}^{(j-1)} (\overline{A}^{(j)} (\overline{A}^{(j-1)})^{n-1-j} \Big]$$

$$\approx \Big(\prod_{q=-[n-1/2]}^{[n/2]-j-2} [(\overline{A}^{(j)}\overline{A}^{(j-1)})^{-n} (\overline{Z}_{j,q+j+1;\,j-1,q+j}) (\overline{A}^{(j)}\overline{A}^{(j-1)})^{n} \Big)$$

$$\cdot \Big[(\overline{A}^{(j)})^{-(n-1)} (\overline{A}^{(j-1)})^{-n} (\overline{Z}_{j,[\frac{n}{2}]-1;\,j-1,[\frac{n}{2}]-1}) (\overline{A}^{(j-1)})^{n} (\overline{A}^{(j)})^{n-1} \Big]$$

$$\cdot \Big[(\overline{A}^{(j-1)})^{-(n-1)} \overline{Z}_{j,[\frac{n-1}{2}];\,j-1,[\frac{n}{2}]-1} (\overline{A}^{(j-1)})^{n-1} \Big]$$

$$\prod_{q=[n/2]-j+1}^{[n/2]-1} \overline{Z}_{j,q-(n-1-j);\,j-1,q-(n-1-j)-1}.$$

(Here and after, we take $\prod_{q=-[n-1/2]}^{[n/2]-j-1} \overset{\text{def}}{=} 1$, if $-[\frac{n-1}{2}] > [\frac{n}{2}] - j - 2$, e.g. $j = n-2$ or $n-1$, and also

$$\prod_{q=[n/2]-j+1}^{[n/2]-1} \overset{\text{def}}{=} 1 \quad \text{if } \left[\frac{n}{2}\right] - j + 1 > \left[\frac{n}{2}\right] - 1, \quad \text{e.g. } j = 1.$$

Moreover, for $j = 1$ we replace here the conjugate of $\overline{Z}_{j,[n/2]-1;j-1,[n/2]-1}$ by the identity element.)

It is convenient to denote the successive factors in (29) by $d_j'(q)$, that is

$$d'_j(q) = \begin{cases} (\overline{A}^{(j)}\overline{A}^{(j-1)})^{-n}(\overline{Z}_{j,q+j+1;\,j-1,q+j})(\overline{A}^{(j)}\overline{A}^{(j-1)})^{n}, \\ \qquad \text{if } q = -[\frac{n-1}{2}],\dots,[\frac{n}{2}]-j-2 \\ \qquad (\text{and } j \neq n-2 \text{ or } n-1); \\ (\overline{A}^{(j)})^{-(n-1)}(\overline{A}^{(j-1)})^{-n}(\overline{Z}_{j,[\frac{n}{2}]-1;\,j-1,[\frac{n}{2}]-1})(\overline{A}^{(j)})^{n-1}(\overline{A}^{(j-1)})^{n}, \\ \qquad \text{if } q = [\frac{n}{2}]-j \ (\text{and } j \neq n-1) \\ (\overline{A}^{(j-1)})^{-(n-1)}(\overline{Z}_{j,-[\frac{n-1}{2}];\,j-1,[\frac{n}{2}]-1})(\overline{A}^{(j-1)})^{n-1} \\ \qquad \text{if } q = [\frac{n}{2}]-j; \\ (\overline{Z}_{j,q-(n-1-j);\,q-(n-1-j)-1}), \\ \qquad \text{if } q = [\frac{n}{2}]-j+1,\dots,[\frac{n}{2}]-1 \ (\text{and } j \neq 1). \end{cases}$$

Thus (29) becomes; $\overline{C}_j \approxeq \prod_{q=-[n-1/2]}^{[n/2]-1} d'_j(q)$. From Lemma 3′ it follows

$$(30)\qquad \begin{aligned} \overline{C} &= \prod_{j=n-1}^{1} \overline{C}_j \approxeq \prod_{j=n-1}^{1} \prod_{q=-[n-1/2]}^{[n/2]-1} d'_j(q) \\ &\approxeq \prod_{q=-[n-1/2]}^{[n/2]-1} \prod_{j=n-1}^{1} d'_j(q) \approxeq \prod_{q=-[n-1/2]}^{[n/2]-1} S'_q, \end{aligned}$$

where

$$S'_q = \prod_{j=n-1}^{1} d'_j(q).$$

From Lemma 2 it follows that $(\prod_{j=0}^{n-1} \overline{D}^{(j)})^{-1}\overline{C}(\prod_{j=0}^{n-1} \overline{D}^{(j)}) \approxeq \overline{C}$. Hence

$$(31)\qquad \left(\prod_{j=0}^{n-1} \overset{\circ}{\overline{D}}{}^{(j)}\right)^{n-1} \cdot \overline{C}^n \approxeq \overline{C} \cdot \left(\left(\prod_{j=0}^{n-1} \overset{\circ}{\overline{D}}{}^{(j)}\right)\overline{C}\right)^{n-1}.$$

It is easy to see that $\forall j = 1,2,\dots,n-1$, $\overline{C\overset{\circ}{D}{}^{(j)}}\,\overline{C}^{-1} = \overset{\circ}{\overline{D}}{}^{(j-1)}$. Now moving in the expression of the right side of (31) each element $\overline{D}^{(n-1)}$ to the left end we get from (31) the expression $((\prod_{j=0}^{n-2} \overset{\circ}{\overline{D}}{}^{(j)})\overline{C})^n \approxeq ((\prod_{j=n-2}^{0} \overset{\circ}{\overline{D}}{}^{(j)})\overline{C})^n$

Now from (28) we obtain

$$(28')\qquad \begin{aligned}\Delta^2 &= \Delta^2(B[\Pi_2^{-1}(a), \Pi_2^{-1}(a) \cap C[2]) \\ &= \left(\Big(\prod_{j=n-2}^{0} \overset{\circ}{\overline{D}}{}^{(j)}\Big)\overline{C}\right)^n.\end{aligned}$$

Denote $\mathcal{L} = (\prod_{j=n-2}^{0} \overset{\circ}{\overline{D}}{}^{(j)})\overline{C}$. Using (30) we get

$$(32)\qquad \mathcal{L} \approx \left(\prod_{j=n-2}^{0} \overset{\circ}{\overline{D}}{}^{(j)}\right) \cdot \prod_{q=-[n-1/2]}^{[n/2]-1} S'_q \approx \prod_{k=n-2}^{0} \overset{\circ}{\overline{D}}{}^{(k)} \overline{S}_k,$$

where

$$\overline{S}_0 = S'_{[n/2]-1} \text{ and } \forall k = 1, \dots, n-2,$$

$$\overline{S}_k = \left(\prod_{k'=k-1}^{0} \overline{D}^{(k')}\right)(S'_{[n/2]-k-1})\left(\prod_{k'=k-1}^{0} \overline{D}^{(k')}\right)^{-1}.$$

Denote by $d_j(0) = d'_j([\frac{n}{2}]-1)$, and for $k = 1, \dots, n-2$

$$d_j(k) = \left(\prod_{k'=k-1}^{0} \overline{D}^{(k')}\right)\left(d'_j\left(\left[\frac{n}{2}\right]-k-1\right)\right)\left(\prod_{k'=k-1}^{0} \overline{D}^{(k')}\right)^{-1}.$$

Clearly, $\overline{S}_k = \prod_{j=n-1}^{1} d_j(k)$. Let us describe more explicitly the factors $d_j(k)$.

1. $j \le k-1$ (and $k \ne 0$ or 1).

$$\begin{aligned}&d_j(k) = \\ &\left(\prod_{k'=k-1}^{0} \overline{D}^{(k')}\right) \\ &(\overline{A}^{(j)}\overline{A}^{(j-1)})^{-n}(\overline{Z}_{j,[\frac{n}{2}]-k+j;\,j-1,[\frac{n}{2}]-k+j-1})(\overline{A}^{(j)}\overline{A}^{(j-1)})^{n} \\ &\left(\prod_{k'=k-1}^{0} \overline{D}^{(k')}\right)^{-1} \\ &= Z_{j,[\frac{n}{2}]-k+j;\,j-1,[\frac{n}{2}]-k+j-1};\end{aligned}$$

2. $j = k$ (and $k \neq 0$).

$$d_k(k) =$$
$$\left(\prod_{k'=k-1}^{0} \overline{D}^{(k')}\right)$$
$$(\overline{A}^{(k)})^{-(n-1)}(\overline{A}^{(k-1)})^{-n}(\overline{Z}_{k,[\frac{n}{2}]-1;\,k-1,[\frac{n}{2}]-1})(\overline{A}^{(k-1)})^{n}(\overline{A}^{(k)})^{n-1}$$
$$\left(\prod_{k'=k-1}^{0} \overline{D}^{(k')}\right)^{-1}$$
$$= (\overline{A}^{(k)})^{-(n-1)}(\overline{Z}_{k,[\frac{n}{2}]-1;\,k-1,[\frac{n}{2}]-1})(\overline{A}^{(k)})^{n-1};$$

3. $j = k+1$

$$d_{k+1}(k) =$$
$$\left(\prod_{k'=k-1}^{0} \overline{D}_{k'}\right)$$
$$(\overline{A}^{(k)})^{-(n-1)}(\overline{Z}_{k+1,-[\frac{n-1}{2}];\,k,[\frac{n}{2}]-1})(\overline{A}^{(k)})^{n-1}$$
$$\left(\prod_{k'=k-1}^{0} \overline{D}^{(k')}\right)^{-1}$$
$$= (\overline{A}^{(k)})^{-(n-1)}(\overline{Z}_{k+1,-[\frac{n-1}{2}];\,k,[\frac{n}{2}]-1})(\overline{A}^{(k)})^{n-1};$$

4. $j \geq k+2$ (and $k \neq n-2$)

$$d_j(k) =$$
$$\left(\prod_{k'=k-1}^{0} \overline{D}^{(k')}\right)$$
$$(\overline{Z}_{j,j-(k+1)-[\frac{n-1}{2}];\,j-1,j-(k+2)-[\frac{n-1}{2}]})$$
$$\left(\prod_{k'=k-1}^{0} \overline{D}^{(k')}\right)^{-1}$$
$$= \overline{Z}_{j,j-(k+1)-[\frac{n-1}{2}];\,j-1,j-(k+2)-[\frac{n-1}{2}]}.$$

We see in particular that for $j \leq k-1$ and $j \geq k+2$, $d_j(k)$ commute with

$\overline{A}^{(k)}$. Using this remark we can write:

$$\overline{S}_k = \prod_{j=n-1}^{1} d_j(k) \cong (\overline{A}^{(k)})^{-(n-1)} .$$

$$\begin{aligned}(33)\qquad &\Bigg[\prod_{j=n-1}^{k+2} \overline{Z}_{j,j-(k+1)-[\frac{n-1}{2}];\, j-1,j-(k+2)-[\frac{n-1}{2}]} \\ &\cdot \overline{Z}_{k+1,-[\frac{n-1}{2}];\, k,[\frac{n}{2}]-1} \cdot \overline{Z}_{k,[\frac{n}{2}]-1;\, k-1;\, [\frac{n}{2}]-1} \\ &\cdot \prod_{j=k-1}^{1} \overline{Z}_{j,[\frac{n}{2}]-k+j;\, j-1,[\frac{n}{2}]-k+j-1}\Bigg](\overline{A}^{(k)})^{n-1}\end{aligned}$$

(where $\prod_{j=n-1}^{k+2} \cdots$ we write only when $k \neq n-2$, $\prod_{j=k-1}^{1} \cdots$ only when $k \neq 0$ or 1 and the factor $\overline{Z}_{k,[n/2]-1;\, k-1,[n/2]-1}$ is included only when $k \neq 0$).

It is easy to see that the expression in brackets in (33) is such that any two neighboring factors in it correspond to chords (segments of straight lines) which are neighbors (e.g. the right end of the chord of the left neighbor coincides with the left end of the cord of the right neighbor.)

Thus this expression corresponds to a polygon $\prod(k)$ with the set of vertices

$$\begin{aligned}\alpha(k) = \{&\overline{y}_{j,j-(k+1)-[\frac{n-1}{2}],\, j=n-1,\ldots,k+2};\ \overline{y}_{k+1,-[\frac{n-1}{2}]}, \\ &\overline{y}_{k,[\frac{n}{2}]-1}, \overline{y}_{j'-1,[\frac{n}{2}]-k+j'-1},\ j' = k-1,\ldots,1\},\end{aligned}$$

(that is, this expression has the form $A(\prod(k), \alpha(k))$ (see p. 206)). Using a cyclic shift in $A(\prod(k), \alpha(k))$ we can write

$$\begin{aligned}
\overline{S}_k \approxeq\ & (\overline{A}^{(k)})^{-(n-1)} [\overline{Z}_{k,[\frac{n}{2}]-1,k-1,[\frac{n}{2}]-1} \\
& \cdot \prod_{j=k-1}^{1} \overline{Z}_{j,[\frac{n}{2}]-k+j;\, j-1,[\frac{n}{2}]-k+j-1} \\
& \cdot \overline{Z}_{0,[\frac{n}{2}]-k;\, n-1,[\frac{n}{2}]-k-1} \\
& \cdot \prod_{j=n-1}^{k+2} \overline{Z}_{j,j-(k+1)-[\frac{n-1}{2}];\, j-1,j-(k+2)-[\frac{n-1}{2}]}] (\overline{A}^{(k)})^{n-1} \\
\approxeq\ & T_k \cdot \left[\left(\prod_{j=k-1}^{0} \overline{Z}_{j,[\frac{n}{2}]-k+j;\, j-1,[\frac{n}{2}]-k+j-1} \right) \right. \\
& \left. \cdot \prod_{j=n-1}^{k+2} \overline{Z}_{j,j-(k+1)-[\frac{n-1}{2}];\, j-1,j-(k+2)-[\frac{n-1}{2}]} \right],
\end{aligned}$$

where

$$T_k = (\overline{A}^{(k)})^{-(n-1)} (\overline{Z}_{k,[\frac{n}{2}]-1;\, k-1,[\frac{n}{2}]-1}) (\overline{A}^{(k)})^{n-1}$$

(and $\overline{y}_{-1,l} = \overline{y}_{n-1,l}$).

Let $\overline{\alpha}(k) = \{\overline{y}_{j,[n/2]-k+j},\ j = k-1,\ldots,0;\ \overline{y}_{j,j-(k+1)-[n-1/2]},\ j = n-1,\ldots,k+2;\ \overline{y}_{k+1,-[n-1/2]}\}$ and $\overline{\prod}(k)$ be a polygon with the vertices set equal to $\overline{\alpha}(k)$. Using a cyclic shift in $A(\overline{\prod}(k), \overline{\alpha}(k))$ and introducing new notations $\tilde{y}_l$ for the points of $N[2]$ by $\overline{y}_{jq} = \tilde{y}_{j(n-1)+q+[n-1/2]}$ and $\tilde{Z}_{l,l'} = \langle \tilde{y}_l, \tilde{y}_{l'} \rangle$ we can write

$$\overline{S}_k \approxeq T_k \cdot \prod_{l=n-3}^{0} \tilde{Z}_{ln+n-1-k,(l+1)\cdot n+n-1-k}.$$

Let $\tilde{E}_k = \prod_{l=n-3}^{0} \tilde{Z}_{ln+n-1-k,(l+1)n+n-1-k}$. So we have $\overline{S}_k \approxeq T_k \cdot \tilde{E}_k$ and from (32)

$$\mathcal{L} \approxeq \prod_{k=n-2}^{0} (\overset{\circ}{\tilde{D}}{}^{(k)} \cdot T_k \cdot \tilde{E}_k). \tag{34}$$

Consider now $\overset{\circ}{\tilde{D}}{}^{(k)} \cdot T_k$. We have

$$\overline{D}^{(k)} \cdot T_k = (\overline{A}^{(k)})^n \cdot [(\overline{A}^{(k)})^{-(n-1)} \overline{Z}_{k,[n/2]-1;\, k-1,[n/2]-1} (\overline{A}^{(k)})^{n-1}]$$
$$= \left(\prod_{r=[n/2]-1}^{-[n-1/2]} \overline{Z}_{k,r;\,k,r-1} \right)^{n-1}$$
$$= \left(\prod_{r'=n-2}^{0} \tilde{Z}_{k(n-1)+r',k(n-1)+r'-1} \right)^{n-1}$$

(see Lemma 1′). Denote by $\hat{A}^{(k)} = \prod_{r'=n-2}^{0} \tilde{Z}_{k(n-1)+r',k(n-1)+r'-1}$ by $\hat{D}^{(k)} = (\hat{A}^{(k)})^{n-1}$ and by $\hat{D}^{\circ(k)}$ the (product)-expression $\overset{\circ}{\overline{D}}{}^{k} \cdot T_k$. Now $\mathcal{L} \approx \prod_{k=n-2}^{0}(\hat{D}^{\circ(k)} \cdot \tilde{E}_k) \approx (\prod_{k=n-2}^{0} \tilde{E}_k) \cdot \prod_{k=n-2}^{o}[(\prod_{k'=k}^{0} \tilde{E}_{k'})^{-1}(\hat{D}^{\circ(k)}) \circ (\prod_{k'=k}^{0} \tilde{E}_{k'})]$. From Lemma 1‴ it follows that

$$\prod_{k'=k}^{0} E_{k'} \approx \prod_{k'=k}^{0} \prod_{l=n-3}^{0} \tilde{Z}_{ln+n-1-k',(l+1)n+n-1-k'}$$
$$(35) \qquad \approx \prod_{l=n-3}^{0} \prod_{k'=k}^{0} \tilde{Z}_{ln+n-1-k',(l+1)n+n-1-k'}$$
$$\approx \prod_{l=n-3}^{0} \prod_{m=n-1-k}^{n-1} \tilde{Z}_{ln+m,(l+1)n+m}.$$

Because $\hat{D}^{\circ(k)}$ corresponds to the sequence $\{\tilde{Z}_{k(n-1)+r',\, k(n-1)+r'-1},$ $r' = n-2, \ldots, 0\}$, we have that $(\prod_{k'=k}^{0} \tilde{E}_{k'})^{-1}(\hat{D}^{\circ(k)})(\prod_{k'=k}^{0} \tilde{E}_{k'})$ corresponds (by the same way) to the sequence

$$(36) \left\{ \left(\prod_{k'=k}^{0} \tilde{E}_{k'} \right)^{-1} \tilde{Z}_{k(n-1)+r',k(n-1)r'-1)} \left(\prod_{k'=k}^{0} \tilde{E}_{k'} \right), r' = n-2, \ldots, 0 \right\}.$$

Using (35) and observing that in (35) $m \in (n-1-k, \ldots, n-1)$ we can easily see that for $r' \geq k+1$, that is when $1 \leq r'-k \leq n-2-k$, the braid $\tilde{Z}_{k(n-1)+r',k(n-1)+r'-1} = \tilde{Z}_{kn+r'-k,kn+r'-k-1}$ commutes with $\prod_{k'=k}^{0} \tilde{E}_{k'}$. For $r' = k$ we get from (35)

$$\left(\prod_{k'=k}^{0}\tilde{E}_{k'}\right)^{-1}\tilde{Z}_{kn,kn-1}\left(\prod_{k'=k}^{0}\tilde{E}_{k'}\right)$$

$$=\left(\prod_{l=n-3}^{0}\prod_{m=n-1-k}^{n-1}\tilde{Z}_{ln+m,(l+1)n+m}\right)^{-1}$$

$$\tilde{Z}_{kn,(k-1)n+n-1}$$

$$\left(\prod_{l=n-3}^{0}\prod_{m=n-1-k}^{n-1}\tilde{Z}_{ln+m,(l+1)n+m}\right)$$

$$=\left(\prod_{m=n-1-k}^{n-1}\tilde{Z}_{(k-1)n+m,kn+m}\right)^{-1}$$

$$\tilde{Z}_{kn,(k-1)n+n-1}$$

$$\left(\prod_{m=n-1-k}^{n-1}\tilde{Z}_{(k-1)n+m,kn+m}\right)$$

$$=\tilde{Z}^{-1}_{(k-1)n+n-1,kn+n-1}\cdot\tilde{Z}_{kn,(k-1)n+n-1}\cdot\tilde{Z}_{(k-1)n+n-1,kn+n-1}$$

$$=\tilde{Z}_{kn,kn+n-1}.$$

Now let $r'\in(0,1,\ldots,k-1)$. Then

$$\left(\prod_{k'=k}^{0}E_{k'}\right)^{-1}\tilde{Z}_{kn+r'-k,kn+r'-k-1}\left(\prod_{k'=k}^{0}\tilde{E}_{k'}\right)$$

$$=\left(\prod_{l=n-3}^{0}\prod_{m=n-1-k}^{n-1}\tilde{Z}_{ln+m,(l+1)n+m}\right)^{-1}$$

$$\tilde{Z}_{(k-1)n+n-k+r',(k-1)n+n-k-1+r'}$$

$$\cdot\left(\prod_{l=n-3}^{0}\prod_{m=n-1-k}^{n-1}\tilde{Z}_{ln+m,(l+1)n+m}\right)$$

$$=\left(\prod_{m=n-1-k}^{n-1}\tilde{Z}_{(k-1)n+m,kn+m}\right)^{-1}$$

$$\tilde{Z}_{(k-1)n+n-k+r',(k-1)n+n-k-1+r'}$$

$$\cdot\left(\prod_{m=n-1-k}^{n-1}\tilde{Z}_{(k-1)n+m,kn+m}\right)$$

$$=\tilde{Z}_{kn+n-k+r',kn+n-k-1+r'}.$$

Using the last remarks we see that (36) is actually equal to the following sequence

$$\{\tilde Z_{kn+r'-k,kn+r'-k-1}, r' = n-2,\dots,k+1;\ \tilde Z_{kn,kn+n-1};\\ \tilde Z_{kn+n-k+r',kn+n-k-1+r'},\ r' = k-1,\dots,0\}. \tag{37}$$

Denote by $\hat{\prod}(k)$ a polygon with the set of vertices: $\hat\alpha(k) = \{\tilde y_{kn+m'}, = m = 0,\dots,n-1\}$. Clearly (37) (= (36)) is a sequence of $n-1$ successive edges of $\hat{\prod}(k)$. Let $\tilde A^{(k)} = A(\hat{\prod}(k), \hat\alpha(k))$ (see p. 206). Using a $(n-2-k+1)$-cyclic shift in $\tilde A^{(k)}$ we see that

$$\tilde A^{(k)} = \prod_{m=n-1}^{1} \tilde Z_{kn+m,kn+m-1}.$$

Denote by $\tilde D^{(k)} = (\tilde A^{(k)})^{n-1}$ and by $\tilde D^{\circ(k)}$ a decomposition of $\tilde D^{(k)}$ given by Lemma 1′. Lemma 1″ says that this decomposition is independent (up to equivalence) of a cyclic shift in $\tilde A^{(k)} = A(\hat{\prod}(k), \hat\alpha(k))$. Thus

$$\tilde D^{\circ(k)} \approxeq (\prod_{k'=k}^{0} \tilde E_{k'})^{-1}(\hat D^{\circ(k)})(\prod_{k'=k}^{0} \tilde E_{k'})$$

and

$$\mathcal{L} \approxeq \left(\prod_{k=n-2}^{0} \tilde E_k\right) \cdot \prod_{k=n-2}^{0} \tilde D^{\circ(k)}$$

Using (28′) we obtain now that the horizontal braid monodromie for $X[2]$ could be written in the form

$$\begin{aligned}\Delta^2 &= \Delta^2\left(B[\Pi_2^{-1}(a), \Pi_2^{-1}(a) \cap C[2]\right)\\ &= \mathcal{L}^n \approxeq \left(\left(\prod_{k=n-2}^{0} \tilde E_k\right) \cdot \prod_{k=n-2}^{0} \tilde D^{\circ(k)}\right)^n.\end{aligned} \tag{38}$$

It follows from (35) that $\prod_{k=n-2}^{0} \tilde E_k \approxeq \prod_{l=n-3}^{0} \prod_{m=l}^{n-1} \tilde Z_{ln+m,(l+1)n+m}$.

Denote by $Z^*_{ln+m,(l+1)n+m} = (\tilde{A}^{(l)}\tilde{A}^{(l+1)})^{n-1-m}\tilde{Z}_{ln+n-1,(l+1)n+n-1}\cdot(\tilde{A}^{(l)}\tilde{A}^{(l+1)})^{-(n-1-m)}$, $m = n-1, n-2, \dots, 1, 0$. From Lemma 2′ we obtain that

$$\prod_{m=1}^{n-1}\tilde{Z}_{ln+m,(l+1)n+m} \approxeq \prod_{m=n-1}^{1} Z^*_{ln+m,(l+1)n+m}.$$

Thus

$$\prod_{k=n-2}^{0}\tilde{E}_k \approxeq \prod_{l=n-3}^{0}\prod_{m=n-1}^{1}\tilde{Z}^*_{ln+m,(l+1)n+m}.$$

Denoting by $\tilde{A} = \prod_{k=n-2}^{0}\tilde{A}^{(k)}$, $\tilde{D} = \prod_{k=n-2}^{0}\tilde{D}^{(k)} = (\tilde{A})^{n-1}$ and $\tilde{D}^\circ = \prod_{k=n-2}^{0}\tilde{D}^{\circ(k)}$ we rewrite (38) in the form:

$$(38')\qquad \Delta^2 = \left(\left(\prod_{l=n-3}^{0}\prod_{m=n-1}^{1} Z^*_{ln+m,(l+1)n+m}\right)\cdot\tilde{D}^\circ\right)^n$$

Now we can write (using Lemma 3′) and denoting by $Q_0 = \prod_{l=n-3}^{0}\tilde{Z}_{ln+n-1,(l+1)n+n-1}$

$$\begin{aligned}
&\left(\left(\prod_{l=n-3}^{0}\prod_{m=n-1}^{1} Z^*_{ln+m,(l+1)n+m}\right)\cdot\tilde{D}^\circ\right)^n \\
&\qquad \approxeq \left(\left(\prod_{m=n-1}^{1}\prod_{l=n-3}^{0} Z^*_{ln+m,(l+1)n+m}\right)\cdot\tilde{D}^\circ\right)^n \\
(39)\qquad &\qquad \approxeq \left(\left(\prod_{m=n-1}^{1}[\tilde{A}^{n-1-m}(Q_0)\tilde{A}^{-(n-1-m)}]\right)\cdot\tilde{D}^\circ\right)^n \\
&\qquad \approxeq \left(\prod_{j=0}^{n-2}\left(\left(\prod_{m=n-1-j}^{1}[\tilde{A}^{n-1-m}(Q_0)\tilde{A}^{-(n-1-m)}]\right)\cdot\tilde{D}^\circ\right.\right. \\
&\qquad\qquad \prod_{m=n-1}^{n-1-j}[\tilde{A}^{n-1-m}(Q_0)\tilde{A}^{-(n-1-m)}]\cdot\tilde{D}^\circ
\end{aligned}$$

$$\approx \left(\prod_{j=0}^{n-2}\left(\left(\prod_{m=n-1-j}^{1}[\tilde{A}^{n-1-m}(Q_0)\tilde{A}^{-(n-1-m)}]\right)\right.\right.$$

$$\left.\left.\cdot\left(\prod_{m=n-1}^{n-1-j}[\tilde{D}\tilde{A}^{n-1-m}(Q_0)\tilde{A}^{-(n-1-m)}]\right)\cdot\tilde{D}^\circ\right)\cdot\tilde{D}^\circ\right.$$

$$\approx \left(\prod_{j=0}^{n-2}\left(\left(\prod_{m=n-1-j}^{1}[\tilde{A}^{n-1-m}(Q_0)\tilde{A}^{-(n-1-m)}]\right)\right.\right.\tag{39}$$

$$\left.\left.\cdot\left(\prod_{m=n-1}^{n-1-j}[\tilde{A}^{2(n-1)-m}(Q_0)\tilde{A}^{-(2(n-1)-m)}]\cdot\tilde{D}^\circ\right)\right)\cdot\tilde{D}^\circ\right.$$

$$\approx \left(\prod_{j=0}^{n-2}\left(\left(\prod_{r=0}^{n-1}[\tilde{A}^{j+r}(Q_0)\tilde{A}^{-(j+r)}]\right)\cdot\tilde{D}^\circ\right)\right)\cdot\tilde{D}^\circ.$$

But

$$\prod_{r=0}^{n-1}[\tilde{A}^{j+r}(Q_0)\tilde{A}^{-(j+r)}]$$

$$\approx \prod_{r=0}^{n-1}\prod_{l=n-3}^{0}[\tilde{A}^{j+r}(\tilde{Z}_{ln+n-1,(l+1)n+n-1})\tilde{A}^{-(j+r)}]$$

$$\approx \prod_{l=n-3}^{0}\prod_{r=0}^{n-1}[\tilde{A}^{j+r}(\tilde{Z}_{ln+n-1,(l+1)n+n-1})\tilde{A}^{-(j+r)}]$$

(the last equivalence follows from Lemma 3′). From Lemma 2 we obtain that

$$\prod_{r=0}^{n-1}[\tilde{A}^{j+r}(\tilde{Z}_{ln+n-1,(l+1)n+n-1})\tilde{A}^{-(j+r)}]$$

$$\approx \prod_{r=0}^{n-1}[\tilde{A}^{r}(\tilde{Z}_{ln+n-1,(l+1)n+n-1})\tilde{A}^{-r}]$$

$$\approx \prod_{q=n-1}^{0} Z^*_{ln+q,(l+1)n+q}$$

and so

$$\prod_{r=0}^{n-1}[\tilde{A}^{j+r}(Q_0)\tilde{A}^{-(j+r)}] \approx \prod_{l=n-3}^{0}\prod_{q=n-1}^{0} Z^*_{ln+q,(l+1)n+q}.$$

Going back to (39) we get

$$\left(\left(\prod_{l=n-3}^{0}\prod_{m=n-1}^{1} Z^*_{ln+m,(l+1)n+m}\right)\cdot\tilde{D}^{\circ}\right)^n$$

$$\approx \left(\prod_{j=0}^{n-2}\left(\prod_{l=n-3}^{0}\prod_{q=n-1}^{0} Z^*_{ln+q,(l+1)n+q}\right)\cdot\tilde{D}^{\circ}\right)\cdot\tilde{D}^{\circ}$$

$$\approx \left(\left(\prod_{l=n-3}^{0} P^*_l\right)\cdot\tilde{D}^{\circ}\right)^{n-1}\cdot\tilde{D}^{\circ},$$

where

$$P^*_l \stackrel{\text{def}}{=} \prod_{q=n-1}^{0} Z^*_{ln+q,(l+1)n+q}.$$

It is easy to check that $\forall k = n-2,\ldots,1,0$

$$\left(\prod_{l=n-3}^{0} P^*_l\right)\tilde{D}^{\circ(k)}\left(\prod_{l=n-3} P^*_l\right)^{-1} \approx \tilde{D}^{\circ(k-1)},$$

where $\tilde{D}^{\circ(-1)} \stackrel{\text{def}}{=} \tilde{D}^{\circ(n-2)}$. This means that

$$\left(\prod_{l=n-3}^{0} P^*_l\right)\tilde{D}^{\circ}\left(\prod_{l=n-3}^{0} P^*_l\right)^{-1} \approx \tilde{D}^{\circ}$$

and so

$$\left(\left(\prod_{l=n-3}^{0} P^*_l\right)\cdot\tilde{D}^{\circ}\right)^{n-1}\cdot\tilde{D}^{\circ} \approx (\tilde{D}^{\circ})^n\cdot\left(\prod_{l=n-3}^{0} P^*_l\right)^{n-1}.$$

So finally we get the following expression for the horizontal braid monodromie on Step 2:

(40) $$\Delta^2 = \Delta^2\left(B\left[\prod_{2}^{-1}(a), \prod_{2}^{-1}(a)\cap C[2]\right]\right) = (\tilde{D}^{\circ})^n\cdot\left(\prod_{l=n-3}^{0} P^*_l\right)^{n-1}.$$

(40) almost immediately gives a normal form for the horizontal braid monodromie of $X[3]$ related to $(24)_3$.

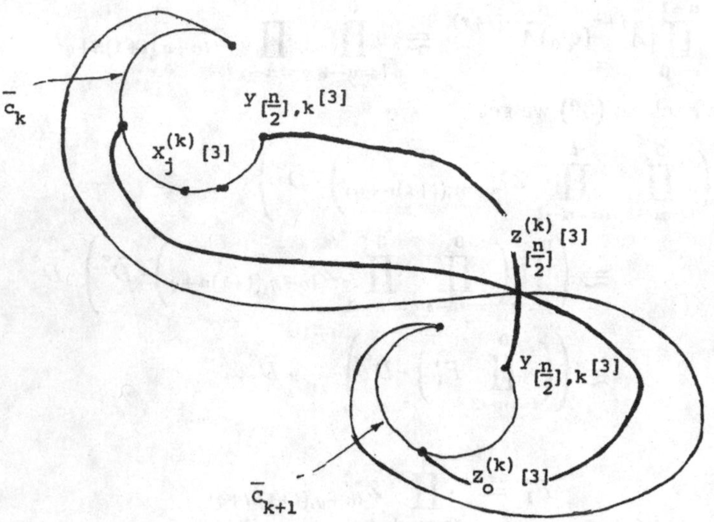

Figure 9

Step 3. Introduce the following notations (see Figure 9):

$$X_j^{(k)}[3] = \langle y_{jk}[3], y_{j+1k}[3]\rangle, \quad k = 0, 1, \ldots, n-2, \; j = 0, 1, \ldots, n-2,$$

$$A^{(k)}[3] = \prod_{j=n-2}^{0} X_j^{(k)}[3], \quad D^{(k)}[3] = (A^{(k)}[3])^{n-1},$$

$\mathring{D}^{(k)}[3]$ = (factorization of $D^{(k)}[3]$ given by Lemma 1′ (product of elements of degree one, two and three)), $\mathring{D}[3] = \prod_{k=n-2}^{0} \mathring{D}^{(k)}[3]$, $Z^{(k)}_{[\frac{n}{2}]}[3]$ = braid (half-twist) corresponding to the path $\langle y_{[\frac{n}{2}],k}[3], y_{[\frac{n}{2}],k+1}[3]\rangle$, $k = 0, \ldots, n-3$, $Z_j^{(k)}[3] = (A^{(k+1)}[3])^{[\frac{n}{2}]-j}(A^{(k)}[3])^{[\frac{n}{2}]-j}(Z^{(k)}_{n-1}[3])$ $(A^{(k)}[3])^{-([\frac{n}{2}]-j)}, (A^{(k+1)}[3])^{-([\frac{n}{2}]-j)}$, $P_k[3] = \prod_{j=n-1}^{0} Z_j^{(k)}[3]$.

An explicit description of a transition $N[2] \to N[3]$ given on p. 246 and the formula (40) give the following normal form of h.b.m. of $X[3]$ (related to $(24)_3$)

$$\Delta^2 = \Delta^2\Big(B[\Pi_2^{-1}(a), \Pi_2^{-1}(a) \cap C[3]\Big)$$

(41)
$$= (\overset{\circ}{D}[3])^n \cdot \Big(\prod_{k=n-3}^{0} P_k[3]\Big)^{n-1}$$

Step 4. Introducing notations $\overset{\circ}{D}^{(k)}[4]$, $\overset{\circ}{D}[4]$, $Z_j^{(k)}[4]$, $P_k[4]$, similar to $\overset{\circ}{D}^{(k)}[3]$, $\overset{\circ}{D}[3]$, $Z_j^{(k)}[3]$, $P_k[3]$, and using Lemma 1″ and Lemma 2, we get from (41) a similar normal form of h.b.m. of $X[4]$ related to $(24)_4$, namely:

$$\Delta^2 = \Delta^2\Big(B\Big[\Pi_2^{-1}(a), \Pi_2^{-1}(a) \cap C[4]\Big]\Big)$$

(42)
$$= (\overset{\circ}{D}[4])^n \cdot \Big(\prod_{k=n-3}^{0} P_k[4]\Big)^{n-1}$$

$$\approx \prod_{k=0}^{n-2} (\overset{\circ}{D}^{(k)}[4])^n \cdot \Big(\prod_{k=n-3}^{0} P_k[4]\Big)^{n-1}$$

Recall that

$$N[4] = \{y_{jk}[4] \approx (1 - \frac{1}{n-1}\alpha_0^{n/n-1}\varsigma_j')\mu_k),$$
$$j = 0, 1, \ldots, n-1, \quad k = 0, 1, \ldots, n-2\},$$

$\varsigma_j' = \exp(\pi i(\frac{2j+1}{n}))$, $\mu_k = \exp(2\pi\frac{ik}{n-1})$. (see p. 244). Denote by

$$Z_{jj'}^{(k)} = \langle y_{jk}[4], y_{j'k}[4]\rangle, \quad X_j^{(k)}[4] = Z_{j,j+1}^{(k)}[4],$$

$$\beta^{(k)}[4] = \Big(\prod_{j=0}^{n-3} X_j^{(k)}[4]\Big)(X_{n-2}^{(k)}[4])\Big(\prod_{j=0}^{n-3} X_j^{(k)}[4]\Big)^{-1},$$

$$A_{n-2}^{(k)}[4] = \prod_{j=1}^{n-3} X_j^{(k)}[4], \quad D_{n-2}^{(k)}[4] = (A_{n-2}^{(k)}[4])^{n-1},$$

$$\overset{\circ}{D}{}_{n-2}^{(k)}[4] \approx (X_1^{(k)}[4])^3 \prod_{j=3}^{n-2}\Big(\Big(\prod_{l=1}^{j-2}(Z_{j,l}^{(k)}[4])^2\Big) \cdot (X_{j-1}^{(k)}[4])^3\Big)$$

(expression given by Lemma 1(i),

$$\hat{Z}^{(k)}_{n-1,j}[4] = (\beta^{(k)}[4])(Z^{(k)}_{0,j}[4])(\beta^{(k)}[4])^{-1}.$$

Now applying Proposition 1 to $\mathring{D}^{(k)}[4]$ we get:

$$(43)\quad \begin{aligned}(\mathring{D}^{(k)}[4])^n \equiv \Big[&(X_0^{(k)}[4])^3 \cdot ((\beta^{(k)}[4])(X_0^{(k)}[4])(\beta^{(k)}[4])^{-1})^3 \\ &\cdot ((\beta^{(k)}[4])^2(X_0^{(k)}[4])(\beta^{(k)}[4])^{-2})^3 \\ &(\prod_{l=2}^{n-2}((Z^{(k)}_{0l}[4])^2 \cdot (\hat{Z}^{(k)}_{n-1,j}[4])^2) \cdot (D^{(k)}_{n-2}[4])\Big]^{n-2} \\ &\cdot \underbrace{\beta^{(k)}[4] \cdot \ldots \cdot \beta^{(k)}[4]}_{n \text{ times}}.\end{aligned}$$

Denote by $\overline{Z}^{(k)}_{n-1}[4]$ the braid (half-twist) corresponding to the path $\langle y_{n-1,k}[4], 0\rangle \cup \langle 0, y_{n-1,k+1}[4]\rangle$, for $j = 1, \ldots, n-2$ by

$$\overline{Z}^{(k)}_j[4] = (Z^{(k)}_{n-1,j}[4] \cdot Z^{(k+1)}_{n-1,j}[4])^{-1}(Z^{(k)}_{n-1}[4])(Z^{(k)}_{n-1,j}[4] \cdot Z^{(k+1)}_{n-1,j}[4])$$

and by $\overline{Z}^{(k)}_0[4] = (D^{(k)}[4] \cdot D^{k+1}[4])(Z^{k}_{n-1}[4])(D^{(k)}[4] \cdot D^{k+1}[4])^{-1}$ (see Figure 10).

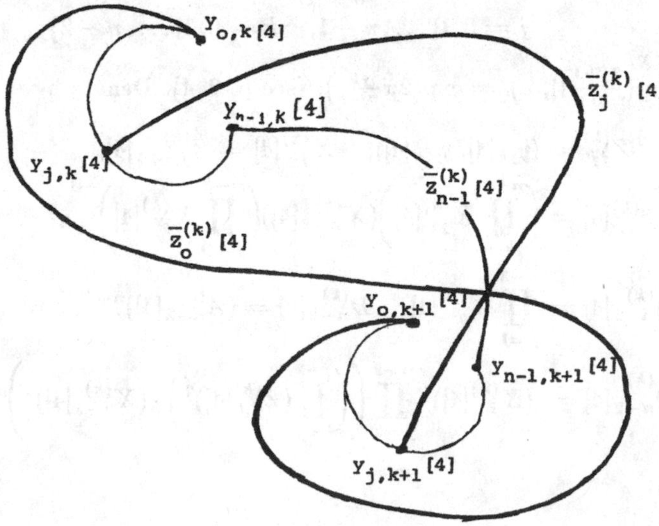

Figure 10

It is easily follows from Lemmas 2 and 2′ that

$$(44)\qquad P_k[4] \cong \left(\prod_{j=1}^{n-2} \overline{Z}_j^{(k)}[4]\right) \cdot \overline{Z}_{n-1}^{(k)}[4] \cdot \overline{Z}_0^{(k)}[4].$$

Step 5. Replacing in (42) each $(\mathring{D}^{(k)}[4])^n$ and each $P_k[4]$ by expressions given by (43) and (44) we will get an expression for the horizontal braid monodromie which in the transition process $N[4] \to N[5]$ (Step 5, see p. 247) becomes a normal form for the horizontal braid monodromie of $X[5]$ related to $[24]_5$. Namely, let $X_l^{(k)}[5] = \langle y_{lk}[5], y_{l+1,k}[5]\rangle$, $l = 0, 1, \ldots, n-4$, $k = 0, 1, \ldots, n-2$,(see p. 245), $A^{(k)}[5] = \prod_{l=n-4}^{0} X_l^{(k)}[5]$, $D_{n-2}^{(k)}[5] = (A^{(k)}[5])^{n-1}$, $\mathring{D}_{n-2}^{(k)}[5]$ be a factorization of $D_{n-2}^{(k)}[5]$ (as a product of cubes and squares) given by Lemma 1(i), V_k be the braid (half twist) corresponding to the path $\langle y_{n-3,k}[5], y_{0,k+1}[5]\rangle$, for $j = 0, 1, \ldots, n-3$,

$$Z_j^{(k)}[5] = (A^{(k)}[5])^{-j}(A^{(k+1)}[5])^{n-3-j}(V_k)(A^{k+1}[5])^{-(n-3-j)}(A^{(k)}[5])^j$$

$$G_k[5] = \prod_{j=n-3}^{0} Z_j^{(k)}[5],$$

$$\beta^{(k)}[5] = \langle \overline{y}_{0k}[5], \overline{y}_{1k}[5]\rangle$$

for $\tau = 0, 1$,

$$Y_{\tau,n-3}^{(k)}[5] = \langle \overline{y}_{\tau k}[5], \overline{y}_{n-3,k}[5]\rangle,$$

$$Y_{\tau,j}^{(k)}[5] = (A^{(k)}[5])^{n-3-j}(Y_{\tau,n-3}^{(k)}[5])(A^{(k)}[5])^{-(n-3-j)} (j = 0, \ldots, n-3),$$

$P_0^{(k)}[5]$ (resp. $P_1^{(k)}[5]$) be the braid (half twist) corresponding to $\overline{Z}_{n-1}^{(k)}[4]$ (resp. to $\overline{Z}_0^{(k)}[4]$) in the transition process of Step 5. Clearly, in this transition $\mathring{D}_{n-2}^{(k)}[5]$ corresponds to $\mathring{D}_{n-2}^{(k)}[4]$, the expression $G_k[5]$ to $\prod_{j=1}^{n-2} Z_j^{(k)}[4]$, $\beta^{(k)}[5]$ to $\beta^{(k)}[4]$, $Y_{1,n-3}^{(k)}[5]$ to $X_0^{(k)}[4]$, $Y_{1l}^{(k)}[5]$ to $Z_{0l}[4]$, $(l = 2, \ldots, n-2)$, $Y_{0l}^{(k)}[5]$ to $\hat{Z}_{n-1,l}[4]$.

Now from (42), (43), (44) we get the following expression for the horizontal braid monodromie of $X[5]$ (related to $(24)_5$):

$$\begin{aligned}\Delta^2 &= \Delta^2(B[\Pi_2^{-1}(a), \Pi_2^{-1}(a) \cap C[5]])\\ &= \Bigg[\prod_{k=0}^{n-2} ((Y_{1,n-3}^{(k)}[5])^3((\beta^{(k)}[5])(Y_{1,n-3}^{(k)}[5])(\beta^{(k)}[5])^{-1})^3\\ &\quad \cdot ((\beta^{(k)}[5])^2(Y_{1,n-3}^{(k)}[5])(\beta^{(k)}[5])^{-2})^3\\ &\quad \cdot \prod_{l=n-4}^{0} ((Y_{1l}^{(k)}[5])^2(Y_{0l}^{(k)}[5])^2) \cdot \overset{\circ}{D}{}_{n-2}^{(k)}[5]\Bigg]^{n-2}\\ &\quad \cdot \left(\prod_{k=n-3}^{0} (G_k[5] \cdot P_0^{(k)}[5] \cdot P_1^{(k)}[5])\right)^{n-1} \cdot \left(\prod_{i'=0}^{n-2} \beta^{(k)}[5]\right)^n. \end{aligned} \tag{45}$$

It is easy to see that (45) is a "ready to collapse" normal form coinciding with the expression given in (22).[5]

In steps 6 and 7 the expression (45) will not have any essential changes.

---

[5]It follows from Lemma 3 that $\left(\prod_{k=n-3}(G_k[5] \cdot P_0^{(k)}[5] \cdot P_1^{(k)}[5])\right)^{n-1} = \left((\prod_{k=0}^{n-3}(G_k[5] \cdot P_0^{(k)}[5] \cdot P_1^{(k)}[5])\right)^{n-1}$.

References

[1] Garside, F. A., "The braid group and other groups," Quart. J. Math. Oxford 20, No. 78, pp. 235–254, (1969).

[2] Moishezon, B.G., "Stable branch curves and braid monodromies," in Lecture Notes in Math., vol. 862 (1981), pp. 107–192.

[3] Persson, U., "Chern invariants of surfaces of general type," Comp. Math. (1981).

Received August 10, 1982.

Professor B. Moishezon
Department of Mathematics
Columbia University
New York, New York 10027.

Towards an Enumerative Geometry of the Moduli Space of Curves

David Mumford

Dedicated to Igor Shafarevitch on his 60th birthday

Introduction

The goal of this paper is to formulate and to begin an exploration of the enumerative geometry of the set of all curves of arbitrary genus g. By this we mean setting up a Chow ring for the moduli space $\mathcal{M}_g$ of curves of genus g and its compactification $\overline{\mathcal{M}}_g$, defining what seem to be the most important classes in this ring and calculating the class of some geometrically important loci in $\overline{\mathcal{M}}_g$ in terms of these classes. We take as a model for this the enumerative geometry of the Grassmannians. Here the basic classes are the Chern classes of the tautological or universal bundle that lives over the Grassmannian, and the most basic cycles are the loci of linear spaces satisfying various Schubert conditions: the so-called Schubert cycles. However, since Harris and I have shown that for g large, $\mathcal{M}_g$ is not unirational [H-M] it is not possible to expect that $\mathcal{M}_g$ has a decomposition into elementary cells or that the Chow ring of $\mathcal{M}_g$ is as simple as that of the Grassmannian. But in the other direction, J. Harer [Ha] and E. Miller [Mi] have strong results indicating that at least the low dimensional homology groups of $\mathcal{M}_g$ behave nicely. Moreover, it appers that many geometrically natural cycles are all expressible in terms of a small number of basic classes.

More specifically, the paper is divided into 3 parts. The goal of the first part is to define an intersection product in the Chow group of $\overline{\mathcal{M}}_g$. The problem is that due to curves with automorphisms, $\overline{\mathcal{M}}_g$ is singular, but in a mild way. In fact it is a "Q-variety", locally the quotient of a smooth variety by a finite group. If it were *globally* the quotient of a smooth variety by a finite group, it would be easy to define a product in $A\cdot(\overline{\mathcal{M}}_g)\otimes \mathbb{Q}$. Instead we have used the fact that $\overline{\mathcal{M}}_g$ is globally the quotient of a Cohen-Macaulay variety by a finite group, plus many of the ideas of Fulton and MacPherson, and especially a strong use of both the Grothendieck and Baum-Fulton-MacPherson forms of the Riemann-Roch theorem to achieve this goal. To handle an arbitrary Q-variety, Gillet has proposed using higher K-theory ($H^n(K^n) \cong A^n$) and this may well be the right technique.

The goal of the second part is to introduce a sequence of "tautological" classes $\kappa_i \in A^i(\overline{\mathcal{M}}_g) \otimes \mathbb{Q}$, derive some relations between them, and calculate the fundamental class of certain subvarieties, such as the hyperelliptic locus, in terms of them. Again the Grothendieck Riemann-Roch theorem is one of the main tools. Some of these results have been found independently by E. Miller [Mi], and it seems reasonable to guess, in view of the results of Harer and Miller (op. cit.), that in low codimensions $H^i(\mathcal{M}_g) \otimes \mathbb{Q}$ is a polynomial ring in the κ_i.

Finally, to make the whole theory concrete, we work out $A^{\cdot}(\overline{\mathcal{M}}_2)$ completely in Part III. An interesting corollary is the proof, as a consequence of general results only, that $\mathcal{M}_2$ is affine. It seems very worthwhile to work out $A^{\cdot}(\overline{\mathcal{M}}_g)$ or $H^{\cdot}(\mathcal{M}_g)$ for other small values of g, in order to get some feeling for the properties of these rings and their relation to the geometry of $\overline{\mathcal{M}}_g$. The techniques of Atiyah-Bott [A-B] may be very useful in doing this.

Part I: Defining a Chow Ring of the Moduli Space

§1. Fulton's Operational Chow Ring

If X is any quasi-projective variety, Fulton and Fulton-MacPherson have defined in two papers ([F1], p. 157, [F-M], p. 92) two procedures to attach to X a kind of Chow cohomology theory: a ring-valued contravariant functor. We combine here the 2 definitions taking the simplest parts of them in a way that is adequate for our applications. The theory also becomes substantially simpler if we take the *coefficients for our cycles to be* $\mathbb{Q}$, and we assume that *char* $k = 0$. This is the case we are interested in, *so we will restrict ourselves to this case henceforth.* We call the resulting ring $opA^{\cdot}(X)$. To form this ring, take:

generators : *elts*(f, α),

$f: X \longrightarrow Y$ *a morphism*

Y *smooth, quasi-projective*

α *a cycle on* Y

relations : $(f,\alpha) \sim 0$ if, for all $g\colon Z \longrightarrow X$, we have:
$(f \circ g)^*\alpha$ rationally equivalent to 0 on Z
via the induced map

$$A^k(Y) \xrightarrow{(f\circ g)^*} A_{n-k}(Z)$$

$(k = \text{codim of } \alpha, n = \dim of Z)$

Equivalently, we may define

$$opA^{\cdot}(X) = \text{Image}\left\{ \varinjlim_{(X \xrightarrow{f} Y)} A^{\cdot}(Y) \longrightarrow \prod_{(Z \xrightarrow{g} X)} \text{End}(A_{\cdot}(Z))\right\}$$

where the map is given by cap product

$$A^k(Y) \times A_l(Z) \xrightarrow{\cap} A_{l-k}(Z)$$

(cf. [F1], §2). This makes it clear that $op A^{\cdot}$ is a ring and a contravarient functor and that $opA^{\cdot}(X)$ acts on $A_{\cdot}(X)$ by cap product. If X is smooth, then $opA^{\cdot}(X) \cong A^{\cdot}(X)$.

Moreover, as in Fulton [F1], §3.2, for all coherent sheaves $\mathcal{F}$ with finite projective resolutions, we can define the Chern classes $c_k(\mathcal{F}) \in opA^k(X)$, (by resolution of $\mathcal{F}$, twisting and pull-back of Schubert cycles from maps of X to Grassmannians).

Using a resolution of X, we can give a very simple description of the relations in $opA^{\cdot}(X)$:

Proposition 1.1. *If $\pi\colon \tilde{X} \longrightarrow X$ is a resolution of X, then*

$$(f,\alpha) \sim 0 \Leftrightarrow (f \circ \pi)^*(\alpha) = 0 \quad \text{in } A^k(\tilde{X}),$$

i.e.,

$$opA^{\cdot}(X) \subset A^{\cdot}(\tilde{X}).$$

Proof. We must show that if

$$g\colon Z \longrightarrow X$$

is any test morphism, then $l(f\circ\pi)^*\alpha$ rationally equivalent to 0 on $\tilde{X}$ implies $l'(f \circ g)^*\alpha$ rationally equivalent to 0 on Z for some l'. But by taking a suitable subvariety of $Z \times_X \tilde{X}$ we get a diagram

$$\begin{array}{ccccc} \tilde{Z} & \xrightarrow{\tilde{g}} & \tilde{X} & & \\ p \downarrow & & \downarrow \pi & & \\ Z & \xrightarrow{g} & X & \xrightarrow{f} & Y \end{array}$$

where p is proper, surjective, generically finite of degree l''. Therefore

$$\begin{aligned} l''(f \circ g)^* \alpha &= p_*((f \circ g \circ p)^* \alpha) \\ &= p_*(\tilde{g}^*((f \circ \pi)^* \alpha)) \end{aligned}$$

hence

$$l.l''(f \circ g)^* \alpha = p_*(\tilde{g}^*(l(f \circ \pi)^* \alpha)) \sim 0.$$

This uses the formula:

$$(*) \quad \text{For all} \quad \begin{array}{ccc} \tilde{Z} & \xrightarrow{\tilde{h}} & Y \\ p \downarrow & & \| \\ Z & \xrightarrow[h]{} & Y \end{array} \quad \begin{array}{l} Y \text{ smooth, } \alpha \text{ a cycle on } Y, \\ p \text{ proper, surjective, generically finite of degree } d \end{array}$$

$$p_* \tilde{h}^*(\alpha) \sim d h^*(\alpha).$$

(See [F1], §2.2, part 2 of lemma). Q.E.D.

In fact, we can say more:

Proposition 1.2. *In the situation of Prop. 1.1, the image of* $opA^k(X)$ *in* $A^k(\tilde{X})$ *is contained in the subgroup of* $A^k(\tilde{X})$ *generated by irreducible subvarieties* W *of* $\tilde{X}$ *such that* $W = \pi^{-1}(\pi(W))$.

Proof. Let (f, α) be a generator of $op A^k(X)$, where $f: X \longrightarrow Y$ is a morphism and Y is smooth, quasi-projective. Let $p = f \circ \pi$, and

$$Y_k = \{y \in Y \mid \dim p^{-1}(y) \geq k\}.$$

Then use the moving lemma on Y to represent α as a cycle $\sum n_i W_i$, whose components W_i meet properly all the Y_k. Then each component W_{ij} of $p^{-1}(W_i)$ meets the open set of $\tilde{X}$ where π is an isomorphism and

$p: \tilde{X} \longrightarrow p(\tilde{X})$ is equidimensional. Therefore, $p^*\alpha$ is represented by a combination of the W_{ij} with suitable multiplicities and each W_{ij} satisfies $W_{ij} = \pi^{-1}(\pi(W_{ij}))$. Q.E.D.

There is also another natural way to give generators for $opA^{\cdot}(X)$:

Proposition 1.3. *Using rational coefficients in $K(X)$ too, the homomorphism*

$$ch : K(X) \longrightarrow opA^{\cdot}(X)$$

is surjective, hence $opA^{\cdot}(X)$ can be defined to be

$$\begin{aligned} \text{Image}\,[K(X) &\rightarrow A^{\cdot}(\tilde{X})] \\ \mathcal{E} &\mapsto ch(\pi^*\mathcal{E}) \end{aligned}$$

($\pi: \tilde{X} \longrightarrow X$ a resolution of X).

Proof. It is well known that for any smooth quasi-projective Z, ch: $gr\,K(Z) \longrightarrow A^{\cdot}(Z)$ is a graded isomorphism, hence taking the total Chern characters, ch: $K(Z) \longrightarrow A^{\cdot}(Z)$ is also an isomorphism[1]. Therefore ch: $K(X) \longrightarrow opA^{\cdot}(X)$ is surjective.

$opA^{\cdot}(X)$ has a much subtler *covariance* for certain morphisms f, whose existence is tied up with the version of the Grothendieck-Riemann-Roch theorems for $opA^{\cdot}$. The result is this:

Theorem 1.4 (Fulton). *Let $f: X \longrightarrow Y$ be a projective local-complete-intersection morphism. Define $Td_f \in opA^{\cdot}(X)$ in the usual way. Then there is a homomorphism*

$$f_*: opA^{\cdot}(X) \longrightarrow opA^{\cdot}(Y)$$

such that

---

[1]This sounds a bid odd, but it is perhaps clarified by the observation that for any rank r and dimension n, there are universal polynomials P_k such that for all vector bundles $\mathcal{E}$ of rank r on n-dimensional varieties,

$$c_k(\mathcal{E}) = P_k(ch\,\mathcal{E}, ch\,\Lambda^2\mathcal{E}, \ldots, ch\,\Lambda^r\mathcal{E})$$

where $ch\,\mathcal{E}$ is the *total* Chern character, and these elements lie in any cohomology ring with the usual Chern formalism and rational coeffcients.

1) *for all cartesian diagrams*

$$\begin{array}{ccc} X' & \xrightarrow{h} & X \\ {\scriptstyle f'}\downarrow & & \downarrow{\scriptstyle f} \\ Y' & \xrightarrow[g]{} & Y \end{array}$$

and all $\alpha \in A_.(Y')$, $\beta \in opA^{\cdot}(X)$

$$f_*\beta \cap \alpha = f'_*(\beta \cap [f]_{Y'}(\alpha)) \quad \text{in } A_.(Y')$$

($[f]_{Y'}$ defined as in Fulton-MacPherson [F-M], p. 95).

2) *for all locally free sheaves* $\mathcal{E}$ *on* X,

$$ch(f_!\mathcal{E}) = f_*(ch\,\mathcal{E}.Td_f).$$

This is proven in Fulton [F2], Ch.18: here $opA^{\cdot}$ is a possibly larger Chow Ring in which (1) is the *definition* of f_*. In this ring (2) is proven, and (2) shows that $f_*\beta$ actually lies in the subring $opA^{\cdot}$ considered here.

§2. Q-Varieties and $\overline{\mathcal{M}}_g$

The moduli space $\overline{\mathcal{M}}_g$ of stable curves is an example of a variety which is locally in the étale topology a quotient of a smooth variety by a finite group. The approach we take to defining a Chow ring for $\overline{\mathcal{M}}_g$ is best studied in this more general context. Because these varieties are quite close to the objects introduced by Matsusaka [Ma], we shall call them quasi-projective Q-varieties. We define a quasi-projective Q-variety to be:

1) a quasi-projective variety X,
2) a finite atlas of charts:

$$\begin{array}{c} X_\alpha \\ \downarrow \\ X_\alpha/G_\alpha \\ \downarrow{\scriptstyle p'_\alpha} \\ X \end{array} \qquad (p_\alpha : X_\alpha \to X)$$

where p'_α is étale, G_α is a finite group acting, faithfully on a quasi-projective *smooth* X_α and

$$X = \bigcup_\alpha (\mathrm{Im}\ p_\alpha),$$

3) The charts should be compatible in the sense that for all α, β, let

$$X_{\alpha\beta} = \text{normalization of } X_\alpha \times_X X_\beta.$$

Then the projections

$$\begin{array}{ccc} & X_{\alpha\beta} & \\ \swarrow & & \searrow \\ X_\alpha & & X_\beta \end{array}$$

should be étale.

Here, of course, a new chart can be added if it satisfies the compatibility conditions (3) with the old ones. For any such Q-variety, we can normalize X in a Galois extension of its function field $k(X)$ containing copies of the field extensions $k(X_\alpha)$ for all α. This leads to a covering $p\colon \tilde{X} \longrightarrow X$ with a group G acting faithfully on $\tilde{X}$ and $X = \tilde{X}/G$. The fact that $k(\tilde{X}) \supset k(X_\alpha)$ leads to a factorization of p locally:

$$\begin{array}{ccccl} \tilde{X} & \supset & \tilde{X}_\alpha & & \tilde{X}_\alpha \text{ open in } \tilde{X} \text{ stabilized by} \\ \Big\downarrow p & & \downarrow & & H'_\alpha \subset G, \\ & & \tilde{X}_\alpha/H_\alpha \cong X_\alpha & & H_\alpha = \text{normal subgroup of } H'_\alpha \\ & & \swarrow p_\alpha & & \text{with } G_\alpha \cong H'_\alpha/H_\alpha. \\ X & & & & \end{array}$$

Q-varieties come in various different grades. The best ones are those such that for some atlas, $\tilde{X}$ can be chosen to be itself smooth. Not so nice, but still amendable to the techniques we shall use are those where $\tilde{X}$ can be chosen to be Cohen-Macaulay. We call these Q-varieties *with a smooth global cover* and with a *Cohen-Macaulay global cover* respectively.

Another important concept is that of a *Q-sheaf on a Q-variety* X. By this, we do not mean a coherent sheaf on X, but rather a family of coherent sheaves $\mathcal{F}_\alpha$ on X_α, plus isomorphisms

$$\mathcal{F}_\alpha \otimes_{\mathcal{O}_{X_\alpha}} \mathcal{O}_{X_{\alpha\beta}} \cong \mathcal{F}_\beta \otimes_{\mathcal{O}_{X_\beta}} \mathcal{O}_{X_{\alpha\beta}}$$

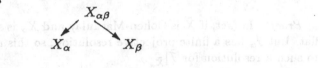

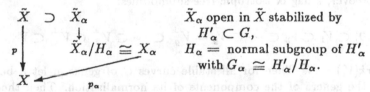

compatible on the triple overlaps. Note that such a coherent sheaf pulls back by tensor product to a family of coherent sheaves on $\tilde{X}_\alpha$, which glue together to one coherent sheaf $\tilde{\mathcal{F}}$ on $\tilde{X}$ on which G acts. Therefore, equivalently we can define a coherent sheaf on the Q-variety X to be a coherent G-sheaf $\tilde{\mathcal{F}}$ on $\tilde{X}$ such that for all α, $\tilde{\mathcal{F}}|_{\tilde{X}_\alpha}$ with its H'_α-action is the pull-back of a coherent sheaf on X_α. The importance of $\tilde{X}$ being Cohen-Macaulay is illustrated by the simple fact:

Proposition 2.1. *If $\tilde{X}$ is Cohen-Macaulay, then for any coherent sheaf $\mathcal{F}$ on the Q-variety X, $\tilde{\mathcal{F}}$ has a finite projective resolution.*

Proof. In fact, if $\tilde{X}$ is Cohen-Macaulay and X_α is smooth, $\tilde{X}_\alpha \to X_\alpha$ is flat. But $\mathcal{F}_\alpha$ has a finite projective resolution, so this resolution pulls back to such a resolution for $\tilde{\mathcal{F}}|_{\tilde{X}_\alpha}$. Q.E.D.

Consider the case of the moduli space $\overline{\mathcal{M}}_g$. Choose an integer $n \geq 3$ prime to the characteristic. Fix a free $\mathbb{Z}/n\mathbb{Z}$-module V of rank $2g$ with an alternating non-degenerate form

$$e: V \times V \longrightarrow \mu_n.$$

Fix, moreover, a flag of isotropic free submodules:

$$(0) \subset V_1 \subset V_2 \subset \cdots \subset V_g = V_g^\perp \subset \cdots \subset V_2^\perp \subset V_1^\perp \subset V$$

where $rk(V_i) = i$. Then for all stable curves C of genus g, let h be the sum of the genera of the components of its normalization. Then there is an isomorphism

$$H^1(C, \mathbb{Z}/n\mathbb{Z}) \cong V_{g-h}^\perp$$

such that the form

$$H^1 \times H^1 \overset{\cup}{\to} H^2 \xrightarrow{\text{f'al class}} \mu_n$$

corresponds to e. We may consider the auxiliary moduli space

$$\left(\overline{\mathcal{M}}_g^{(n,h)}\right)' = \text{set of pairs } (C, \phi), C \text{ stable curve,}$$

$$\phi : V_{g-h}^\perp \xhookrightarrow{\phi \text{ injective}} H^1(C, \mathbb{Z}/n\mathbb{Z}) \text{ a sympl. map}$$

which can be constructed by standard arguments. Inside this space, define an *open* subset by:

$$\overline{\mathcal{M}}_g^{(n,h)} = \text{those pairs } (C,\phi) \text{ such that every automorphism } \alpha\colon C \longrightarrow C \text{ fixes the submodule } \operatorname{Im}\phi \subset H^1(C,\mathbb{Z}/n\mathbb{Z}) \text{ and, if } \alpha \neq 1_C, \text{ then } \alpha \text{ acts non-trivially on } \operatorname{Im}\phi.$$

Since the pairs (C,ϕ) in this subset have no automorphisms, $\overline{\mathcal{M}}_g^{(n,h)}$ is smooth and represents the universal deformation space of any curve occuring in it. Note that every curve C occurs in the space $\overline{\mathcal{M}}_g^{(n,h)}$ such that $g+h = rk\, H^1(C)$ (see [D-M], Th. 1.13). Next consider the finite groups

$$G = Sp(V,\mathbb{Z}/n\mathbb{Z})$$

$$\cup$$

$$H'_h = (\text{stabilizer of } V_{g-h})$$

$$\cup$$

$$H_h = \begin{pmatrix}\text{elements which}\\ \text{act identically}\\ \text{on } V_{g-h}^{\perp}\end{pmatrix}$$

$$G_h = H'_h/H_h = (\text{induced group of automorphisms of } V_{g-h}^{\perp}).$$

Then G_h acts on $\overline{\mathcal{M}}_g^{(n,h)}$ and we have canonical morphisms

$$\begin{array}{c} \overline{\mathcal{M}}_g^{(n,h)} \\ \downarrow \\ \overline{\mathcal{M}}_g^{(n,h)}/G_h \\ \downarrow p'_h \\ \overline{\mathcal{M}}_g \end{array} \qquad (p_h \colon \overline{\mathcal{M}}_g^{(n,h)} \to \overline{\mathcal{M}}_g \text{ the composite})$$

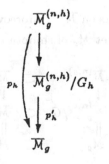

I claim that p'_h is étale. In fact, if $(C,\phi) \in \overline{\mathcal{M}}_g^{(n,h)}$, then $\mathrm{Aut}(C)$ can be identified with a subgroup of G_h and *formally* near $[C,\phi]$, the isomorphism

$$\overline{\mathcal{M}}_g^{(n,h)} \xleftarrow[\text{formal isom.}]{} \mathrm{Def}\,(C)$$

commutes with the action of $\mathrm{Aut}(C)$.

Therefore we have a diagram

$$\begin{array}{ccc}
\overline{\mathcal{M}}_g^{(n,h)} & \longleftarrow & \mathrm{Def}(C) \\
\downarrow & & \downarrow \\
\overline{\mathcal{M}}_g^{(n,h)}/\,\mathrm{Aut}(C) & \longleftarrow & \mathrm{Def}\,(C)/\,\mathrm{Aut}(C) \\
{\scriptstyle p}\downarrow & & \Big\downarrow \\
\overline{\mathcal{M}}_g^{(n,h)}/G_h & & \\
{\scriptstyle q}\downarrow & & \\
\overline{\mathcal{M}}_g & \longleftarrow & \mathrm{Spec}\,\hat{\mathcal{O}}_{\overline{\mathcal{M}}_g,[C]}
\end{array}$$

where the horizontal arrows are formal isomorphisms. Now $\mathrm{Def}\,(C)/\,\mathrm{Aut}(C)$ i.e., Spec of the $\mathrm{Aut}(C)$-invariants in the complete local ring representing the deformations of C, is isomorphic to Spec of the complete local ring of $\overline{\mathcal{M}}_g$ at [C]. Thus the morphism indicated by $q \circ p$ in the diagram is étale at $[C,\phi]$. Therefore so is q.

This proves that the atlas $\{p_h\colon \overline{\mathcal{M}}_g^{(n,h)} \longrightarrow \overline{\mathcal{M}}_g\}$ puts a structure of Q-variety on $\overline{\mathcal{M}}_g$. In this setting, what is the variety $\tilde{X}$ dominating all the charts? This is

$$\overline{\mathcal{M}}_g^{(n)} = (\text{normalization of } \overline{\mathcal{M}}_g \text{ in this field extension } \overline{\mathcal{M}}_g^{(n,g)}).$$

Note that the full group G acts on $\overline{\mathcal{M}}_g^{(n)}$. Moreover, $\overline{\mathcal{M}}_g^{(n,g)}$ is the open subset of $\overline{\mathcal{M}}_g^{(n)}$ lying over the open set $\mathcal{M}_g$ of smooth curves.

$$\begin{array}{cccc}
p^{-1}(\mathcal{M}_g) = \overline{\mathcal{M}}_g^{(n,g)} & \subset & \overline{\mathcal{M}}_g^{(n)} & \\
\downarrow & & \downarrow & \\
\overline{\mathcal{M}}_g^{(n,g)}/G & \subset & \overline{\mathcal{M}}_g^{(n)}/G & \Big) \, p \\
\| & & \| & \\
\mathcal{M}_g & \subset & \overline{\mathcal{M}}_g &
\end{array}$$

What we call $\overline{\mathcal{M}}_g^{(n,h)}$, for $h < g$, however, is recovered by dividing an open subset of $\overline{\mathcal{M}}_g^{(n)}$ by H_h.

Life would be particularly simple if $\overline{\mathcal{M}}_g^{(n)}$ were smooth. However, it is not, nor is it known whether the normalization of $\overline{\mathcal{M}}_g^{(n)}$ in any finite extension field is smooth. However, fortunately $\overline{\mathcal{M}}_g$ does have a Cohen-Macaulay global cover, because:

Proposition 2.2. *$\overline{\mathcal{M}}_g^{(n,g)} \subset \overline{\mathcal{M}}_g^{(n)}$ is a toroidal embedding, i.e., $\overline{\mathcal{M}}_g^{(n)}$ is formally isomorphic to $\mathbb{A}^{3g-3}$ modulo an abelian group acting diagonally, and with $\overline{\mathcal{M}}_g^{(n)} - \overline{\mathcal{M}}_g^{(n,g)}$ isomorphic to the image of a union of coordinate hyperplanes in $\mathbb{A}^{3g-3}$.*

Proof. At every point of $\overline{\mathcal{M}}_g^{(n)}$, $\overline{\mathcal{M}}_g^{(n)}$ is a Galois covering of one of the smooth varieties $\overline{\mathcal{M}}_g^{(n,h)}$ with group H_h. Note that H_h is abelian of order prime to char(k). The covering is ramified only on $\overline{\mathcal{M}}_g^{(n,h)} - p_h^{-1}(\mathcal{M}_g)$. Since $\overline{\mathcal{M}}_g^{(n,h)}$ is formally the universal deformation space of some curve C, this is formally a ramified cover of $\mathbb{A}^{3g-3}$, ramified only in coordinate hyperplanes. But if char$(k) \not| \, n$, the n-cyclic extensions $k[[x_1, \ldots, x_{3g-3}]]$ ramified only over the ideals (x_i) are all given by

$$\left(\prod_{i\in I} x_i^{a_i}\right)^{1/n}, \quad I \subset \{1, \ldots, 3g-3\}.$$

Thus the covering in hand is sandwiched between $k[[x_1, \ldots, x_{3g-3}]]$ and $k[[x_1^{1/n}, \ldots, x_{3g-3}^{1/n}]]$, hence, by Galois theory, is as described.

Corollary 2.3. *$\overline{\mathcal{M}}_g^{(n)}$ is Cohen-Macaulay.*

§2b. Q-Stacks

Unfortunately, the concept of Q-variety, although adequate to deal with $\overline{\mathcal{M}}_g$, $g \geq 3$, or with any moduli variety whose general object has no automorphisms, breaks down for $\overline{\mathcal{M}}_2$ and $\overline{\mathcal{M}}_{1,1}$ where the general object has automorphism group $\mathbb{Z}/2\mathbb{Z}$. Consider, for instance, $\overline{\mathcal{M}}_2$. Let $\mathcal{M}_2^\circ \subset \overline{\mathcal{M}}_2$

be the open set of smooth curves C such that $\mathrm{Aut}(C) \cong \mathbb{Z}/2\mathbb{Z}$. Then, although $\mathcal{M}_2^0$ gives a local deformation space for its curves, it does not carry a universal family of curves. And if $\mathcal{M}_2'$ is an étale cover of $\mathcal{M}_2^0$ carrying some family

$$p\colon \mathcal{C}' \longrightarrow \mathcal{M}_2'$$

the sheaf $\mathbb{E} = p_* \Omega^1_{\mathcal{C}'/\mathcal{M}'}$ on $\mathcal{M}_2'$ will not be a Q-sheaf. In fact, to compare $p_1^*\mathbb{E}$, $p_2^*\mathbb{E}$ on $\mathcal{M}_2' \times_{\mathcal{M}_2} \mathcal{M}_2'$, we want an isomorphism of the 2 families

$$\begin{array}{ccc} \mathcal{C}' \times_{\mathcal{M}_2} \mathcal{M}_2' & \cong & \mathcal{M}_2' \times_{\mathcal{M}_2} \mathcal{C}' \\ & \searrow \quad \swarrow & \\ & \mathcal{M}_2' \times_{\mathcal{M}_2} \mathcal{M}_2' & \end{array}$$

and although these families are fibrewise isomorphic, the isomorphism is not unique and may not globalize.

To deal with this, one must use some variant of the ideal of stack (cf. [D-M], §4). The most natural thing is to replace the normalization of $X_\alpha \times_X X_\beta$ by a scheme $X_{\alpha\beta}$ which must be given as part of the data and cannot be derived from the rest. $X_{\alpha\beta}$ should map to $X_\alpha \times_X X_\beta$ and given $x \in X_{\alpha\beta}$, $y \in X_{\beta\gamma}$ with the same projection to X_β, a "composition" $x \circ y \in X_{\alpha\gamma}$ should be defined. A point $x \in X_{\alpha\beta}$ lying over $u \in X_\alpha$, $v \in X_\beta$ should be thought of as meaning an isomorphism from the object C_u corresponding to u to the object C_v corresponding to v.

Definition 2.4. A *Q-stack* is a collection of quasi-projective varieties and morphisms:

$$\coprod X_{\alpha\beta} \underset{p_2}{\overset{p_1}{\rightrightarrows}} \coprod X_\alpha \xrightarrow{p} X$$

(with an arrow $\coprod X_\alpha \to \coprod X_{\alpha\beta}$)

X_α, $X_{\alpha\beta}$ smooth, X normal, p_1, p_2 étale

$$\begin{aligned} \coprod X_\alpha &\to X \text{ surjective} \\ X_{\alpha\beta} &\to X_\alpha \times_X X_\beta \text{ surjective, finite} \\ p_i \circ \epsilon &= \text{identity} \end{aligned}$$

plus morphisms[2]:

$$X_{\alpha\beta} \times_{X_\beta} X_{\beta\gamma} \xrightarrow{\circ} X_{\alpha\gamma}$$
$$X_{\alpha\beta} \xrightarrow{-1} X_{\beta\alpha}$$

making $\coprod X_{\alpha\beta}$ into a pseudo-group (i.e., $\circ$ is associative where defined, $^{-1}$ is its inverse and ϵ is an identity).

It is an interesting exercise in categorical style constructions to show that this collection of data can be derived from a finite group G acting on a normal variety $\tilde{X}$, plus open sets $\tilde{X}_\alpha \subset \tilde{X}$ stabilized by $H_\alpha \subset G$, very much as above:

$$\begin{array}{ccc} \tilde{X} & \supset & \tilde{X}_\alpha \\ \Big\downarrow p & & \Big\downarrow \\ & & \tilde{X}_\alpha/H_\alpha = X_\alpha \\ & \swarrow p_\alpha & \\ X = \tilde{X}/G & & \end{array}$$

and satisfying:

(2.5) a) *For all* $x \in \tilde{X}_\alpha \cap g(\tilde{X}_\beta)$ *and all* $h \in H_\alpha$ *such that* $h(x) = x$ *then* $g^{-1}hg \in H_\beta$.
b) H_α *acts faithfully on* X_α.

Then define

$$X_{\alpha\beta} = \coprod_{\substack{g = \text{repres. of double} \\ \text{cosets } H_\alpha \backslash G / H_\beta}} \tilde{X}_\alpha \cap g(\tilde{X}_\beta)/H_\alpha \cap gH_\beta g^{-1}$$

$p_1 =$ natural map $\tilde{X}_\alpha \cap g(\tilde{X}_\beta)/H_\alpha \cap gH_\beta g^{-1} \to \tilde{X}_\alpha/H_\alpha$

$p_2 =$ the map induced by g^{-1}

$$\tilde{X}_\alpha \cap g(\tilde{X}_\beta)/H_\alpha \cap gH_\beta g^{-1} \to \tilde{X}_\beta/H_\beta$$

and if

$$x \in \tilde{X}_\alpha \cap g(\tilde{X}_\beta) \text{ maps to } \bar{x} \in X_{\alpha\beta}$$
$$y \in \tilde{X}_\beta \cap g'(\tilde{X}_\gamma) \text{ maps to } \bar{y} \in X_{\beta\gamma}$$

---

[2]The maps $\circ$ can also be introduced by giving as extra data more of a semi-simplicial variety:

$$\coprod X_{\alpha\beta\gamma} \rightrightarrows \coprod X_{\alpha\beta}$$

in the usual way.

so that

$$g^{-1}x = hy, \qquad h \in H_\beta$$

then let

$$\bar{x} \circ \bar{y} = (\text{image of } x \in \tilde{X}_\alpha \cap ghg'(\tilde{X}_\gamma) \text{ in } X_{\alpha\gamma}).$$

The object so constructed is a Q-variety *if G acts faithfully on $\tilde{X}$*, but in general only a Q-stack.

Note that $\overline{\mathcal{M}}_2$ and $\overline{\mathcal{M}}_{1,1}$ are in a natural way Q-stacks. We let $\tilde{\mathcal{M}}_2$, $\tilde{\mathcal{M}}_{1,1}$ be the normalization of $\overline{\mathcal{M}}_2, \overline{\mathcal{M}}_{1,1}$ in the level n covering, some $n \geq 3$ and let $G = Sp(4, \mathbb{Z}/n\mathbb{Z})$, $SL(2, \mathbb{Z}/n\mathbb{Z})$ resp. The open sets X_α and subgroups H_α are defined exactly as in the case $g \geq 3$ treated above. The most general chart is any $X_\alpha \to \overline{\mathcal{M}}_2$ (resp. $\overline{\mathcal{M}}_{1,1}$) such that X_α comes with a family of the corresponding curves over it which represents locally everywhere the universal deformation space. Given 2 charts X_α, X_β, with families C_α, C_β, then $X_{\alpha\beta}$ is by definition:

$$\text{Isom}\,(C_\alpha, C_\beta) = \{(x, y, \phi) \mid x \in X_\alpha, y \in X_\beta, \phi \text{ an isom. of } C_{\alpha,x} \text{ with } C_{\beta,y}\}$$

Finally morphisms between Q-stacks X, Y are given by sets of morphisms and commuting diagrams:

$$\begin{array}{ccccc}
\coprod X_{\alpha\beta} & \rightrightarrows & \coprod X_\alpha & \to & X \\
\downarrow{\scriptstyle f_{\alpha\beta}} & & \downarrow{\scriptstyle f_\alpha} & & \downarrow{\scriptstyle f} \\
\coprod Y_{\alpha\beta} & \rightrightarrows & \coprod Y_\alpha & \to & Y
\end{array}$$

provided the atlas for X is suitably refined. For suitable $\tilde{X}$ and $\tilde{Y}$ the morphism will be induced by a morhism

$$\tilde{f} : \tilde{X} \longrightarrow \tilde{Y}$$

which is equivariant with respect to a homomorphism $G_X \to G_Y$ of the finite groups acting on $\tilde{X}, \tilde{Y}$. However, if the Q-stack X is already presented as $\tilde{X}/G_X$ for one $\tilde{X}$, one may have to pass to a bigger covering before $\tilde{f}$ will be defined. This gives a diagram

$$\begin{array}{ccc}
\tilde{X}' & \overset{\tilde{f}}{\dashrightarrow} & \tilde{Y} \\
\downarrow & & \\
\tilde{X} & & \downarrow \\
\downarrow & & \\
\tilde{X}'/G'_X = \tilde{X}/G_X = X & \xrightarrow{f} & Y = \tilde{Y}/G_Y.
\end{array}$$

Among morphisms of Q-stacks, the simplest class consists of those that satisfy:

$$(2.6)\quad \left| \begin{array}{l} \forall \alpha, \text{ let } X'_{\alpha\alpha} = \{x \in X_{\alpha\alpha} \mid f_{\alpha\alpha}(x) = \epsilon_Y(f_\alpha(p_1(x)))\}. \\ \text{Then } X'_{\alpha\alpha} \text{ acts freely on } X_\alpha \end{array} \right.$$

These are the morphisms whose fibres are bona fide varieties, not just Q-varieties or Q-stacks. For such morphisms, it is possible to choose $\tilde{X}, \tilde{Y}$ with the same finite group G acting and $\tilde{f}$ G-equivariant:

$$\begin{array}{ccc} & \tilde{X} \xrightarrow{\tilde{f}} \tilde{Y} & \\ & \downarrow \quad\;\; \downarrow & \\ \tilde{X}/G = & X \;\rightarrow\; Y & = \tilde{Y}/G \end{array}$$

such that, moreover, locally, $\tilde{X}_\alpha \cong (\tilde{X}_\alpha/H_\alpha) \times_{(\tilde{Y}_\alpha/H_\alpha)} \tilde{Y}_\alpha$. The fibres of $\tilde{f}$ are then the "true fibres" of f. The typical example of this is

$$\begin{array}{ccc} \overline{\mathcal{C}}_g^{(n)} & \rightarrow & \overline{\mathcal{M}}_g^{(n)} \\ \downarrow & & \downarrow \\ \overline{\mathcal{C}}_g & \rightarrow & \overline{\mathcal{M}}_g \end{array}$$

where

$\overline{\mathcal{C}}_g =$ moduli space of pairs (C, x), C a stable curve, $x \in C$

$\cong$ moduli space of pairs (C, x), a 1-pointed stable curve[3]

$\overline{\mathcal{C}}_g^{(n)} =$ normalization of $\overline{\mathcal{C}}_g$ in the covering defined by the moduli space of triples (C, x, ϕ), C a smooth curve, $x \in C$ and ϕ a level n structure on C.

Such morphisms may be called *representable* morphisms of Q-stacks. We do not need to develop the theory of Q-stacks for our applications, so we stop at merely these definitions.

---

[3] An n-pointed stable curve $(C, x_1, \ldots, x_n)$ is a reduced, connected curve C with at most ordinary double points plus n distinct smooth points $x_i \in C$ such that every smooth rational component E of C contains at least 3 points which are either x_i's or double points of C.

§3. The Chow Group for Q-Varieties with Cohen-Macaulay Global Covers

We want to study a quasi-projective Q-variety $\{p_\alpha: X_\alpha \longrightarrow X\}$ of dimension n such that the big global cover $\tilde{X}$ is Cohen-Macaulay. We use the notation of §2, esp. $X = \tilde{X}/G$. We also choose a resolution of singularities

$$\pi: \tilde{X}^* \longrightarrow \tilde{X}.$$

Then we assert:

Theorem 3.1. *With the above hypotheses, there is a canonical isomorphism γ between the Chow group of X and G-invariants in the operational Chow ring of $\tilde{X}$, (as usual after extending the coefficients to $\mathbb{Q}$):*

$$\gamma: A_{n-k}(X) \stackrel{\approx}{\rightarrow} opA^k(\tilde{X})^G, \quad 0 \le k \le n.$$

This key result does two important things for us:

a) it defines a ring structure on $A.(X)$,

b) for all Q-sheaves $\mathcal{F}$ on the Q-variety X, we can define Chern classes

$$c_k(\mathcal{F}) \in A.(X).$$

I don't know if these things can be done if we drop the hypothesis that $\tilde{X}$ is Cohen-Macaulay. My guess is that this hypothesis can be dropped, but more powerful tools seem to be needed to treat this case.

Proof. The first step is to define, for all subvarieties $Z \subset X$, an element $\gamma(Z) \in opA^\cdot(\tilde{X})^G \otimes \mathbb{Q}$. We use the local covers $p_\alpha: X_\alpha \longrightarrow X$ and let $p_\alpha^{-1}(Z)$ be the *reduced* subscheme of X_α with support $p_\alpha^{-1}(Z)$. Then $\{\mathcal{O}_{p_\alpha^{-1}(Z)}\}$ is a Q-sheaf on the Q-variety X. As above, let p factor locally

$$\tilde{X}_\alpha \stackrel{q_\alpha}{\rightarrow} \tilde{X}_\alpha/H_\alpha \cong X_\alpha \stackrel{p_\alpha}{\rightarrow} X$$

and lift $p_\alpha^{-1}(Z)$ to its *scheme-theoretic* inverse image $q_\alpha^*(p_\alpha^{-1}(Z))$. Define

$$\tilde{Z} = \left\{\begin{array}{l} G\text{-invariant subscheme of } \tilde{X} \text{ supported} \\ \text{on } p^{-1}(Z) \text{ such that } \tilde{Z}|_{\tilde{X}_\alpha} = q_\alpha^*(p_\alpha^{-1}(Z)) \end{array}\right\}.$$

Note that because $\tilde{X}$ is Cohen-Macaulay, and X_α is smooth, q_α is *flat*. Since $O_{p_\alpha^{-1}(Z)}$ has a finite projective resolution on X_α, this implies that O_Z has a finite projective resolution on $\tilde{X}$. Therefore, by Fulton's theory [F1], the Chern classes $c_k(O_{\tilde{Z}})$ are defined in $opA^k(\tilde{X})$. Next define the "ramification index":

$$e(Z) = \text{order of the stabilizer in } G_\alpha \text{ of almost all points of } p_\alpha^{-1}(Z).$$

Here α is any index such that $p_\alpha^{-1}(Z) \neq \phi$: the definition does not depend on α. Then let

$$\gamma(Z) = \frac{(-1)^{k-1}e(Z)}{(k-1)!} \cdot c_k(O_{\tilde{Z}}), \quad \text{if } k = \operatorname{codim} Z,$$
$$(\text{or} \quad = e(Z)ch_k(O_{\tilde{Z}}), \text{ using lemma 3.3 below}).$$

An important point in the study of $\tilde{Z}$ is that the family of subschemes $\{p_\alpha^{-1}(Z)\}$ can be simultaneously resolved:

Theorem 3.2 (Hironaka). *For all subvarieties $Z \subset X$, there is a birational map $\pi: Z^* \longrightarrow Z$, Z^* normal, such that for all α*

$$\left(p_\alpha^{-1}(Z) \times_Z Z^*\right)_{nor}$$

is smooth.

This is a Corollary of Hironaka's strong resolution theorem, giving a resolution compatible with the pseudo-group of all local analytic isomorphisms between open sets in the original variety: see [H], Theorem 7.1, p. 164. One may proceed as follows: first resolve $p_{\alpha_1}^{-1}(Z)$ as in [H]. Since $\left((\tilde{X}_{\alpha_1}/H_\alpha) \times_X (\tilde{X}_{\alpha_1}/H_{\alpha_1})\right)_{nor}$ is étale over $\tilde{X}_{\alpha_1}/H_{\alpha_1}$ this equivalence relation extends to one on the resolution $p_{\alpha_1}^{-1}(Z)^*$, hence there is a blow-up $\pi_1: Z_1 \longrightarrow Z$ such that $p_{\alpha_1}^{-1}(Z)^* \cong \left(p_{\alpha_1}^{-1}(Z) \times_Z Z_1\right)_{nor}$. Secondly, resolve $\left(p_{\alpha_2}^{-1}(Z) \times_Z Z_1\right)_{nor}$ by a blow-up over $X - p(X_{\alpha_1})$ so as not to affect the first step. Again, descend this to a further blow-up (π_2: $Z_2 \rightarrow Z_1 \rightarrow Z$) of Z. Eventually, we get the needed resolution.

Lemma 3.3. *For all Q-sheaves $\mathcal{F}$ on the Q-variety X, let $S \subset X$ be the support of $\mathcal{F}$. Then in $opA^\cdot(\tilde{X})$,*

a) $c_k(\tilde{\mathcal{F}}) = 0$ *if* $k <$ codim S

b) *if* $k =$ codim S, *and* $S_1, \ldots, S_n$ *are the codimension* k *components of* S, *then*

$$c_k(\tilde{\mathcal{F}}) = \sum_{i=1}^{n} l_i c_k(O_{\tilde{S}_i})$$

where l_i *is the length of the stalk of* $\mathcal{F}_\alpha$ *at the generic point of* $p_\alpha^{-1}(S_i)$ *when* S_i *meets* $p(\tilde{X}_\alpha)$.

Proof. This results from an application of Fulton's Grothendieck-Riemann-Roch Theorem 1.4 and Hironaka's resolution 3.2. By the usual dévissage, reduce the lemma to the case where S is irreducible of codimension k, and $\tilde{\mathcal{F}}$ is an $O_{\tilde{S}}$-module. Let $S^* \to S$ be a "resolution" as in 3.2. This gives a family of resolutions

$$S_\alpha^* = \left(p_\alpha^{-1}(S) \times_S S^*\right)_{nor} \xrightarrow{\pi_\alpha} p_\alpha^{-1}(S) \subset \tilde{X}_\alpha / H_\alpha$$

which are local complete intersection (or l.c.i.) morphisms. Therefore, by fibre product with the flat morphism $\tilde{X}_\alpha \to \tilde{X}_\alpha / H_\alpha$,

$$S_\alpha^* \times_{(\tilde{X}_\alpha/H_\alpha)} \tilde{X}_\alpha \to \tilde{X}_\alpha$$

are l.c.i. morphisms. These glue together to an l.c.i. morphism

$$\tilde{S}^* \xrightarrow{\pi} \tilde{X}$$

such that $\tilde{\mathcal{F}}$ is a $\pi_* O_{\tilde{S}^*}$-module. Let $\tilde{\mathcal{F}}^* = \pi^*(\tilde{\mathcal{F}})$. Then by 1.4:

$$ch\, \pi_!\left(\tilde{\mathcal{F}}^*\right) = \pi_*(ch\, \tilde{\mathcal{F}}^* \cdot Td_\pi).$$

Now if $i > 0$, $R^i \pi_*(\tilde{\mathcal{F}}^*)$ are Q-sheaves on the Q-variety X with supports properly contained in S, so by induction we can assume

$$c_l(R^i \pi_* \tilde{\mathcal{F}}^*) = 0 \quad \text{if} \quad l \le k, i > 0.$$

Therefore

$$ch_l(\pi_! \tilde{\mathcal{F}}^*) = ch_l(\pi_* \tilde{\mathcal{F}}^*) = ch_l(\tilde{\mathcal{F}}) \quad \text{if} \quad l \le k.$$

But $\pi_*: opA^\cdot(\tilde{S}^*) \longrightarrow opA^\cdot(\tilde{X})$ raises the codimension of a cycle by k. So

$$ch_l(\tilde{\mathcal{F}}) = 0 \quad \text{if} \quad l < k$$

and

$$ch_k(\tilde{\mathcal{F}}) = (\text{generic rank of } \tilde{\mathcal{F}}^* \text{ as free } \mathcal{O}_{\tilde{S}^*}\text{-module}) \cdot \pi_*(1)$$
$$= \big(\text{length of } \mathcal{F}_\alpha \text{ at generic point of } p_\alpha^{-1}(S)\big) \cdot ch_k(\mathcal{O}_{\tilde{S}}).$$

Q.E.D.

Lemma 3.4. *If two cycles $\sum n_i Z_i$, $\sum m_i W_i$ on X are rationally equivalent, then*

$$\sum n_i \gamma(Z_i) = \sum m_i \gamma(W_i) \quad \text{in } opA^{\cdot}(\tilde{X}).$$

Proof. Let L be an ample line bundle on X. Rational equivalence on X may be defined by requiring that for all subvarieties $Y \subset X$ and all $s_1, s_2 \in \Gamma(Y, L^n \otimes \mathcal{O}_Y)$, if D_i is the divisor of zeroes of s_i on X, then

$$D_1 \underset{rat}{\sim} D_2.$$

So to prove the lemma, it will suffice to prove that for all $s \in \Gamma(Y, L^n \otimes \mathcal{O}_Y)$, if $D = \sum n_i Z_i$ is the divisor of zeroes of s, then

$$\sum n_i \gamma(Z_i) = e(Y) \cdot [ch_k(\mathcal{O}_{\tilde{Y}}) - ch_k(L^{-n} \otimes \mathcal{O}_Y)].$$

To see this, use the exact sequence of Q-sheaves

$$0 \to L^{-n} \otimes \mathcal{O}_{\tilde{Y}} \overset{\otimes s}{\to} \mathcal{O}_{\tilde{Y}} \to \mathcal{O}_{\tilde{D}} \to 0$$

where $\tilde{D} \subset \tilde{X}$ is the scheme of zeroes of s in $\tilde{Y}$ and the local calculation[4] that on $p^{-1}(Y)$:

$$(*) \qquad (\text{Divisor of zeroes of } p_\alpha^*(s) \text{ on } p_\alpha^{-1}(Y)) = \sum n_i \frac{e(Z_i)}{e(Y)} p_\alpha^{-1}(W_i)$$

---

[4]We use the lemma that if a finite group G acts faithfully on a variety Y and ϕ is a G-invariant function on Y, zero on a subvariety $W \subset Y$ of codimension 1, and $\overline{\phi}$ is the induced function on Y/G, then:

$$\text{ord}_W(\phi) = \#\{g \in G \mid g = id. \text{ on } W\} \cdot \text{ord}_{W/G}(\overline{\phi}).$$

(n.b., $p_\alpha^{-1}(Y)$, $p_\alpha^{-1}(W_i)$ are the *reduced* inverse images). Therefore

$$\begin{aligned} ch_k O_{\tilde{Y}} - ch_k L^{-n} \otimes O_{\tilde{Y}} &= ch_k O_{\tilde{D}} \\ &= \frac{(-1)^{k-1}}{(k-1)!} c_k O_{\tilde{D}} \\ &= \sum_i \begin{pmatrix} \text{length of } O_{D_\alpha} \text{ at} \\ \text{gen.pt. of } p_\alpha^{-1}(W_i) \end{pmatrix} \cdot \frac{(-1)^{k-1}}{(k-1)!} c_k O_{\tilde{W}_i} \quad \text{by (3.3)} \\ &= \sum_i \operatorname{ord}_{p_\alpha^{-1}(W_i)}(p_\alpha^* s) \cdot \frac{\gamma(W_i)}{e(W_i)} \quad \text{by def}^{\underline{n}} \text{ of } \gamma \\ &= \frac{1}{e(Y)} \sum n_i \gamma(W_i) \quad \text{by } (*). \end{aligned}$$

This proves (3.4), which shows that γ factors:

$$\gamma\colon A_.(X) \longrightarrow opA^{\cdot}(\tilde{X})^G.$$

Lemma 3.5. *The composition of maps*

$$A_.(X) \xrightarrow{\gamma} opA^{\cdot}(\tilde{X})^G \xrightarrow{\cap[\tilde{X}]} A_.(\tilde{X})^G \xrightarrow{p_*} A_.(X)$$

is multiplication by n, the degree of p.

Proof. To prove this we use another Riemann-Roch theorem: the version of Baum-Fulton-MacPherson [BFM]. This says that there is a natural transformation $\tau\colon K_o(Z) \longrightarrow A_.(Z)$ for all varieties Z such that

$$\begin{array}{ccc} K^o(Z)\otimes K_o(Z) & \overset{\otimes}{\rightarrow} & K_o(Z) \\ \downarrow ch\otimes\tau & & \downarrow \tau \\ opA^{\cdot}(Z)\otimes A_.(Z) & \overset{\cap}{\rightarrow} & A_.(Z) \end{array}$$

commutes. By the lemma, p. 129 of [BFM] and dévissage, τ satisfies:

> *for all $\mathcal{F}$ with support $\cup Z_i$ of codimension k, $\tau(\mathcal{F})$ has codimension k and*
>
> $$\tau(\mathcal{F})_k = \text{class of} \sum \begin{pmatrix} \text{length of } \mathcal{F} \text{ at} \\ \text{gen. pt. of } Z_i \end{pmatrix} [Z_i].$$

We apply this to $Z = \tilde{X}$ and $\mathcal{F} = O_{\tilde{Z}}$ where $Z \subset X$ is a subvariety. It follows that

$$ch(O_{\tilde{Z}}) \cap \tau(O_{\tilde{X}}) = \tau(O_{\tilde{Z}}).$$

Therefore if $k =$ codimension Z,

$$\begin{aligned} p_*\big(\gamma(Z) \cap [\tilde{X}]\big) &= e(Z) \cdot p_*\big(ch_k(O_{\tilde{Z}}) \cap [\tilde{X}]\big) \\ &= e(Z) \cdot p_*([\tilde{Z}]) \\ &= e(Z) \cdot [\tilde{Z} : Z] \cdot \text{class of } Z \\ &= n \cdot \text{class of } Z. \end{aligned}$$

Q.E.D.

Lemma 3.6. *If $\pi: \tilde{X}^* \longrightarrow \tilde{X}$ is a resolution, and $Z \subset X$ is a subvariety of codimension k such that for all components Z_i of $p^{-1}(Z)$, $\pi^{-1}(Z_i)$ is irreducible of codimension k, then $\pi^*\big(\gamma(Z)\big)$ is represented by a cycle*

$$c \cdot \sum \pi^{-1}(Z_i), \quad \text{some } c \in \mathbb{Q}, c > 0.$$

Proof. Let $U \subset \tilde{X}$ be the open set over which π is an isomorphism. Then $Z_i \cap U \neq \phi$, all i. Now

$$\begin{aligned} \pi^*\big(\gamma(Z)\big) &= e(Z) \cdot \pi^*\big(ch_k(O_{\tilde{Z}})\big) \\ &= e(Z) \cdot \Big(\sum (-1)^l ch_k\big(\text{tor}_l(O_{\tilde{Z}}, O_{\tilde{X}^*})\big)\Big). \end{aligned}$$

But these tor_l are supported on proper subsets of $\pi^{-1}(Z_i)$, hence have no k^{th} Chern character. Therefore:

$$\begin{aligned} \pi^*\big(\gamma(Z)\big) &= e(Z) \cdot ch_k(O_{\tilde{Z}} \otimes O_{\tilde{X}^*}) \\ &= e(Z) \cdot \text{class of } \pi^{-1}(\tilde{Z}) \end{aligned}$$

by the Riemann-Roch theorem on $\tilde{X}^*$. Q.E.D.

Corollary 3.7. *γ is bijective.*

Proof. 3.5, 3.6 and 1.2.

This proves the theorem. A few comments can be made on the ring structure that this introduces in $A.(X)$. First of all, suppose W_1, W_2 are two cycles on X that intersect properly. Then the product $[W_1], [W_2]$ in the above ring structure can also be defined directly by assigning suitable multiplicities to the components of $\operatorname{Supp} W_1 \cap \operatorname{Supp} W_2$. In fact, define:

$$W_1 \cdot W_2 = \sum_{\substack{\text{comp.} U \text{ of} \\ \operatorname{Supp} W_1 \cap \operatorname{Supp} W_2}} i(W_1 \cap W_2; U) \cdot U$$

where if $p_\alpha^{-1}(U) \neq \phi$, then

$$i(W_1 \cap W_2; U) = \frac{e(W_1) \cdot e(W_2)}{e(U)} \cdot i(p_\alpha^{-1}(W_1) \cap p_\alpha^{-1}(W_2); p_\alpha^{-1} U).$$

Note that the intersection multiplicity on the right is taken on the smooth ambient variety $\tilde{X}_\alpha / H_\alpha$, hence is defined, e.g., by

$$\sum_l (-1)^l \begin{pmatrix} \text{length at gen. pt.} \\ \text{of } p_\alpha^{-1} U \end{pmatrix} (\operatorname{tor}_l(\mathcal{O}_{p_\alpha^{-1} W_1}, \mathcal{O}_{p_\alpha^{-1} W_2})).$$

The proof that this is the same as the product in $opA^{\cdot}(\tilde{X})$ is straightforward, i.e.,

$$\begin{aligned}
\gamma(Z_1) \cdot \gamma(Z_2) &= e(Z_1)e(Z_2) ch_{k_1}(\mathcal{O}_{\tilde{Z}_1}) \cdot ch_{k_2}(\mathcal{O}_{\tilde{Z}_2}) \\
&= e(Z_1)e(Z_2) ch_{k_1+k_2}(\mathcal{O}_{\tilde{Z}_1} \overset{\mathrm{L}}{\otimes} \mathcal{O}_{\tilde{Z}_2}) \\
&\quad (\overset{\mathrm{L}}{\otimes} \text{ means take tensor product of projective resol.}) \\
&= \sum_U e(Z_1)e(Z_2) i(p_\alpha^{-1}(W_1) \cap p_\alpha^{-1}(W_2); p_\alpha^{-1} U) \cdot ch_{k_1+k_2}(\mathcal{O}_{\tilde{U}}) \\
&= \sum \frac{e(Z_1)e(Z_2)}{e(U)} i(p_\alpha^{-1}(W_1) \cap p_\alpha^{-1}(W_2); p_\alpha^{-1} U) \cdot \gamma(U).
\end{aligned}$$

This product could be introduced directly without relating it to the product in $opA^{\cdot}(\tilde{X})$. This has been done by Matsusaka in his book "Theory of Q-varieties", [Ma], where associativity and other standard formulae are proven. The missing ingredient, however, is the moving lemma. This follows as a Corollary of the isomorphism of $A.(X)$ with $opA^{\cdot}(\tilde{X})$, i.e., by representing a cycle on X as the projection from $\tilde{X}$ of the Chern class of a sheaf with finite resolution. In particular, I don't know any way to get a moving lemma unless some $\tilde{X}$ is Cohen-Macaulay.

Henceforth, in the study of the Chow rings of Q-varieties we shall identify $A_{n-k}(X)$ and $opA^k(\tilde{X})^G$ via the map γ, and write this as $A^k(X)$ just like the k-codimension piece of the Chow ring of an ordinary non-singular variety. This does not usually lead to any confusion, except with regard to the concept of the *fundamental class* of a subvariety $Y \subset X$. The important thing to realize here is that there are really two different notions of fundamental class, differing by a rational number, and both are important. Thus for all Y of codimension k, we will write

$$[Y] = \text{class of the cycle } Y \text{ in the Chow group } A_{n-k}(X) = A^k(X)$$

and

$$[Y]_Q = \text{the class } ch_k(O_{\tilde{Y}}) \text{ in } opA^k(\tilde{X})^G = A^k(X).$$

Since we are using the identification γ, we have:

$$[Y]_Q = \frac{1}{e(Y)} \cdot [Y].$$

When one makes calculations of intersections in local charts $\tilde{X}_\alpha/H_\alpha$, then one is verifying an identity between classes $[Y]_Q$. But when one has a rational equivalence between cycles on X, one has an identity between $[Y]$'s: e.g., if X is unirational, then for all points $P_1, P_2 \in X$,

$$[P_1] = [P_2],$$

but the point classes $[P]_Q$ are fractions $1/e(P)$ of the basic point class $[P] \in A^{\cdot}(X)$.

If X is a Q-stack, exactly the same theorem holds and we have an isomorphism

$$\gamma\colon A_{\cdot}(X) \xrightarrow{\approx} opA^{\cdot}(\tilde{X})^G.$$

The only difference is that a subgroup $Z \subset G$ acts identically on $\tilde{X}$. If $\#Z = z$, then the effect of this is merely to modify the ring structure on $A^{\cdot}(X)$ as follows. Let W_1, W_2 be cycles on X and consider:

i) the *Q-variety* structure on X given by the action of G/Z on $\tilde{X}$, and the multiplication $W_1 \cdot_{var} W_2$,

ii) the Q-*stack* structure on X given by the action of G on $\tilde{X}$, and the multiplication $W_1 \cdot_{st} W_2$.

Then by the moving lemma plus the formula above for proper intersections, it follows:

$$W_1 \cdot_{st} W_2 = z.W_1 \cdot_{var} W_2.$$

In particular, the identity in the Chow ring of a Q-stack X is $[X]_Q$, not $[X]$.

The Chow ring for Q-varieties, or more generally Q-stacks, has good contravariant functorial properties. We consider morphisms of Q-stacks with global Cohen-Macaulay covers:

$$X \xrightarrow{f} Y$$

as defined in §2. Then I claim:

Proposition 3.8. *There is a canonical ring homomorphism*

$$f^* : A^{\cdot}(Y) \to A^{\cdot}(X)$$

satisfying:

i) $f_*(a.f^*b) = f_*a.b$ *(f_* being defined from $A.(X)$ to $A.(Y)$ as usual),*
ii) $f^*(c_k\mathcal{E}) = c_k(f^*\mathcal{E})$ *for all Q-vector bundles $\mathcal{E}$ on Y*
iii) *if W is a subvariety of Y such that codim $f^{-1}(W) =$ codim W then*

$$f^*([W]) = \text{class of} \sum_{\substack{\text{comp. } V_k \\ \text{of } f^{-1}(W)}} i_k \cdot [V_k]$$

where i_k is calculated on suitable charts $f_\alpha : X_\alpha \longrightarrow Y_\alpha$ by pull-backs in the smooth case adjusted by $e(W)/e(V_k)$.

Proof. Although the moving lemma plus (iii) provides us with the simplest formula for f^*, to see that f^* is well-defined, we use $opA^{\cdot}$. There is one complication. X and Y have global Cohen-Macaulay covers $\tilde{X}, \tilde{Y}$ but f may not lift to $\tilde{f}: \tilde{X} \longrightarrow \tilde{Y}$. Instead, we may have to 'refine' $\tilde{X}$:

$$\begin{array}{ccc} \tilde{X}' & & \\ {\scriptstyle q}\downarrow & \searrow & \\ \tilde{X} & & \tilde{Y} \\ {\scriptstyle p}\downarrow & & \downarrow \\ X & \xrightarrow{f} & Y \end{array}$$

and then $\tilde{X}'$ may not longer be Cohen-Macaulay. Still, once one has one Cohen-Macaulay $\tilde{X}$ with which to set up the theory, one proves that

$$(3.9) \qquad opA^{\cdot}(\tilde{X})^{G_X} \xrightarrow{q^*} opA^{\cdot}(\tilde{X}')^{G'_X}$$

is an isomorphism, hence f^* may be defined by:

$$A.(Y) \underset{\gamma_Y}{\approx} opA^{\cdot}(\tilde{Y})^{G_Y} \underset{\tilde{f}_*}{\rightarrow} opA^{\cdot}(\tilde{X}')^{G'_X} \approx opA^{\cdot}(\tilde{X})^{G_X} \approx A.(X).$$

To check (3.9), use

$$\begin{array}{ccccc} A.(X) & \underset{\gamma_X}{\approx} opA^{\cdot}(\tilde{X})^{G_X} & \underset{q^*}{\rightarrow} & opA^{\cdot}(\tilde{X}')^{G'_X} \hookrightarrow & A^{\cdot}(\tilde{X}'^*) \\ & \nwarrow & & & \downarrow \cap[\tilde{X}'] \\ & & {\scriptstyle (p\circ q)_*} & & A.(\tilde{X}') \end{array}$$

($\tilde{X}'^* =$ resolution of $\tilde{X}'$) and argue as in lemmas 3.5 and 3.6. There is one hitch: namely, in 3.5, we get

$$\begin{aligned} (p\circ q)_*\big(q^*(\gamma(Z)) \cap [\tilde{X}']\big) &= e(Z)(p\circ q)_*\big(q^*(ch_k(O_{\tilde{Z}})) \cap [\tilde{X}']\big) \\ &= e(Z)(p\circ q)_*\big(ch_k(O_{\tilde{Z}} \overset{\mathbf{L}}{\otimes} O_{\tilde{X}'}) \cap [X']\big) \end{aligned}$$

where $\overset{\mathbf{L}}{\otimes}$ means take a resolution of $O_{\tilde{Z}}$ and tensor it with $O_{\tilde{X}'}$. But now if $\tilde{X}'$ is not Cohen-Macaulay, $O_{\tilde{Z}} \overset{\mathbf{L}}{\otimes} O_{\tilde{X}'}$ will not be a resolution of some $O_{\tilde{Z}'}$ and we get instead:

$$\begin{aligned} &= e(Z)\sum(-1)^l (p\circ q)_*\big(ch_k \operatorname{tor}_l(O_{\tilde{Z}}, O_{X'}) \cap [\tilde{X}']\big) \\ &= e(Z)\cdot(p\circ q)_*\Big(\sum_n i_n \cdot [\tilde{Z}_n]\Big) \end{aligned}$$

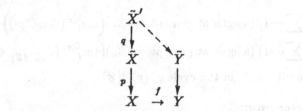

where $\tilde{Z}_l$ are the components of $(p \circ q)^{-1} Z$ and

$$\begin{aligned} i_n &= \sum (-1)^l (\text{length at gen. pt. of } \tilde{Z}_n)\Big(\operatorname{tor}_l^{\mathcal{O}_{\tilde{X}}}(\mathcal{O}_{\tilde{Z}}, \mathcal{O}_{\tilde{X}'})\Big) \\ &= \sum (-1)^l (\text{length at gen. pt. of } \tilde{Z}_n)\Big(\operatorname{tor}_l^{\mathcal{O}_{X_\alpha}}(\mathcal{O}_{p_\alpha^{-1}(Z)}, \mathcal{O}_{\tilde{X}'})\Big) \\ &= \text{mult. of } \tilde{Z}_n \text{ in the cycle } q_\alpha'^{*}(p_\alpha^{-1}(Z)) \end{aligned}$$

where we factor $p \circ q$:

$$\begin{array}{ccc} \tilde{X}' & \supset & \tilde{X}'_\alpha \\ \Big\downarrow & & \Big\downarrow {\scriptstyle q'_\alpha} \\ & & X_\alpha \\ \Big\downarrow & \swarrow {\scriptstyle p_\alpha} & \\ X & & \end{array}$$

Thus

$$\begin{aligned} e(Z)(p \circ q)_* \Big(\sum_n i_n [\tilde{Z}_n]\Big) & \\ &= e(Z) p_{\alpha,*}(\text{class of } q'_{\alpha,*}(q_\alpha'^{*}(p_\alpha^{-1} Z))) \\ &= \deg q'_\alpha \cdot e(Z) \cdot p_{\alpha,*}(p_\alpha^{-1} Z) \\ &= n.\text{class of } Z. \end{aligned}$$

f^* being defined, the rest of the proof is straightforward.

For representable morphisms $f\colon X \longrightarrow Y$ of Q-stacks, there is a further important compatibility. For such f, let

$$\tilde{f}\colon \tilde{X} \longrightarrow \tilde{Y}$$

be a G-equivariant morphism such that $X = \tilde{X}/G$, $Y = \tilde{Y}/G$ and

(3.9) $$\tilde{X}_\alpha \cong X_\alpha \times_{Y_\alpha} \tilde{Y}_\alpha$$

as in §2*b*. Then we have:

Proposition 3.10. *If the morphism f on local charts*

$$f_\alpha\colon X_\alpha \longrightarrow Y_\alpha$$

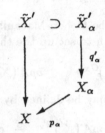

is a local complete intersection and $\tilde{Y}$ is Cohen-Macaulay, then $\tilde{f}$ is l.c.i. and the diagram

$$\begin{array}{ccc} opA^*(\tilde{X})^G & \xrightarrow{\tilde{f}_*} & opA^{\cdot}(\tilde{Y})^G \\ \gamma \uparrow & & \gamma \uparrow \\ A_{\cdot}(X) & \xrightarrow{f_*} & A_{\cdot}(Y) \end{array}$$

commutes.

Proof. Let W be a codimension k subvariety of X such that W is generically finite over $f(W)$. Then we must check

$$\tilde{f}_*(\gamma W) = [W : fW] \cdot \gamma(fW),$$

i.e.,

$$e(X) \cdot \tilde{f}_*(ch_k(O_{\tilde{W}})) = [W : fW] \cdot e(fW) \cdot ch_k(O_{f\tilde{W}}).$$

But by Riemann-Roch for $\tilde{f}$,

$$\begin{aligned} \tilde{f}_*(ch_k(O_{\tilde{W}})) &= ch_k(\tilde{f}_* O_{\tilde{W}}) \cdot \\ &= [W_\alpha : f_\alpha(W_\alpha)] \cdot ch_k(O_{f\tilde{W}}) \end{aligned}$$

because $\tilde{f}_*(O_{\tilde{W}})$ is generically a locally free $O_{f\tilde{W}}$-algebra of length $[W_\alpha : f_\alpha W_\alpha]$ by (3.9). But now use:

$$[W_\alpha : W].[W : fW] = [W_\alpha : f_\alpha W_\alpha] \cdot [f_\alpha W_\alpha : fW]$$

and

$$\begin{aligned} [W_\alpha : W] \cdot e(W) &= [X_\alpha : X] \cdot e(X) \\ &= [G : H_\alpha] \\ &= [Y_\alpha : Y] \cdot e(Y) \\ &= [f(W)_\alpha : fW] \cdot e(fW) \end{aligned}$$

and the equality of the coefficients follows.

Part II: Basic Classes in the Chow Ring of the Moduli Space

§4. Tautological Classes

Whenever a variety or topological space is defined by some universal property, one expects that by virtue of its defining property, it possesses certain cohomology classes called tautological classes. The standard example is a Grassmannian, e.g., the Grassmannian Grass of k-planes in $\mathbb{C}^n$. By its very definition, there is a universal bundle E on Grass of rank k, and this induces Chern classes $c_l(E)$, $1 \leq l \leq k$, in both the cohomology ring of Grass and the Chow ring of Grass. These two rings are, in fact, isomorphic and generated as rings by $\{c_l(E)\}$. Moreover, one gets tautological relations from the fact that E is a sub-bundle of the trivial bundle $\mathbb{C}^n \times$ Grass. This gives an exact sequence:

$$0 \to E \to \mathcal{O}^n \to F \to 0, \quad F \text{ a bundle of rank } n-k,$$

hence

$$\left(1 + c_1(E) + \cdots + c_k(E)\right)_l^{-1} = 0, \quad l > n-k.$$

As is well known, these are a complete set of relations for the cohomology and Chow rings of Grass.

We shall begin a program of the same sort for the Chow ring (or cohomology ring) of $\overline{\mathcal{M}}_g$. Our purpose is merely to identify a natural set of tautological classes and some tautological relations. To what extent these lead to a presentation of either ring is totally unclear at the moment.

The natural place to start is with the universal curve over $\overline{\mathcal{M}}_g$. This is the same as the coarse moduli space of 1-pointed stable curves (C, P) (see Knudsen [K], Harris-Mumford [H-M]), which we call $\overline{\mathcal{M}}_{g,1}$ or $\overline{\mathcal{C}}_g$ alternatively. $\overline{\mathcal{C}}_g$ is a Q-variety, too, and everything we have said about $\overline{\mathcal{M}}_g$ applies to $\overline{\mathcal{C}}_g$ too. The morphism $\overline{\mathcal{C}}_g \to \overline{\mathcal{M}}_g$ is a representable morphism, and via level n structures, we have a covering $\tilde{\mathcal{C}}_g$ of $\overline{\mathcal{C}}_g$ and a morphism:

$$\pi \colon \tilde{\mathcal{C}}_g \longrightarrow \tilde{\mathcal{M}}_g$$

which is a flat, proper family of stable curves, with a finite group G acting on both, and $\overline{\mathcal{C}}_g = \tilde{\mathcal{C}}_g/G$, $\overline{\mathcal{M}}_g = \tilde{\mathcal{M}}_g/G$. If $g \geq 3$, then

$$\tilde{\mathcal{C}}_g = (\text{normalization of } \overline{\mathcal{C}}_g \times_{\overline{\mathcal{M}}_g} \tilde{\mathcal{M}}_g),$$

but if $g = 2$, the generic curve has automorphisms, $Sp(4, \mathbb{Z}/n\mathbb{Z})$ does not act faithfully on $\tilde{\mathcal{M}}_2$, and $\tilde{\mathcal{C}}_2$ is a double cover of this normalization. In any case, $\overline{\mathcal{C}}_g$ has a Q-sheaf $\omega_{\overline{\mathcal{C}}_g/\overline{\mathcal{M}}_g}$ represented by the invertible sheaf $\omega_{\tilde{\mathcal{C}}_g/\tilde{\mathcal{M}}_g}$ on $\tilde{\mathcal{C}}_g$. Henceforth, whenever we talk of sheaves on $\overline{\mathcal{C}}_g$ or $\overline{\mathcal{M}}_g$ we shall mean *Q-sheaves* and they are always represented by usual coherent sheaves on $\tilde{\mathcal{C}}_g$ and $\tilde{\mathcal{M}}_g$ with G-action. Furthermore, we shall make calculations in $A^{\cdot}(\overline{\mathcal{M}}_g)$ and $A^{\cdot}(\overline{\mathcal{C}}_g)$ by implicitly identifying these with $opA^{\cdot}(\tilde{\mathcal{M}}_g)^G$ and $opA^{\cdot}(\tilde{\mathcal{C}}_g)^G$.

Now define the *tautological classes:*

$$K_{\overline{\mathcal{C}}_g/\overline{\mathcal{M}}_g} = c_1(\omega_{\overline{\mathcal{C}}_g/\overline{\mathcal{M}}_g}) \in A^1(\overline{\mathcal{C}}_g)$$
$$\kappa_l = \left(\pi_* K^{l+1}_{\overline{\mathcal{C}}_g/\overline{\mathcal{M}}_g}\right) \in A^l(\overline{\mathcal{M}}_g)$$
$$\mathbb{E} = \pi_*(\omega_{\overline{\mathcal{M}}_g/\overline{\mathcal{M}}_g}) : \text{a locally free } Q\text{-sheaf of rank } g \text{ on } \overline{\mathcal{M}}_g$$
$$\lambda_l = c_l(\mathbb{E}), \quad 1 \leq l \leq g.$$

I believe that the κ_l are the natural tautological classes to consider on $\overline{\mathcal{M}}_g$. On the other hand, the λ_l are the natural classes for abelian varieties. Let me sketch this link, which will not be used subsequently. In fact, if

$$\mathcal{A}_g^* = \begin{pmatrix}\text{Satake's compactification of the moduli} \\ \text{space of principally polarized abelian varieties}\end{pmatrix}$$

then there is a natural morphism

$$t\colon \overline{\mathcal{M}}_g \longrightarrow \mathcal{A}_g^*$$

carrying the point $[C]$ to the point of $\mathcal{A}_g^*$ defined by the Jacobian of C. This morphism lifts to a G-equivariant morphism

$$\tilde{t}\colon \tilde{\mathcal{M}}_g \longrightarrow \tilde{\mathcal{A}}_g$$

where $\tilde{\mathcal{A}}_g$ is a suitable toroidal compactification of the level n covering of $\mathcal{A}_g$: see Namikawa [N]. Moreover, $\tilde{\mathcal{A}}_g$ carries a universal family

$$\pi\colon \tilde{\mathcal{G}}_g \longrightarrow \tilde{\mathcal{A}}_g$$

of semi-abelian group schemes, i.e., $\tilde{\mathcal{G}}/\tilde{\mathcal{A}}$ is a group scheme whose fibres are extensions of abelian varieties by algebraic tori $(\mathbb{C}^*)^h$. The family $\tilde{\mathcal{G}}_g$ pulls back on $\tilde{\mathcal{M}}_g$ to the family of Jacobians and generalized Jacobians of $\tilde{\mathcal{C}}_g$. Over $\tilde{\mathcal{A}}_g$, define

$$\mathbb{E}' = \Omega^1_{\tilde{\mathcal{C}}_g/\tilde{\mathcal{A}}_g}|_{\text{0-section}}, \text{ a locally free sheaf of rank } g$$
$$\lambda'_l = c_k(\mathbb{E}'), \quad 1 \leq l \leq g.$$

Then it follows that

$$\tilde{t}^*\mathbb{E}' \cong \mathbb{E}$$

and

$$\tilde{t}^*\lambda'_l = \lambda_l.$$

The class $K_{\overline{\mathcal{C}}_g/\overline{\mathcal{M}}_g}$ played a central role in the basic paper [A] of Arakelov, who proved the essential case of:

Theorem 4.1 (Arakelov). *The divisor $K_{\overline{\mathcal{C}}_g/\overline{\mathcal{M}}_g}$ is numerically effective on $\overline{\mathcal{C}}_g$, i.e., for all curves $C \subset \overline{\mathcal{C}}_g$,*

$$\deg_C K_{\overline{\mathcal{C}}_g/\overline{\mathcal{M}}_g} \geq 0.$$

Proof. In fact, Arakelov proved that for all normal surfaces F fibred in stable curves over a smooth curve C, $\omega_{F/C}$ is ample on F. This implies that for all curves $C \subset \tilde{\mathcal{C}}_g$ such that $\pi(C) \cap \mathcal{M}_g \neq \phi$,

$$\deg_C K_{\tilde{\mathcal{C}}_g/\tilde{\mathcal{M}}_g} > 0.$$

Now suppose $C \subset \tilde{\mathcal{C}}_g$ and $\pi(C) \subset \overline{\mathcal{M}}_g - \mathcal{M}_g$.

Case 1: $\pi(C) =$ one pt. Then $\deg_C K > 0$ because ω is ample on all fibres of $\tilde{\mathcal{C}}_g \to \tilde{\mathcal{M}}_g$.

Case 2: $d\pi|_C \equiv 0$, i.e., C is in the locus Sing $\mathcal{C}$ of double points of the fibres. But Sing $\mathcal{C}$ has an étale double cover Sing′ $\mathcal{C}$ parametrizing pairs consisting of a double point of a fibre of π and a branch through this point. By residue

$$\omega_{\tilde{\mathcal{C}}_g|\tilde{\mathcal{M}}_g} \otimes \mathcal{O}_{\text{Sing}'\mathcal{C}} \cong \mathcal{O}_{\text{Sing}'\mathcal{C}}$$

so $\deg_C K = 0$.

Case 3: Other. After a suitable case change

$$C' \to \pi(C) \subset \tilde{M}_g$$

we can assume that the pull-back family $\tilde{C}_g \times_{\tilde{M}_g} C'$ is obtained by glueing several generically smooth stable families $Y_\alpha \to C'$ along a set of sections $t_{\alpha\beta}\colon C' \longrightarrow Y_\alpha$. Lying over C there will be a curve C'' contained in one of the Y_α's, say Y_{α_0}, mapping onto C' and not equal to $t_{\alpha_0\beta}(C')$, any β. The pull-back of $\omega_{\tilde{C}_g/\tilde{M}_g}$ to Y_α will be equal to $\omega_{Y_\alpha/C'}(\sum_\beta t_{\alpha_0,\beta}(C'))$ and, by Arakelov, this will have non-negative degree on C' if genus $C' \geq 2$. If genus $C' = 0$ or 1, it is easy to check that this is still the case. Q.E.D.

Corollary 4.2. *The classes κ_l are numerically effective, i.e., for all subvarieties $W \subset \overline{M}_g$ of dimension l,*

$$(W.\kappa_l) \geq 0.$$

Proof. $K_{\overline{C}/\overline{M}}$ numerically effective implies $K^{l+1}_{\overline{C}/\overline{M}}$ numerically effecitve (see [K1]), hence $\pi_*(K^{l+1}_{\overline{C}/\overline{M}})$ is numerically effective.

In fact, κ_1 is ample, see [M], §5.

§5. Tautological Relations via Grothendieck-Riemann-Roch

Grothendieck's Riemann-Roch theorem (G-R-R) is, in many cases, tailor-made to find relations among tautological classes. For example, see Atiyah-Bott [A-B], §9. We can compute the classes λ_k in terms of the classes κ_k. To do this, we apply the G-R-R to the morphism

$$\pi\colon \tilde{C}_g \longrightarrow \tilde{M}_g.$$

This gives us

$$ch\, \pi_!\, \omega_{\overline{C}/\overline{M}} = \pi_*\big(ch\, \omega_{\overline{C}/\overline{M}}\, Td^\vee(\Omega^1_{\overline{C}/\overline{M}})\big).$$

Here we use the notaion $Td^\vee(\mathcal{E})$ to write the universal multiplicative polynomial in the Chern classes of $\mathcal{E}$ such that for line bundles L,

$$\begin{aligned} Td^\vee(L) &= \frac{\lambda}{e^\lambda - 1}, \quad \lambda = c_1(L) \\ &= 1 - \frac{1}{2}\lambda - \sum_{k=1}^{\infty}(-1)^{k-1}\frac{B_k}{(2k)!}\lambda^{2k}, \end{aligned}$$

(i.e., the usual $Td(L)$ is $\lambda/1-e^{-\lambda}$ or $1+\frac{1}{2}\lambda+\cdots$). Since $R^1\pi_*\omega_{\overline{C}/\overline{M}} \cong O_{\overline{M}}$, this means:

$$ch\,\mathbb{E} = 1 + \pi_*(e^K\, Td^\vee(\Omega^1_{\overline{C}/\overline{M}})).$$

Now use the exact sequence:

$$0 \to \Omega^1_{\overline{C}/\overline{M}} \to \omega_{\overline{C}/\overline{M}} \to \omega_{\overline{C}/\overline{M}} \otimes O_{\mathrm{Sing}\,\overline{C}} \to 0$$

(compare [M], pf. of 5.10). Let $\mathrm{Sing}'\,\overline{C}$ be the double cover of $\mathrm{Sing}\,C$ consisting of singular points plus branches: as a Q-variety, it is an étale double cover, i.e., the map between the charts

$$(\mathrm{Sing}'\,\overline{C})_\alpha \to (\mathrm{Sing}\,\overline{C})_\alpha$$

which are local universal deformation spaces, is étale. Then via residue

$$\omega_{\overline{C}/\overline{M}} \otimes O_{\mathrm{Sing}'\overline{C}} \cong O_{\mathrm{Sing}'\overline{C}}.$$

Therefore:

$$\begin{aligned} ch\,\mathbb{E} &= 1 + \pi_*(e^K \cdot Td^\vee(\omega_{\overline{C}/\overline{M}}) \cdot Td^\vee(O_{\mathrm{Sing}\,\overline{C}})^{-1}) \\ &= 1 + \pi_*(e^K \cdot \frac{K}{e^K - 1} + [Td^\vee(O_{\mathrm{Sing}\,\overline{C}})^{-1} - 1]) \end{aligned}$$

since K intersects any cycle on $\mathrm{Sing}\,\overline{C}$ in zero. Now use the lemma:

Lemma 5.1. *There is a universal power series P such that for all $i: Z \longrightarrow X$, an inclusion of a smooth codimension two subvariety in a smooth variety,*

$$(Td^\vee O_Z)^{-1} - 1 = i_*[P(c_1N, c_2N)]$$

where N is the normal bundle I_Z/I_Z^2.

Proof. In fact

$$(Td^\vee O_Z)^{-1} = 1 + (\text{polyn. in } ch_k(O_Z), \quad k \geq 1)$$

and by G-R-R for i, $ch_k(O_Z)$ is i_* of a polynomial in c_1N, c_2N.

To compute this polynomial P, say $Z = D_1 \cdot D_2$. Then use

$$0 \to O_X(-D_1 - D_2) \to O_X(-D_1) \oplus O_X(-D_2) \to O_X \to O_Z \to 0.$$

This gives us

$$Td^\vee \mathcal{O}_Z = \left(Td^\vee \mathcal{O}_X(-D_1)\right)^{-1} \cdot \left(Td^\vee \mathcal{O}_X(-D_2)\right)^{-1} \cdot Td^\vee \mathcal{O}_X(-D_1 - D_2)$$
$$= \left(\frac{-D_1}{e^{-D_1}-1}\right)^{-1} \cdot \left(\frac{-D_2}{e^{-D_2}-1}\right)^{-1} \cdot \left(\frac{-D_1-D_2}{e^{-D_1-D_2}-1}\right).$$

Thus

$$D_1 D_2 \cdot P(D_1 + D_2, D_1 \cdot D_2) = Td^\vee(\mathcal{O}_Z)^{-1} - 1$$
$$= \frac{D_1}{1-e^{-D_1}} \cdot \frac{D_2}{1-e^{-D_2}} \cdot \frac{1-e^{-D_1-D_2}}{D_1+D_2} - 1$$
$$= \frac{1}{D_1+D_2} \cdot \left[D_1 \cdot \left(\frac{D_2}{1-e^{-D_2}} - 1\right) + D_2 \cdot \left(\frac{D_1}{1-e^{-D_1}} - 1\right) - D_1 \cdot D_2\right]$$
$$= \frac{D_1 D_2}{D_1+D_2} \cdot \sum_{k=1} \frac{(-1)^{k-1} B_k}{(2k)!} \left(D_1^{2k-1} + D_2^{2k-1}\right)$$

So

$$P(D_1 + D_2, D_1 \cdot D_2) = \sum_{k=1}^{\infty} \frac{(-1)^{k-1} B_k}{(2k)!} \left(\frac{D_1^{2k-1} + D_2^{2k-1}}{D_1 + D_2}\right)$$
$$= \frac{1}{12} - \frac{1}{720}\left((D_1+D_2)^2 - 3D_1 D_2\right) + \frac{1}{30,240}\left((D_1+D_2)^4 - 5D_1D_2(D_1+D_2)^2 + 5D_1^2 D_2^2\right) + \cdots$$

Therefore

$$ch\,\mathbb{E} = 1 + \pi_*\left(\frac{K}{1-e^{-K}}\right) + (\pi \circ i)_* P(c_1 N, c_2 N).$$

Now Sing $\overline{\mathcal{C}}$ breaks up into pieces depending on whether the double point disconnects the fibre in which it lies or not, and if it does, what the genera are of the two pieces. Thus:

$$\operatorname{Sing} \overline{\mathcal{C}} = \coprod_{0 \le h \le [g/2]} \Delta_h^*$$

where Δ_0^* are the non-disconnecting double points and if $h \geq 1$, Δ_h^* are the points for which one piece has genus h. Moreover, looking at the two pieces, one sees that

$$\Delta_h^* \cong \overline{\mathcal{C}}_h \times \overline{\mathcal{C}}_{g-h} \quad \text{if} \quad 1 < h < g/2$$

while

$$\begin{aligned} \Delta_{g/2}^* &\cong \overline{\mathcal{C}}_{g/2} \times \overline{\mathcal{C}}_{g/2}/(\mathbb{Z}/2\mathbb{Z}) \text{ if } \quad g \text{ is even} \\ \Delta_0^* &\cong \overline{\mathcal{M}}_{g-1,2}/(\mathbb{Z}/2\mathbb{Z}) \end{aligned}$$

where $\overline{\mathcal{M}}_{g-1,2}$ is the space of stable curves with two ordered points P_1, P_2 and $\mathbb{Z}/2\mathbb{Z}$ permutes either the two factors or the two points. In fact, specifying a branch too, we get:

$$\begin{aligned} \operatorname{Sing}' \overline{\mathcal{C}} &= \coprod_{0 \leq h \leq [g/2]} \Delta_h' \\ \Delta_h' &\cong 2 \text{ copies of } \overline{\mathcal{C}}_h \times \overline{\mathcal{C}}_{g-h} \quad 1 \leq h \leq g/2 \\ &\cong \overline{\mathcal{C}}_{g/2} \times \overline{\mathcal{C}}_{g/2} \quad \text{if } h = g/2 \\ &\cong \overline{\mathcal{M}}_{g-1,2} \quad \text{if } h = 0. \end{aligned}$$

Let K_1, K_2 be the divisor classes defined

a) on $\overline{\mathcal{C}}_h \times \overline{\mathcal{C}}_{g-h}$ by $K_1 = p_1^* K_{\overline{\mathcal{C}}_h/\overline{\mathcal{M}}_h}$, $K_2 = p_2^* K_{\overline{\mathcal{C}}_{g-h}/\overline{\mathcal{M}}_{g-h}}$

b) on $\overline{\mathcal{M}}_{g-1,2}$ by K_i = conormal bundle at the ith point.

Writing out $ch\,\mathbb{E}$ finally we get

$$(5.2) \qquad \begin{aligned} ch\,\mathbb{E} = g + \sum_{l=1}^{\infty} \frac{(-1)^{l+1} \cdot B_l}{(2l)!} \cdot \\ \left[\kappa_{2l+1} + \frac{1}{2} \sum_{h=0}^{g-1} i_{h,*}(K_1^{2l-2} - K_1^{2l-3} \cdot K_2 + \cdots + K_2^{2l-2}) \right]. \end{aligned}$$

Here we have expanded $K/1 - e^{-K}$ and used the fact that $\pi_* K$ is $(2g-2)$ times the fundamental class of $\overline{\mathcal{M}}_g$. The morphism i_h is

$$\begin{aligned} &i_0 \colon \overline{\mathcal{M}}_{g-1,2} \to \operatorname{Sing} \overline{\mathcal{C}} \to \overline{\mathcal{M}}_g \\ &i_h \colon \overline{\mathcal{C}}_h \times \overline{\mathcal{C}}_{g-h} \to \operatorname{Sing} \overline{\mathcal{C}} \to \overline{\mathcal{M}}_g, \quad 1 \leq h \leq g-1. \end{aligned}$$

Note that i_0 and $i_{g/2}$ have degree 2 and the other i_h's are repeated twice in the sum: hence the factor 1/2. Moreover, we have evaluated the normal bundle to $\operatorname{Sing}\overline{\mathcal{C}}$ in $\overline{\mathcal{C}}_g$ as the direct sum of the tangent bundle to the two branches of the curve at the singular point:

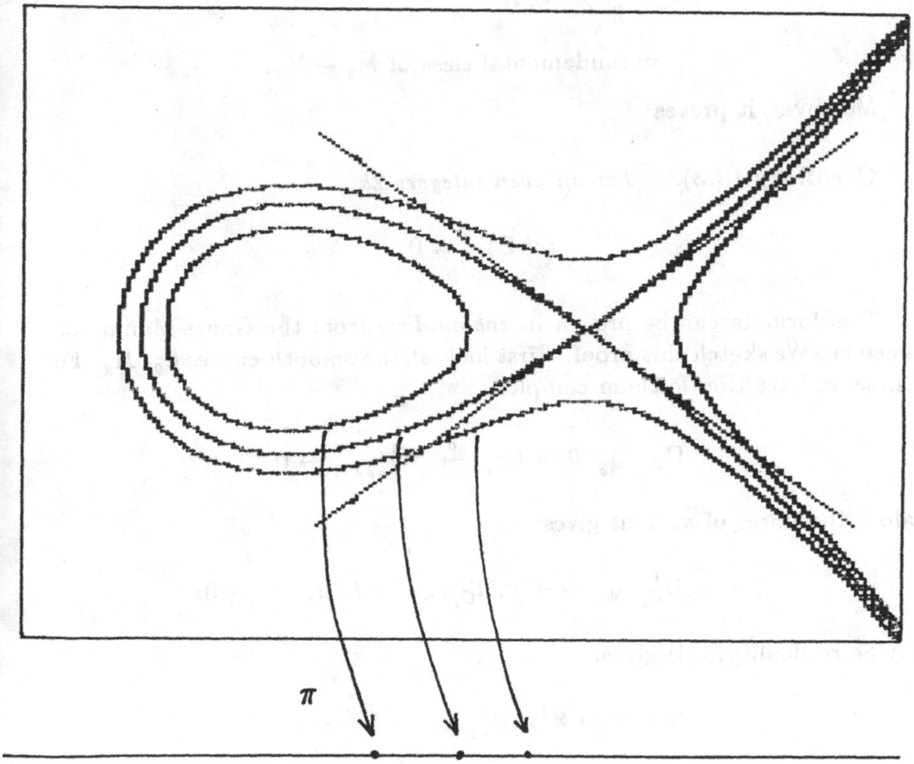

(In a transversal to $\operatorname{Sing}\overline{\mathcal{C}}$, $\overline{\mathcal{C}}_g/\overline{\mathcal{M}}_g$ looks like $xy = t$, and the tangent bundle to the x, y-surface at $(0,0)$ is the sum of the tangent line to the branch $x = 0$ and to the branch $y = 0$.)

The formula (5.2) specializes in codimension 1 to the formula of [M], p. 102:

$$\lambda_1 = c_1(\mathbb{E}) = \frac{1}{12}(\kappa_1 + \delta)$$

where

$$\begin{aligned}\delta &= \frac{1}{2}\sum_{h=0}^{g-1} i_{h,*}(1) \\ &= \text{fundamental class of } \overline{\mathcal{M}}_g - \mathcal{M}_g.\end{aligned}$$

Moreover, it proves

Corollary (5.3). *For all even integers $2k$,*

$$(ch\,\mathbb{E})_{2k} = 0.$$

This formula can be proven in *cohomology* from the Gauss-Manin connection. We sketch this proof. First look at the smooth curves $\mathcal{C}_g/\mathcal{M}_g$. For these we have the DeRham complex

$$\Omega^{\cdot}_{\mathcal{C}_g/\mathcal{M}_g} : 0 \to \mathcal{O}_{\mathcal{C}_g} \xrightarrow{d} \Omega^1_{\mathcal{C}_g/\mathcal{M}_g} \to 0$$

along the fibres of π. This gives:

$$0 \to \pi_*\Omega^1_{\mathcal{C}_g/\mathcal{M}_g} \to \mathbb{R}^1\pi_*\Omega^{\cdot}_{\mathcal{C}_g/\mathcal{M}_g} \to R^1\pi_*\mathcal{O}_{\mathcal{C}_g} \to 0.$$

By Serre duality, this gives:

$$0 \to \mathbb{E} \to \mathbb{R}^1\pi_*\Omega^{\cdot}_{\mathcal{C}_g/\mathcal{M}_g} \to \mathbb{E}^{\vee} \to 0.$$

The vector bundle in the middle has rank $2g$, is isomorphic to $R^1\pi_*\mathbb{C}$ and possesses the Gauss-Manin connection. Therefore its Chern classes are zero and over $\mathcal{M}_g$:

$$c(\mathbb{E}) \cdot c(\mathbb{E}^{\vee}) = 1. \tag{5.4}$$

This identity can be extended to $\overline{\mathcal{M}}_g$ if we use the complex

$$\omega^{\cdot}_{\overline{\mathcal{C}}_g/\overline{\mathcal{M}}_g} : 0 \to \mathcal{O}_{\overline{\mathcal{C}}_g} \xrightarrow{d} \omega_{\overline{\mathcal{C}}_g/\overline{\mathcal{M}}_g} \to 0$$

from which we get the sequence:

$$(5.5)\qquad 0 \to \mathbb{E} \to \mathbb{R}^1\pi_*\omega^{\cdot}_{\overline{\mathcal{C}}_g/\overline{\mathcal{M}}_g} \to \mathbb{E}^{\vee} \to 0.$$

Although the Gauss-Manin connection does not extend regularly to $\mathbb{R}^1\pi_*\omega^{\cdot}_{\overline{\mathcal{C}}_g/\overline{\mathcal{M}}_g}$, it has regular singularities with a polar part which is nilpotent. This is enough to conclude that its Chern classes zero, extending (5.5) to $\overline{\mathcal{M}}_g$. This means equivalently that

$$ch(\mathbb{E}) + ch(\mathbb{E}^{\vee}) = 0$$

or

$$ch(\mathbb{E})_{2k} = 0, \quad k \geq 1.$$

This identity in fact holds on $\tilde{\mathcal{A}}_g$, the toroidal compactification of $\mathcal{A}_g$. It can be deduced, for instance, from the extension of Hirzebruch's proportionality theorem to $\tilde{\mathcal{A}}_g$ (see [M2]).

The conclusion to be drawn from (5.2) and (5.3) is that the even λ_k's are polynomials in the odd ones, and that all the λ_k's are polynomials in the κ_k's and in boundary cycles. Moreover, applying (5.2) in odd degree above g, we can express κ_k for k odd, $k > g$, in terms of lower κ_l's and boundary cycles. We shall strengthen this in the next section, where we find a simpler way to get identities on the κ_k's.

The exact sequence (5.4) is remarkable in another way that reveals something of the nature of $\overline{\mathcal{M}}_g$. Note that $\mathbb{E}$ tends to be a positive bundle: at least $c_1(\mathbb{E})$ is the pull-back of an ample line bundle by a birational map. But it is also a sub-bundle of a bundle with connection, i.e., the DeRham bundle $\mathbb{R}^1\pi_*\omega^{\cdot}$ is unstable yet has a connection.

§6. Tautological Relations via the Canonical Linear System

There is another very different way to get relations on the λ_i and κ_i. For this, we will not try to get the full relations in $A^{\cdot}(\overline{\mathcal{M}}_g)$ as the boundary terms seem to be a bit involved, but instead get the relations in $A^{\cdot}(\mathcal{M}_g)$. Because of the exact sequence:

$$A^{\cdot}(\overline{\mathcal{M}}_g - \mathcal{M}_g) \to A^{\cdot}(\overline{\mathcal{M}}_g) \to A^{\cdot}(\mathcal{M}_g) \to 0$$

this is the same as a relation in $A^{\cdot}(\overline{\mathcal{M}}_g)$ with an undetermined boundary term.

The method is based on the fact that for all smooth curves C, the sheaf ω_C is generated by its global sections.[5]

Now if we let $\tilde{C}_g/\tilde{M}_g$ temporarily stand for the family of *smooth* stable curves, i.e., replace $\tilde{C}_g$ by $\pi^{-1}(\mathcal{M}_g)$, then we have an exact sequence:

$$0 \to \mathcal{F} \to \pi^*\pi_*\omega_{\tilde{C}_g/\tilde{M}_g} \to \omega_{\tilde{C}_g/\tilde{M}_g} \to 0$$

where all these sheaves are Q-sheaves and $\mathcal{F}$ is locally free of rank $g-1$. Taking Chern classes, we get:

$$c(\mathcal{F}) = \pi^*(1+\lambda_1+\cdots+\lambda_g)\cdot(1+K_{\tilde{C}_g/\tilde{M}_g})^{-1}.$$

Using the fact that $c_n(\mathcal{F})=0$ if $n \geq g$, this says:

$$(K^n_{\tilde{C}/\tilde{M}}) - \pi^*(\lambda_1)\cdot(K^{n-1}_{\tilde{C}/\tilde{M}}) + \cdots + (-1)^g\pi^*(\lambda_g)\cdot(K^{n-g}_{\tilde{C}/\tilde{M}}) = 0$$

for all $n \geq g$. Taking π_*, this means

$$\begin{aligned} \kappa_{n-1}-\lambda_1\cdot\kappa_{n-2}+\cdots+(-1)^g\lambda_g\cdot\kappa_{n-g-1} &= 0 \quad \text{if } n \geq g+2\\ \kappa_g-\lambda_1\cdot\kappa_{g-1}+\cdots+(-1)^g\lambda_g\cdot(2g-2) &= 0 \quad \text{if } n = g+1\\ \kappa_{g-1}-\lambda_1\cdot\kappa_{g-2}+\cdots+(-1)^{g-1}\lambda_{g-1}\cdot(2g-2) &= 0 \quad \text{if } n = g. \end{aligned}$$

---

[5]If C is a singular stable curve, then one can show that $\Gamma(\omega_C)$ generates the subsheaf of ω_C of sections which are zero

i) at all double points P for which $C-P$ is disconnectd,

ii) on all components E_0 of C which are isomorphic to $\mathbf{P}^1$ and such that all double

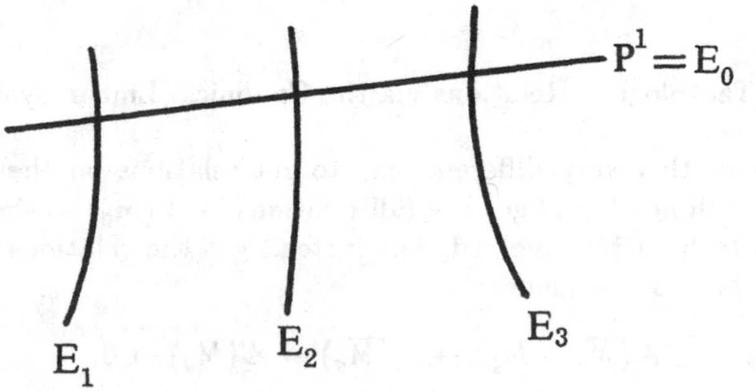

points P on E_0 are disconnecting double points.

Corollary 6.2. *For all g, all the classes λ_i, κ_i restricted to $A^{\cdot}(\mathcal{M}_g)$ are polynomials in $\kappa_1, \kappa_2, \ldots, \kappa_{g-2}$.*

Proof. This is clear except for κ_g, κ_{g-1}. Here we must combine the above relations with (5.2). There are two cases depending on whether g is even or odd. Recalling that

$$ch_n \mathbb{E} = \frac{(-1)^{n-1} \cdot c_n(\mathbb{E})}{(n-1)!} + \text{polyn. in lower Chern class}$$

it follows that if $g = 2k$ or $2k-1$, then

$$\frac{(-1)^{k+1} \cdot B_k}{(2k)!} \kappa_{2k-1} = ch_{2k-1}\mathbb{E} = \frac{\lambda_{2k-1}}{(2k-2)!} + (\text{polyn. in lower } \lambda\text{'s}).$$

If $g = 2k$, we want to show that the 2 equations

$$\begin{cases} (-1)^{k+1} B_k \cdot \kappa_{2k-1} = 2k \cdot (2k-1)\lambda_{2k-1} + \text{lower} \\ \kappa_{2k-1} = 2 \cdot (2k-1)\lambda_{2k-1} + \text{lower} \end{cases}$$

have independent leading terms, and if $g = 2k-1$, then we want to do the same with

$$\begin{cases} (-1)^{k+1} B_k \cdot \kappa_{2k-1} = 2k \cdot (2k-1)\lambda_{2k-1} + \text{lower} \\ \kappa_{2k-1} = 4(k-1) \cdot \lambda_{2k-1} + \text{lower.} \end{cases}$$

This follows, however, by inspection if $k \leq 10$ and for larger k by the estimate:

$$B_k = \frac{2 \cdot (2k)!}{(2\pi)^{2k}} \zeta(2k) > \frac{2 \cdot (2k/e)^{2k}}{(2\pi)^{2k}} = 2 \cdot \left(\frac{k}{e \cdot \pi}\right)^{2k} > 2 \cdot k \quad \text{if } k \geq 11.$$

Q.E.D.

With this approach, the first relation between $\kappa_1, \ldots, \kappa_{g-2}$ that we get occurs in codimension $g+1$ or $g+2$. One should, however, get the relation $\kappa_1^2 = 0$ in $A^{\cdot}(\mathcal{M}_3)$ so we clearly do not have all the relations on the κ_i's and λ_i's yet. It does seem reasonable to conjecture, however, that $\kappa_1, \ldots, \kappa_{g-2}$ have no relations up to something *like* codimension g, e.g., g −(small constant).

§7. The Tautological Classes via Arbarello's Flag of Subvarieties of $\mathcal{M}_g$

We want to consider the following subsets of $\mathcal{C}_g$ and $\mathcal{M}_g$:

$$W_l^* = \{C, x \in \mathcal{C}_g \mid h^o(\mathcal{O}_C(l \cdot x)) \geq 2\}$$
$$= \left\{ \begin{array}{r} C, x \in \mathcal{C}_g \mid \exists \text{ a morphism } \pi\colon c \to \mathbb{P}^1 \text{ of degree} \\ d \leq l \text{ with } \pi^{-1}(\infty) = d.x \end{array} \right\}$$
$$W_l = \pi(W_l^*) \subset \mathcal{M}_g$$

where $2 \leq l \leq g$. Thus W_g^* =locus of Weierstrass points in $\mathcal{C}_g$, W_{g-1} = curves with an exceptional Weierstrass of one of the two simplest types, and W_2 = hyperelliptic curves. Note that:

$$\mathcal{C}_g \supset W_g^* \supset W_{g-1}^* \supset \cdots \supset W_2^*$$
$$\mathcal{M}_g = W_g \supset W_{g-1} \supset \cdots \supset W_2.$$

I first heard of this flag from E. Arbarello who proposed (see [Arb][6]) that they might be used as a ladder to climb from the reasonably well-known space W_2 to the still mysterious $\mathcal{M}_g$.

Let me first recall and sketch the proof of the following well-known facts:

Proposition 7.1. *W_l^* is irreducible of codimension $g - l + 1$ and $W_l^* - W_{l-1}^*$ is an open dense subset smooth in the local charts for $\mathcal{C}_g$, i.e., the local deformation space for the pairs (C, x).*

Sketch of proof. Firstly, W_l^* is a determinantal subvariety of $\tilde{\mathcal{C}}_g$. In fact, consider

$$\begin{array}{ccccc} & & \tilde{\mathcal{C}}_g \times_{\tilde{\mathcal{M}}_g} \tilde{\mathcal{C}}_g \supset \Delta, \text{ diagonal} & & \\ & {\scriptstyle p_1}\swarrow & & \searrow{\scriptstyle p_2} & \\ \tilde{\mathcal{C}}_g & & & & \tilde{\mathcal{C}}_g \\ & {\scriptstyle \pi_1}\searrow & & \swarrow{\scriptstyle \pi_2} & \\ & & \tilde{\mathcal{M}}_g & & \end{array}$$

Let

$$\mathcal{F}_l = R^1 p_{2,*}(\mathcal{O}_{\tilde{\mathcal{C}} \times \tilde{\mathcal{C}}}(l\Delta)).$$

Then over a point $[C, x] \in \tilde{\mathcal{C}}_g$,

$$\mathcal{F}_l \otimes \mathbf{k}([C, x]) \cong H^1(C, \mathcal{O}_C(l.x)).$$

---

[6] Unfortunately, the proof of Theorem 3.27 in [Arb] is incomplete as it stands.

Now if $[C,x] \notin W_l^*$, $h^0(O_C(lx)) = 1$ and $h^1(O_C(lx)) = g-l$, while if $[C,x] \in W_l^*$, both numbers are bigger. Thus $\mathcal{F}_l$ is locally free of rank $g-l$ on $\tilde{C}_g - W_l^*$ and not locally free anywhere on W_l^*. But look at the sequence:

$$0 \to O_{\tilde{C}\times\tilde{C}} \to O_{\tilde{C}\times\tilde{C}}(l\Delta) \to O_{\tilde{C}\times\tilde{C}}(l\Delta)/O_{\tilde{C}\times\tilde{C}} \to 0$$

which gives us:

$$(7.2) \qquad 0 \to p_{2,*}(O_{\tilde{C}\times\tilde{C}}(l\Delta)/O_{\tilde{C}\times\tilde{C}}) \overset{\alpha}{\to} R^1p_{2,*}(O_{\tilde{C}\times\tilde{C}}) \to \mathcal{F}_l \to 0.$$

The first sheaf is locally free of rank l, the second locally free of rank g, hence

$$W_l^* = \{[C,x] \in \tilde{C}_g \mid rk_{[C,x]}(\alpha) < l\}.$$

Thus the codimension of W_l^* is at most $g-l+1$. But describing $W_l^* - W_{l-1}^*$ as the set of l-fold covers of $\mathbb{P}^1$, totally ramified at ∞, one gets the *upper bound* $2g+l_1-3$ on $\dim W_{l_1}^* - W_{l_1-1}^*$ for all l_1, hence the same upper bound on $\dim W_l^*$. Comparing the two, it follows that $\operatorname{codim} W_l^*$ is *exactly* $g-l+1$ and W_l^* is determinantal as well as that $W_l^* - W_{l-1}^*$ is dense in W_l^*. The irreducibility of $W_l^* - W_{l-1}^*$ is a classical result of Lüroth, describing all l-fold covers of $\mathbb{P}^1$ as branched covers with a standard set of transpositions.

The smoothness of $W_l^* - W_{l-1}^*$ in the universal deformation space may be checked by the following calculation: let f have an l-fold pole at $x \in C$ and make an infinitesimal deformation C of C over $\mathbb{C}[\epsilon]$ by glueing open sets $U_\alpha \times \operatorname{Spec} \mathbb{C}[\epsilon]$ via a 1-cocycle $D_{\alpha\beta}$ of derivations zero at x.. Then f lifts to a rational function on C with l-fold pole at x if there are functions g_α with l-fold poles at x and:

$$(1+\epsilon D_{\alpha\beta})(f+\epsilon g_\alpha) = f + \epsilon g_\beta.$$

This means that $\{D_{\alpha\beta}f\} \in H^1(C, O(lx))$ is zero. But $D_{\alpha\beta}f = \langle D_{\alpha\beta}, df\rangle$ is the image:

$$(7.3) \qquad \{D_{\alpha\beta}\} \in H^1(C, T_C(-x)) \xrightarrow[\langle\ ,df\rangle]{} H^1(C, O_C(lx)).$$

Note that $H^1(C, T_C(-x))$ is the tangent space to the universal deformation space of (C,x). Moreover (7.3) is dual to the injective map

$$H^0(C, O(2K_C + x)) \xleftarrow[\otimes df]{} H^0(C, O(K_C - lx))$$

hence (7.3) is *surjective*, i.e., the subscheme of the universal deformation space where f lifts is smooth of codimension $h^1(\mathcal{O}_C(lx)) = g - l$.

In order to work out the fundamental class of W_l^*, it is convenient to split up (7.2) into pieces as follows. Starting with

$$0 \to \mathcal{O}_{\tilde{C}\times\tilde{C}}((l-1)\Delta) \to \mathcal{O}_{\tilde{C}\times\tilde{C}}(l\Delta) \to \mathcal{O}_{\tilde{C}\times\tilde{C}}(l\Delta)/\mathcal{O}_{\tilde{C}\times\tilde{C}}((l-1)\Delta) \to 0$$

$$\Vert$$

$$\mathcal{O}_\Delta \otimes p_2^*\mathcal{O}_{\tilde{C}}(-lK_{\tilde{C}/\tilde{\mathcal{M}}})$$

we get via $R^{\cdot}p_{2,*}$:

$$0 \to \mathcal{O}_{\tilde{C}}(-lK_{\tilde{C}/\tilde{\mathcal{M}}}) \xrightarrow{\beta} \mathcal{F}_{l-1} \to \mathcal{F}_l \to 0. \tag{7.4}$$

It follows that on $\tilde{C}_g - W_{l-1}^*$ where $\mathcal{F}_{l-1}$ is locally free:

$$\begin{aligned} W_l^* &= \{[C,x] \in \tilde{C}_g \mid \beta_{[C,x]} = 0\} \\ &= \text{zeroes of the section } \beta' \in \Gamma(\tilde{C}_g, \mathcal{F}_{l-1}(lK_{\tilde{C}/\tilde{\mathcal{M}}})). \end{aligned}$$

Moreover, on the universal deformation space of $[C,x]$, this section β' vanishes to 1st order along W_l^*: in fact, the differential of β' at a point of W_l^* is a map

$$\begin{array}{ccc} T_{[C,x],\mathcal{C}_\alpha} & \longrightarrow & \mathcal{F}_{l-1}(lk)\otimes \mathrm{k}([C,x]) \\ \Vert & & \Vert \\ H^1(C, T_C(-x)) & & H^1(C, \mathcal{O}_C(lx))\otimes(\mathfrak{m}_x/\mathfrak{m}_x^2)^{\otimes l} \end{array}$$

which is readily seen to be the surjective map (7.3) (the factor $(\mathcal{M}_x/\mathcal{M}_x^2)^l$ is hidden in (7.3) in the choice of f). Thus

$$[W_l^*]_Q = c_{g-l+1}(\mathcal{F}_{l-1}(lK_{\tilde{C}/\tilde{\mathcal{M}}})), \quad \text{on } \tilde{C}_g - W_{l-1}^*. \tag{7.5}$$

But W_{l-1}^* has codimension $g - l + 2$ so

$$A^{g-l+1}(C_g) \cong A^{g-l+1}(C_g - W_{l-1}^*).$$

Thus (7.5) holds as an equation in $A^{g-l+1}(C_g)$, hence in $opA^{g-l+1}(\tilde{C}_g)$. Now let's calculate the fundamental class of W_l^*:

$$\begin{aligned} [W_l^*]_Q &= c_{g-l+1}(\mathcal{F}_{l-1}(lK)) \\ &= c_{g-l+1}(\mathcal{F}_l(lK)) \\ &= c_{g-l+1}(\mathcal{F}_l). \end{aligned}$$

Here we have abbreviated $K_{\bar{\mathcal{C}}/\bar{\mathcal{M}}}$ to K, and the last equality follows from the general fact

$$c_n(\mathcal{G}(D)) =$$
$$c_n(\mathcal{G}) + (r-n+1)D.c_{n-1}(\mathcal{G}) + \binom{r-n+2}{2} D^2.c_{n-2}(\mathcal{G}) + \cdots + \binom{r}{n} \cdot D^n$$

($r =$ generic rank $\mathcal{G}$), whence

$$c_{r+1}(\mathcal{G}(D)) = c_{r+1}(\mathcal{G}), \text{ all divisors } D.$$

But now

$$\begin{aligned} c(\mathcal{F}_l) &= c(\mathcal{F}_{l-1}).(1-lK)^{-1} \\ &= c(\mathcal{F}_{l-2}).(1-(l-1)K)^{-1}.(1-lK)^{-1} \\ &\cdots \\ &= c(\mathcal{F}_0).(1-K)^{-1}.(1-2K)^{-1}.\cdots.(l-lK)^{-1} \\ &= \pi_2^*(c(R^1\pi_{1,*}\mathcal{O}_{\bar{\mathcal{C}}/\bar{\mathcal{M}}})).(1-K)^{-1}.\cdots.(1-lK)^{-1} \\ &= \pi_2^*(1-\lambda_1+\lambda_2-\cdots+(-1)^g\lambda_g).(1-K)^{-1}.\cdots.(1-lK)^{-1}. \end{aligned}$$

Thus

$$(7.6)\quad \begin{aligned} [W_l^*]_Q =& (g-l+1)^{st} \text{ component of} \\ & \pi^*(1-\lambda_1+\lambda_2-\cdots+(-1)^g\lambda_g).(1-K)^{-1}.\cdots.(1-lK)^{-1}. \end{aligned}$$

If we define W_l *as a cycle* as $\pi_*(W_l^*)$, we get also

$$(7.7)\quad \begin{aligned} [W_l]_Q =& (g-l)^{th}\text{-component of} \\ & (1-\lambda_1+\lambda_2-\cdots+(-1)^g\lambda_g).\pi_*[(1-K)^{-1}.\cdots.(1-lK)^{-1}]. \end{aligned}$$

This shows that $[W_l]_Q$ is a polynomial in the tautological classes κ_j. Presumably the coefficient of κ_l is always non-zero and hence we can solve for the κ_l's in terms of the classes $[W_j]$, but this looks like a messy calculation.

Let's work out the hyperelliptic locus $\mathcal{H}$ as an example. Note that

$$W_2^* \to W_2$$

is a covering of degree $2g+2$ because for all hyperelliptic curves C, there are exactly $2g+2$ points x such that $h^0(\mathcal{O}_C(2x)) \geq 2$ — namely the Weierstrass points. Thus

$$
\begin{aligned}
[\mathcal{H}]_Q &= \frac{1}{2g+2}[W_2]_Q \\
&= \frac{1}{2g+2}\left\{\begin{matrix}(g-2)^{nd} \text{ component of} \\ (1-\lambda_1+\cdots+(-1)^g\lambda_g)\cdot\pi_*((1-K)^{-1}\cdot(1-2K)^{-1})\end{matrix}\right\} \\
&= \frac{1}{2g+2}\{(2^g-1)\kappa_{g-2}-(2^{g-1}-1)\lambda_1\cdot\kappa_{g-3}+\cdots+ \\
&\quad (-1)^{g-3}\cdot 7\cdot\lambda_{g-3}\cdot\kappa_3+(-1)^{g-2}\cdot(6g-6)\lambda_{g-2}\}.
\end{aligned}
$$

Finally every hyperelliptic curve has an automorphism of order 2, so

$$
\begin{aligned}
[\mathcal{H}] &= 2\cdot[\mathcal{H}]_Q \\
&= \frac{1}{g+1}\{(2^g-1)\kappa_{g-2}-\cdots+(-1)^{g-2}(6g-2)\lambda_{g-2}\}.
\end{aligned}
$$

Part III: The Case g = 2

§8. Tautological Relations in Genus 2

First of all, let's specialize the calculations of Part II to the case $g=2$ and see what we have. From $\mathbb{E}$, we get 2 elements

$$
\begin{aligned}
\lambda_1 &\in A^1(\overline{\mathcal{M}}_2) \\
\lambda_2 &\in A^2(\overline{\mathcal{M}}_2)
\end{aligned}
$$

and because $ch_2(\mathbb{E})=0$, we get

$$\lambda_2 = \lambda_1^2/2.$$

From K on $\overline{\mathcal{C}}_2$, we get

$$\kappa_i \in A^i(\overline{\mathcal{M}}_2), \qquad i=1,2,3.$$

The calculations of §5 give us the relation:

$$\lambda_1 = \frac{1}{12}(\kappa_1+\delta). \tag{8.1}$$

Here $\overline{\mathcal{M}}_2 - \mathcal{M}_2$ has 2 components Δ_0 and Δ_1, Δ_0 the closure of the locus of irreducible singular curves, Δ_1 the locus of singular curves $C_1 \cup C_2$, $C_1 \cap C_2 =$ one pt., $p_a(C_1) = p_1(C_2) = 1$. By definition

$$\delta = [\Delta_0]_Q + [\Delta_1]_Q.$$

We shall write δ_0 for $[\Delta_0]_Q$ and δ_1 for $[\Delta_1]_Q$. We don't need (5.2) in codimension 3 but it provides an interesting check on the calculations later. It gives:

$$\frac{1}{6}\lambda_1^3 - \frac{1}{2}\lambda_1\lambda_2 = ch_3\mathbb{E} = -\frac{1}{720}\left[\kappa_3 + \frac{1}{2}\cdot\sum_{h=0}^{1} i_{h,*}\left((K_1+K_2)^2 - 3K_1\cdot K_2\right)\right]$$

or

$$60\lambda_1^3 = \kappa_3 + \frac{1}{2}\sum_{h=0}^{1} i_{h,*}(K_1^2 - K_1K_2 + K_2^2).$$

In §10*b* we shall work out these terms numerically and check this.

We can refine the calculations of §6 by working out the boundary term too. It is easy to see that if C is a stable curve of genus 2, ω_C is generated by its global sections, unless $C = C_1 \cup C_2$, $C_1 \cap C_2 = \{P\}$, in which case $\Gamma(\omega_C)$ generates $m_P \cdot \omega_C$. Therefore, working over the whole of $\tilde{C}_2$ we get:

$$0 \to \mathcal{F} \to \pi^*\pi_*\omega_{\bar{C}_2/\bar{\mathcal{M}}_2} \to I_{\Delta_1^*}\cdot\omega_{\bar{C}_2/\bar{\mathcal{M}}_2} \to 0$$

(following the notation of §5). $I_{\Delta_1^*}$ has two generators at every point, so its projective dimension is 1, i.e., $\mathcal{F}$ is locally free, hence invertible. Now use:

$$0 \to I_{\Delta_1^*}\cdot\omega \to \omega \to \omega\otimes O_{\Delta_1^*} \to 0$$

and the fact that via residue, ω is trivial on the double cover Δ_1' of Δ_1^*, hence ω^2 is trivial on Δ_1^*. It follows that

$$c(\mathcal{F}) = \pi^*(1+\lambda_1+\lambda_2)\cdot(1+K_{\bar{C}_2/\bar{\mathcal{M}}_2})^{-1}\cdot c(O_{\Delta_1^*}).$$

A useful lemma that we can use here is:

Lemma 8.2. *If $Y \subset X$ is a local complete intersection of codimension 2, and $i\colon Y \to X$ is the inclusion, then*

$$c(O_Y) = 1 - i_*\left(c(I_Y/I_Y^2)^{-1}\right).$$

Proof. If $Y = D_1.D_2$ globally, then this formula is easily checked. But by the G-R-R,

$$c(\mathcal{O}_Y) = 1 + i_*\,(\text{univ. polyn. in } c_1(I/I^2), c_2(I/I^2))$$

and the universal polynomial must be $c(I/I^2)^{-1}$ because the two are equal whenever $Y = D_1.D_2$.

As in §5, $\Delta_1' \cong \overline{\mathcal{M}}_{1,1} \times \overline{\mathcal{M}}_{1,1}$, i.e., there is a 2-1 map:

$$i_1 \colon \overline{\mathcal{M}}_{1,1} \times \overline{\mathcal{M}}_{1,1} \longrightarrow \Delta_1^*$$

and $i_1^*(I/I^2) = K_1 + K_2$. Thus

$$c(\mathcal{O}_{\Delta_1^*}) = 1 - \frac{1}{2} \cdot i_{1,*}((1 - K_1 - K_2)^{-1}).$$

Thus

$$c(\mathcal{F}) = \pi^*(1 + \lambda_1 + \lambda_2) \cdot \\ (1 - K + K^2 - K^3 + K^4) \cdot \left(1 - \frac{1}{2} \cdot i_{1,*}(1 + K_1 + K_2 + K_1 \cdot K_2)\right).$$

In particular,

$$0 = c_2(\mathcal{F}) = \pi^*\lambda_2 - K.\pi^*\lambda_1 + K^2 - [\Delta_1^*]_Q.$$

Multiplying this by K and K^2, we get even simpler formulae:

$$0 = K \cdot \pi^*\lambda_2 - K^2 \cdot \pi^*\lambda_1 + K^3$$
$$0 = K^2 \cdot \pi^*\lambda_2 - K^3 \cdot \pi^*\lambda_1 + K^4.$$

Taking π_*, this gives

(8.3)
$$\begin{aligned} \kappa_1 &= 2\lambda_1 + \delta_1 \\ \kappa_2 &= \kappa_1 \cdot \lambda_1 - 2\lambda_2 = \lambda_1 \cdot (\lambda_1 + \delta_1) \\ \kappa_3 &= \kappa_2\lambda_1 - \kappa_1\lambda_2 = \frac{1}{2}\lambda_1^2(\delta_1). \end{aligned}$$

Combining (8.1) and (8.3), we see that both κ_1 and λ_1 are expressible in terms of δ_0, δ_1:

$$10\lambda_1 = \delta_0 + 2\delta_1 \tag{8.4}$$

$$5\kappa_1 = \delta_0 + 7\delta_1. \tag{8.5}$$

As κ_1 is ample, this implies the well-known fact that $\mathcal{M}_2$ is affine!

This relation (8.4) has a very simple analytic proof. Consider the modular form of weight 10 on Siegel's space $\mathcal{H}_2$ given by

$$f(Z) = \left[\prod_{a,b \text{ even}} \theta\begin{bmatrix} a \\ b \end{bmatrix}(0, Z) \right]^2$$

(Each θ has weight 1/2 and there are ten even a, b's.) It vanishes on $\mathcal{H}_2$ precisely when

$$\gamma Z = \begin{pmatrix} Z_1 & 0 \\ 0 & Z_2 \end{pmatrix}, \quad \text{some } \gamma \in Sp(4, \mathbb{Z})$$

and then to order 2. At the principal cusp

$$\begin{pmatrix} i\infty & w \\ w & z \end{pmatrix}$$

it has the form

$$\begin{aligned} &(\text{unit}) \cdot \theta\begin{bmatrix} 1 & 0 \\ 0 & 0 \end{bmatrix}^2 \cdot \theta\begin{bmatrix} 1 & 1 \\ 0 & 0 \end{bmatrix}^2 \cdot \theta\begin{bmatrix} 1 & 0 \\ 0 & 1 \end{bmatrix}^2 \cdot \theta\begin{bmatrix} 1 & 1 \\ 1 & 1 \end{bmatrix}^2 \\ &= (\text{unit}) \cdot (e^{\pi i (1/2) \Omega_{11}})^8 \\ &= \text{unit} \cdot e^{2\pi i \Omega_{11}} \end{aligned}$$

i.e., it vanishes to order 1. Thus f defines a section of $(\Lambda^2 \mathbb{E})^{\otimes 10}$ whose zeroes in $\tilde{\mathcal{M}}_2$ are $2\tilde{\Delta}_1 + \tilde{\Delta}_0$. This reproves (8.4).

§9. Generators of $A.(\overline{\mathcal{M}}_2)$.

We use the exact sequence:

$$A.(Y) \to A.(X) \to A.(X - Y) \to 0$$

($Y \subset X$ closed subvariety) to get generators of $A_.(\overline{\mathcal{M}}_2)$. Recall that $\mathcal{M}_2$ is known from Igusa's results [I] to be isomorphic to $\mathbb{C}^3$ modulo $\mathbb{Z}/5\mathbb{Z}$ acting by

$$(x, y, z) \mapsto (\varsigma x, \varsigma^2 x, \varsigma^3 y).$$

Then

$$A_.(\mathbb{C}^3) \to A_.(\mathcal{M}_2)$$

is surjective, hence $A_k(\mathcal{M}_2) = (0)$, if $k < 3$. Thus

$$A_k(\Delta_0) \oplus A_k(\Delta_1) \to A_k(\overline{\mathcal{M}}_2)$$

is surjective if $k < 3$. In particular, $A_2(\overline{\mathcal{M}}_2)$ is generated by δ_0 and δ_1.

Define the dimension 1 subsets:

$$\Delta_{00} = \left\{ \begin{matrix} \text{closure of curve in } \overline{\mathcal{M}}_2 \text{ parametrizing} \\ \text{irreducible rational curves with 2 nodes} \end{matrix} \right\}$$

$\Delta_{01} = \Delta_0 \cap \Delta_1$
$=$ Curve in $\overline{\mathcal{M}}_2$ parametrizing curves $C_0 \cup C_2$, where $C_1 \cap C_2 = \{x\}$, C_1 is elliptic or rational with one node and C_2 is rational with one node

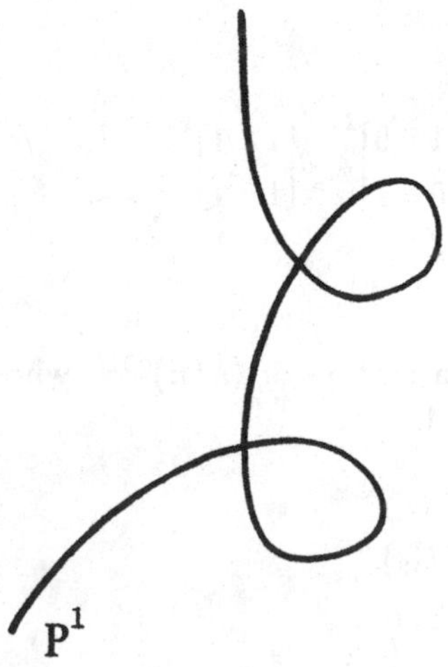

curves in Δ_{00}

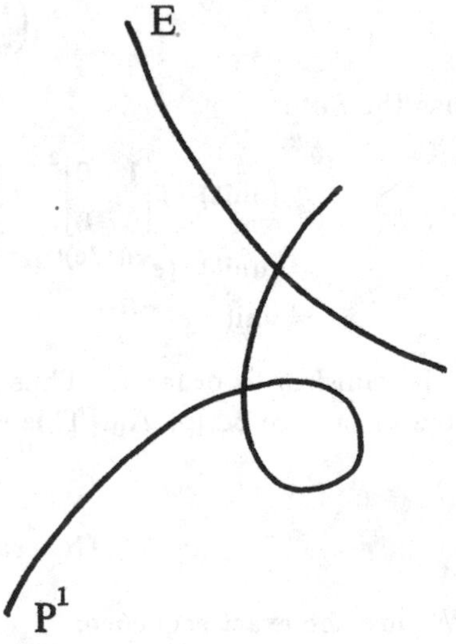

curves in Δ_{01}

Note that Δ_{00} contains, besides the irreducible curves illustrated, the two reducible curves

C_{001}:

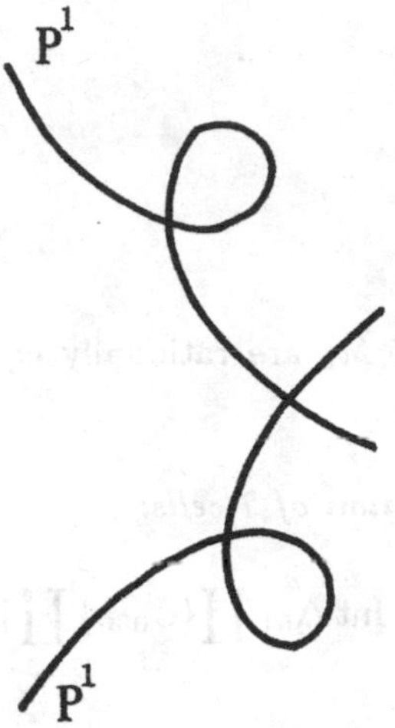

C_{000}:

$\operatorname{Int}\Delta_1 = (\Delta_1 - \Delta_{01})$ is the locus of curves $C_1 \cup C_2$ where $C_1 \cap C_2 = \{x\}$, C_1, C_2 smooth elliptic. It is isomorphic to Symm ${}^2 \mathcal{M}_{1,1}$, i.e., to the product of the affine j-line by itself mod the involution interchanging the factors. Therefore it is coordinatized by $j(C_1) + j(C_2)$, $j(C_1).j(C_2)$:

$$\operatorname{Int}\Delta_1 \cong \mathbb{C}^2.$$

Moreover, $\operatorname{Int}\Delta_0 = \Delta_0 - (\Delta_{00} \cup \Delta_{01})$ is the locus of irreducible elliptic curves with one node, i.e., the space $\mathcal{M}_{1,2}$ of triples (E, x_1, x_2), $x_1, x_2 \in E$, $x_1 \neq x_2$ mod the involution interchanging the 2 points. Write all elliptic curves as

$$y^2 = x(x-1)(x-\lambda), \quad \lambda \neq 0, 1$$

and take $x_1 =$ pt. at ∞, $x_2 = (x, y)$. Interchanging x_1, x_2 carries (x, y) to $(x, -y)$. So we get a surjective map

$$\{(x, \lambda) \mid \lambda \neq 0, 1\} \to \operatorname{Int}\Delta_0.$$

Putting this together

$$A_k(\Delta_1 - \Delta_{01}) = (0), \quad k < 2$$
$$A_k(\Delta_0 - (\Delta_{00} \cup \Delta_{01})) = (0), \quad k < 2.$$

Thus $A_1(\overline{\mathcal{M}}_2)$ is generated by:

$$\delta_{00} = [\Delta_{00}]_Q$$

and

$$\delta_{01} = [\Delta_{01}]_Q.$$

Finally, $\overline{\mathcal{M}}_2$ is unirational so all points of $\overline{\mathcal{M}}_2$ are rationally equivalent and we have proven:

Proposition 9.1. *$\overline{\mathcal{M}}_2$ is the disjoint union of 7 cells:*

$$\overline{\mathcal{M}}_2 = \mathcal{M}_2 \coprod \operatorname{Int}\Delta_0 \coprod \operatorname{Int}\Delta_1 \coprod \operatorname{Int}\Delta_{00} \coprod \operatorname{Int}\Delta_{01} \coprod \{C_{000}\} \coprod \{C_{001}\}.$$

Correspondingly, $A^{\cdot}(\overline{\mathcal{M}}_2)$ is generated by

a) *1 in codimension 0,*
b) *δ_0, δ_1 in codimension 1,*
c) *δ_{00}, δ_{01} in codimension 2,*
d) *the class $[x]$ of a point in codimension 3: call this p.*

Note that by the results of §8, λ_1 and κ_1 are also generators in codimension 1. We shall see that all the above cycles are independent. This will follow as a Corollary once we work out the multiplication table for these cycles.

§10. Multiplication in $A^{\cdot}(\overline{\mathcal{M}}_2)$

We shall prove:

Theorem 10.1. *The ring $A^{\cdot}(\overline{\mathcal{M}}_2)$ has a $\mathbb{Q}$-basis consisting of* 1, δ_0,

δ_1, δ_{00}, δ_{01}, *p and multiplication table:*

$$\delta_0^2 = \frac{5}{3}\delta_{00} - 2\delta_{01}$$

$$\delta_0 \cdot \delta_1 = \delta_{01}$$

$$\delta_1^2 = -\frac{1}{12}\delta_{01}$$

$$\delta_0 \cdot \delta_{00} = -\frac{1}{4}p$$

$$\delta_0 \cdot \delta_{01} = \frac{1}{4}p$$

$$\delta_1 \cdot \delta_{00} = -\frac{1}{8}p$$

$$\delta_1 \cdot \delta_{01} = -\frac{1}{48}p$$

An easier way to describe the ring structure is via λ_1. Using the identities $10\lambda_1 = \delta_0 + 2\delta_1$, we can describe the multiplication by:

$$a)\ \delta_0 \cdot \delta_1 = \delta_{01}$$

$$b)\ \delta_{00} \cdot \delta_1 = \frac{1}{8}p$$

$$c)\ \delta_{00} \cdot \lambda_1 = 0$$

$$d)\ \delta_1 \cdot \lambda_1 = \frac{1}{12}\delta_{01}$$

$$e)\ \delta_0 \cdot \lambda_1 = \frac{1}{6}\delta_{00}.$$

The reader can check that these are equivalent to the relations of the Theorem.

Relations (a) and (b) are proper intersections of cycles and are proved by the explicit formula of §3: thus (a) follows because the lifts of δ_0, δ_1 to the universal deformation space of a curve $C \in \Delta_{01}$ are smooth divisors meeting transversely in the smooth curve lifting Δ_{01}. And for (b), $\Delta_{00} \cap \Delta_1$ is the one curve C_{001} whose automorphism group has order 8. In the universal deformation space of C_{001}, Δ_{00} and Δ_1 lift to a smooth curve and surface meeting transversely, so

$$\delta_{00}.\delta_1 = [C_{001}]_Q = \frac{1}{8}p.$$

c) is an immediate consequence of the general theory of Knudsen [K] or of the fact that δ_{00} is blown down to a point in the Satake compactification $\mathcal{A}_2^*$ of $\mathcal{A}_2$. To prove (d), consider

$$i_1 \colon \overline{\mathcal{M}}_{1,1} \times \overline{\mathcal{M}}_{1,1} \longrightarrow \Delta_1 \subset \overline{\mathcal{M}}_2.$$

We check that $i_1^*(\lambda_1 - \frac{1}{12}\delta_0) = 0$, hence $(\lambda_1 - \frac{1}{12}\delta_0).\delta_1 = 0$. But

$$i_1^*(\lambda_1^{(2)}) = p_1^*(\lambda_1^{(1)}) + p_2^*(\lambda_1^{(1)})$$

where, for the sake of clarity, we write

$$\begin{aligned} \lambda_1^{(2)} &= \text{the class } \lambda_1 \text{ in } A^{\cdot}(\overline{\mathcal{M}}_2) \\ \lambda_1^{(1)} &= \text{the class } \lambda_1 \text{ in } A^{\cdot}(\overline{\mathcal{M}}_{1,1}). \end{aligned}$$

This is simply because on curves $E_1 \cup E_2$, E_i elliptic, $E_1 \cap E_2 =$ one point,

$$\Gamma(E_1 \cup E_2, \omega_{E_1 \cup E_2}) \cong \Gamma(E_1, \omega_{E_1}) \oplus \Gamma(E_2, \omega_{E_2}).$$

Moreover,

$$i_1^*(\delta_0^{(2)}) = p_1^*(\delta^{(1)}) + p_1^*(\delta^{(1)}).$$

But in $A^{\cdot}(\overline{\mathcal{M}}_{1,1})$, the relation

$$\lambda_1 = \frac{1}{12}\delta$$

holds. This is well known and is just the specialization to genus 1 of the theory of §5. Or else it may be seen using the elliptic modular form Δ of weight 12 with a simple pole at the cusp. Finally, to prove (e), consider:

$$i_o\colon \overline{\mathcal{M}}_{1,2} \longrightarrow \Delta_0 \subset \overline{\mathcal{M}}_2.$$

Then $i_{0,*}(1_{\overline{\mathcal{M}}_{1,2}}) = 2\delta_0$. One should be careful here to note that the presence of automorphisms generically on $\overline{\mathcal{M}}_2$ does not affect this: in fact

$$\begin{aligned} i_{0,*}(1_{\overline{\mathcal{M}}_{1,2}}) &= i_{0,*}([\overline{\mathcal{M}}_{1,2}]) \\ &= [\Delta_0] \quad \text{(because } \overline{\mathcal{M}}_{1,2} \to \Delta_0 \text{ is birational)} \\ &= 2\delta_0 \quad \text{(because } \mathrm{Aut}(C) = \mathbb{Z}/2\mathbb{Z}, [C] \in \Delta_0 \text{ generic).} \end{aligned}$$

Therefore

$$\begin{aligned} \lambda_1.\delta_0 &= \frac{1}{2}\lambda_1 \cdot i_{0,*}(1_{\overline{\mathcal{M}}_{2,1}}) \\ &= \frac{1}{2} i_{0,*}(i_0^*(\lambda_1)). \end{aligned}$$

Now let

$$\pi\colon \overline{\mathcal{M}}_{1,2} \longrightarrow \overline{\mathcal{M}}_{1,1}$$

be the natural projection. Note that $\overline{\mathcal{M}}_{1,1}$ is the j-line and $\overline{\mathcal{M}}_{1,2}$ is the universal family over the j-line of elliptic curves mod automorphisms. Then

$$i_0^*(\lambda_1^{(2)}) = \pi^*(\lambda_1^{(1)})$$

by Knudsen's theory. This corresponds to the fact that if E' is elliptic with one node P, and E is the normalization of E', then there is a canonical sequence

$$0 \to \Gamma(\omega_E) \to \Gamma(\omega_{E'}) \overset{res}{\to} \mathrm{k}(P) \to 0$$

hence $\Lambda^2(\Gamma(\omega_{E'})) \cong \Gamma(\omega_E)$. Therefore

$$\begin{aligned} \lambda_1.\delta_0 &= \frac{1}{2} i_{0,*}\Big(\pi^*(\lambda_1^{(1)})\Big) \\ &= \frac{1}{24} i_{0,*}\Big(\pi^*(\delta^{(1)})\Big). \end{aligned}$$

But $\pi^*(\delta^{(1)}) = [\tilde{\Delta}], [\tilde{\Delta}] \subset \overline{\mathcal{M}}_{1,2}$ being the closure of the locus of triples (C, x_1, x_2), C a rational curve with a node, x_1, x_2 distinct smooth points of C. $\tilde{\Delta}$ maps birationally to Δ_{00} in $\overline{\mathcal{M}}_2$, and the automorphism group of the generic rational curve with 2 nodes is $(\mathbb{Z}/2\mathbb{Z})^2$, hence:

$$\begin{aligned}\lambda_1.\delta_0 &= \frac{1}{24}[\Delta_{00}] \\ &= \frac{1}{6}\delta_{00}.\end{aligned}$$

Q.E.D.

§10*b*. A Check

An interesting check that these Q-stack-theoretic calculations are OK is to evaluate all terms in the identity

$$60\,\lambda_1^3 = \kappa_3 + \frac{1}{2}\sum_{h=0}^{1} i_{h,*}(K_1^2 - K_1K_2 + K_2^2)$$

obtained in §8. Using Theorem 10.1, one finds

$$60\,\lambda_1^3 = \frac{1}{48}p.$$

Using (8.3) plus theorem 10.1, one finds

$$\kappa_3 = \frac{1}{1152}p.$$

To calculate

$$i_{0,*}(K_1^2 - K_1K_2 + K_2^2)$$

let $\pi\colon \overline{\mathcal{M}}_{1,2} \longrightarrow \overline{\mathcal{M}}_{1,1}$ be the natural map and in $\tilde{\mathcal{M}}_{1,2}$ consider the points $\pi^{-1}([E])$, i.e., representing (E, x_1, x_2), with E fixed. Up to automorphisms of E, x_1 can be normalized to be the identity. Letting x_2 vary, we parametrize this subset of $\tilde{\mathcal{M}}_{1,2}$ by E itself, and describe the universal family of triples (E, x_1, x_2) as $E \times E$ over E with x_1 being given by $s_1(x) = e(x)$, x_2 by the diagonal $s_2(x) = (x, x)$:

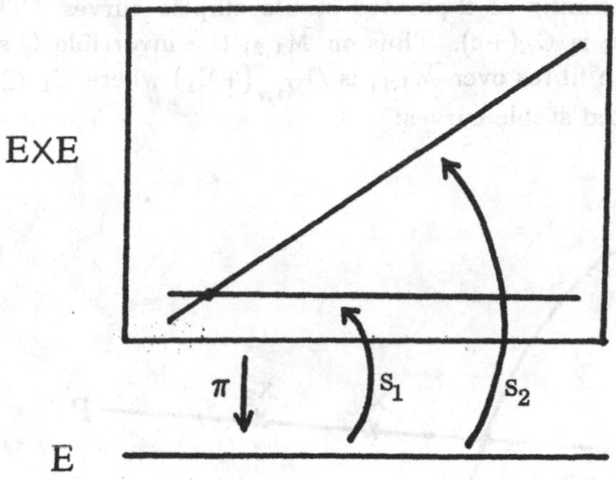

However, this allows $x_1 = x_2$ over $e \in E$, where we should have instead $E \cup \mathbb{P}^1$, $x_1, x_2 \in \mathbb{P}^1 - E \cap \mathbb{P}^1$, $x_1 \neq x_2$. Thus we must blow up $(e, e) \in E \times E$, getting:

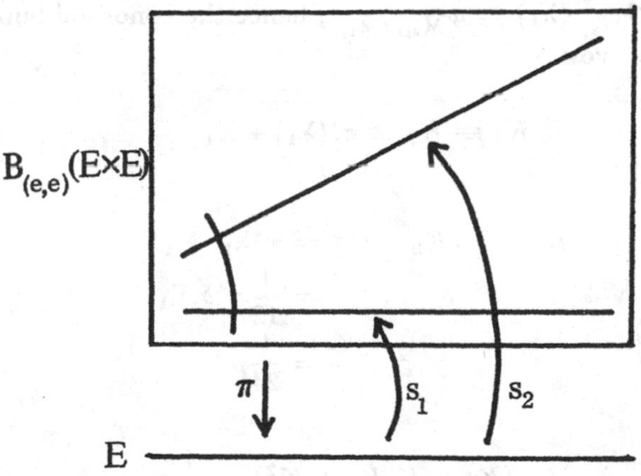

This is now a family of 2-pointed stable elliptic curves. The conormal bundle K_i to s_i is $\mathcal{O}_E(+e)$. Thus on $\overline{\mathcal{M}}_{1,2}$, the invertible Q-sheaf $\mathcal{O}(K_i)$, restricted to the fibres over $\overline{\mathcal{M}}_{1,1}$, is $\mathcal{O}_{\overline{\mathcal{M}}_{1,2}}(+\Sigma_1)$ where $\Sigma_1 \subset \overline{\mathcal{M}}_{1,2}$ is the locus of 2-pointed stable curves:

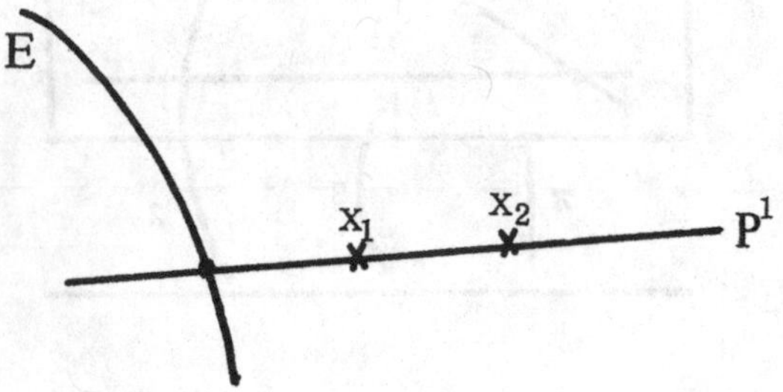

Therefore $K_1 \equiv K_2 \equiv \pi^*(A) + \Sigma_1$, for some Q-divisor class A on $\overline{\mathcal{M}}_{1,1}$. But *along* $\sum_1$, a canonical coordinate can be put on $\mathbb{P}^1$ making $E \cap \mathbb{P}^1 = \{\infty\}$, $x_1 = 0$, $x_2 = 1$, hence $\mathcal{O}_{\overline{\mathcal{M}}_{1,2}}(K_i)$, restricted to Σ_1, is trivial. Now $\pi^*\mathcal{O}_{\overline{\mathcal{M}}_{1,1}}(\lambda_1) \cong \omega_{\overline{\mathcal{M}}_{2,1}/\overline{\mathcal{M}}_{1,1}}$, hence the conormal bundle to Σ_1 is $\mathcal{O}(\lambda_1)$. This proves:

$$K_1 \equiv K_2 \equiv \pi^*(\lambda_1) + \Sigma_1.$$

Therefore

$$\begin{aligned} K_1^2 - K_1K_2 + K_2^2 &= \pi^*\lambda_1 \cdot \Sigma_1 \\ &= \frac{1}{12}\pi^*\delta.\Sigma_1 \\ &= \frac{1}{24}p \end{aligned}$$

hence

$$\frac{1}{2}i_{0,*}(K_1^2 - K_1K_2 + K_2^2) = \frac{1}{48}p.$$

Finally to calculate

$$i_{1,*}(K_1^2 - K_1K_2 + K_2^2)$$

note that on $\overline{\mathcal{M}}_{1,1} \times \overline{\mathcal{M}}_{1,1}$, $K_1^2 = K_2^2 = 0$, and K_i is the pull back from $\overline{\mathcal{M}}_{1,1}$ of λ_1. Since $\lambda_1 = \frac{1}{24}p$,

$$\begin{aligned}\frac{1}{2}i_{1,*}(K_1^2 - K_1K_2 + K_2^2) &= \frac{1}{2}i_{1,*}(-\frac{1}{24}\cdot\frac{1}{24}p)\\ &= -\frac{1}{1152}p.\end{aligned}$$

This checks!

References

[A] Arakelov, S., *Families of algebraic curves with fixed degeneracies*, Izv. Akad. Nauk, *35* (1971).

[Arb] Arbarello, E., *Weierstrass points and moduli of curves*, Comp. Math., *29* (1974), pp. 325–342.

[A-B] Atiyah, M., and Bott, R., *The Yang-Mills equations over Riemann surfaces*, to appear.

[B-F-M] Baum, P., Fulton, W., and MacPherson, R., *Riemann-Roch for singular varieties*, Publ. I.H.E.S., *45* (1975), pp. 101–145.

[D-M] Deligne, P., and Mumford, D., *The irreducibility of the space of curves of given genus*, Publ. I.H.E.S. *36* (1969) pp. 75–109.

[F1] Fulton, W., *Rational equivalence on singular varieties*, Publ. I.H.E.S. *45* (1975), pp. 147–167.

[F2] Fulton, W., *Intersection Theory*, Springer-Verlag, 1983.

[F-M] Fulton, W. and MacPherson, R., *Categorical framework for the study of singular spaces*, Memoirs A.M.S. 243, (1981).

[Ha] Harer, J., The second homology group of the mapping class group of an orientable surface, to appear.

[H-M] Harris, J., and Mumford, D., *On the Kodaira dimension of the moduli space of curves*, Inv. Math., *67* (1982), pp. 23–86.

[H] Hironaka, H., *Bimeromorphic smoothing of a complex-analytic space*, Acta. Math. Vietnamica, *2* (1977), pp. 103–168.

[I] Igusa, J.I., *Arithmetic theory of moduli for genus two*, Annals of Math., *72* (1960), pp. 612–649.

[K] Knudsen, F., *The projectivity of the moduli space of stable curves*, Math. Scand. to appear.

[Kl] Kleiman, S., *Towards a numerical theory of ampleness*, Annals of Math., *84* (1966), pp. 293–344.

[Ma] Matsusaka, T., *Theory of Q-varieties*, Publ. Math. Soc. Japan, *8* (1964)

[Mi] Miller, E., *The homology of the mapping class group of surfaces*, in preparation.

[M] Mumford, D., *Stability of projective varieties*, L'Ens, Math., *24* (1977), pp. 39-110.

[M2] Mumford, D., *Hirzebruch's proportionality principle in the non-compact case*, Inv. math., *42* (1977), pp. 239–272.

[N] Namikawa, Y., *A new compactification of the Siegel space and degeneration of abelian varieties I and II*, Math. Annalen, *221* (1976), pp. 97–141, 201–241.

Received July 1, 1982

Professor David Mumford
Department of Mathematics
Harvard University
Cambridge, Massachusetts 02138

Schubert Varieties and the Variety of Complexes

C. Musili and C. S. Seshadri

Dedicated to Professor I.R. Shafaravich on his 60th birthday

Introduction

De Concini and Strickland [3] have obtained interesting results on the "variety of complexes" (e.g. the Cohen-Macaulay nature of its irreducible components) by introducing certain "standard monomials" in its study (see also Kempf [6] for an earlier study of the variety of complexes and [3] for other references).

The starting point of this investigation is the observation that the variety of complexes is a nice open subset of a union of Schubert varieties in $Sl(n)/Q$, where Q is a (non-maximal) parabolic subgroup. It is than natural to expect that the results of [3] could be deduced from the standpoint of the general stadard monomial theory for classical groups, developed in [9]. In fact, this is what is done in this paper.

The variety of complexes provides a good example of the standard monomial theory for Schubert varieties, for what is called the *mixed case* in [9]. In [9], detailed proofs are given for the mixed case, only for a certain class of Schubert varieties which are called *special* and the Schubert varieties represented by the irreducible components of the variety of complexes are *not* special. Besides, in treating the variety of complexes, one is obliged to study scheme-theoretic unions and intersections of Schubert varieties (in the mixed case) in a systematic way, which is not done in [9]. Hence in Part A, *we present a standard monomial theory, valid for unions of arbitrary Schubert varieties in* $Sl(n)/B$ (cf. Theorem 5.5 and Corollary 5.6)[1]. Our treatment here is related to Remark 15.4 of [9] and brings greater clarity to the proofs presented in Part B of [9].

[1] Detailed proofs of the standard monomial theory on a G/Q, as announced in §16 of [9], will appear shortly. (cf. Remark 5.7)

The crucial result in Part B is that the variety of complexes (with its canonical scheme theoretic structure) is an open subset of a closed subscheme of $Sl(n)/Q$ (as above), defined by a bunch of Plücker coordinates, such that among these the ones coming from one fixed Grassmannian, define a union of Schubert varieties in this Grassmannian (see Lemma 10.1). This, with the results of Part A (see Theorem 6.1), shows that the variety of complexes is an open subset of a union of Schubert varieties in $Sl(n)/Q$, endowed with its *canonical reduced structure*. The Cohen-Macaulay nature of the irreducible components of the variety of complexes is now a consequence of a result of Huneke and Lakshmibai ([f], see also Remark 7.4) or Remark 7.3. It can be seen easily that the standard monomials of [3] are the same as given by the general theory of [9].

A generalisation of the variety of complexes is the scheme M defined by $XY = YX = 0$ (X and Y are say $(n \times n)$ matrices). This scheme occurs naturally in the study of the moduli of vector bundles (to be precise torsion-free sheaves) on projective curves, whose only singularities are ordinary double points (see the lectures of the second author to appear in Astérisque). Strickland [17] has extended the method of [3] to this case. An interesting question is to relate the study of M to that of Schubert varieties.

We are thankful to V. Lakshimibai for many fruitful discussions, especially in relation to §7.

Part A
Schubert Varieties in $Sl(n)/B$

§1. Notation

Let $G = Sl(n+1)$ be the special linear group of rank n over an algebraicaly closed field k of *arbitrary characteristic*. Let T (resp. B) be the subgroup of diagonal (resp. upper triangular) matrices. Let Δ be the system of roots of G relative to T and Δ^+, the set of positive roots relative to B and S, the set of simple roots. Let $\pi_1, \ldots, \pi_n$ be the fundamental weights. For $\pi \in X(T)$ (= the character group of T), write $\pi = \sum a_i\pi_i$, $a_i \in \mathbf{Z}^+$. Let $L(\pi)$ denote the line bundle on the flag variety G/B so that $H^o(G/B, L(\pi)) = \{f\colon G \longrightarrow k / f(gb) = \pi(b)f(g)$, for all $g \in G$ and $b \in B\}$. Recall that π is *dominant* (written $\pi \geq 0$) iff $a_i \geq 0$ for all i iff

$H^o(G/B, L(\pi)) \neq (0)$. Further, π is *ample* (written $\pi > 0$) iff $a_i > 0$ for all i.

Let $Q \supseteq B$ be any parabolic subgroup of G. Let $W_Q = N_Q(T)/T$ be the Weyl group of Q and write $W = W_G$. For $\theta \in W/W_Q$, let $X_Q(\theta)$ be the Schubert variety in G/Q associated to θ, i.e., $X_Q(\theta) =$ the closure of $B\theta Q/Q$ endowed with the canonical reduced structure. We have the Bruhat partial order $\geq$ on W/W_Q, namely, $\theta_1 \geq \theta_2$ if $X_Q(\theta_1) \supseteq X_Q(\theta_2)$. For $Q = B$, we write $X(\theta)$ for $X_B(\theta)$, $\theta \in W$.

Let $P \supseteq B$ be a maximal parabolic subgroup of G, associated to a fundamental weight π (or a simple root α). Recall that P (or π or α) is known (cf. [8, §2]) to be *minuscule*, in the sense that

$$(\pi, \beta^*) = 2(\pi, \beta)/(\beta, \beta) \leq 1$$

for all positive roots β, where (,) denotes a W-invariant inner product on $X(T)$. Or, equivalently, all the weights of the fundamental representation V_π are *extremal*, i.e. they are of the form $\tau(\pi)$, $\tau \in W$; and are parametrised by $\tau \in W/W_P$.

§2. Standard Monomials

We *fix an ordering* $\pi_1, \ldots, \pi_n$ of the fundamental weights.

Let P_i denote the maximal parabolic subgroup of G corresponding to the fundamental weight π_i, $1 \leq i \leq n$. We write $W_i = W_{P_i}$. Fix an $\mathbf{a} = (a_1, \ldots, a_n) \in (\mathbf{Z}^+)^n$. By a *Young diagram of type* $\mathbf{a}$ or multidegree $\mathbf{a}$, we mean an element $(\theta_{ij}) \in \prod_{1 \leq i \leq n} (W/W_i)^{a_i}$, $1 \leq j \leq a_i$, $1 \leq i \leq n$. (If $a_r = 0$ for any r, $1 \leq r \leq n$, the symbols θ_{r-} are understood to be vacuous).

Let $\tau \in W$ and $\mathbf{a}$ as above. We say that a Young diagram (θ_{ij}) *is standard on* $X(\tau)$ *(or relative to* τ*) and written* $\tau \geq (\theta_{ij})$ if there exists a chain (φ_{ij}), $\varphi_{ij} \in W$, $1 \leq j \leq a_i$, $1 \leq i \leq n$, called a *defining chain* for (θ_{ij}) relative to τ, such that

(i) each φ_{ij} in W is a lift for θ_{ij} in W/W_i and

(ii) $\tau \geq \varphi_{11} \geq \cdots \geq \varphi_{1a_1} \geq \varphi_{21} \geq \cdots \geq \varphi_{ij} \geq \varphi_{ij+1} \geq \cdots \geq \varphi_{ia_i} \geq \varphi_{i+11} \geq \cdots \geq \varphi_{na_n}$ (in W).

Now we recall the following (cf. [9, §11, Cor. 11.1 and 11.2′]): Let (θ_{ij}) be a standard Young diagram on $X(\tau)$. Then, there exists a unique *maximal* (resp. *minimal*) *defining chain* (φ_{ij}^{+}) (resp. φ_{ij}^{-}) for (θ_{ij}) relative to τ, in the obvious sense that if (φ_{ij}) is a defining chain for (θ_{ij}), then we have

$$\tau \geq \varphi_{ij}^{+} \geq \varphi_{ij} \geq \varphi_{ij}^{-} \text{ for all } i \text{ and } j.$$

Furthermore, we note that the maximal defining chain (φ_{ij}^{+}) depends on τ but the minimal one (φ_{ij}^{-}) depends *only* on (θ_{ij}) (i.e., not on τ).

Now we recall the following results of standard monomial theory on the Grassmannian. These are in [12]; see also [10, p. 434] and [8, §3].

Theorem 2.1. *Let $L_i = L(\pi_i)$ be the ample generator to* Pic G/P_i, $1 \leq i \leq n$. *Then, there exists a basis $\{p_\theta\}$ of $H^o(G/P_i, L_i)$, parametrised by $\theta \in W/W_i$, $1 \leq i \leq n$, ($\{p_\theta\}$ are the Plücker coordinates) such that*

(1) *p_θ is a weight vector of weight $\chi(\theta) = -\theta(\pi_i)$,*

(2) *for each $\tau \in W/W_i$, the restriction of p_θ to $X_{P_i}(\tau)$ is not zero iff $\tau \geq \theta$ in W/W_i,*

(3) *$\{p_\theta \mid \tau \geq \theta\}$ is a basis of $H^o(X_{P_i}(\tau), L_i)$,*

c) *the closed subscheme $p_\tau = 0$ in $X_{P_i}(\tau)$ is reduced and is the union of all the Schubert divisors in $X_{P_i}(\tau)$.*

Let $\pi = \sum_i a_i \pi_i$, $\mathbf{a} = (a_i)(\mathbf{Z}^+)^n$. To each Young diagram (θ_{ij}) of multidegree $\mathbf{a}$, we define the monomial $p(\theta_{ij}) \in H^o(G/B, L(\pi))$ by

$$p(\theta_{ij}) = \prod_{1 \leq i \leq n} \prod_{1 \leq j \leq a_i} p_{\theta_{ij}}.$$

Note that $p(\theta_{ij})$ is a weight vector of weight

$$\chi(\theta_{ij}) = -\sum_i \sum_j \theta_{ij}(\pi_i).$$

We say that the monomial $p(\theta_{ij})$ is standard on $X(\tau)$, $\tau \in W$, if (θ_{ij}) is a standard Young diagram on $X(\tau)$, i.e., $\tau \geq (\theta_{ij})$.

§3. Pieri's Formula

Let $Ch(G/B)$ denote the Chow ring of G/B. For $\tau \in W$, let $[X(\tau)]$ denote the element in $Ch(G/B)$ determined by the Schubert variety $X(\tau)$, $\tau \in W$. One knows that the elements $[X(\tau)]$, $\tau \in W$, form a basis of $Ch(G/B)$, considered as a Z-module. Let $P \supseteq B$ be a maximal parabolic subgroup of G and π (resp. α) be the associated fundamental weight (resp. simple root). One notes that the codimension one subvariety $X(s_\alpha w_o)$ of G/B is the inverse image of the unique codimension one Schubert subvariety H_p of G/P. Consider now the multiplication $[X(\tau)] \cdot [X(s_\alpha w_o)]$ in the ring $Ch(G/B)$. This element is of the form $\sum d_i[X(\tau_i)]\, d_i \neq 0$. Then by intersection theory it follows that $X(\tau_i)$ is of codimension one in $X(\tau)$; further, by a formula of Chevalley (cf. [1] and [4]), it follows that if we write

$$\tau_i = s_{\beta_i}\tau = \tau_{\delta_i}; \quad \beta_i, \delta_i \in \Delta^+$$

we have

$$d_i = (\pi, \delta_i^*) = |(\pi, \tau^{-1}(\delta_i)^*)| = |(\tau(\pi), \beta_i^*)|$$

Since all the fundamental weights of $G(= Sl(n+1))$ are *minuscule*, it follows that $0 \leq d_i \leq 1$. Thus we have

$$[X(\tau)] \cdot [X(s_\alpha w_o)] = \sum_i [X(\tau_i)] \qquad (*)$$

where on the right side τ_i runs over a certain subset of Schubert subvarieties of codimension one in $X(\tau)$ (this subset could be empty, namely when the left side represents the zero element of $Ch(G/B)$).

Let $X_P(\theta)$, $\theta \in W/W_P$, be the image of $X(\tau)$ in G/P under the canonical map η: $G/B \longrightarrow G/P$. Suppose that $\dim X_P(\theta) > 0$. Then one knows that $\{X_P(\theta)\} \cap \{p_\theta = 0\}$ (cf. (4) of Theorem 2.1) is precisely the union of *all* the Schubert subvarieties in G/P of codimension one in $X_P(\theta)$. It follows, then that the intersection $\{X(\tau)\} \cap \{p_\theta = 0\}$ in G/B is *proper* (now we consider $\{p_\theta = 0\}$ as a subset of G/B) and that it is a union of Schubert varieties, for we have

$$X(\tau) \cap \{p_\theta = 0\} = X(\tau) \cap \left(\eta^{-1}(X_P(\theta) \cap \{p_\theta = 0\})\right)$$

One knows that $\{p_\theta = 0\}$ is linearly equivalent to $X(s_\alpha w_o)$. Hence from $(*)$ it follows that *set theoretically*, we have

$$X(\tau) \cap \{p_\theta = 0\} = \bigcup_i X(\tau_i), \quad \tau_i \text{ as in } (*) \qquad (**)$$

Further, from $(*)$ it follows that the intersection multiplicity of the left side of $(**)$ along $X(\tau_i)$ is 1. Now one knows that the Schubert variety $X(\tau)$ is *normal* (cf. Th.2, [16]). Hence, by standard commutative algebra, it follows that the scheme theoretic intersection $X(\tau)\cap\{p_\theta=0\}$ is *reduced.* Since a scheme theoretic union of reduced schemes is reduced, it follows that $(**)$ holds scheme theoretically. This is referred to as *(the scheme theoretic) Pieri's formula.*

Remark 3.1. The image of $X(\tau)\cap\{p_\theta = 0\}$ in G/P is precisely the union of all the Schubert subvarieties of $X_P(\theta)$ of codimension one. However, it is not true[2] that the image of every $X(\tau_i)$ (irreducible component of $X(\tau)\cap\{p_\theta=0\}$) in $X_P(\theta)$ is of codimension one in $X_P(\theta)$.

§4. Some Properties of Standard Monomials

Definitions. Let $\varphi_1,\varphi_2,\ldots,\varphi_s\in W$ and $Y=X(\varphi_1)\cup\cdots\cup X(\varphi_s)$ be the schematic union of Schubert varieties. (Note that Y is reduced). Let us define for $\mathbf{a}=(a_i)\in(\mathbf{Z}^+)^n$.

$$\text{(i)}\quad S(Y,\mathbf{a})=\left\{\begin{array}{l}\text{Set of standard monomials of } Y \text{ of}\\ \text{multidegree } \mathbf{a}\text{, i.e., the set of all}\\ \text{standard monomials of multidegree } \mathbf{a}\\ \text{on } X(\varphi_t) \text{ for some } t, 1\le t\le s.\end{array}\right\}$$

(ii) $s(Y,\mathbf{a})=\#S(Y,\mathbf{a})=$ cardinality of $S(Y,\mathbf{a})$,
(iii) $h^o(Y,\mathbf{a})=\dim H^o(Y,L(\pi))$, $\pi=\sum_i a_i\pi_i$.

Lemma 4.1. *The set of standard monomials on Y is linearly independent; in particular, for all $\mathbf{a}=(a_1,\ldots,a_n)\in(\mathbf{Z}^+)^n$, we have*

$$h^o(Y,L(\pi))\ge s(Y,\mathbf{a}),\quad \pi=\sum_i a_i\pi_i.$$

Proof. We have $Y=X(\varphi_1)\cup\cdots\cup X(\varphi_s)$, $\varphi_i\in W$. The proof is by induction on a_1: suppose $a_1=0$. Let $Q=P_2\cap\cdots\cap P_n$ and let $X(\varphi_i')$ be

[2]This was pointed out to us by Lakshmibai.

the image of $X(\varphi_i)$ under $\eta: G/B \longrightarrow G/Q$. Let

$$Y' = X(\varphi'_1) \cup \cdots \cup X(\varphi'_s) = \eta(Y).$$

We find that any standard monomial of multi-degree $(0, a_2, \ldots, a_n)$ on $X(\varphi_i)$ can be canonically identified with a standard monomial of multi-degree $(a_2, \ldots, a_n)$ on $X(\varphi'_i)$. This is true for every $i = 1, \ldots, s$. Thus to prove the result for Y, it suffices to prove it for Y'. But Y' is a union of Schubert varieties in G/Q, and so by decreasing induction on the rank of Q, we may assume that the result is true for Y'. (In Case,

$$a_2 = a_3 = \cdots = a_{n-1} = 0;$$

$Q = P_n$ is a maximal parabolic subgroup and the result is well known. In fact, our proof includes even this case). Thus we may assume $a_1 \geq 1$. We may also assume, by induction, that

$$h^o(Y, L(\pi')) \geq s(Y, \mathbf{a}') \text{ where } \mathbf{a}' = (a_1 - 1, a_2, \ldots, a_n).$$

Let $p(\lambda^k_{ij})$, $1 \leq k \leq t$, be a minimal set of linearly dependent standard monomials on Y, where $\lambda_{ij} \in W/W_i$, $1 \leq j \leq a_i$, $1 \leq i \leq n$; $a \leq k \leq t$; and

$$p(\lambda^k_{ij}) = \prod_{1 \leq i \leq n} \prod_{1 \leq j \leq a_i} p_{\lambda^k_{ij}}$$

Let (θ^k_{ij}) be a defining chain for $p(\lambda^k_{ij})$. We note that for each k, $\varphi_l \geq \theta^k_{11}$ for some l, $1 \leq l \leq s$. We distinguish two cases:

Case (1). Let $\lambda^1_{11} = \lambda^2_{11} = \cdots = \lambda^t_{11} = \lambda$, say. By definition, each $p(\lambda^k_{ij})$, $1 \leq k \leq t$, is standard on a Schubert variety Y_l which is an irreducible component of Y. Let $Z = Y_1 \cup \cdots \cup Y_t$. Observe that $p(\lambda^k_{ij})$, $1 \leq k \leq t$, is standard on Z.

Write $p(\lambda^k_{ij}) = p_\lambda p'(\lambda^k_{ij})$ in an ovbious way and observe that the $p'(\lambda^k_{ij})$ are standard monomials of multi-degree $\mathbf{a}' = (a_1 - 1, a_2, \ldots, a_n)$ on Z and hence are linearly independent (by induction). Since Z is reduced and $p_\lambda \neq 0$ on Y_k $(1 \leq k \leq t)$, we find that a non-trivial dependency relation for the $p(\lambda^k_{ij})$ gives one for the $p'(\lambda^k_{ij})$, a contradiction.

Case (2). Let λ be a minimal element in the set $\{\lambda^1_{11}, \ldots, \lambda^t_{11}\}$; say $\lambda = \lambda^1_{11} = \cdots = \lambda^{t_o}_{11}$ and $\lambda \not\geq \lambda^{t_o+1}_{11}, \ldots, \lambda^t_{11}$ (Note that $1 \leq t_o < t$).

Let $Y_o = X(\theta^1_{11} \cup \cdots \cup X(\theta^{t_o}_{11})$. By definition, $\eta_1(Y_o) = X(\lambda)$ in G/P_1, where $\eta_1 \colon G/B \longrightarrow G/P_1$ and $p_{\lambda^k_{11}} \equiv 0$ on $X(\lambda)$ for $t_o + 1 \leq k \leq t$. Hence we get

$$p(\lambda^k_{ij}) \equiv 0 \quad \text{on} \quad Y_o \quad \text{for} \quad t_o + 1 \leq k \leq t.$$

Moreover, $p(\lambda^k_{ij})$, $1 \leq k \leq t_o$, are standard on Y_o and are linearly independent on Y_o by Case (1). Now start with a dependency relation on Y, say

$$\alpha_1 p(\lambda^1_{ij}) + \cdots + \alpha_t p(\lambda^t_{ij}) = 0.$$

Note that $\alpha_k \neq 0$ for *every* k by choice of t. Restricting this relation to Y_o, we get the relation on Y_o

$$\alpha_1 p(\lambda^1_{ij}) + \cdots + \alpha_{t_o}(\lambda^{t_o}_{ij}) = 0$$

which is a contradiction. This completes the proof.

Remark 4.2. The method of proof of the above lemma can be adapted to yield a proof of Theorem 11.3 in 9, pp. 343–345, valid also for a union of Schubert varieties.

Lemma 4.3. *Let $\varphi_i, \psi_j \in W$, $1 \leq i \leq s$, $1 \leq j \leq t$; $Y_1 = \bigcup_i X(\varphi_i)$ and $Y_2 = \bigcup_j X(\psi_j)$ and $\mathbf{a} = (a_i) \in (\mathbf{Z}^+)^n$. Then we have*

$$s(Y_1 \cup Y_2, \mathbf{a}) = s(Y_1, \mathbf{a}) + s(Y_2, \mathbf{a}) - s((Y_1 \cap Y_2)_{\text{red}}, \mathbf{a}).$$

Proof. Recall that $(Y_1 \cap Y_2)_{\text{red}}$ is a union of Schubert varieties, say for suitable $\sigma_{ij} \in W$,

$$(Y_1 \cap Y_2)_{\text{red}} = \bigcup_{i,j} X(\sigma_{ij})(\varphi_i, \psi_j \geq \sigma_{ij})$$

We see that we have only to check that if $p(\theta_{ij})$ is a stadard monomial of multidegree $\mathbf{a}$ on $X(\varphi_\alpha)$ and also on $X(\psi_\beta)$ for some α, β with $1 \leq \alpha \leq s$ and $1 \leq \beta \leq t$, then $p(\theta_{ij})$ is standard on $X(\sigma_{\alpha\beta})$ for some $\sigma_{\alpha,\beta}$. Look at the *minimal* defining chain (λ^-_{ij}) for the standard Young diagram (θ_{ij}), which depends only on (θ_{ij}) but not on φ_α or ψ_β. We have

$$\varphi_\alpha \geq \lambda^-_{11} \text{ and } \psi_\beta \geq \lambda^-_{11} \quad (\text{in } W).$$

But then we know that $\varphi_\alpha, \psi_\beta \geq \sigma_{\alpha,\beta} \leq \lambda_{11}^-$ for some $\sigma_{\alpha,\beta}$ and hence $p(\theta_{ij})$ is standard on $X(\sigma_{\alpha\beta})$, as required.

Lemma 4.4. *Let τ, τ_i be as in (**) of §3. Suppose moreover that $P = P_1$. Let $\mathbf{a} = (a_1, \ldots, a_n) \in (\mathbf{Z}^+)^n$ be such that $a_1 \geq 1$. Let $\mathbf{a}' = (a_1 - 1, a_2, \ldots, a_n)$. Then we have*

$$s\big(X(\tau), \mathbf{a}\big) - s\big(X(\tau), \mathbf{a}'\big) = s\Big(\bigcup\nolimits_{i=1}^{r} X(\tau_i), \mathbf{a}\Big).$$

Proof. It is clear that by the mapping $q \to p_\tau q$, we can identify $S(X(\tau), \mathbf{a}')$ with the subset of elements of $S\big(X(\tau), \mathbf{a}\big)$ which begin with p_τ. It is also clear, since $X(\tau_i)$ are the maximal lifts in $X(\tau)$ of the set of all Schubert divisors of $X_p(\theta)$, that the subset of elements of $S\big(X(\tau), \mathbf{a}\big)$ which do not begin with p_τ can be identified with $S\big(\bigcup_{i=1}^{r} X(\tau_i), \mathbf{a}\big)$. The lemma is now an immediate consequence.

Remark 4.5. A posteriori, Lemma 4.4 would also hold when P is any P_i, $1 \leq i \leq r$, not merely for $i = 1$. A direct "combinatorial proof" of this fact would be of interest (keeping in mind that the ordering of the fundamental weights remains the same as $\pi_1, \ldots, \pi_n$ above).

§5. The Basis Theorem

If Z is a closed subscheme of G/B, let us denote by $\mathbf{O}_Z(a)$ and $\mathbf{O}_Z(m)$ the invertible sheaves on Z, associated respectively to the line bundle $L(\pi)|_Z$ and $L(m\rho)|_Z$, where $\pi = \sum a_i \pi_i$ (cf. §1), $\rho = \sum \pi_i$ and $m \in \mathbf{Z}$. Note that $\mathbf{O}_Z(1)$ is a very ample sheaf on Z. If Y is a union of Schubert varieties in G/B (endowed with its canonical reduced structure), we write $S(Y, m\rho)$, $s(Y, m\rho)\ldots$ etc. for $S\big(Y, (m, \ldots, m)\big)\ldots$ etc.

Lemma 5.1. *Let Y_1, Y_2 be respectively unions of Schubert subvarieties of G/B. Suppose that for $m \gg 0$, we have*

$$h^o(Y_i, m\rho) = s(Y_i, m\rho), \qquad i = 1, 2.$$

Thus we have the following:

(a) *the scheme-theoretic intersection* $Y_1 \cap Y_2$ *is reduced*

(b) $h^o(Y_1 \cup Y_2, m\rho) = s(Y_1 \cup Y_2, m\rho)$; *and* $h^o(Y_1 \cap Y_2, m\rho) = s(Y_1 \cap Y_2, m\rho)$ *for* $m \gg 0$.

Proof. We have the following exact sequence (as sheaves of $\mathbf{O}_{G/B}$ modules)

$$0 \to \mathbf{O}_{Y_1 \cup Y_2} \to \mathbf{O}_{Y_1} \oplus \mathbf{O}_{Y_2} \to \mathbf{O}_{Y_1 \cap Y_2} \to 0$$

where $Y_1 \cap Y_2$ (resp. $Y_1 \cup Y_2$) denotes the scheme theoretic intersection (resp. union). This gives the exact sequence

$$0 \to \mathbf{O}_{Y_1 \cup Y_2}(m) \to \mathbf{O}_{Y_1}(m) \oplus \mathbf{O}_{Y_2}(m) \to \mathbf{O}_{Y_1 \cap Y_2}(m) \to 0$$

Writing the cohomology exact sequence and using Serre's vanishing theorem, we get

$$(1) \qquad \begin{cases} h^o(Y_1 \cap Y_2, m\rho) = h^o(Y_1, m\rho) + h^o(Y_2, m\rho) - h^o(Y_1 \cup Y_2, m\rho) \\ \text{for } m \gg 0. \end{cases}$$

We have by lemma 4.1

$$h^o(Y_1 \cup Y_2, m\rho) \geq s(Y_1 \cup Y_2, m\rho).$$

Thus using the hypothesis of the lemma, we get for $m \gg 0$

$$(2) \qquad \text{RHS of (1)} \leq s(Y_1, m\rho) + s(Y_2, m\rho) - s(Y_1 \cup Y_2, m\rho).$$

Now by Lemma 4.3, we have for $m \gg 0$

$$\text{RHS of (2)} = s(Y_1 \cap Y_2)_{\text{red}}, m\rho).$$

Thus we conclude that

$$(3) \qquad h^o(Y_1 \cap Y_2, m\rho) \leq s((Y_1 \cap Y_2)_{\text{red}}, m\rho).$$

On the other hand, we see that if $Y_1 \cap Y_2$ is *not* reduced

$$(4) \qquad h^o(Y_1 \cap Y_2, m\rho) > h^o((Y_1 \cap Y_2)_{\text{red}}, m\rho), \quad m \gg 0.$$

Since we have also (by Lemma 4.1)

$$s((Y_1 \cap Y_2)_{\text{red}}, m\rho) \leq h^o((Y_1 \cap Y_2)_{\text{red}}, m\rho)$$

we see that (4) implies that

$$h^o(Y_1 \cap Y_2, m\rho) > s((Y_1 \cap Y_2)_{\text{red}}, m\rho) \quad \text{for } m \gg 0.$$

This contradicts (3) so that it follows that $Y_1 \cap Y_2$ is *reduced*. The last assertions are immediate.

Notation 5.2 We denote by S_t the st of *all* reduced subschemes X of G/B such that every irreducible component of X is a Schubert variety of dimension $\leq t$.

Lemma 5.3. *Suppose that for every Schubert variety $X \in S_d$, we have*

$$(*) \qquad S(X, \mathbf{a}) = h^o(X, L(\pi)), \quad \pi = \sum a_i\pi_i, \quad \mathbf{a} = (a_i), \quad a_i \geq 0.$$

Then we have the following:

(i) *for any $Y_1, Y_2 \in S_d$, $Y_1 \cap Y_2$ is reduced,*
(ii) *the assertion $(*)$ holds for any $Y \in S_d$.*

Proof. Suppose that $Y \in S_d$. Then by induction on the number of components of Y, it follows, by an immediate extension of the argument of Lemma 5.1, that $h^o(Y, m\rho) = s(Y, m\rho)$ for $m \gg 0$. Then, again by Lemma 5.1, the assertion (i) above follows. Thus, it remains only to prove the assertion (ii) above.

We observe that (ii) obviously holds for S_o. We now prove (ii) by induction on t, i.e., le $t \leq d$ and suppose that (ii) holds for every $X \in S_l$, $l < t$; then we shall show that (ii) also holds for evey $X \in S_t$. We shall make another inductive argument, namely, if $C(X)$ denotes the number of irreducible components of $X \in S_t$, we suppose that (ii) is true for $X \in S_t$ such that $C(X) \leq (r-1)$. We now take $X \in S_t$ such that $C(X) = r$ and will show that (ii) holds for X. This would prove the lemma. Let then

$$X = X_1 \cup \cdots \cup X_r$$

where X_i are the *distinct* irreducible components of X. We set

$$Y_1 = X_1 \cup \cdots \cup X_{r-1} \text{ and } Y_2 = X_r$$

By (i) $Y_1 \cap Y_2$ is reduced and we get the following exact sequence of $\mathcal{O}_{G/B}$-modules:

$$(1) \qquad 0 \to \mathcal{O}_X(\mathbf{a}) \to \mathcal{O}_{Y_1}(\mathbf{a}) \oplus \mathcal{O}_{Y_2}(\mathbf{a}) \to \mathcal{O}_{Y_1 \cap Y_2}(\mathbf{a}) \to 0.$$

We see that $Y_1 \cap Y_2 \in S_{t-1}$ since X_i are the distinct irreducible components of X. By our inductive hypothesis, (ii) holds for Y_1, Y_2 and $Y_1 \cap Y_2$. Hence $H^o(Y_1 \cap Y_2, L(\pi))$ $(\pi = \sum a_i \pi_i)$ has a basis formed of standard monomials. This implies that the canonical mapping

$$H^o(\mathcal{O}_{Y_1}(\mathbf{a}) \oplus \mathcal{O}_{Y_2}(\mathbf{a})) \to H^o(\mathcal{O}_{Y_1 \cap Y_2}(\mathbf{a}))$$

is *surjective*. Writing the cohomology exact sequence of (1), we get then

$$(2) \qquad h^o(Y_1, L(\pi)) + h^o(Y_2, L(\pi)) = h^o(X, L(\pi)) + h^o(Y_1 \cap Y_2, L(\pi)).$$

We have, on the other hand

$$h^o(Y_i, L(\pi)) = s(Y_i, \mathbf{a}), \qquad i = 1, 2$$
$$h^o(Y_1 \cap Y_2, L(\pi)) = s(Y_1 \cap Y_2, \mathbf{a}).$$

Hence form (2) we get

$$(3) \qquad h^o(X, L(\pi)) = s(Y_1, \mathbf{a}) + s(Y_2, \mathbf{a}) - s(Y_1 \cap Y_2, \mathbf{a}).$$

By Lemma 4.3

$$\text{RHS of (3)} = s(Y_1 \cup Y_2, \mathbf{a}) \quad (X = Y_1 \cup Y_2).$$

Hence we get that

$$h^o(X, L(\pi)) = s(X, \mathbf{a}).$$

This proves Lemma 5.3.

Remark 5.4. In Lemma 5.3, if in addition to (*), we suppose also that for every Schubert variety $X \in S_d$

$$H^i(X, L(\pi)) = 0, \qquad i > 0, \pi \geq 0,$$

then the same proof gives also, in addition,

(iii) $H^i(Y, L(\pi)) = 0, \quad i > 0, \pi \geq 0$ and $Y \in S_d$

We shall now state and prove the main basis theorem:

Theorem 5.5. *Let X be a union of Schubert subvarieties of G/B (endowed with its canonical reduced structure). Then we have (note $G = Sl(n+1)$):*

$$(*) \qquad h^o(X, L(\pi)) = s(X, \mathbf{a}), \quad \pi \geq 0, \pi = \sum a_i \pi_i, \mathbf{a} = (a_i),$$

i.e., $H^o(X, L(\pi))$ has a basis consisting of its standard monomials.

Proof. We prove the theorem by induction on $\dim X$. We observe that if $\dim X = 0$, the theorem is also true. We suppose that the theorem is true for every $Y \in S_t$ where $t < d = \dim X$. By Lemma 5.3, it suffices to prove $(*)$ for the case when X is a Schubert variety. Suppose then that $X = X(\tau)$, $\tau \in W$.

Let now $\pi = \sum a_i \pi_i (\pi \geq 0)$ be such that $a_k = 0$, $k < i$. Let Q_i be the parabolic subgroup $Q_i = P_i \cap P_{i+1} \cdots \cap P_n$ and g: $G/B \longrightarrow G/Q_i$ the canonical morphism of G/B onto G/Q_i. We see that $L(\pi)$ descends to a line bundle on G/Q_i, which we again denote by the same letter $L(\pi)$. Let Y be the Schubert variety in G/Q_i, which is the image of X (under g). We see that standard monomials on X of type $\mathbf{a}$ are precisely *standard monomials* on Y of type $\mathbf{a}$ (this is an obvious generalization of standard monomials on Schubert varieties in G/B given in §1). One knows also that Y is *normal* (cf. [16], Th.2). From this it follows, without much difficulty, that if f: $X \longrightarrow Y$ is the canonial morphism, we have $f_* \mathbf{O}_X = \mathbf{O}_Y$ (there is an open subset of X, namely the big cell which is a "product" over the big cell in Y, which shows that the function field of Y is algebraically closed in that of X etc. see also [16]). This fact implies that $H^o(X, L(\pi)) \xrightarrow{\sim} H^o(Y, L(\pi))$. Thus it suffices to prove the theorem for the Schubert variety Y in G/Q_i. Suppose that $i = n$ so that $Q_n = P_n$. Then we are reduced to the case of one maximal parabolic group, where $(*)$ is known (cf. Theorem 2.1). We do one more inductive argument, i.e., assume $(*)$ for $\pi = \sum a_i \pi_i$, $(\pi \geq 0)$ of the form

$$a_k = 0, \quad k < (i+1)$$

or what can be termed as assuming $(*)$ for Schubert varieties in G/Q_k, $k \geq (i+1)$. Now if we are very strict we work in G/Q_i or one observes that *we can without loss of generality, work in G/B, assuming that* $(*)$ *holds when* $a_1 = 0$.

Let $X_{P_1}(0)$ be the image of $X(\tau)$ in G/P_1 and $X(\tau_i)$ the Schubert divisors in $X(\tau) \cap \{p_\theta = 0\}$ (cf. §3). Let $H(\tau)$ be the union of the Schubert varieties $X(\tau_i)$

$$H(\tau) = \bigcup_i X(\tau_i).$$

Then by the scheme-theoretic Pieri's formula (cf. §3), we get the exact sequence

$$(1) \qquad 0 \to \mathbf{O}_{X(\tau)}(-1, 0, 0, \ldots, 0) \to \mathbf{O}_{X(\tau)} \to \mathbf{O}_{H(\tau)} \to 0.$$

This yields the exact sequence

$$(2) \qquad 0 \to \mathbf{O}_{X(\tau)}(\mathbf{a}') \to \mathbf{O}_{X(\tau)}(\mathbf{a}) \to \mathbf{O}_{H(\tau)}(\mathbf{a}) \to 0.$$

where $\mathbf{a}' = (a_1 - 1, a_2, \ldots, a_n)$. By our induction hypothesis, we have $s(H(\tau), \mathbf{a}) = h^o(H(\tau), \mathbf{a})$ which implies, in particular, that the canonical mapping

$$H^o(X(\tau), L(\pi)) \to H^o(H(\tau), L(\pi))$$

is *surjective*. Writing the cohomology exact sequence of (1), we get the exact sequence

$$0 \to H^o(X(\tau), L(\pi')) \to H^o(X(\tau), L(\pi)) \to H^o(H(\tau), L(\pi)) \to 0$$

where $\pi' = (a_1 - 1)\pi_1 + a_2\pi_2 + \cdots + a_n\pi_n$. This gives

$$(3) \qquad h^o(X(\tau), L(\pi)) = h^o(X(\tau), L(\pi')) + h^o(H(\tau), L(\pi)).$$

Now we prove $(*)$ by induction on a_1. If $a_1 = 0$, as we observed above, $(*)$ holds. Let then $a_1 > 0$. Then we can suppose that

$$h^o(X(\tau), L(\pi')) = s(X(\tau), \mathbf{a}'), \ \pi' \text{ as above.}$$

Besides, by our induction hypothesis, we have also (as observed above)

$$h^o(H(\tau), L(\pi)) = s(H(\tau), \mathbf{a}).$$

Hence (3) gives

$$(4) \qquad h^o(X(\tau), L(\pi)) = s(X(\tau), \mathbf{a}') + s(H(\tau), \mathbf{a}).$$

On the other hand, by Lemma 4.4 we have

$$s(X(\tau), \mathbf{a}) = s(X(\tau), \mathbf{a}') + s(H(\tau), \mathbf{a}) \tag{5}$$

which implies that

$$s(X(\tau), \mathbf{a}) = h^o(X(\tau), L(\pi)).$$

This proves (*) for $X(\tau)$ and concludes the proof of the theorem.

Corollary 5.6. *Let X_1, X_2 be respectively unions of Schubert varieties in G/B. Then the scheme theoretic intersection $X_1 \cap X_2$ is reduced.*

Proof. This is an immediate consequence of Lemma 5.3 and Theorem 5.5.

Remark 5.7. A sketch of proof is given in [9] for the generalization of Theorem 5.5 to that of *any* Schubert variety in G/Q, where G is semi-simple and Q is a parabolic subgroup of *classicl type* (see Theorem 16.2, [9]). We find that there is a gap in the proof we had in mind and this is the step (iv) in the proof of Theorem 16.2, [9]. Unlike the case of G/P, P a *maximal* parabolic subgroup (or the class of *special* Schubert varieties in Part B of [9]), step (ii) (in the proof of Theorem 16.2, [9]) does *not* suffice to imply generation by standard monomials, the difficulty being that (in the mixed case) there are *weakly standard monomials* which need *not* be standard. However, a proof of the results of §16 of [9] can be given by following a different approach and it will appear shortly.

§6. Ideal Theory of Schubert Varieties

If P_i are the maximal parabolic subgroups of G associated to the fundamental weights π_i, $1 \leq i \leq n$, we have the Plücker embedding of the Grassmannian, namely

$$G/P_i \to H^o(G/P_i, L(\pi_i))$$

whose homogeneous coordinate ring R_i is $\bigoplus_{r \geq 0} H^o(G/P_i, L(r\pi_i))$. By the "diagonal embedding" of the flag variety

$$G/B \to \prod_{1 \leq i \leq n} G/P_i$$

we consider the closed immersion

$$G/B \to \prod_{1\leq i\leq n} \mathbf{P}^{m_i} = Z, \quad \mathbf{P}^{m_i} = \mathbf{P}(H^o(G/P_i, L(\pi_i)).$$

We denote by L_i the ample generator of Pic $\mathbf{P}^{m_i}$. The restriction of L_i to G/P_i is $L(\pi_i)$. We denote by S the *multi-homogeneous coordinate ring of* Z i.e.,

$$\begin{cases} S = \bigoplus_{\mathbf{a}} S_{\mathbf{a}}, \quad \mathbf{a} = (a_1, \ldots, a_n), \qquad a_i \geq 0 \\ S_{\mathbf{a}} = H^o(Z, L_1^{a_1} \otimes \cdots \otimes L_n^{a_n}). \end{cases}$$

We have

$$S = k[x_0^1, \ldots, x_{m_1}^1; \ldots; x_0^i, \ldots, x_{m_i}^i, \ldots]$$

$$\mathbf{A} = \prod_i \mathbf{A}^{m_i+1} = \operatorname{Spec} S$$

$$\mathbf{A}^{m_i+1} = \operatorname{Spec} k[x_0^i, \ldots, x_{m_i}^i].$$

We denote by T the torus group $T = (t_1, \ldots, t_n)$, $t_i \in G_m$. We have a canonical action of T on $\mathbf{A}$, namely multiplication by t_i on the component $\mathbf{A}^{m_i+1}$. We denote by $\mathbf{A}^o$, the open subscheme formed of points $x = (x_i)$, $x_i \in \mathbf{A}^{m_i+1}$ such that $x_i \neq 0$ for all i. Then T operates freely on $\mathbf{A}^o$ and Z identifies with the orbit space $\mathbf{A}^o/T$.

Let X be a closed subscheme of the multi-prodjective space Z. We denote by $I(X)$ the ideal of S generated by *all* $f \in S_{\mathbf{a}}$ (for varying $\mathbf{a}$) such that f vanishes on X (f considered, canonically as above, as a section of a line bundle on Z). We call $I(X)$ *the ideal of* X *(in* S*)*. Obviously, $I(X)$ is a *multigraded ideal* in S. If $\hat{X} = \operatorname{Spec} S/I(X)$, we call $\hat{X}$ *the multicone over* X. On the other hand, let J be a multigraded ideal in S. We can then canonically associate a closed subscheme $V(J)$ of $Z (V(J) =$ "variety" associated to $J)$ in the following way. The ideal J determines a T stable sheaf $\tilde{J}$ of ideals in $\mathbf{A}$. The restriction of $\tilde{J}$ to $\mathbf{A}^o$ goes down to a sheaf of ideals in Z and hence defines a closed subscheme $V(J)$ of Z. More concretely, Z can be covered by open subsets of the form $U_1 \times \cdots \times U_n$, where U_i is the open subset of $\mathbf{P}^{m_i}$ defined by $x_{k_i}^i \neq 0$, i.e., $U_i = \operatorname{Spec}[x_0^i/x_{k_i}^i, x_1^i/x_{k_i}^i, \ldots]$ and the restriction of $V(J)$ to $U_1 \times \cdots \times U_n$ is given by the ideal (in the coordinate ring of $U_1 \times \cdots \times U_n$), generated by elements of the form

$$F/\prod_{1\leq i\leq n} (x_{k_i}^i)^{a_i}, \qquad F \in S_{\mathbf{a}}.$$

The mapping $J \to V(J)$ has the property that a union of ideals is taken to the corresponding scheme theoretic intersection (in Z) and an intersectin of ideals to the corresponding scheme theoretic union. We observe that $J \subseteq I(V(J))$. Note however that if J_1, J_2 are two multigraded ideals of S such that $V(J_1) = V(J_2)$, it does *not* necessarily follow that $(J_1) = (J_2)$ for all but a finite number of **a**.

Let R be the multigraded ring, defined by

$$R_{\mathbf{a}} = H^{\circ}(G/B, L(\pi)), \quad \pi = \sum a_i\pi_i, \quad a_i \geq 0.$$

One knows that the canonical mapping

$$H^{\circ}(Z, L_1^{a_1} \otimes \cdots \otimes L_n^{a_n}) \to H^{\circ}(G/B, L(\pi)), \quad \pi \text{ as above}$$

is *surjective* (for example, by Theorem 5.5). Hence we see that the ideal $I(G/B)$ of G/B is precisely the kernel of the canonical homomorpism $S \to R$. Let now I be a multigraded ideal of R. Then $V(I)$ is a closed subscheme of G/B; conversely, if X is a closed subscheme of G/B, the ideal in R generated by all multihomogeneous $f \in R$, vanishing on X is called the ideal of X (in R) and is the image of $I(X)$ under the canonical map $S \to R$. We denote this ideal in R by the same letter.

Let now J_i be a graded ideal of the homogeneous coordinate ring R_i of G/P_i and $\tilde{J}_i$ the multigraded ideal of R generated by J_i. We denote (as usual) by $V(J_i)$ the closed subscheme G/P_i defined by J_i. Then we observe that if η: $G/B \longrightarrow G/P_i$ is the canonical projection, we have

$$V(\tilde{J}_i) = \eta^{-1}(V(J_i))$$

in the scheme theoretic sense. This is simple to check, for example, use the concrete description, given above, for the restriction of $V(\tilde{J}_i)$ to affine open subsets of a suitable covering of Z.

Recall that a subset T^i of W/W_i ($W_i = W_{P_i}$) is called a right half space, if for $\alpha, \beta \in W/W_i$, $\alpha \in T^i$ and $\beta \geq \alpha$, then $\beta \in T^i$ (cf. [12] and [14]). The following result is crucial for the "variety of complexes".

Theorem 6.1. *Let J be an ideal of R generated by Plücker coordinates p_φ, $\varphi \in T$, where $T = \bigcup_{1 \leq i \leq n} T^i$ such that T^i is a right half space in W/W_i (some T^i could be empty). Then the closed subscheme $V(J)$ of G/B is reduced and is in fact a union of Schubert varieties.*

Proof. Let J_i (resp. $\tilde{J}_i$) be the ideal in the homogeneous coordinate ring R_i (resp. R) of G/P_i generated by p_φ, $\varphi \in T^i$. One knows that the closed subscheme $V(J_i)$ of G/P_i is *reduced* and is a union of Schubert varieties (in G/P_i); in fact R_i/J_i is itself reduced (cf. [12]). Then obviously $\eta^{-1}(V(J_i))$ is *reduced*, η being the canonical mapping $G/B \to G/P_i$. As we observed above, we have $V(\tilde{J}_i) = \eta^{-1}(V(J_i))$, so that $V(\tilde{J}_i)$ is reduced and is a union of Schubert varieties in G/B. Now one has

$$V(J) = V(\tilde{J}_1) \cap V(\tilde{J}_2) \cap \cdots \cap V(\tilde{J}_n).$$

Now by Corollary 5.6, it follows that $V(J)$ is reduced and is a union of Schubert varieties in G/B.

Remark 6.2. If Q is any parabolic subgroup of G, then, as we did above for the case of B, we can define the multigraded ring of G/Q. We see also that Theorem 6.1 has an obvious extension to this case.

Remark 6.3. Note that we do *not* claim that $R/V(J)$ or $R/V(\tilde{J}_i)$ is *reduced.* This is probably false, unlike the case of G/P, P a maximal parabolic subgroup. Examples of other difficulties, which crop up in the mixed case, are as follows:

(i) Let X_1, X_2 be two Schubert varieties in G/B. Then

$$I(X_1 \cap X_2) = I(X_1) \cup I(X_2)?$$

(ii) Let $X = X(\tau)$ be a Schubert variety in G/B and let $X_P(\theta)$ be the image of X in G/P, P a maximal parabolic subgroup. Let $R(\tau) = R/I(X(\tau))$. Let $H(\tau) = X(\tau) \cap \{p_\theta = 0\}$. Let $I_\tau(H(\tau))$ be the ideal of $H(\tau)$ in $R(\tau)$, i.e., the image of $I(H(\tau))$ in R_τ. Then

$$I_\tau(H(\tau)) = p_\theta R(\tau)?$$

If $f \in R(\tau)$ $(\mathbf{a} = (a_1, \ldots, a_n))$ such that $a_1 > 0$, then by applying Theorem 5.5, it follows that $f = p_\theta g$, $g \in R(\tau)$. However, it seems very likely that $I_\tau(H(\tau)) \cap R(\tau)_{\mathbf{a}} \neq 0$ with $a_1 = 0$, in which case the above equality does not hold. Thus, while the ideal theory of Schubert varieties in G/B is as good as in the case of G/P (P a maximal parabolic subgroup), it doesn't seem to extend in the same manner to the *multicone over* G/B.

Remark 6.4. A closed subscheme X of G/B is said to be of *product type*, if it is of the form $V(J)$ as in Theorem 7.1. We shall now show that a Schubert *variety* in G/B is always of product type. This is a consequence of the following more general

Lemma 6.5. *Let G be a semi-simple algebraic group, B a Borel subgroup and P_i the maximal parabolic subgroups containing B. (one knows that $B = \bigcap_{1 \le i \le n} P_i$). Let η_i: $G/B \longrightarrow G/P_i$ be the canonical morphisms. For a Schubert variety $X(\theta)$ in G/B set $X'_i = \eta(X(\theta))$ and $X''_i = \eta_i^{-1}(X'_i)$. Then we have (set theoretically)*

$$X(\theta) = \bigcap_{1 \le i \le n} X''_i = \begin{cases} \text{Intersection of the pullback of} \\ \text{the projections of } X(\theta) \text{ in } G/P_i \end{cases}$$

Proof. Let $\theta_i \equiv \theta \,(\mathrm{mod}\, W_i)$ so that $X'_i = X_{P_i}(\theta_i)$. Let $\varphi_i \in W$ be such that $X''_i = X(\varphi_i)$, $1 \le i \le n$. Now we observe that $\exists\, \sigma_i \in W_i$ such that

$$\varphi_i = \theta\sigma_i \text{ with } l(\varphi_i) = l(\theta) + l(\sigma_i), \qquad 1 \le i \le n.$$

But then by (Lemma 3.11, p. 115, [8]), we find that if $\varphi \in W$ is such $\varphi \le \varphi_i$ for all i, then $\exists \sigma_o$ with $\varphi \le \theta\sigma_o$ and $\sigma_o \le \sigma_i$ for all i. Hence $\sigma_o \in W_i$ for all i. But $\bigcap_{1 \le i \le n} W_i = \{\mathrm{id}\}$ and so $\sigma_o = \mathrm{id}$. This means that if $X(\varphi) \subseteq \bigcap X(\varphi_i)$, then $X(\varphi) \subseteq X(\theta)$. Hence $X(\theta) = \bigcap_i X(\varphi_i) = \bigcap_i X''_i$, as required.

Remark 6.6. One notes that an intersection of subschemes of G/B *product type* is of *product type* and that the subscheme $H(\tau)$ of $X(\tau)$ (see (ii) of Remark 6.3 above) is of product type. An arbitrary union of Schubert varieties in G/B is perhaps not of product type.

§7. Vanishing Theorems and Cohen-Macaulay Properties

Theorem 7.1. *Let X be a union of Schubert varieties in G/B, endowed with its canonical reduced structure. Then we have*

$$H^i(X, L(\pi)) = 0, \qquad i > 0, \pi \ge 0 \tag{$*$}$$

Proof. In view of Theorem 5.5 and Remark 5.4, we see that it suffices to prove $(*)$ when X is a Schubert *variety* $X(\tau)$, $\tau \in W$. Because of

Theorem 5.5, the canonical mapping

$$H^o(G/B, L(\pi)) \to H^o(X, L(\pi)), \quad \pi \geq 0$$

is *surjective*. Then, under this hypothesis, (*) is known for a Schubert variety (cf. Theorem 2 and Remark 5, [16]). Now (*) can also be proved in a more direct way by writing exact sequences. For this we consider the following exact sequence (the exact sequence (2) in the proof of Theorem 5.5)

$$(1) \qquad 0 \to \mathbf{O}_{X(\tau)}(\mathbf{a}') \to \mathbf{O}_{X(\tau)}(\mathbf{a}) \to \mathbf{O}_{H(\tau)}(\mathbf{a}) \to 0$$

where $\pi = \sum a_i\pi_i$ and $(\mathbf{a}') = (\mathbf{a}') = (a_1 - 1, a_2, \ldots, a_n)$. Setting $\pi' = \sum a'_i\pi_i$, by a suitable induction hypothesis, we can suppose that

(a) $H^i(X(\tau), L(\pi')) = 0, \qquad i > 0$
(b) $H^i(H(\tau), L(\pi)) = 0, \qquad i > 0$

Then writing the cohomology sequence for (1), (*) would follow: but there is one crucial point, namely that we may *not* have $\pi' \geq 0$. But in this case, there exists a "$\mathbf{P}^1$ fibration of $X(\tau)$ of degree -1" (cf. [8]) and (a) would follow in this case (another argument would be that the case $\pi \geq 0$ and $\pi' \not\geq 0$ occurs when $a_1 = 0$ and this case could be treated by taking the image of $X(\tau)$ in G/Q, Q a suitable parabolic subgroup, on the same lines as in the proof of theorem 5.5; the crucial point in this proof is to show that the cohomologies are preserved when we take this image and this type of argument is done in [16]). We skip the details.

REMARK 7.2. Consider Pieri's formula (**) of §3, namely

$$X(\tau) \cap \{p_\theta = 0\} = \bigcup X(\tau_i) = H(\tau).$$

Suppose that we can index the irreducible components of $X(\tau_i)$ of $H(\tau)$, say $1 \leq i \leq r$, in such a way that[3]

$$(+) \qquad X(\tau_j) \cap \left(\bigcup_{1 \leq i \leq (j-1)} X(\tau_i) \right)$$

---

[3]Lakshmibai has proved (+), deducing it as a consequence of the *shellability* of W/W_Q, Q being any parabolic subgroup of G.

is of pure codimension one in $X(\tau_j), j, 1 \le j \le r$. Then one can prove the following:

$$(**) \qquad H^i(X(\tau), L(-\pi)) = 0, \quad \begin{cases} 0 \le i < \dim X(\tau) \\ \pi > 0 \end{cases}$$

The proof of $(**)$ follows easily by writing the cohomology exact sequence of the exact sequence (1) and the "$\mathbf{P}^1$-fibration argument" in the proof of Theorem 7.1. It also uses exact sequences connecting unions and intersections of the irreducible components of $H(\tau)$ (see the proof of Th. 7.1, [8]). We skip the details.

REMARK 7.3. By Theorems 7.1 and Remark 7.2 we see that if $L(\pi)$ is any ample line bundle on G/B (note that $L(\pi)$ is very ample) and $\widehat{X(\tau)}$ denotes the cone over $X(\tau)$ defined by the projective imbedding of $X(\tau)$, associated to the very ample line bundle $L(\pi), \widehat{X(\tau)}$ is Cohen-Macaulay. This is a consequence of the well-known theorems connecting vanishing theorems and the Cohen-Macaulay nature of $\widehat{X(\tau)}$.

REMARK 7.4. Huneke and Lakshmibai [5] have shown that the multigraded ring $R(\tau)$ associated to a Schubert variety $X(\tau)$ of G/B is Cohen-Macaulay. This result could also be deduced as a consequence of the results of §5. The main obsrvation is that for *a suitable choice of a maximal parabolic subgroup P of G*, we have indeed (with notations as above)

$$(*) \qquad I_r(H(\tau)) = p_\theta \cdot R(\tau) \quad \text{(compare (ii) of Remark 6.3).}$$

In fact, if $\tau = ws_\alpha$ with α simple and $\dim X(w) = \dim X(\tau) - 1$, $P =$ the maximal parabolic subgroup corresponding to α, we see easily that $X(w)$ is an irreducible component of $H(\tau)$, and the projections of $X(w)$ and $X(\tau)$ in G/P', for every maximal parabolic subgroup P', $P' \ne P$, coincide. Hence the projections of $X(w)$ and $X(\tau)$ in G/Q coincide, Q being the intersection of all the maximal parabolic subgroups P', $P \ne P'$. Suppose now that $f \in R(\tau)_{\mathbf{a}}$, $\mathbf{a} = (a_1, \ldots, a_n)$ and $f \equiv 0$ on $H(\tau)$ (in particular, on $X(w)$). If $a_1 = 0$, we have, in fact, $f \equiv 0$ on $X(\tau)$, since f "comes from G/Q". On the other hand, as we saw in (ii) of Remark 6.3, if $a_1 \ne 0$, $f \in p_\theta R(\tau)$. This proves $(*)$ above.

By induction, we can suppose that the multigraded ring of $H(\tau)$, namely $R(\tau)/p_\theta R(\tau)$ is Cohen-Macaulay (using (+) of Remark 7.2 and a simple "patching up" argument, see for example p. 162, [12]). Now (*) implies that $R(\tau)$ is Cohen-Macaulay.

Part B
Variety of Complexes

§8. Definitions

Let $V_1, \ldots, V_{r+1}$ be a sequence of finite dimensional vector spaces over a field k; $\dim V_i = n_i$, $1 \le i \le r+1$. Let

$$\mathbf{A} = \bigoplus_{1 \le i \le r} \operatorname{Hom}(V_i, V_{i+1})$$

be the affine space whose coordinate ring A is the polynomial ring

$$A = k[Y^{(1)}, \ldots, Y^{(r)}]$$

where each $Y^{(i)}$ denotes an $n_{i+1} \times n_i$ matrix of indeterminates over k, $1 \le i \le r$. Let $C \subset \mathbf{A}$ be the closed subscheme of "complexes", i.e.,

$$C = \{(f_1, \ldots, f_r) \mid V_1 \xrightarrow{f_1} V_2 \xrightarrow{f_2} \cdots \to V_r \xrightarrow{f_r} V_{r+1}; \quad f_i \text{ linear and } f_{i+1} \circ f_i = 0, 1 \le i \le r-1.\}$$

In other words, the ideal I of C in A is generated by the quadratic forms, namely, the entries of the products of the matrices $Y^{(i+1)} \cdot Y^{(i)}$, $1 \le i \le r-1$. Let $B = A/I$ be the coordinate ring of C.

§9. Schubert Varieties and the Variety of Complexes

Let $G = Gl(n, k)$ where $n = n_1 + \cdots + n_{r+1}$. Let Q be the parabolic subgroup of G of the shape:

$$Q = \left\{ \begin{pmatrix} A_1 & * & * & \cdot & \cdot & * \\ 0 & A_2 & * & * & \cdot & * \\ 0 & 0 & A_3 & * & \cdot & * \\ \cdot & \cdot & \cdot & \cdot & \cdot & \cdot \\ \cdot & \cdot & \cdot & \cdot & \cdot & \cdot \\ 0 & 0 & \cdot & \cdot & 0 & A_{r+1} \end{pmatrix} : A_i \in Gl(n_i, k) \quad 1 \le i \le r+1 \right\}$$

i.e., with entries $x_{kl} = 0$ for $k \geq m_i + 1$ and $l \leq m_i$ where $m_1 = n_1$, $m_2 = n_1 + n_2, \ldots, m_{r+1} = n = n_1 + \cdots + n_{r+1}$. Recall that G/Q is the variety of partial flags of type $(m_1, \ldots, m_r)$ in an n-dimensional vector space. Let Z be the unipotent radical of $\mathbf{Q} = w_o Q w_o^{-1}$, the parabolic subgroup of G, opposite to Q; w_o being the Weyl involution of G. In other words, Z is the affine space

$$Z = \left\{ \begin{pmatrix} I_1 & 0 & \cdot & \cdot & \cdot & 0 \\ * & I_2 & 0 & \cdot & \cdot & 0 \\ * & * & I_3 & 0 & \cdot & 0 \\ \cdot & \cdot & \cdot & \cdot & \cdot & \cdot \\ \cdot & \cdot & \cdot & \cdot & \cdot & \cdot \\ * & * & \cdot & \cdot & * & I_{r+1} \end{pmatrix} : I_i = Id \in G(n_i, k) \quad 1 \leq i \leq r+1 \right\}.$$

We recall the well-known fact that the restriction of the natural morphism $\epsilon\colon G \longrightarrow G/Q$ to Z is an open immersion. The open affine space $\epsilon(Z)$ is called the *opposite big cell* of G/Q. (Often we identify Z with $\epsilon(Z)$).

Let $P_1, \ldots, P_r$ be *all* the (standard) maximal parabolic subgroups of G containing Q (so that $Q = \bigcap_i P_i$). We have, for $1 \leq i \leq r$,

$$P_i = \left\{ \begin{pmatrix} A_i & * \\ 0 & B_i \end{pmatrix} : \begin{matrix} A_i \in Gl(m_i) \\ B_i \in Gl(n - m_i) \end{matrix} \right\}.$$

Via the Plücker embedding of G/P_i in a projective space, say $\mathbf{P}_i$, we have the multi-projective embedding of G/Q, namely,

$$G/Q \to \prod_{1 \leq i \leq r} G/P_i \to \prod_{1 \leq i \leq r} \mathbf{P}_i.$$

Identifying G/Q with its image on $\prod_i \mathbf{P}_i$, we can describe the natural morphism $\epsilon\colon G \longrightarrow G/Q$ to be

$$\epsilon\colon G \longrightarrow \prod_i \mathbf{P}_i$$

where for a matrix $g \in G$, the multi-Plücker coordinaes of $\epsilon(g)$ are simply all the $m_i \times m_i$ minors of the first m_i columns of g, $1 \leq i \leq r$. It follows that the restriction of ϵ to Z is an isomorphism and thus we embed Z in the multi-projective space $\prod_i \mathbf{P}_i$. (It can be seen that

$$Z = (G/Q) \cap (Z_1 \times \cdots \times Z_r)$$

where Z_i is the opposite big cell of G/P_i, $1 \leq i \leq r$). *We can thus speak of a Plücker coordinate as a function on Z.*

For a Schubert variety S in G/Q, the affine open subset $S \cap Z$ of S (or the closed subst $S \cap Z$ of the opposite big cell Z of G/Q) is called the *opposite big cell of* S. This is an abuse of language since $S \cap Z$ is *not a cell* itself but only a closed subset of the cell Z of G/Q.

We identify the variety of complexes C with the closed subscheme of Z defined by

$$(1)\quad C = \left\{ \begin{pmatrix} I_1 & 0 & \cdot & \cdot & \cdot & 0 \\ Y^{(1)} & I_2 & 0 & \cdot & \cdot & 0 \\ 0 & Y^{(2)} & I_3 & 0 & \cdot & 0 \\ \cdot & \cdot & \cdot & \cdot & \cdot & \cdot \\ \cdot & \cdot & \cdot & \cdot & \cdot & \cdot \\ 0 & \cdot & \cdot & 0 & Y^{(r)} & I_{r+1} \end{pmatrix} : \begin{array}{c} I_i = Id \in Gl(n_i, k) \\ 1 \leq i \leq r \\ T^{(i+1)} \cdot Y^{(i)} = (0); \\ 1 \leq i \leq r-1 \\ Y^{(i)} \text{ is } n_{i+1} \times n_i \text{ matrix} \\ 1 \leq i \leq r \end{array} \right\}.$$

As with Z, C is identified with a locally closed subscheme of $\prod_i \mathbf{P}_i$ via the morphism ϵ.

§10. Plücker Coordinates Vanishing on C

As usual, identifying the Weyl group $W = W(Gl(n))$ with the symmetric group on the n symbols $\{1, \ldots, n\}$, we find that the Weyl group $W_i = W(P_i)$ of P_i is the subgroup leaving the subsets $\{1, \ldots, m_i\}$ and $\{m_i + 1, \ldots, n\}$ stable and consequently, we have

$$W/W_i = \{\mathrm{j}^{(i)} = (j_1, \ldots, j_{m_i}) : 1 \leq j_1 < \cdots < j_{m_i} \leq n\}.$$

Then we can identify the coset space W/W_Q to be the set

$$W/W_Q = \{(\mathrm{j}^{(1)}, \ldots, \mathrm{j}^{(r)}) : \mathrm{j}^{(1)} \subset j^{(2)} \cdots \subset \mathrm{j}^{(r)}; \mathrm{j}^{(i)} \in W/W_i\}.$$

Recall that we have identified the Plücker coordinates on G/P_i as functions on Z (see §8 above). Now we have the crucial

Lemma 10.1. *The ideal I of the variety of complexes C (in the coordinate ring of) is generated by all the Plücker coordinates p_{j}, where*

$\mathbf{j} = (j_1, \ldots, j_{m_t}) \in W/W_t$, $1 \leq t \leq r$ *and* $\mathbf{j}$ *is not of the form*

$$(2) \qquad \begin{cases} (a)\ j_s = s & \text{for } 1 \leq s \leq m_{t-1} \text{ if } t \geq 2 \text{ and} \\ (b)\ j_{m_t} \leq m_{t+1}. \end{cases}$$

or equivalently $\mathbf{j}$ *is of the form*

$$(2)' \qquad \begin{cases} \text{either } (a)\ j_s \neq s \text{ for some } s, 1 \leq s \leq m_{t-1} \text{ if } t \geq 2 \\ \text{or } (b)\ j_{m_t} > m_{t+1}. \end{cases}$$

Proof. As a closed subscheme of Z, C is defined by the equations as in (1), i.e., by the ideal spanned by

$$(3) \qquad \begin{cases} (a)\ Y^{(i+1)} \cdot Y^{(i)}, & 1 \leq i \leq r-1 \\ (b)\ x_{kl} \text{ for } k \geq m_{i+1}, & l \leq m_i, \quad 1 \leq i \leq r. \end{cases}$$

We have to show that the ideal spanned by the elements in (2)′ is the same as that spanned by the elements in (3) in the coordinate ring of Z. Let us denote by D the closed subscheme of Z defined by (3) (b), i.e.,

$$D = \left\{ \begin{pmatrix} I_1 & 0 & \cdot & \cdot & \cdot & 0 \\ Y^{(1)} & I_2 & 0 & \cdot & \cdot & 0 \\ 0 & Y^{(2)} & I_3 & 0 & \cdot & 0 \\ 0 & 0 & Y^{(3)} & I_4 & \cdot & \cdot \\ \cdot & \cdot & \cdot & \cdot & \cdot & \cdot \\ \cdot & \cdot & \cdot & \cdot & \cdot & \cdot \end{pmatrix} : \begin{matrix} I_i = Id \in Gl(n_i) \\ Y^{(i)} = n_{i+1} \times n_i \text{ matrix} \\ 1 \leq i \leq r \end{matrix} \right\}$$

We see that D is an *affine space.* We first claim that D is defined by the ideal in (2)′ (b), i.e., the ideal generated by (2)′ (b) is the same as that generated by (3)(b). It is rather immediate that the elements in (2)′(b) vanish on D, i.e., if we take the sub-matrix of D formed by the first m_t columns, then the determinant of the $(m_t \times m_t)$ minor associated to a Plücker coordinate as in (2)′(b) is zero. Conversely, let

$$k \geq (m_{i+1} + 1), \quad l \leq m_i; \quad 1 \leq i \leq r \text{ and}$$
$$\mathbf{j} = (1, \ldots, \hat{l}, \ldots, m_i, k)$$

Then as a function on Z, p is the minor of the submatrix of Z formed by the first m_i columns and rows corresponding to $\mathbf{j}$. Then we see easily that

$$p = \begin{cases} \pm x_{kl} & \text{if } \quad l = m_i \\ \pm x_{kl} + \text{ terms involving } x_{kj},\ j \geq l+1 \text{ if } m_i > l. \end{cases}$$

By a decreasing induction on l, it follows that the elements of (3)(b) are in the ideal spanned by (2)′(b). This concludes the proof that the ideal of D is spanned by the elements of (2)′(b).

Now to conclude the proof of the lemma, it suffices to show that in the coordinate ring of D, the ideal generated by (3)(a) is the same as that generated by (2)′(a). In fact, we show that

$$(*)_\delta \qquad \begin{cases} \text{the ideal generated by } Y^{(t)} \cdot Y^{(t-1)}, \quad t \leq \delta \\ = \text{ the ideal generated by the elements } (2)'(a), \quad t \leq \delta. \end{cases}$$

We prove $(*)$ by induction on δ so that we assume $(*)_\delta$ is true for $j \leq (\delta - 1)$. We see that it suffices to prove $(*)_\delta$ in the coordinate ring of $\overline{D}$, where $\overline{D}$ is the closed subscheme of D defined by the ideal generated by the entries of $Y^{(t)} \cdot Y^{(t-1)}$, $t \leq (\delta - 1)$. We have then only to show that (in the coordinate ring of $\overline{D}$) the ideal generated by the entries of $Y^{(\delta)} \cdot Y^{(\delta-1)}$ is the same as the one generated by $p_{\mathbf{j}}$

$$(4) \qquad \begin{cases} \mathbf{j} \in W/W_\delta \\ j_s \neq s, \text{ for some } s, m_{\delta-2} \leq s \leq m_{\delta-1}. \end{cases}$$

In $\overline{D}$, we have the following relation (product of an $(m_{\delta+1} \times m_\delta)$ matrix by an $(m_\delta \times m_\delta)$ matrix)

$$(5) \begin{pmatrix} I_1 & 0 & \cdot & 0 \\ Y^{(1)} & I_2 & \cdot & 0 \\ \cdot & \cdot & \cdot & \cdot \\ 0 & \cdot & Y^{(\delta-1)} & I_\delta \\ 0 & \cdot & 0 & Y^{(\delta)} \end{pmatrix} \begin{pmatrix} I_1 & 0 & \cdot & 0 \\ -Y^{(1)} & I_2 & \cdot & 0 \\ \cdot & \cdot & \cdot & \cdot \\ 0 & 0 & -Y^{(\delta-1)} & I_\delta \end{pmatrix} = \begin{pmatrix} I_1 & 0 & \cdot & 0 \\ 0 & I_2 & \cdot & 0 \\ \cdot & \cdot & \cdot & \cdot \\ \cdot & \cdot & \cdot & \cdot \\ 0 & 0 & 0 & NY_\delta \end{pmatrix}$$

where $N = -Y^{(\delta)} Y^{(\delta-1)}$.

i.e., $g_\delta u_\delta = h_\delta$, say. Now, $\det u_\delta = 1$ so that the determinant of an $(m_\delta \times m_\delta)$ minor of g_δ is the same as the determinant of the corresponding $(m_\delta \times m_\delta)$ minor of h_δ. Let Y_{kl} be an element of the matrix $Y^{(\delta)} \cdot Y^{(\delta-1)}$ sitting as an element of the matrix h_δ as above. Then we easily see that

$$(6) \qquad k \geq (m_\delta + 1); \quad m_{\delta-2} \leq l \leq m_{\delta-1}.$$

Let us now set

$$(7) \qquad \mathbf{j} = (\text{complement of } l \text{ in } (1, \ldots, m) \cup \{k\}).$$

Then it can be seen without much difficulty that

$$(8) \qquad \pm y_{kl} = \begin{cases} \text{determinant of the } (m_\delta \times m_\delta) \text{ minor of } h_\delta \\ \text{corresponding to } \mathbf{j}. \end{cases}$$

In fact, more generally we have the following rule (a variant of which is used in [11]). Consider the $(m \times r)$ matrix, $m = n + r$ of the form

$$J = \begin{pmatrix} I \\ A \end{pmatrix} \quad \begin{matrix} I = (r \times r) \text{ identity matrix} \\ A = (n \times r) \text{ matric.} \end{matrix}$$

Let $p_{(i),(j)}$ = determinant of a minor A formed by rows and columns which correspond respectively to the indices $(i) = (i_1, \ldots, i_k)$, $(j) = (j_1, \ldots, j_k)$, $k \leq r$. Let $p_{(\lambda)}$ be the determinant of an $(r \times r)$ minor of J whose row indices are $(\lambda) = (\lambda_1, \ldots, \lambda_r)$ (column indices being of course $1, \ldots, r$). Then we have

$$(9) \qquad \begin{cases} p_{(\lambda)} = \pm p_{(i),(j)} \\ (\lambda) = (r + (i)) \cup (\text{complement of } (j) \text{ in } (1, \ldots, r) \\ \qquad\qquad \text{arranged in increasing order.} \end{cases}$$

We see that $\mathbf{j}$ in (7) is of the type as in (4). This shows that the ideal generated by the entries of $Y^{(\delta)} \cdot Y^{(\delta-1)}$ is contained in the ideal generated by p, where $\mathbf{j}$ is of type (4). Now, more generally, it is easily seen that *all* the elements p of type (4) are obtained by taking determinandts of suitable minors of the product matrix $Y^{(\delta)} \cdot Y^{(\delta-1)}$ (or we can also see that the p of type (4) vanish on C by applying the above rule). This shows that the ideal generated by the entries of $Y^{(\delta)} \cdot Y^{(\delta-1)}$ is the same as the one generated by the p of type (4). This concludes the proof of the lemma.

Now we prove the main result of this section, namely:

Theorem 10.2. *The variety of complexes C is reduced and its irreducible components are the opposite big cells of suitable Schubert varieties in G/Q. In fact, the closure of C in G/Q is a subscheme of G/Q of product type (cf. Remarks 6.4 and 6.2, Part A).*

Proof. This is an immediate consequence of Lemma 10.1 and Theorem 6.1 (see also Remark 6.2). For $1 \leq i \leq r$, let

$$T^i = \{\mathbf{j} = (j_1, \ldots, j_{m_i}) \in W/W_i; \quad \mathbf{j} \text{ is of the form } (2)'\}.$$

We observe that each T^i is a right half space (see §7, Part A). Let $T = \bigcup_i T^i$ and J the ideal of R generated by the Plücker coordinates P_φ, $\varphi \in T$. Then by Theorem 6.1 (Part A), the closed subscheme $V(J)$ of G/B is reduced and is a union of Schubert varieties (and of product type, see Remarks 7.4 and 7.2, Part A). By Lemma 10.1, the restriction of $V(J)$ to the opposite big cell Z is precisely C. This proves Theorem 10.2.

As an immediate consequence of Remark 7.3 or Remark 7.4, Part A, we have the following.

Corollary 10.3. *The irreducible components S of the variety of complexes C are normal and Cohen-Macaulay. Further, the restrictions of standard monomials (on Z) to C (resp. S) generate the coordinate ring of C (resp. S) and the non-zero standard monomials on S form a basis for the coordinate ring of S.*

§11. Irreducible Components of C

Let $\mathbf{k} = (k_1, \ldots, k_r) \in (\mathbf{Z}^+)^r$ be such that $k_i \leq \min(n_i, n_{i+1})$, $k_i + k_{i+1} \leq n_{i+1}$, $1 \leq i \leq r$ (with $k_{r+1} = 0$). Let $C(\mathbf{k})$ be the closed subscheme of C defined by the ideal $I(\mathbf{k})$ generated by I (the ideal of C) and the minors of $Y^{(i)}$ of size $k_i + 1$, $1 \leq i \leq r$. Let $B(\mathbf{k}) = A/I(\mathbf{k})$ be the coordinate ring of $C(\mathbf{k})$. We note that $C(\mathbf{k})$ is the set of points $(f_1, \ldots, f_r) \in C$ such that rank $f_i \leq k_i$, $1 \leq i \leq r$.

From definitions, it follows easily that $C(\mathbf{k}) \subseteq C(\mathbf{k}')$ if and only $\mathbf{k} \leq \mathbf{k}'$ (i.e., $k_i \leq k_i'$ for all $i = 1, \ldots, r$), and further, set-theoretically, we have $C = \bigcup C(\mathbf{k})$. Thus, the irreducible components of C are the $C(\mathbf{k})$ corresponding to the maximal elements $\mathbf{k}$ for the partial order above.

Now we have the following.

Theorem 11.1. *Each $C(\mathbf{k})$ is reduced and irreducible. In fact, $C(\mathbf{k})$ is the opposite big cell of a Schubert variety $X(\mathbf{k})$ in G/Q where $X(\mathbf{k})$ is defined by the element $w(\mathbf{k}) = (\mathbf{w}^{(1)}, \ldots, \mathbf{w}^{(r)}) \in W/W_Q$ with*

$$\mathbf{w}^{(1)} = (k_1 + 1, \ldots, m_1; m_2 - k_1 + 1, \ldots, m_2)$$

and, for $2 \leq t \leq r$,

$$\mathbf{w}^{(t)} = (1, \ldots, m_{t-1}; m_{t-1} + k_t + 1, \ldots, m_t; m_{t+1} - k_t + 1, \ldots, m_{t+1}).$$

In particular, $C(\mathbf{k})$ is normal and Cohen-Macaulay.

Proof. Proceeding exactly as for the proof of Lemma 10.1 and Theorem 10.2, we conclude the following.

(a) The ideal $I(\mathbf{k})$ of $C(\mathbf{k})$ in Z is generated by the Plücker coordinates $P, \mathbf{j} \in T = \bigcup_i T^{(i)}$, say, where

$$T_k^{(i)} = \{\mathbf{j} \in W/W_i \mid \mathbf{w}^{(i)} \not\geq \mathbf{j}\}, \qquad 1 \leq i \leq r.$$

To see this, we have only to observe that each of the projections $\epsilon^i(X(\mathbf{k}))$ in G/P_i is simply the determinantal subvariety of the affine space $Y^{(i)}$ of rank at most k_i, which in turn is the opposite big cell of the Schubert variety $X_i'(\mathbf{k}) = X(\mathbf{x}^{(i)})$ in G/P_i (cf. [13, §7, p. 139]).

(b) Let $X_i''(\mathbf{k}) = \epsilon_i^{-1}(X_i'(\mathbf{k}))$. Then we conclude that the Schubert variety $X(\mathbf{k}) = X(w(\mathbf{k}))$ in G/Q is simply the reduced variety $\bigcap_i X_i''(\mathbf{k})$ and the ideal of the opposite big cell $S(\mathbf{k})$ of $X(\mathbf{k})$ in Z is $C(\mathbf{k})$. Hence $C(\mathbf{k}) = S(\mathbf{k})$ which is reduced and irreducible, as required.

Remark 11.2. For $\mathbf{k} = (k_1, \ldots, k_r)$ as above, we have

$$\dim(C(\mathbf{k})) = \sum_{1 \leq i \leq r+1} (n_i - k_i)(k_{i-1} + k_i) \text{ with } k_o = k_{r+1} = 0.$$

This can be seen in several ways. For instance, one can apply a classical formula to find the length of a permutation $\sigma = (a_1, \ldots, a_n)$ in terms of the simple transpositions $s_i = (i, i+1), 1 \leq i \leq n-1$; to get

$$l(\sigma) = \frac{1}{2}n(n-1) - \left(\sum_i a_i'\right)$$

where $a_i' = \#\{a_j \mid a_j < a_i, j = 1, \ldots i-1\}$ with $a_1' = 0$. We apply this formula to the $w(\mathbf{k}) \in W/W_Q$ lifting $w(\mathbf{k})$ to its minimal representative in W. *Or*, we observe easily that $C(\mathbf{k})$ contains, as an open set, the algebraic set $(U \times T)$, where T is the product of the flag varieties

$$T = \prod_{1 \leq i \leq r} F(k_i, k_i + k_{i+1}, n_{i+1})$$

and for each $t \in T$, $U \times \{t\}$ being the affine space

$$U \times \{t\} = \mathrm{Hom}(V_1, W_1^2) \times \mathrm{Hom}(W_2^2/W_1^2, W_1^3) \times \cdots \times \\ \times \cdots \times \mathrm{Hom}(W_2^{r-1}/W_1^{r-1}, W_1^r)$$

where $t = ((W_1^1, W_1^2, V_2), \ldots, (W_1^{r-1}, W_2^{r-1}, V_r), (W_1^r, W_1^r, V_{r+1}))$ with each $W_1^i \subset W_2^i \subset V_{i+1}$ being a flag of type $(k_i, k_i + k_{i+1}, n_{i+1})$ in the vector space $V_{i+1} = (\dim V_{i+1} = n_{i+1})$. Now we see that

$$\dim(U \times T) = \sum_{1 \leq i \leq r+1} (n_i - k_i)(k_{i-1} + k_i)$$

with $k_o = k_{r+1} = 0$. This completes the proof.

References

[1] C. C. Chevalley, Sur les décompositions cellulaires de espaces G/B, (unpublished manuscript), 1958.

[2] C. DeConcini and V. Lakshmibai, Arithmetic Cohen-Macaulayness and arithmetic normality of Schubert varieties, Amer. J. of Math., *103* (1981), 835–850.

[3] C. DeConcini and E. Strickland, On the variety of complexes, Advances in Math., *41* (1981) 57–77.

[4] M. Demazure, Désingularisation des variétés de Schubert généralisées, Ann. Sci. École Norm. Sup., 7 (1974), 53–88.

[5] C. Huneke and V. Lakshmibai, Arithmetic Cohen-Macaulayness and normality of the multi-cones over Schubert varieties in $Sl(n)/B$, preprint.

[6] G. R. Kempf, Images of homogeneous vector bundles and variety of complexes, Bull. A.M.S., *81* (1975), 900–901.

[7] V. Lakshmibai, C. Musili and C. S. Seshadri, Cohomology of line bundles on G/B, Ann. Sci. École Norm. Sup., 7 (1974), 89–138.

[8] V. Lakshmibai, C. Musili and C. S. Seshadri, Geometry of G/P-III (Standard monomial theory for a quasi-miniscule P), Proc. Ind. Acad. Sci., *87* (1978), 93–177.

[9] V. Lakshmibai, C. Musili and C. S. Seshadri, Geometry of G/P-IV (Standard monomial theory for classical types), Proc. Ind. Acad. Sci., *88* (1979), 279–362.

[10] V. Lakshmibai, C. Musili and C. S. Seshadri, Geometry of G/P, Bull. A.M.S., (New Series), *1* (1979), 432–435.

[11] V. Lakshmibai, C. Musili and C. S. Seshadri, Geometry of G/P-II (The work of DeConcini and Procesi and the basic conjectures), Proc. Ind. Acad. Sci., *87* (1978), 1–54.

[12] C. Musili, Postulation formula for Schubert varieties, J. Ind. Math. Soc., *36*, (1972), 143–171.

[13] C. Musili, Some properties of Schubert varieties, J. Ind. Math. Soc., *38* (1974), 131–145.

[14] C. Musili and C. S. Seshadri, Standard monomial theory, Séminaire d'Algèbre Paul Dubreil et Marie-Paule Malliavin proceedings, Paris 1980, 33 ème Année), 441–476.

[15] C. S. Seshadri, Geometry of G/P-I (Standard monomial theory for a minuscule P), "C. P. Ramanujam : A Tribute", 207, (Springer-Verlag), Published for the Tata Institute of Fundamental Research, Bombay, 1978, 207–239.

[16] C. S. Seshadri, Standard monomial theory and the work of Demazure, Symposia in Math., 1 (1981) (Proceedings of a Symposium on Algebraic and Analytic Varieties, Tokyo), to appear.

[17] E. Strickland, On the conormal bundle of the Determinantal variety, preprint.

Received July 21, 1982

Professor C. Musili
School of Mathematics
University of Hyderabad
Central University, P.O.
Hyderabad 500134, India

Professor Conjeeveram S. Seshadri
Tata Institute for Fundamental Research
Homi Bhabha Road
Bombay 400005, India

A Crystalline Torelli Theorem for Supersingular K3 Surfaces

Arthur Ogus

To I.R. Shafarevich

Introduction

I like to argue that crystalline cohomology will play a role in characteristic p analogous to the role of Hodge theory in characteristic zero. One aspect of this analogy is that the F-crystal structure on crystalline cohomology should reflect deep geometric properties of varieties. This should be especially true of varieties for which the "p-adic part" of their geometry is the most interesting, as seems often to be true of supersingular varieties in the sense of Shioda [26]. For example, in [19 §6] I proved that supersingular abelian varieties of dimension at least two are determined up to isomorphism by the F-crystal structure (and trace map) on H^1_{cris}, just as abelian varieties over $\mathbb{C}$ are determined by the Hodge structure on H^1_{DR}.

One of the most impressive triumphs of Hodge theory is the proof [21] by Piatetski-Shapiro and Shafarevich of the Torelli theorem for $K3$ surfaces over $\mathbb{C}$, which asserts that a $K3$ surface is determined up to isomorphism by the Hodge structure (and quadratic form) on H^2_{DR}. In [19], I conjectured that a supersingular $K3$ surface in characteristic $p > 2$ should be determined by the F-crystal structure and quadratic form on H^2_{cris}, and obtained some partial results, including a proof for Kummer surfaces. I stated "... since the period space is *compact*, it seems reasonable to hope that supersingular $K3$'s cannot degenerate in any serious way. This would prove that the period space is in fact a fine moduli space of (rigidified) supersingular $K3$ surfaces." Recently, Rudakov and Shafarevich have proved such a nondegeneracy result [23], and say in their introduction, "It is known that this implies that (supersingular $K3$ surfaces) are determined by their periods."

It is the purpose of this note to justify these claims. The first step is to make precise formulations of the results we want to prove. The first two formulations are taken from [19] and restated near the beginning of §1

(Theorems I and II). Before proving them, however, it is necessary to give a third formulation (Theorems III, at the end of §1). Here there are two main points: to define a suitable notion of a "marking" of a supersingular $K3$ surface, and to define a period space $\mathcal{P}$ which keeps track of both the crystalline periods and the ample cone in the Neron-Serevi group. (The latter is necessary because we have found it desirable to work with unpolarized $K3$ surfaces, along the lines pursued by Burns and Rapoport in their proof [7] of the Torelli theorem for Kahlerian $K3$ surfaces.) Next we study families of marked $K3$ surfaces and show that there is a fine moduli space $\mathcal{S}$ classifying them, using Artin's construction techniques for algebraic spaces. As is well-known, $\mathcal{S}$ is not separated, because of the possibility of "elementary modifications" of families of $K3$'s (called "elementary transformations" by Burns and Rapoport [7]).

Following the outline of [19], we construct a period map $\pi\colon \mathcal{S} \longrightarrow \mathcal{P}$ and verify that it is étale and separated. The theorem of Rudakov and Shafarevich [23] can be used to show that in fact π is essentially proper. Finally, we use the Torelli theorem of [19] for supersingular Kummer surfaces to find enough points p in $\mathcal{P}$ for which $card\{\pi^{-1}(p)\} = 1$. It is then easy to check that this implies that π is an isomorphism.

In the final section of this paper, written after a conversation with S. Bloch, we show that his work on the "mysterious functor" allows one to prove that the inner product on the crystalline cohomology of a $K3$ surface in characteristic 2 is even, as conjectured by Shafarevich and Deligne. We also give a new proof of the crystalline discriminant formula for $K3$'s using Bloch's very deep results.

It goes almost without saying that the work of many Russian mathematicians (Kulikov, Rudakov, and of course especially Shafarevich) has been fundamental in countless places in the theory of $K3$ surfaces, and in particular to the completion of this paper. It also seems hardly necessary to recall that Artin began the systematic investigation of supersingular $K3$ surfaces in [4], and even suggested a notion of "periods" (based on flat, rather than crystalline, cohomology).

The research for this paper was made possible by a grant from the Alfred P. Sloan Foundation to the Institute for Advanced Study. I am grateful to them and to the many mathematicians there whose conversation helped stimulate my work on it. Special thanks go to M. Raynaud, R. Miranda, D. Morrison, and P. Deligne. I also want to thank P. Blass for translating a preprint version of [23] for the $K3$ surface seminar held at the Institute in 1981.

§ 1. Statements of the Torelli Theorem

We begin by changing slightly the definition of a $K3$ crystal given in [19]. Recall that k is an algebrically closed field of characteristic $p > 2$.

(1) *Definition.* A *$K3$ crystal over k* consists of a triple $(H, (\ |\ ), \Phi)$, where H is a free W-module of rank 22, $(\ |\ ): H \otimes H \longrightarrow W$ is a symmetric bilinear form, and $\Phi: H \longrightarrow H$ is an F_W-linear endomorphism, satisfying the following conditions:

(1.1) The map $(\ |\ ): H \to H^\vee =: Hom_W[H, W]$ induced by $(\ |\ )$ is an isomorphism.

(1.2) For $x, y \in H$ we have $(\Phi(x) \mid \Phi(y)) = p^2(x \mid y)$.

(1.3) The Hodge number $h^0(H) = 1$, i.e., $\Phi \otimes id_k$ has rank one.

(1.4) The discriminant $\left(\frac{H}{p}\right)$ of $(H, (\ |\ ), \Phi)$ [20], is -1. (Equivalently, if λ is a basis for $\Lambda^{22} H$ satisfying $\Phi(\lambda) = p^{22}\lambda$, then $(\lambda \mid \lambda) = -1$.)

It is now known that the F-crystal associated to any $K3$ surface satisfies the above conditions. The proof of (1.4) is subtle, cf. §4, (or [19] for the supersingular case).

(1.5) *Remark.* Every $K3$ crystal is isomorphic to the F-crystal of some $K3$ surface. If H is such a crystal, its Newton polygon is determined by its smallest slope λ, where $\lambda = 1 - 1/k$ and $k \in \{1, \ldots, 10, \infty\}$. If $k \neq \infty$, then the Newton and Hodge polygons touch at two points, and by a theorem of Katz [11] H splits into a direct sum of F-crystals: $H = H_\lambda \oplus H_1 \oplus H_\mu$, where H_λ has pure slope λ and $\mu = 2 - \lambda$. It is easy to check that H_λ and H_μ are totally isotropic and orthogonal to H_1, and the inner product defines an isomorphisms of W-modules $H_\lambda \to Hom[H_\mu, W]$. Moreover, H_λ has Hodge numbers $h^0 = 1$, $h^1 = h - 1$, $h^i = 0$ for $i > 1$, so there is a V_λ: $H_\lambda \longrightarrow H_\lambda$ with $\Phi_\lambda \circ V_\lambda = V_\lambda \circ \Phi_\lambda = \cdot p$. As is well known, such a crystal is uniquely determined up to isomorphism by λ. The map $H_\lambda \stackrel{\sim}{\rightarrow} Hom[H_\mu, W]$ sends pV_λ to Φ_μ^{tr}, and we conclude that the isomorphism class of $(H_\lambda \oplus H_\mu, \Phi_\lambda \oplus \Phi_\mu, (\ |\ )|_{H_\lambda \oplus H_\mu})$ is uniquely determined by λ. Finally, H_1 is a twist of a unit root crystal, and $(\ |\ )|_{H_1}$ is perfect. This implies that the isomorphism class of $(H_1, \Phi_1, (\ |\ )|_{H_1})$ is determined by its rank and discriminant $\left(\frac{H_1}{p}\right)$. But $H = H_1 \oplus H_\lambda \oplus H_\mu$, with H_λ

and H_μ totally isotropic, so $\left(\frac{H_1}{p}\right) = \left(\frac{H}{p}\right)\left(\frac{-1}{p}\right)^h = -\left(\frac{-1}{p}\right)^h$. Thus the isomorphism class of a $K3$ crystal with $h \neq \infty$ is uniquely determined by h. Since Artin has shown [4] that all values of h actually occur, our remark is certainly true (and rather trivial) if $h \neq \infty$. When $h = \infty$, we say that the crystal is supersingular; this case is not at all trivial and is a consequence of Theorem III below.

Our main goal is the following theorem and its variants.

Theorem I. *If X and Y are supersingular $K3$ surfaces and if there exists an isomorphism: $H^2_{cris}(X/W) \to H^2_{cris}(Y/W)$ compatible with Φ and $(\ |\ )$, then X and Y are isomorphic.*[1]

Recall that it is quite easy to "calculate" whether or not two supersingular $K3$ crystals are isomorphic; they have been completely classified in [19].

In fact, we shall prove a stronger version of Theorem I which will allow us to find the isomorphisms between supersingular $K3$ surfaces. This requires looking at the Neron-Severi group, which will play the role of integral homology.

(1.6) *Definition.* A $K3$ *lattice* is a free abelian group N of rank 22 with an even symmetric bilinear form $(\ |\ )$ with the following properties:

(a) $\operatorname{disc}(N \otimes \mathbb{Q}) = -1$ in $\mathbb{Q}^*/\mathbb{Q}^{*2}$.

(b) The signature of $(N \otimes \mathbb{R})$ is $(1, 21)$.

(c) $Cok : (\ |: N \to N^\vee$ *is annihilated by* p.

(1.7) Let us recall the following facts from [4] and [19]. If N is a $K3$ lattice, $\operatorname{disc}(N) = -p^{2\sigma_0}$, where $1 \leq \sigma_0 \leq 10$, and σ_0 determines the isomorphism class of N. If $N_1 =: N/pN^\vee$, then N_1 is an $\mathbb{F}_p$-vector space of rank $22 - 2\sigma_0$ and $(\ |\ )$ induces a nondegenerate (perfect) bilinear form $(\ |\ )_1$ on N_1; this form is not neutral, that is, its maximal totally isotropic subspaces have rank $10 - \sigma_0$. The form $(\ |\ )$ on $pN^\vee \subseteq N$ is divisible by p; dividing it by p we get a nonneutral nondegenerate bilinear

[1] This depends on Theorem (3.1) of Rudakov and Shafarevich below, which at present seems to require the restriction $p \geq 5$. We make this assumption from now on.

form on the $2\sigma_0$-dimensional $\mathbb{F}_p$-vector space $N_0 =: pN^\vee/pN$. If $\Lambda \subseteq N_0$ is totally isotropic, then $N_\Lambda =: \{x \in N \otimes \mathbb{Q} : px \in N \text{ and } \overline{px} \in \Lambda\}$ is also a $K3$ lattice, and $\sigma_0(N_\Lambda) = \sigma_0(N) - \operatorname{rank}(\Lambda)$. Dually, if $\Lambda \subseteq N_1$ is totally isotropic, then $N^\Lambda =: \{x \in N : \overline{x} \in Ann(\Lambda)\}$ is again a $K3$ lattice, and $\sigma_0(N^\Lambda) = \sigma_0(N) + \operatorname{rank}(\Lambda)$. These facts make it easy to "compute" with $K3$ lattices. The Neron-Severi group of a supersingular $K3$ surface is a $K3$ lattice, and it can be proved that any $K3$ lattice occurs in this way.

In [19] it is proved that Theorem I follows from:

Theorem II. *If X and Y are supersingular $K3$ surfaces and if θ: $NS(X) \longrightarrow NS(Y)$ is an isomorphism, then θ is induced by a (unique) isomorphism $Y \to X$ if it satisfies the following conditions:*

(i) *It is compatible with the quadratic forms.*

(ii) *It takes effective classes to effective classes.*

(iii) *There is a commutative diagram:*

$$\begin{array}{ccc} NS(X) & \xrightarrow{\theta} & NS(Y) \\ c_1 \downarrow & & \downarrow c_1 \\ H^2_{cris}(X/W) & \to & H^2_{cris}(Y/W) \end{array}$$

We shall rigidify our $K3$ surfaces in such a way so that σ_0 is allowed to change. This approach was introduced in [19] for $K3$ crystals, and we follow the same idea for surfaces. Let us fix a $K3$ lattice N with $\sigma_0(N) = a$ as a reference.

(1.8) *Definition.* An *N-marking* on a $K3$ surface X/k is a map η: $N \longrightarrow NS(X)$, compatible with the bilinear forms. An *isomorphism of N-marked $K3$ surfaces $\theta : (X, \eta) \to (Y, \varsigma)$* is an isomorphism θ: $X \longrightarrow Y$ such that $NS(\theta) \circ \varsigma = \eta$.

It is clear from (1.7) that a supersingular $K3$-surface X/k admits an N-marking η iff $\sigma_0\big(NS(X)\big) \leq a$. We can therefore deduce Theorem II from a similar theorem for marked surfaces. If (X, η) is an N-marked $K3$ surface, we recall from [19] the following construction of its "periods." The composite of η: $N \longrightarrow NS(X)$ with the Chern class map: c_{DR}: $NS(X) \longrightarrow H^2_{DR}(X/k)$ induces a map: $\overline{\eta}$: $N \otimes k \longrightarrow H^2_{DR}(X/k)$; we let $Ker_{(X,\eta)} =: Ker(\overline{\eta})$. It is proved in [19] that $Ker_{(X,\eta)} \subseteq N_0 \otimes k$

and is totally isotropic of rank a. Let ϕ: $N_0 \otimes k \longrightarrow N_0 \otimes k$ be $id_{N_0} \otimes F_k$ and let $K_{(X,\eta)} =: \phi^{-1}(Ker_{(X,\eta)})$. Recall also from [19] that $K_{(X,\eta)}$ is a "characteristic subspace of $N_0 \otimes k$," i.e., it is totally isotropic of rank a and $\phi(K_{(X,\eta)}) \cap K_{(X,\eta)}$ has rank $a-1$. In fact, $K_{(X,\eta)}$ determines $H^2_{cris}(X/W)$ and $NS(X)$, as the following result of [19] explains.

(1.9) Proposition. *With the above notations, the map $\eta : N \to NS(X)$ and the crystalline Chern class map $c_{cris} : NS(X) \to H^2_{cris}(X/W)$ together induce a map $\eta_{cris} : N \otimes W \to H^2_{cris}(X/W)$. With the obvious identifications induced by η_{cris}, we have:*

$$H^2_{cris}(X/W) \cong \{x \in N \otimes W \otimes \mathbb{Q} : px \in N \otimes W \text{ and } \overline{px} \in K_{(X,\eta)}\} \text{ and}$$
$$NS(X) \cong N_{(K,\eta)} =: \{x \in N \otimes \mathbb{Q} : px \in N \text{ and } \overline{px} \in K_{(X,\eta)}\} = \{x \in N \otimes \mathbb{Q} : \eta_{cris}(x) \in H^2_{cris}(X/W)\}.$$

If (X,η) is an N-marked $K3$ surface, let $NE_{(X,\eta)} \subseteq N_K \subseteq N \otimes \mathbb{Q}$ denote the subset consisting of those elements λ of N_K such that $\eta(\lambda)$ contains an effective divisor. It is clear that if θ: $(X,\eta) \longrightarrow (Y,\varsigma)$ is an isomorphism of N-marked $K3$ surfaces, then $K_{(X,\eta)} = K_{(Y,\varsigma)}$ and $NE_{(X,\eta)} = NE_{(Y,\varsigma)}$.

Theorem II′. *If (X,η) and (Y,ς) are N-marked $K3$-surfaces and if $K_{(X,\eta)} = K_{(Y,\varsigma)}$ and $NE_{(X,\eta)} = NE_{(Y,\varsigma)}$, then there is a unique isomorphism θ: $(X,\eta) \longrightarrow (Y,\varsigma)$.*

It is clear from Proposition (1.9) that Theorem II and Theorem II′ are equivalent for these surfaces X with $\sigma_0(NS(X)) \leq a$.

It turns out to be slightly more convenient to work with the ample cone than the set of effective divisors, although of course these are equivalent. We shall need some elementary facts about the set of all possible ample cones. In characteristic zero, at least, these facts are well known, but I find the references usually given rather inaccurate. Since it is easy to provide direct proofs, I prefer to do so.

In what follows we let N denote an arbitrary $K3$ lattice (or in fact any free abelian group with an even nondegenerate symmetric bilinear form of signature $(1,n)$, $n > 0$). Let $\Delta_N =: \{\delta \in N : (\delta \mid \delta) = -2\}$, and $V_N =: \{x \in N \otimes \mathbb{R} : (x \mid x) > 0 \text{ and } (x \mid \delta) \neq 0 \text{ for all } \delta \in \Delta_N\}$. If $\delta \in \Delta_N$, let r_δ be the reflection: $x \to x + (x \mid \delta)\delta$, and let R_N be the subgroup of $Aut(N)$ generated by $\{r_\delta \colon \delta \in \Delta_N\}$. We denote by $\pm R_N$ the subgroup of $Aut(N)$ generated by R_N and $-id$.

(1.10) Proposition. *The subset V_N of $N \otimes \mathbb{R}$ is open, and each of its connected components meets N. The group $\pm R_N$ operates simply and transitively on the set C_N of connected components of V_N.*

Proof. The first claim is easy. Suppose $\lambda \in V_N$; we must prove that there is a neighborhood U of λ in $N \otimes \mathbb{R}$ contained in V_N. Since $\delta^\perp$ is closed in $N \otimes \mathbb{R}$, it is enough to show that there is a neighborhood U' of λ such that $\{\delta \in \Delta_N : \delta^\perp \text{ meets } U'\}$ is finite. Since $N \subseteq N \otimes \mathbb{R}$ is discrete, this set will be finite as soon as it is bounded.

Let $u =: \lambda/\sqrt{(\lambda \mid \lambda)}$, and for any $x \in V$ let $x' =: (x \mid u)$ and $x'' =: x - x'u$. Since the form $(\ \mid\ )$ is negative definite on $u^\perp$ and positive definite on $\mathbb{R}u$, the inner product $\langle x \mid y\rangle =: x'y' - (x'' \mid y'')$ is positive definite on $N \otimes \mathbb{R}$ and hence defines a Euclidean norm $\|\ \|$. If $\delta \in \Delta_N$, $-2 = (\delta \mid \delta) = \delta' \cdot \delta' + (\delta'' \mid \delta'') = |\delta'|^2 - \|\delta'\|^2$, so $\|\delta''\| \leq |\delta'| + 2$. If $z \in N \otimes \mathbb{R}$ and $\delta \cdot z = 0$, then $\delta' \cdot z' + (\delta'' \mid z'') = 0$, and hence:

$$|\delta' \cdot z'| \leq \|\delta''\|\,\|z''\| \leq \left(|\delta'| + 2\right)\|z''\|$$

and

$$|\delta'|(|z'| - \|z''\|) \leq 2\,\|z''\|.$$

If z is near λ, $|z'|$ is near $\|\lambda\|$ and $\|z''\|$ is small, and this gives a bound on $|\delta'|$ and hence on $\|\delta\|$.

This shows that V_N is open. A Baire category argument implies that it is not empty. Since V_N is open, each of its connected components is open and hence meets $N \otimes \mathbb{Q}$. But each connected component α of V_N is bounded by hyperplanes, and hence if $x \in \alpha$ and $r \in \mathbb{R}^+$, $rx \in \alpha$. This implies that α meets N.

The proof of the statement about R_N is somewhat more involved. Along the way we shall prove some additional results which perhaps have independent interest. We often write $x \cdot y$ for $(x \mid y)$ in what follows.

Note that if $\lambda \in V_N$ and $x \in N$ with $x \cdot x \geq -2$, then $x \cdot \lambda \neq 0$ unless $x = 0$, by definition of V_N if $x \cdot x = -2$ and by the index assumption if $x \cdot x \geq 0$. Choose a connected component α of V_N and let λ be an element of $\alpha \cap N$. Define NE_α to be the submonoid of N generated by $\{x \in N : x \cdot x \geq -2 \text{ and } x \cdot \lambda > 0\}$. (Clearly it depends only on α, not on λ.) If $x \in N$ and $x \cdot x \geq -2$, $\pm x \in NE_\alpha$.

(1.10.1) *If $x \in NE_\alpha$ and $x \cdot \lambda \leq 0$, then $x = 0$. Any element of NE_α can be written as a sum of irreducible elements, and any irreducible element x satisfies $x \cdot x \geq -2$. If x and y are distinct and irreducible, $x \cdot y \geq 0$.*

Proof. The first statement is obvious. If $x \in NE_\alpha$ is not irreducible, we have $x = x' + x''$ with $x', x'' \in NE_\alpha$ and nonzero. Hence $0 < \lambda \cdot x' < \lambda \cdot x$, and $0 < \lambda \cdot x'' < \lambda \cdot x$, so by induction on $\lambda \cdot x$ the result is proved. Since any element of NE_α can by definition be written as a positive linear combination of elements satisfying $x \cdot x \geq -2$, any irreducible element must satisfy this inequality. If x and y are distinct and irreducible, let $z = x - y$. Then $z \cdot z = x \cdot x - 2x \cdot y + y \cdot y \geq -2(x \cdot y + 2)$. Thus if $x \cdot y < 0$, $z \cdot z \geq -2$, hence $\lambda \cdot z \neq 0$ and $\pm z \in NE_\alpha$. Each of these possibilities leads to a contradiction of the irreducibility of x or of y.

(1.10.2) *Suppose $e \in \Delta_N$ is irreducible in NE_α and $x \in N$ satisfies $x \cdot x \geq -2, x \in NE_\alpha$. Then $r_e(x) \in NE_\alpha$ unless $x = e$.*

Proof. If $x \cdot e \geq 0$, this is obvious. If $x \cdot e = -m$ with $m > 0$, let $z_i =: x - ie$. Then $z_i z_i = x \cdot x - 2i(i - m) \geq -2$ if $0 \leq i \leq m$, and hence $\pm z_i \in NE_\alpha$. Note that $z_i \neq 0$ unless $i = 1$ and $x = e$. Since $z_0 = x \in NE_\alpha$ and $e = z_i - z_{i+1}$ is irreducible, we see by induction that $z_m = r_e(X) \in NE_\alpha$.

Let R_N^α be the subgroup of R_N generated by $\{r_e : e$ is irreducible in $NE_\alpha\}$.

(1.10.3) *If $x \cdot x \geq 0$ and belongs to NE_α, then there is an $r \in R_N^\alpha$ such that $r(x)$ belongs to the closure of α. If $x \cdot x = -2$ and belongs to NE_α, there is an $r \in R_N^\alpha$ such that $r(x)$ is irreducible in NE_α.*

Proof. If $x \cdot x \geq 0$ and $x \cdot \lambda > 0$, then $x \cdot y \geq 0$ for every $y \in NE_\alpha$ with $y \cdot y \geq 0$. Thus, x belongs to the closure of α iff $x \cdot \delta \geq 0$ for every $\delta \in \Delta_N \cap NE_\alpha$, and it suffices to consider irreducible such δ. If $x \notin \overline{\alpha}$, there exists an irreducible e with $e \cdot x < 0$, and then $r_e(x)$ still belongs to NE_α by (1.10.2). Since $0 < \lambda \cdot r_e(x) < \lambda \cdot x$, we conclude the proof by induction.

If $x \cdot x = -2$ and x is not irreducible, write $x = \sum n_i x_i$ with each x_i irreducible, $n_i > 0$. Then $-2 = \sum n_i x_i \cdot x$ so $x_j \cdot x < 0$ for some j. Since $x_j \cdot x = \sum_{i \neq j} n_i x_i \cdot x_j + n_j x_j \cdot x_j$, $x_j \cdot x_j < 0$, hence $x_j \cdot x_j = -2$. Thus, there exists an irreducible e with $x \cdot e < 0$. Again we see that $r_e(x) \in NE_\alpha$ by (1.10.2), and we conclude by induction on $\lambda \cdot x$ as before.

(1.10.4) *The groups R_N^α and R_N are equal.*

Proof. If $\delta \in \Delta_N$, $r_\delta = r_{-\delta}$, so R_N is generated by $\{\delta \in \Delta_N \cap NE_\alpha\}$. Choose $r \in R_N^\alpha$ such that $r(\delta)$ is irreducible in NE_α; then

$$r_\delta = r^{-1} r_{r(\delta)} r \in R_N^\alpha.$$

(1.10.5) *If $r \in R_N$ and x and $r(x)$ belong to $\overline{\alpha} \cap N$, then $r(x) = x$.*

Proof. By the previous result, any $r \in R_N$ can be written as a product $r_{e_1} \ldots r_{e_n}$, where each e_i is irreducible in NE_α. Let $|r|$ denote the minimum n for all such expressions. The proof is by induction on $|r|$. If $|r| = 0$, $r = id$, and the result is trivial.

I claim that if $e \in \Delta_N$ is irreducible in NE_α and $r \in R_N$ is such that $r(e) \notin NE_\alpha$, then $|r \circ r_e| < |r|$. Certainly $n =: |r| > 0$; write $r = r_1 r_2 \ldots r_n$, where $r_i = r_{e_i}$ and e_i is irreducible. Let $\delta_i =: r_{i+1} \ldots r_n(e)$, for $0 \leq i \leq n$. For each i, $\delta_i \cdot \delta_i = -2$, so $\pm\delta_i \in NE_\alpha$. Moreover, $\delta_n = e \in NE_\alpha$ and $\delta_0 = r(e) \notin NE_\alpha$. Thus the largest i such that $r_i(\delta_i) \notin NE_\alpha$ is in $[1, n]$, and $\delta_i \in NE_\alpha$. By (1.10.2), $\delta_i = e_i$. Let $\tilde{r} = r_{i+1} \ldots r_n$; then $\tilde{r}(e) = e_i$, so $\tilde{r} r_e \tilde{r}^{-1} = r_{\tilde{r}(e)} = r_i$. Then

$$r r_e = (r_1 \ldots r_{i-1} r_i \tilde{r}) r_e = r_1 \ldots r_{i-1} \tilde{r} r_e \tilde{r}^{-1} \tilde{r} r_e = r_1 \ldots r_{i-1} r_{i+1} \ldots r_n.$$

This proves the claim.

Next I claim that if $x \in \overline{\alpha}$ and $r \in R_N$, then $r(x) - x \in NE_\alpha$. We prove this by induction on $|r|$, the case $|r| = 0$ being trivial. If $|r| \geq 1$, write $r = \tilde{r} r_e$ with $e \in NE_\alpha$ irreducible and $|\tilde{r}| < |r|$. Then if $m =: e \cdot x$, $m \geq 0$ since $x \in \overline{\alpha}$, and $r(x) = \tilde{r}(x) + m\tilde{r}(e)$. Since $\tilde{r}(x) - x \in NE_\alpha$ by our induction hypothesis, it is sufficient to note that the previous claim implies that $\tilde{r}(e) \in NE_\alpha$.

Suppose now that x and $r(x)$ belong to $\overline{\alpha}$. Then let $y =: r(x) - x$, which belongs to NE_α. Then:

$$y \cdot x + y \cdot r(x) = 2y \cdot x + y \cdot y = (y + x) \cdot (y + x) - x \cdot x = r(x) \cdot r(x) - x \cdot x = 0.$$

Since x and $r(x)$ both belong to $\overline{\alpha}$ and y belongs to NE_α, $y \cdot x$ and $y \cdot r(x)$ are nonnegative, so in fact $y \cdot x = y \cdot r(x) = 0$, and hence we also have $y \cdot y = 0$.

If $x \cdot x > 0$, $(\ |\ )$ is negative definite on $x^\perp$, so $y = 0$ and $x = r(x)$. If $x \cdot x = 0$, the form $(\ |\ )$ induces on $x^\perp/\mathbb{R}x$ is negative definite, and so $y = ax$ for some real number a. This implies that $r(x) = b(x)$ for some $b \in \mathbb{R}$. But x and $r(x)$ belong to N, so b is an integer, and since in fact r induces an isomorphisms $N \to N$, $b = \pm 1$. Since x and $r(x)$ both belong to $\overline{\alpha}$, we must have $x = r(x)$.

We immediately deduce:

(1.10.6) *Every $\pm R_N$-orbit of $\{x \in N : x \cdot x \geq 0\}$ meets $\overline{\alpha}$ exactly once.*

Now the proposition is almost proved. It is clear that $\pm R_N$ acts transitively on C_N; we have only to check that its action is simple. This follows from:

(1.10.7) *If $x \in \alpha$ and $g \in \pm R_N$ and $g(x) \in \alpha$, then $g = id$.*

Proof. The set $\{x \in N \otimes R : x \cdot x > 0\}$ has two connected components, preserved by R_N and reversed by $-id$. Thus if x, g, and α are as above, g must belong to R_N. Now (1.10.5) tells us that in fact $g(x) = x$. But now if x' is any element of α, $g(x')$ and $g(x)$ must lie in the same connected component of V_N, i.e. α, and hence it is also true that $g(x') = x'$. Since α is open and $g\,|_\alpha = id$, $g = id$.

(1.10.8) *Remark.* It follows from the above arguments that any element δ of $\Delta_N \cap NE_\alpha$ can be written as a linear combination: $\delta = \sum m_i \delta_i$, where $m_i > 0$ and $\delta_i \in \Delta_N \cap NE_\alpha$ is irreducible. We prove this by induction on $(\delta \cdot \lambda)$. If δ is not irreducible, there exists an irreducible element e of $\Delta_N \cap NE_\alpha$ with $\delta \cdot e < 0$; then by (1.10.2), $r_e(\delta) \in \Delta_N \cap NE_\alpha$ and $\delta = r_e(\delta) + |\delta \cdot e|e$. Since $r_e(\delta) \cdot \lambda < \delta \cdot \lambda$, $r_e(\delta)$ has an expression of the above form.

(1.10.9) *Remark.* If X is a $K3$ surface and $N =: NS(X)$, let $\lambda \in N$ be the class of an ample divisor H on X. Then if D is a divisor on X, its class $\delta \in N$ lies in NE_α iff D is linearly equivalent to an effective divisor. We check this by a well-known argument. If $D \sim D'$ with D' effective, write D' as a positive linear combination of irreducible curves D'_i; to prove that $\delta \in NE_\alpha$, it suffices to check that each $\delta'_i \in NE_\alpha$. On a $K3$ surface,

$C \cdot C \geq -2$ for any irreducible curve C by the adjunction formula, and since $C \cdot H > 0$, class$(C) \in NE_\alpha$ by the definition of NE_α. Conversely, if $\delta \in NE_\alpha$, then δ can be written as a positive linear combination of elements δ' with $\delta' \cdot \delta' \geq -2$ and $\delta' \cdot \lambda > 0$, and so D is a positive linear combination of divisors D' with these properties. Thus it suffices to show that for such a D', $H^0(X, \mathcal{O}_X(D')) \neq 0$. By the Riemann-Roch theorem we see that $H^0(X, \mathcal{O}_X(\pm D')) \neq 0$, and since $D' \cdot H > 0$, we cannot have $H^0(X, \mathcal{O}_X(-D')) \neq 0$.

Notice that if the class δ of an effective divisor D is irreducible in NE_α, then every element in the complete linear system $|D|$ is irreducible. It would be interesting to determine which divisors have this property. Note that if E is effective and $E \cdot E = -2$, then the class e of E is irreducible in NE_α iff E is an irreducible curve.

We can now reformulate Theorem II′ as follows. If X is a $K3$ surface, it is well known that a line bundle L on X is ample iff $L \cdot L > 0$, $L \cdot E > 0$ for every effective E with $E \cdot E = -2$, and $h^0(L) \neq 0$. This implies that there is a unique connected component α of $V_{NS(X)}$ such that a line bundle L on X is ample iff its class in $NS(X) \otimes \mathbb{R}$ lies in α. Now if (X, η) is an N-marked $K3$ surface and $K =: K(X, \eta) \subseteq N_0 \otimes k$, η induces isomorphisms: $N_K \to NS(X)$, $\Delta_{N_K} \to \Delta_{NS(X)}$, $N \otimes \mathbb{R} \to NS(X) \otimes \mathbb{R}$. Let C_K denote the set of connected components of

$$\{x \in N \otimes \mathbb{R} : x \cdot x > 0,\ x \cdot \delta \neq 0\, \forall\, \delta \in \Delta_{N_K}\}.$$

Then the ample component of $V_{NS(X)}$ corresponds to a well-defined element $\alpha_{(X,\eta)}$ of C_K. It is clear that Theorem II′ is equivalent to:

Theorem II″. *If (X, η) and (Y, ς) are N-marked $K3$ surfaces over k and if $K_{(X,\eta)} = K_{(Y,\varsigma)}$ and $\alpha_{(X,\eta)} = \alpha_{(Y,\varsigma)}$, then there is a unique isomorphisms $\theta: (X, \eta) \longrightarrow (Y, \varsigma)$.*

It is clear from (1.10) that Theorem II″ will imply:

Corollary. *If (X, η) is an N-marked $K3$ surface over k with periods $K \subseteq N_0 \otimes k$ and ample cone $\alpha \in C_N$, let $G_K = \{g \in Aut(N) : gK = K\}$. Then $Aut(X)$ is isomorphic to the subgroup of G_K stabilizing α, and G_K is the semi-direct product of $\pm R_N$ and $Aut(X)$.*

(1.11) *Remark.* It turns out that every supersingular $K3$ surface contains enough -2 curves to determine its ample cone. More precisely, we have the following result (which is not needed for the proof of Theorem II″):

If L is a line bundle on a supersingular $K3$ surface and $L \cdot L > 0$ and $L \cdot E > 0$ for every irreducible effective E with $E \cdot E = -2$, then L is ample.

To prove this we need two lemmas, the first of which we will use again later.

(1.12) Lemma. *If N is any $K3$ lattice and m is any integer, there exists a $\lambda \in N$ with $(\lambda \mid \lambda) = 2m$.*

Proof. First of all let us conside the $K3$ lattice attached to a "super-special" $K3$ surface X, i.e., the Kummer surface attached to $E \times E$, where E is a supersingular elliptic curve. If $x \in E$ is not a 2-torsion point, the strict transform of (x, E) in X is a smooth elliptic curve C, and the strict transform of $(E, 0)$ in X is a smooth rational curve Γ with $C \cdot \Gamma = 1$. Thus, the class γ of $(kC + \Gamma)$ in $NS(X)$ satisfies $(\gamma \mid \gamma) = 2(k-1)$.

Now let N be an arbitrary $K3$ lattice. Let $\overline{\gamma}$ be the image of γ in $NS(X)_1$, which we recall has a (nonneutral) nondegenerate quadratic form, of rank 20. Then $Ann(\overline{\gamma}) \subseteq NS(X)_1$ has rank ≥ 19, and since $\sigma_0(N) - 1 \leq 9$, we can find a totally isotropic subspace Λ of $Ann(\overline{\gamma})$ of rank $\sigma_0(N) - 1$. Then the space $NS(X)^{\Lambda}$ (cf. (1.7)) is a $K3$ lattice with $\sigma_0 = \sigma_0(N)$. Hence $NS(X)^{\Lambda} \cong N$. Since $\gamma \in NS(X)^{\Lambda}$, the lemma follows.

(1.13) Lemma. *If X is any supersingular $K3$ surface there exist effective curves D, E on X such that $D \cdot D = E \cdot E = -2$ and $D \cdot E$ is arbitrarily large.*

Proof. By the previous lemma we know there exists a $\delta \in NS(X)$ with $\delta \cdot \delta = -2$, and it follows that there must also exist an effective irreducible curve D with $D \cdot D = -2$. Now let δ denote the class of D in $NS(X)$.

Choose a large prime l and let $\overline{\delta}$ be the image of δ in $NS(X) \otimes \mathbb{F}_l$. It is easy to see that we can find an $\overline{\epsilon} \in NS(X) \otimes \mathbb{F}_l$ such that $\overline{\epsilon} \cdot \overline{\epsilon} = \overline{\delta} \cdot \overline{\delta}$ and $\overline{\epsilon} \cdot \overline{\delta} = 0$, and a $\overline{g} \in Aut(NS(X) \otimes \mathbb{F}_l)$ such that $\overline{g}(\overline{\delta}) = \overline{\epsilon}$. Since $Aut(NS(X) \otimes \mathbb{Z}_l) \rightarrow Aut(NS(X) \otimes \mathbb{F}_l)$ is surjective, we find a $g \in Aut(NS(X) \otimes \mathbb{Z}_l)$ such that $g(\overline{\delta}) = \overline{\epsilon}$. It is easy to see that we can choose g first of all to have determinant one and then to have spinorial norm one (cf. [19]). Therefore g lifts to the spin group, and by strong approximation we deduce that there is a $g' \in Aut(NS(X))$ such that $g'(\overline{\delta}) = \overline{\epsilon}$ but g' acts as the identity mod p. Then $\epsilon' =: g'(\delta)$ satisfies: $\epsilon' \cdot \epsilon' = -2$, $l \mid (\epsilon' \cdot \delta)$, but $(\epsilon' \cdot \delta) \equiv -2 \bmod p$. In particular, $|\epsilon \cdot \delta| \geq l$. Replacing ϵ' by $-\epsilon'$ if necessary, we may assume that ϵ' is effective. If $(\epsilon' \cdot \delta) > 0$, the lemma is proved. If $(\epsilon' \cdot \delta) < 0$, let $\epsilon =: r_\delta(\epsilon')$; then by (1.10.2), ϵ is still effective, and $\epsilon \cdot \delta = -\epsilon' \cdot \delta > 0$.

Now to prove (1.11): suppose L is as described there, and let λ be its image in $NS(X)$. If we can show that λ contains an effective divisor, then L will be ample. Let λ' be the class of an ample divisor; it is enough to check that $\lambda \cdot \lambda' > 0$, i.e., that λ and λ' lie in the same connected component of $\{x \in N \otimes \mathbb{R} : x \cdot x > 0\}$. By the previous lemma we can find effective δ, ϵ with $\delta \cdot \delta = \epsilon \cdot \epsilon = -2$ and $\delta \cdot \epsilon \geq 3$. Then $\gamma =: \delta + \epsilon$ is effective and $\gamma \cdot \gamma > 0$. By (1.10.8) each of δ and ϵ can be written as a positive linear combination of irreducible curves E with $E \cdot E = -2$, and hence our assumption on L implies tht $\gamma \cdot \lambda > 0$. Since $\gamma \cdot \lambda' > 0$ and $\gamma \cdot \lambda > 0$, all three lie in the same connected component.

(1.14) *Remark.* If X is the Kummer surface attached to $E \times E$ (in any characteristic), then in (1.13) we can take D and E to be irreducible. I am not sure if this is true for all supersinguar $K3$ surfaces. Let me remark that in characteristic zero, we can conclude that in the context of the Burns-Rapoport period space [7], *not every* combination of a partition Δ^+ and a choice of a connected component of $\{x \in H^{1,1}(X, \mathbb{R}) : x \cdot x > 0\}$ can occur in the image of the period mapping.

Now we define an analogue of the Burns-Rapoport period space, which simultaneously keeps track of the "periods" K of an N-marked $K3$ surface and the ample cone in Neron-Severi. Let $\mathcal{M}_N$ denote the k-scheme parameterizing the characteristic subspaces of N_0 introduced in [19]. If S is a k-scheme (or algebraic space), $\underline{\mathcal{M}}_N(S)$ is the set of all $K \subseteq \mathcal{O}_S \otimes N_0$ which are local direct factors, totally isotropic, have rank $\sigma_0(N)$, and for which $F_S^*(K) \cap K \subseteq K$ is a local direct factor of rank $\sigma_0(N) - 1$. If $K \in \underline{\mathcal{M}}_N(S)$, then for each point s of S we set

$$\begin{aligned}
&\Lambda(s) =: N_0 \cap K(s),\\
&N(s) = N_{K(s)} =: \{x \in N \otimes \mathbb{Q} : px \in N \text{ and } \overline{px} \in \Lambda(s)\},\\
&\Delta(s) = \{\delta \in N(s) : \delta \cdot \delta = -2\}, \text{ etc.}
\end{aligned}$$

Note that if s is a specialization of σ, then $\Lambda(\sigma) \subseteq \Lambda(s)$, with equality on a dense open subset of S (because N_0 is finite). It follows that $N(\sigma) \subseteq N(s)$, $\Delta(\sigma) \subseteq \Delta(s)$, and $V(s) \subseteq V(\sigma)$. Thus every connected component of $V(s)$ is contained in a unique connected component of $V(\sigma)$.

(1.15) *Definition.* If $K \in \underline{\mathcal{M}}_N(S)$, an *ample cone for* K is an element α of $\prod\{C_{N(s)} : s \in S\}$ such that $\alpha(s) \subseteq \alpha(\sigma)$ whenever s is a specialization of σ. We denote by $\underline{P}_N$ the functor which assigns to each S the set of pairs (K, α), where $K \in \underline{\mathcal{M}}_N(S)$ and α is an ample cone for K.

We shall call a scheme *almost proper* iff it satisfies the surjectivity part of the valuative criterion (with DVR's as test rings).

(1.16) **Proposition.** *The functor $\underline{P}_N$ is represented by a k-scheme which is locally of finite type and almost proper. The natural map $P_N \to \mathcal{M}_N$ is étale and surjective.*

Proof. Choose a totally subspace Λ of N_0 and let $\mathcal{M}_N^{\Lambda}$ be the open subset of $\mathcal{M}_N$ corresponding to those K such that $K \cap N_0 \subseteq \Lambda$. For each $\alpha \in C_{N_\Lambda}$, let $\mathcal{M}_N^{\Lambda,\alpha}$ be a copy of $\mathcal{M}_N^{\Lambda}$. There is a unique way to extend α to an ample cone for the universal $K\,|_{\mathcal{M}_N^{\Lambda}}$: if $s \in S$, $\Lambda(s) \subseteq \Lambda$ and $V_{N(s)} \subseteq V_{N_\Lambda}$, so $\alpha(s)$ must be the connected component of $V_{N(s)}$ containing α.

If $\Lambda' \subseteq \Lambda$, $\mathcal{M}_N^{\Lambda'} \subseteq \mathcal{M}_N^{\Lambda}$. We take P_N to be the scheme obtained from the disjoint union of all the $\mathcal{M}_N^{\Lambda,\alpha}$ by identifying $\mathcal{M}_N^{\Lambda',\alpha'}$ with the open set $\mathcal{M}_N^{\Lambda',\alpha} \subseteq \mathcal{M}_N^{\Lambda,\alpha}$, provided α and α' agree there. It is clear that P_N represents the functor $\underline{P}_N$ and that the map $P_N \to \mathcal{M}_N$ is étale and surjective.

To prove the almost properness, it is enough to show that $P_N \to \mathcal{M}_N$ is almost proper, since $\mathcal{M}_N$ is proper over k. This is local on $\mathcal{M}_N$, so it suffices to work over each $\mathcal{M}_N^{\Lambda}$. Clearly what we must prove is that if $\Lambda' \subseteq \Lambda$ and $\alpha' \in C_{N_\Lambda}$, then there exists an $\alpha \in C_{N_\Lambda}$ contained in α' — which is obvious.

Our final version of the Torelli theorem is the following, which also includes a "surjectivity" statement.

Theorem III. *The scheme P_N is a fine moduli space for N-marked $K3$ surfaces.*

Of course, to make this precise, we have to define what we mean by a family of N-marked $K3$'s. This will be the focus of the next section.

§ 2. Families of N-Marked K3 Surfaces

This section is devoted to a study of families of N-marked $K3$ surfaces, and in particular to the proof of Theorem (2.7), which asserts the existence of a universal such family over an (abstractly constructed) algebraic space. A key point is the smoothness of this space, which is the content of (2.6).

(2.1) *Definition.* If S is an algebraic space over k, a *family of N-marked $K3$ surfaces over S* is a smooth proper map $f: X \longrightarrow S$ of algebraic spaces, each of whose fibers is a $K3$ surface, together with a morphism of group schemes $\eta: \underline{N}_S \longrightarrow \underline{Pic}_{X/S}$, compatible with the intersection forms.

In the above defintion, $\underline{N}_S$ is the "constant" group scheme on S defined by N, and $\underline{Pic}_{X/S}$ is the relative Picard scheme [2]. An isomorphisms of N-marked $K3$ surfaces over S is defined in the obvious way.

(2.2) Lemma. *If (X,η) is an N-marked $K3$ surface over S,*

$$Aut_S(X,\eta) = \{id\}.$$

Proof. If $S = Spec\ k$, this is proved in [19] (cf. also [5 I, 3.23]). Since Rudakov and Shafarevich have proved that $K3$ surfaces in characteristic p have no vector fields [22] (cf. also [13]), we can conclude that the functor $\underline{Aut}_{X/S}$ is unramified. Hence if $s \in S$ is a k-valued point,

$$Aut_S(X,\eta) \rightarrow Aut_k(X(S), \eta(s))$$

is injective near s. The lemma follows.

(2.3) Lemma. *If S is the spectrum of a noetherian strictly Henselian local ring and (X,η) is an N-marked $K3$ surface over S, there exists a*

$\lambda \in N$ such that $\eta(\lambda)$ is ample on X. Moreover, λ may be chosen so that $(\lambda \mid \lambda)$ is prime to p.

Proof. Since X/S is smooth and S is strictly Henselian, $X \to S$ admits a section and $\underline{Pic}_{X/S}(S) \cong Pic(X)$. Let $s \in S$ be the closed point and choose an ample line bundle L on $X(s)$. The map $\eta_s \colon N \longrightarrow NS(X(s))$ has cokernel killed by p, so there is a $\lambda \in N$ with $\eta_s(\lambda) = L^p$. Moreover, we can choose μ in N such that $(\mu \mid \mu)$ is prime to p; then if a is large, $\eta_s(a\lambda + \mu) = L^{pa} \otimes \eta_s(\mu)$ is still ample on $X(s)$ and has degree prime to p. Replace λ by $a\lambda + \mu$; since $\eta(\lambda)\,|_{X(s)}$ is ample, so is $\eta(\lambda)$ [9, 4.7].

(2.4) *Remark.* The above lemma works just as well if S is the formal spectrum of complete local ring and X is a formal $K3$ surface over S with N-marking η. This implies in particular that $X \mid S$ can be algebraized.

(2.5) **Proposition.** *Let (X, η) and (Y, ς) be two N-marked $K3$ surfaces over S. The functor $\underline{Isom}_{(X,\eta),(Y,\varsigma)}$ is represented by a locally closed subalgebraic space of S.*

Proof. The functor $\underline{Isom}$ assigns to any S-space S' the set of all isomorphisms: $(X \times_S S', \eta_{S'}) \to (Y \times_S S, \varsigma_{S'})$. Its representability follows from [1]. It is clear that $\underline{Isom}$ is a pseudotorsor under $\underline{Aut}_{(X,\eta)/S}$, and hence by (2.2) the map $Isom \to S$ is unramified and radicial. This certainly implies that it is of finite type. To prove that it is a locally closed immersion, we may assume that S is the spectrum of a strictly Henselian local ring, and that the closed fiber of $Isom \to S$ is not empty. In this case we shall show that $Isom \to S$ is proper, hence a closed immersion.

We must verify the valuative criterion of properness. This reduces our assertion to the case in which S is the spectrum of a complete DVR, with closed point s and generic point σ. We must show that if

$$\theta_* : (X(s), \eta(s)) \to (Y(s), \varsigma_s)$$

and

$$\theta_\sigma : (X(\sigma), \eta_\sigma) \to (Y(\sigma), \varsigma_\sigma)$$

are isomorphisms, then there exists a unique isomorphism $\theta\colon (X, \eta) \longrightarrow (Y, \varsigma)$ inducing θ_s and θ_σ. Indeed, since isomorphisms between N-marked $K3$'s are unique if they exist, it suffices to show that (X, η) and (Y, ς) are isomorphic.

By Lemma (2.3), Y is projective, and in fact we can choose $\lambda \in N$ such that $\rho(\lambda)$ is ample on Y. Since $\theta_s: X(s) \longrightarrow Y(s)$ is an isomorphism, $\theta_s^* \varsigma_s(\lambda) = \eta_s(\lambda)$ is ample on $X(s)$, and hence $\eta(\lambda)$ is ample on X. But $\theta_\sigma: X_\sigma \longrightarrow Y_\sigma$ is then an isomorphism of *polarized* $K3$ surfaces, since $\theta_\sigma^* \varsigma_\sigma(\lambda) = \eta_\sigma(\lambda)$, and so by Matsusake-Mumford [14] prolongs to an isomorphism $\theta: X \longrightarrow Y$. Since $Pic(X) \cong Pic(X_\sigma)$ and $Pic(Y) \cong Pic(Y_\sigma)$, θ is necessarily compatible with the N-markings.

The next result is perhaps the crucial step in the proof of Theorem (2.7). It implies the existence of nice families of N-marked $K3$ surfaces. It can be proved using (2.10) (and a cumbersome inductive assumption of theorem II). However, Deligne has suggested a direct proof which severs the last remaining dependence of our study of supersingular $K3$ surfaces on the theory of p-divisible groups.

(2.6) Proposition. *An N-marked $K3$ surface (X_0, η_0) over k has a universal formal deformation $(\mathcal{X}_N, \eta)$ over S_N, where S_N is formally smooth of dimension $\sigma_0(N) - 1$. The Kodaira-Spencer mapping induces a commutative diagram:*

$$\begin{array}{ccccc} N \otimes k & \overset{c_{Hodge}}{\longrightarrow} & H^1(X_0, \Omega^1_{X_0/k}) & \longrightarrow & Cok(c_{Hodge}) \\ & & \downarrow & \swarrow \cong & \\ & & H^2(X_0, \mathcal{O}_{X_0}) \otimes \Omega^1_{S_N/K}(0). & & \end{array}$$

Proof. We know that the parameter space S of the universal formal k-deformation of X is formally smooth of dimension 20, and that the Kodaira-Spencer mapping induces an isomorphism:

$$H^1(X_0, \Omega^1_{X_0/k}) \rightarrow H^2(X_0, \mathcal{O}_{X_0}) \otimes \Omega^1_{S/k}(0).$$

If $L_0 \in Pic(X_0)$, the universal formal k-deformation of (X_0, L_0) has as parameter space S_{L_0}, a closed formal subscheme of S defined by a single equation, and the isomorphisms $m/m^2 \rightarrow \Omega^1_{S/k}(0)$ takes the image of this equation mod m^2 to the image of $c_{Hodge}(L_0)$ under the Kodaira-Spencer map [19]. Thus it is clear that there is a maximal closed formal subscheme S_N of S over which all the elements of $\eta_0(N)$ prolong and that S_N is the parameter space of the universal formal deformation of (X_0, η_0). Moreover, we evidently have a diagram as claimed in the theorem, and it is proved in [19] that $Cok(\eta_0)$ has dimension $\sigma_0(N) - 1$.

To prove the proposition it remains to show that $\dim S_N \geq \sigma_0(N)-1$. We proceed as follows. Let $a =: \sigma_0(N)$ and $b =: 11 - a$; then $N \otimes \mathbb{F}_p \cong N_0 \oplus N_1$, where N_0 and N_1 have ranks $2a$ and $2b$, respectively. Since the intersection form on $N_1 \otimes k$ is perfect, the map:

$$c_{Hodge}: N_1 \otimes k \longrightarrow H^1(X_0, \Omega^1_{X_0/k})$$

is injective. Since

$$c_{Hodge}: N \otimes k \longrightarrow H^1(X_0, \Omega^1_{X_0/k})$$

has rank $20-(a-1) = a+2b-1$, it follows that

$$c_{Hodge}: N_0 \otimes k \longrightarrow H^1(X_0, \Omega^1_{X/k})$$

has rank $a-1$. On the other hand,

$$c_{DR}: N \otimes k \longrightarrow H^2_{DR}(X_0/k)$$

has rank $22 - a = a + 2b$, so c_{DR}: $N_0 \otimes k \longrightarrow H^2_{DR}(X_0/k)$ has rank a. Thus, we can choose an $\mathbb{F}_p$-basis $\{\overline{L}_1 \ldots \overline{L}_{2a}\}$ for N_0 such that

$$\{c_{Hodge}(\overline{L}_1) \ldots c_{Hodge}(\overline{L}_{a-1})\}$$

are linearly independent in $H^1(X_0, \Omega^1_{X_0/k})$ and $\{c_{DR}(\overline{L}_1) \ldots c_{DR}(\overline{L}_a)\}$ are linearly independent in $H^2_{DR}(X_0/k)$, and an $\mathbb{F}_p$-basis $\{\overline{M}_1 \ldots \overline{M}_{2b}\}$ for N_1. Let $\{L_i, M_i\} \subseteq N$ lift $\{\overline{L}_i, \overline{M}_i\}$ and let

$$N' =: span\{L_1, \ldots, L_{a-1}, M_1 \ldots M_{2b}\}.$$

It is clear that there is a maximal closed formal subscheme $S_{N'}$ of S over which the elements of N' prolong, and that $S_{N'} \subseteq S$ is defined by $a-1+2b = 21-a$ equations. Thus, $S_{N'}$ has dimension $\geq a-1$, and since $m_{S_{N'}}/m^2_{S_{N'}} \cong m_{S_N}/m^2_{S_N}$, $S_{N'}$ is in fact formally smooth of dimension $(a-1)$.

The proposition will be proved if we can show that in fact all the elements of N extend to $\mathcal{X}\,|_{S_{N'}}$. Since $S_{N'}$ is formal, it suffices to prove that if $T' \subseteq T \subseteq S_{N'}$ with $T' \subseteq T$ defined by a square zero ideal J and if the claim is true for $X' =: \mathcal{X}\,|_{T'}$ then it is also true for $X =: \mathcal{X}\,|_T$. Recall that there is a commutative diagram [19],

$$\begin{array}{ccc} Pic(X') & \xrightarrow{c_{DR}} & H^2_{DR}(X/T) \\ {\scriptstyle ob}\downarrow & & \downarrow{\scriptstyle edge} \\ H^2(X, JO_X) & \hookrightarrow & H^2(X, O_X) \end{array} \tag{2.6.1}$$

where $ob(L')$ is the obstruction to extending a line bundle L' from X' to X and where the kernel of the edge homomorphism shown is $F^1H^2_{DR}(X/T)$. If $T = Spec\,A$, $H^2_{DR}(X/T)$ is a free A-module of rank 22 equipped with a perfect pairing, and c_{DR} is compatible with the pairings. It induces a map of A-modules:

$$c\colon N \otimes A \cong N_0 \otimes A \oplus N_1 \otimes A \dashrightarrow H^2_{DR}(X/T).$$

The map $N_1 \otimes A \to H^2_{DR}(X/T)$ is still injective when tensored with the residue field k of A, and this implies that $c(N_1 \otimes A)$ is a direct summand of $H^2_{DR}(X/T)$. Let $H^\perp$ denote the orthogonal complement of $c(N_1 \otimes A)$ in $H^2_{DR}(X/T)$; since the quadratic form on $(N_1 \otimes A)$ is perfect, the same is true of the quadratic form on $H^\perp$. Moreover, $H^\perp$ is a free A-module of rank $2a$, and since

$$c(N_1 \otimes A) \subseteq F^1H^2_{DR}(X/T),$$
$$H^\perp \supseteq Ann\,F^1H^2_{DR}(X/T) = F^2H^2_{DR}(X/T).$$

We have to prove that $c(N \otimes A) \subseteq F^1H^2_{DR}(X/T)$, and it suffices to check $c(N_0 \otimes A)$. Clearly $c(N_0 \otimes A) \subseteq H^\perp$, and by assumption the $\mathbb{F}_p$-span N_0' of $(\bar{L}_1 \ldots \bar{L}_{a-1})$ satisfies $c(N_0' \otimes A) \subseteq F^1H^\perp$. Let N_0'' be the $\mathbb{F}_p$-span of $(\bar{L}_1 \ldots \bar{L}_a)$; since the map $N_0'' \otimes A \to H^2_{DR}(X/T)$ is injective when tensored with k, $c(N_0'' \otimes A)$ is a direct summand of $H^2_{DR}(X/T)$. In particular, $c(N_0'' \otimes A) \subseteq H^\perp$ is a totally isotropic direct summand of rank $a = 1/2\,\mathrm{rank}\,(H^\perp)$, and it follows that $c(N_0'' \otimes A) = Ann\,c(N_0'' \otimes A)$.

By assumption, $c(N_0'' \otimes A') \subseteq F^1H^1 \otimes A'$, where $A' =: A/J$, hence $F^2H^\perp \otimes A' = Ann\,F^1H^\perp \otimes A' \subseteq Ann\,c(N_0'' \otimes A') = c(N_0'' \otimes A')$. Let ω be a basis for $F^2H^\perp$ and choose $\lambda \in N_0'' \otimes A$ such that $c(\lambda) \equiv \omega \bmod J$. Since $c(L_1) \ldots c(L_{a-1})$ are independent in $F^1/F^2 \bmod m$, $\{L_1 \ldots L_{a-1}\}$ is a basis for $N_0 \otimes A$. Write $\lambda = \omega + \alpha\varsigma$, with $\alpha \in J$, $\varsigma \in H$; since the quadratic form on N_0'' is divisible by p, $\lambda \cdot \lambda = 0$ so

$$0 = \omega \cdot \omega + (\omega \mid \alpha\varsigma) + \alpha^2(\varsigma \mid \varsigma) = (\omega \mid \alpha\varsigma).$$

But the pairing between $F^1H^\perp$ and $H^\perp/F^1$ is perfect, and therefore $\alpha\varsigma \in F^1H^\perp$. This implies that $\lambda \in F^1$, and since $\{L_1 \ldots L_{a-1}, \lambda\}$ form a basis for $N'' \otimes A$, we conclude that $N'' \otimes A \subseteq F^1H^\perp$. But since the quadratic form on $N_0 \otimes A$ is zero, $c(N_0 \otimes A) \subseteq Ann\, c(N_0'' \otimes A) = c(N_0'' \otimes A)$, and hence $c(N_0 \otimes A) \subseteq F^1H^\perp$.

If T is any algebraic space over k, let $\underline{S}_N(T)$ denote the set of isomorphism classes of N-marked $K3$ surfaces over T. Evidently $\underline{S}_N$ is a functor from the category of k-algebraic spaces to the category of sets. We can summarize the above results in the following statement:

(2.7) Theorem. *The functor $\underline{S}_N$ is representable by an algebraic space S_N over k, which is locally of finite presentation, locally separated, and smooth of dimension $\sigma_0(N) - 1$.*

Proof. This is just a matter of checking Artin's list [2]. First of all, $\underline{S}_N$ is a sheaf for the étale topology. This follows easily from descent theory for algebraic spaces, using the fact that N-marked $K3$'s have no automorphisms, and from the fact that $\underline{Pic}_{X/S}$ is a sheaf for the étale topology. Next, $\underline{S}_N$ is locally of finite presentation, because a $K3$ surface is of finite presentation and N is of finite type. Proposition (2.5) immediately implies that $\underline{S}_N$ is locally separated, and Proposition (2.6) implies that any $\varsigma_0 \in \underline{S}_N(k)$ admits a universal formal deformation. By Remark (2.4), such a deformation is effective. Finally, since the parameter space of each formal universal deformation is normal and of fixed dimension, the "vache" condition (4) of [2, 3.4] is automatic [2, 3.9].

It is important to observe that S_N is not separated. This phenomenon was noted by Burns and Rapoport in their study [7] of the period mapping for Kählerian $K3$ surfaces. We shall need a mild generalization of their notion of an "elementary transformation," which I prefer to call, with David Morrison [16], an "elementary modification." Let $f\colon X \longrightarrow T$ be a family of $K3$-surfaces, with T a normal local scheme. The relative Picard scheme $\underline{Pic}_{X/T}$ is unramified and essentially proper over T, and if γ is a path

[10] from a geometric generic point $\bar{\tau}$ of T to a closed point t, we get a specializaion map:

$$\gamma^*: Pic(X_{\bar{\tau}}) \longrightarrow Pic(X_t).$$

Note that $Im(\gamma^*)$ does not depend on the choice of γ.

(2.8) **Proposition.** *Suppose that Γ is an irreducible curve on X_t with $(\Gamma \mid \Gamma) = -2$ and suppose that the image of Γ in $Pic(X_t)$ is not in $Im(\gamma^*)$. Then there exist a $K3$ surface Y/T and isomorphisms: θ_τ: $X_\tau \longrightarrow Y_\tau$, θ_t: $X_t \longrightarrow Y_t$, such that the following diagram commutes:*

$$\text{(2.8.1)} \qquad \begin{array}{ccccc} Pic(Y_{\bar{\tau}}) & \xrightarrow{\gamma^*} & Pic(Y_0) & \xrightarrow{Pic(\theta_0)} & Pic(X_0) \\ {\scriptstyle Pic(\theta_\tau)}\downarrow & & & \swarrow {\scriptstyle r_\Gamma} & \\ Pic(X_{\bar{\tau}}) & \xrightarrow{\gamma^*} & Pic(X_0) & & \end{array}$$

where r_Γ is the reflection determined by Γ.

Proof. This is well known, but since I was not satisfied with any of the proofs in the literature, it seemed worthwhile to give a sketch. Let $T' \subseteq T$ be the maximal closed subscheme to which $O_X(\Gamma)$ prolongs, a proper Cartier divisor in T. Let L' be the prolongation of $O_X(\Gamma)$ to $X' =: X\,|_{T'}$; it is easy to see that $H^0(X', L')$ is a free $O_{T'}$-module of rank one. Choose a basis s' for $H^0(X', L')$; then s' defines a relative Cartier divisor Γ' over Γ', which is in fact a $\mathbb{P}^1$-bundle. Let J be the ideal of Γ' in X, so that J/J^2 is a vector bundle of rank 2 on Γ'. It is easy to check from the maximality of Γ' and standard obstruction theory that

$$J/J^2 \cong O_{\Gamma'}(1) \oplus O_{\Gamma'}(1)$$

(in the obvious notation).

Now let $\beta; \tilde{X} \to X$ be the blowing up of X along the closed subscheme Γ' and let $\tilde{D} =: \beta^{-1}(\Gamma')$; so $\tilde{D} \cong \Gamma' \times_{T'} \Gamma'$ (with pr_1 corresponding to the map $\beta\,|_{\tilde{D}}$: $\tilde{D} \to \Gamma'$). By results of [3], we can blow down $\tilde{D}$ along the fibers of pr_2 and obtain an algebraic space Y which will be smooth over T. It is clear that there is a natural isomorphism between X and Y over $T - T'$ and in particular over τ. Moreover, the closed fiber $\tilde{X}_t$ of $\tilde{X}$ has two irreducible components: $\tilde{X}_t = X'_t + \tilde{D}_{t'}$ where X'_t is the strict transform of X_t and $\tilde{D}_t$ the closed fiber of $\tilde{D}$. Clearly $X'_t \to X_t$ and $X'_t \to Y_t$ are isomorphisms; this defines θ_t.

To check the commutativity of (2.8) it certainly suffices to prove the analogue in étale cohomology. The scheme $\tilde{X} \subseteq X \times Y$ induces a cohomological correspondence $X \to Y$; evidently X is cohomologically equivalent to $\tilde{X}_t = X'_t + \tilde{D}_t$. The correspondence induced by X'_t is θ_t; using θ_t to identify X_t and Y_t, the correspondence induced by $\tilde{D}_t = \Gamma_t \times \Gamma_t$ sends x to $(x \cdot \Gamma)\Gamma$. This proves the commutativity.

(2.9) *Remark.* If (X, η) is an N-marked $K3$ surface, then Y/T inherits a unique N-structure from the isomorphism θ_T. Moreover, Y/T is automatically projective, and it is easy to see that Y/T is unique up to unique isomorphism.

(2.10) *Remark.*. If X is a supersingular $K3$ surface, its universal formal k-deformation $\mathcal{X}/S$ has a smooth parameter space of dimension 20, and Artin showed [4] that there is a natural formal subscheme S_∞ of S defined by the condition that the formal Brauer group of $\mathcal{X}\mid_S$ has infinite height. Let $S_{ss} =: (S_\infty)_{red}$; we explained in [19] that each irreducible component Σ of S_{ss} is smooth. Moreover, the generic fiber of $\mathcal{X}\mid_\Sigma$ is supersingular and has $\sigma_0 = 10$, and the image of the specialization map $NS(\mathcal{X}_\Sigma) \to NS(X)$ corresponds to a totally isotropic subspace Λ of $N_1(X)$ of rank $10 - \sigma_0(X)$. It follows from (2.6) that all such Λ's actually occur. This answers a question of L. Illusie.

In fact, the Corollary to the Torelli Theorem II″ tells us more. Endow X with the tautological $NS(X)$-marking $id_{NS(X)}$ and let

$$G_K = \{g \in Aut\, NS(X) : gK = K\}.$$

It is not difficult to verify (using the spin group) that G_K acts transitively on the set of all Λ's above. If now Λ and Λ' are two such subspaces we can write $\Lambda' = r\alpha g$, where $\alpha \in Aut(X)$ and r is a product of reflections through irreducible -2 curves. This implies that one can pass from one branch of S_{ss} to any other by applying automorphisms of X and elementary modifications. [Question: do automorphisms suffice?]

§ 3. The Period Mapping

Since S_N is smooth, we can follow the procedure of [19] to define a period map π: $S_N \longrightarrow \mathcal{M}_N$. The N-marking η: $N \longrightarrow \underline{Pic}_{X/S}$ can be composed with the natural Chern class map $\underline{Pic}_{X/S} \to H^2_{DR}(X/S)$, and we obtain a map $N \otimes \mathcal{O}_S \to \underline{H}^2_{DR}(X/S)$. This map is compatible with the connections $id \otimes d_S$ on $N_0 \otimes \mathcal{O}_S$ and the Gauss-Manin connection on $H^2_{DR}(X/S)$. Its kernel ker therefore $= F_S(K)$ for some $K \subseteq N \otimes \mathcal{O}_S$, and it can be checked pointwise that $K \subseteq N_0 \otimes \mathcal{O}_S$ is characteristic.

We must show that π factors naturally through a map $\tilde{\pi}: S_N \longrightarrow \mathcal{P}_N$. If σ is a point of S_N, not necessarily closed, we have a map $\eta: N \longrightarrow \underline{Pic}_{X(\sigma)/k(\sigma)}(k(\sigma))$. For a suitable finite extension $k(\sigma')$ of $k(\sigma)$, $\underline{Pic}_{X(\sigma)/k(\sigma)}(k(\sigma')) = Pic(X(\sigma'))$, and there will be an ample line bundle λ' on $X(\sigma')$. If $k(\sigma'')$ is an extension of $k(\sigma')$, then λ is ample on $X(\sigma')$ iff it becomes ample on $X(\sigma'')$, so we get a well-defined subset of

$$\underline{Pic}_{X(\sigma)/k(\sigma)}(k(\sigma'))$$

corresponding to the ample bundles. This defines a connected component $\alpha(\sigma)$ of $V_{N(\sigma)}$.

If s is a specializatin of σ, and if $\lambda \in N(s) \cong \underline{Pic}_{X/S}(s)$ is ample, it is immediate that $p\lambda \in N(\sigma) \cong \underline{Pic}_{X/S}(\sigma)$ is also ample. Thus, α defines an ample cone for K, in the sense of (1.15).

It is clear that Theorem III (and hence also all our other theorems) follows from:

Theorem III′. *The map $\tilde{\pi}: S_N \longrightarrow \mathcal{P}_N$ is an isomorphism.*

Proof. We may work locally on $\mathcal{P}_N$. Let $\Lambda \subseteq N_0$ be totally isotropic of rank $\sigma_0(N) - 1$, let α be a connected component of C_{N_Λ}, and let $\mathcal{P}' \subseteq \mathcal{P}$ be the corresponding open subset. Set $S' = \tilde{\pi}^{-1}(\mathcal{P}') \subseteq S_N$ and let $\pi': S' \longrightarrow \mathcal{P}'$ be the restriction of $\tilde{\pi}$. As we saw in the proof of (1.16), the sets of the form $\mathcal{P}'$ cover $\mathcal{P}_N$, so it suffices to prove that π' is an isomorhism.

STEP 1: The map $\pi': S' \longrightarrow \mathcal{P}'$ is separated.

Let T be the spectrum of a DVR with generic point τ and closed point t. Suppose (X, η) and (X', η') are two N-marked $K3$ surfaces over T which map to the same point in $\mathcal{P}'(T)$ and which become isomorphic over τ; we must prove that they are isomorphic over T. We may replace T by its strict Henselization. We have then a commutative diagram:

$$\begin{array}{ccccc} Pic\,X(t) & & \overset{sp}{\hookleftarrow} & & Pic\,X(\tau) \\ \wr\!\wr & \nwarrow^{\eta_t} & & \nearrow^{\eta_\tau} & \wr\!\wr \\ N_{K(t)} & \leftarrow & N & \rightarrow & N_{K(\tau)} \\ \wr\!\wr & \swarrow_{\eta'_t} & & \searrow_{\eta'_\tau} & \wr\!\wr \\ Pic\,X(t') & & \underset{sp}{\hookleftarrow} & & Pic\,X'(\tau) \end{array} \quad \theta_t^*$$

where $\theta_\tau\colon X(\tau) \longrightarrow X'(\tau)$ is an isomorphism compatible with the N-structures. Choose $\lambda \in N \cap \alpha$; then since α corresponds to the ample cone in $X(t)$ and in $X'(t)$, $\eta_t(\lambda)$ is ample on $X(t)$ and $\eta'_t(\lambda)$ is ample on $X'(t)$. Then $\eta(\lambda)$ is relatively ample on X and $\eta'(\lambda)$ is relatively ample on X', and $\theta_\tau^*(\eta'_\tau(\lambda)) = \eta_\tau(\lambda)$. In other words, θ_τ is an isomorphism between the generic fibers of two polarized $K3$ surfaces over T and its compatible with the polarizations. It follows from the theorem of Matsusaka–Mumford [14] that θ_τ prolongs to an isomorphism $\theta\colon X \longrightarrow X'$. Since this θ is automatically compatible with the N-marking, the claim is proved.

STEP 2: π' is essentially proper (i.e., satisfies the valuative criterion for properness, with DVR's as test rings).

(3.1) **Theorem** (Rudakov and Shafarevich) [23]: *Let* $R =: k[[t]]$, $K = k((t))$, *and let* X *be a* $K3$ *surface over* K *whose geometric fiber has* $\rho = 22$. *Then there exist a finite extension* R'/R, *a smooth surface* X'/R' *and an isomorphism* $X'_{K'} \cong X_{K'}$.

Now to prove Step 2, let τ be the generic point of $T =: Spec\,R$ and suppose (X, η) is an N-marked $K3$ surface over τ. Let $\langle X, \eta\rangle$ denote the corresponding point of $\underline{S}_N(\tau)$, which we assume to lie in $S'(\tau)$. Assume also that there is a point (K, α) of $\underline{P}'(T)$ mapping to $\bar{\pi}(X, \eta) \in \underline{P}'(\tau)$. We must prove that there is a $\langle Y, \varsigma\rangle \in S'(T)$ such that $\bar{\pi}\langle Y, \varsigma\rangle = (K, \alpha)$ and whose image in $\underline{S}'(\tau)$ is $\langle X, \eta\rangle$. Since $\underline{S}'$ is representable, we may assume that R is complete, and we may replace R by a finite extension if necessary. By (3.1), we may assume that there is a smooth $K3$ surface Y over T extending X. Since T is strictly Henselian, $\underline{Pic}_{Y/T}(T) \cong Pic(Y)$, and since T is a DVR, $Pic(Y) \cong Pic(Y_\tau) = Pic(X)$. Thus the N-marking on X extends (uniquely) to an N-marking ς of Y.

If $(K_Y, \alpha_Y) =: \pi'(Y, \varsigma)$, we have $K_Y(\tau) = K(\tau)$ and $\alpha_Y(\tau) = \alpha(\tau)$. Since $\mathcal{M}'$ is separated, $K_Y = K$. We will see that after applying a finite sequence of elementary modifications (2.8) to Y/T, we can also arrange to have $\alpha_Y = \alpha$. It will be enough to find Y so that $\alpha_Y(t) = \alpha(t)$; note that $\alpha_Y(t)$ and $\alpha(t)$ both are contained in $\alpha(\tau)$.

Choose an ample h on $Y(t)$ and a λ in $\alpha(t) \cap N$. Since $\varsigma_t^{-1}(h)$ and λ both belong to $\alpha(\tau)$, $h \cdot \varsigma_t(\lambda) > 0$, and we can argue by induction on $h \cdot \varsigma_t(\lambda)$. If $\alpha_Y(t) \neq \alpha(t)$, $\varsigma_t(\lambda)$ is not ample on $Y(t)$, and there exists an irreducible curve E on $Y(t)$ with $\varsigma_t(\lambda) \cdot E < 0$. By Hodge index, $E \cdot E < 0$, so $E \cdot E = -2$. Since $\varsigma_\tau(\lambda)$ is ample on $Y(\tau)$, E does not extend to $Y(\tau)$, and we can perform an elementary modification centered at E to obtain a new N-marked $K3$ surface (Y', ς'), as explained in (2.8). If $\theta_t \colon Y'_t \longrightarrow Y_t$ is the isomorphism at the closed fiber, $\varsigma'_t = \theta_t^* \circ r_E \circ \varsigma_t$. Let $h' =: \theta_t^*(h)$, which is ample on Y'_t. We have

$$\begin{aligned} 0 \leq (h' \mid \varsigma_t(\lambda)) &= (h \mid r_E \varsigma_t(\lambda)) \\ &= (h \mid \varsigma_t(\lambda)) + (h \mid E)(E \cdot \varsigma_t(\lambda)) < (h \mid \varsigma_t(\lambda)). \end{aligned}$$

The process must therefore eventually terminate.

STEP 3: π' is étale.

This follows form the fact that $\pi \colon S'_N \longrightarrow \mathcal{M}_N$ is étale, proved as in [19].

STEP 4: Every irreducible component of $\mathcal{P}'$ contains a point ς such that $(\pi')^{-1}(\varsigma)$ consists of a single point.

(3.2) **Lemma.** *If $\tilde{N}$ is a $K3$ lattice with $\sigma_0(\tilde{N}) = 1$, $\mathcal{M}_{\tilde{N}}(k)$ consists of exactly two points, interchanged by the action of $Aut(\tilde{N})$.*

Proof. Since N_0 is a 2-dimensional $\mathbb{F}_p$-vector space equipped with a nondegenerate nonsplit quadratic form, $\mathcal{M}_{\tilde{N}}(k) = \{$isotropic subspaces of $N_0 \otimes k\} = 2$ points. These two points are interchanged by an element of N with determinant -1. Let $\bar{g}_0$ be such an element. Since

$$N \otimes \mathbb{Z}_p \cong T_0 \oplus T_1$$

and $T_0 \otimes \mathbb{F}_p = N_0$, $T_1 \otimes \mathbb{F}_p = N_1$, we can find an element $g_p \in Aut\,(N \otimes \mathbb{Z}_p)$ mapping to $\bar{g}_0$, and we can choose g_p to have determinant one. Let K_p be the subgroup of elements of $Aut(N \otimes \mathbb{Z}_p)$ acting as the identity on N_0 and with determinant one. Since the spinorial norm map $K_p \rightarrow \mathbb{Z}_p^*/\mathbb{Z}_p^{*2}$ is easily seen to be surjective (cf. [19, 7.3]), we may choose g_p to have spinorial norm 1, and hence to lift to the group $\tilde{K}_p =: \{g \in \mathrm{Spin}_{N\otimes\mathbb{Q}}(\mathbb{Q}_p) : \rho(g) \in K_p\}$. By the strong approximation theorem, we can find a $g \in \mathrm{Spin}_{N\otimes\mathbb{Q}}(\mathbb{Q})$ such that $\rho(g) \in Aut(N)$ and $\rho(\tilde{g})$ has the same image as g_p in $Aut(N_0)$.

Now to prove Step 4, recall from the end of the proof of (1.16) that $\mathcal{P}' \to \mathcal{M}^\Lambda$ is bijective. There are precisely two points K of $\mathcal{M}^\Lambda(k)$ such that $\sigma_0(N_K) = 1$, viz., the two points K containing Λ. (They correspond precisely to the two points of $\mathcal{M}_{N_\Lambda}(k)$.) By [19, §4], these parameterize the two connected components of $\mathcal{M}^\Lambda$.

Let X be a superspecial Kummer surface. Then

$$\sigma_0(NS(X)) = 1 = \sigma_0(N_\Lambda),$$

and we can choose an isomorphism $N_\Lambda \rightrightarrows NS(X)$. This defines an N-marking η: $N \longrightarrow NS(X)$, and $K_{(X,\eta)}$ contains Λ. If g is an automorphism of $NS(X)$, g acts on

$$\mathcal{M}_{NS(X)_0}(k) \cong \mathcal{M}_{N_\Lambda}(k) \cong \{K \in \mathcal{M}^\Lambda(k) : K \supseteq \Lambda\}.$$

Lemma (3.2) tells us that this action is transitive. Clearly $(X, g\eta)$ is another N-marked superspecial Kummer surface and $K_{(X,g\eta)} = gK_{(X,\eta)}$. Thus, both points of $\{K \in \mathcal{M}^\Lambda(k) : K \supseteq \Lambda\}$ occur as the periods of an N-marked superspecial Kummer surface.

We must also show that all possible ample cones occur. If $\delta \in N_\Lambda$ and $\delta \cdot \delta = -2$, then the image of δ in $N_\Lambda \otimes \mathbb{F}_p$ is orthogonal to

$$Ker(N_\Lambda \otimes k \to H^2_{DR}(X/k)).$$

Therefore r_δ leaves Ker and hence K invariant. This implies that the group $\pm R_{N_\Lambda}$ stabilizes K, and since it acts transitively on C_{N_Λ} by (1.10), we may choose η so that $(X, \eta) = (K, \alpha)$.

Finally, we recall from [19] that Theorem II is proved for supersingular Kummer surfaces. More precisely, we proved in [19] that theorem II is true if $\sigma_0(NS(X)) \leq 2$. Clearly this implies that the fibers of $\tilde{\pi}$ over points with $\sigma_0 \leq 2$ have cardinality ≤ 1.

We are now almost ready to finish the proof of Theorem III. Indeed, if we knew that π' were of finite type, we could conclude that it is proper (Step 2), hence finite étale (Step 3), and of degree 1 (Step 4). Unfortunately there seems to be no *a priori* way to prove this, so instead we have to make an indirect argument. Clearly the following lemma suffices.

(3.3) **Lemma.** *Let $f\colon S \longrightarrow T$ be a morphism of k-algebraic spaces. Assume that f satisfies the valuative criterion for properness (with respect to DVR's), that f is étale, and that T is irreducible. Then:*

(3.3.1) *If $f^{-1}(t)$ is empty for some closed point t of T, S is empty.*

(3.3.2) *If $f^{-1}(t)$ is a single point for some closed point t of T, f is an isomorphism.*

Proof. The question is local for the étale topology on T, so we may assume that T is an affine scheme. Since S is locally of finite type over k and f is étale, the image $f(S)$ of S in T is open.

We first prove (3.3.1). Assume $f(S)$ is not empty; then it is dense, and there exists a one-dimensional irreducible subscheme T' of T containing t and meeting $f(S)$. If $f'\colon S' \longrightarrow T'$ is obtained by base change, it satisfies the same hypotheses as f.

Let τ' be the generic point of T'; $S_{\tau'} =: f^{-1}(\tau')$ is étale over the field $k(\tau')$. Then $S_{\tau'}$ is a union of the spectra of finite separable extensions of $k(\tau')$, and since $S_{\tau'} \neq \emptyset$ there is such an extension, call it $k(\tau'')$, and a map $\tau'' \to S_{\tau'}$ over τ'. Let R'' be the normalization of the local ring of T' at t in $k(\tau'')$ and let $T'' =: Spec\ R''$. We have a map $T'' \to T'$, which by the valuative criterion lifts to a map $T'' \to S$. The image of the closed point of T'' lies in $f^{-1}(t)$, a contradiction.

Now we can prove (3.3.2). I claim that the diagonal Δ of $S \times_T S$ is closed. This question is local on $S \times_T S$; choosing étale open subsets U and V of S, of finite type over k, we must prove that $U \times_S V$ is closed in $U \times_T V$. This follows from the valuative criterion. On the other hand, since S is étale over T, Δ is also open in $S \times_T S$, and so its complement W is open and closed in $S \times_T S$. It is still locally of finite type, étale, and satisfies the valuative criterion. Since $f^{-1}(t)$ is a single point, the fiber of W over t is empty. By the first part of the lemma, W is empty. This implies that f is universally injective and hence an open immersion. By the valuative criterion again, $S \to T$ is also closed in T (now that we know S is of finite type).

(3.4) *Remark.* In fact, it can be checked that the natural $\mathbb{F}_p$-scheme structure on $\mathcal{P}_N$ makes it the fine moduli space classifying N-marked $K3$ surfaces over $\mathbb{F}_p$. Note that $\mathcal{P}_N$ has no $\mathbb{F}_p$-rational points — there exist no supersingular $K3$ surfaces over $\mathbb{F}_p$, all of whose cycles over $\overline{\mathbb{F}}_p$ are defined over $\mathbb{F}_p$.

§ 4. The Cup Product Pairing on H^2_{cris} (with S. Bloch)

In this section we explain how recent results of Bloch, Gabber, and Kato can be used to obtain information about the bilinear form on the crystalline H^2 of a $K3$ surface. We first answer a question of Deligne and Shafarevich by showing that in characteristic 2, this form is *even*. According to Rudakov and Shafarevich (who proved it for strongly elliptic surfaces), this allows one to formulate and prove a crystalline Torelli theorem for supersingular $K3$ surfaces in characteristic 2 [24].

Let k be an algebraically closed field of characteristic 2 and let X be a smooth proper surface over k. We have a canonical injection:

$$H^2_{cris}(X/W) \otimes k \to H^2_{DR}(X/k)$$

compatible with cup-product, so to decide if $(\ \mid\ )$ on $H^2_{cris}(X/W)$ is even it is enough to study $(\ \mid\ )$ on $H^2_{DR}(X/k)$. It is clear from Poincaré duality that there is a unique element w_X of $H^2_{DR}(X/k)$ such that $(w_X \mid x)^2 = (x \mid x)$ for all $x \in H^2_{DR}(X/k)$.

(4.1) **Problem** (Shafarevich): *Is w_X equal to the first Chern class of Ω^2_X?*

We shall prove that this is indeed the case for $K3$ surfaces, for which c_1 is of course zero. In fact, we shall establish a somewhat more general result which suggests that the problem is likely to have an affirmative answer in general.

Let X/k be a smooth proper scheme, and recall that we have two spectral sequences abutting to $H^*_{DR}(X/k)$, called the Hodge and conjugate spectral sequences. These are compatible with Poincaré duality, and if F is either of the two filtrations and X/k has dimension n, then the pairing

$$H^{n-i}_{DR}(X/k) \times H^{n+i}_{DR}(X/k) \to k$$

identifies $F^j H^{n-i}_{DR}(X/k)$ with $Ann[F^{n-j} H^{n+i}_{DR}(X/k)]$. Furthermore, we have the inverse Cartier isomorphism:

$$C^{-1} : H^q(X, \Omega^p_{X/k}) \to H^q(X, \underline{H}^p(\Omega^{\cdot}_{X/k})).$$

If the Hodge to DeRham spectral sequence of X/k degenerates at E_1, the conjugate spectral degenerates at E_2, and C^{-1} can be regarded as a (semi-linear) isomorphism: $gr^p_{F_{Hodge}} H^{q+p}_{DR}(X/k) \to gr^q_{F_{con}} H^{q+p}_{DR}(X/k)$.

(4.2) **Lemma.** *Assume that the Hodge to DeRham spectral sequence of the smooth proper surface X/k degenerates at E_1. Then*

$$w_X \in F^1_{Hodge} \cap F^1_{con}.$$

If π_{Hodge} (respectively π_{con}) is the projection:

$$F^1 H^2_{DR}(X/k) \to gr^1_F H^2_{DR}(X/k), \quad C^{-1}(\pi_{Hodge}(w_X)) = \pi_{con}(w_X).$$

Proof. To prove that $w_X \in F^1_{Hodge}$, it suffices to check that $(w_X \mid x) = 0$ for all $x \in F^2_{Hodge}$, i.e., that $(x \cdot \mid x) = 0$ for $x \in F^2_{Hodge}$, which is trivial. The same argument works for F^1_{con}. (Note that we have not yet used the degeneration of the spectral sequences.) Since the pairing induced on $gr^1_{F_{con}} H^2_{DR}(X/k)$ is perfect and C^{-1} is an isomorphism, the last claim will be proved if we show that

$$\left(C^{-1}\pi_{Hodge}(w_X) \mid C^{-1}\pi_{Hodge}(x)\right) = \left(\pi_{con} w_X \mid C^{-1}\pi_{Hodge}(x)\right)$$

for all $x \in F^1_{Hodge}$. The left side of this is

$$\left(\pi_{Hodge}(w_X) \mid \pi_{Hodge}(x)\right)^2 = (w_X \mid x)^2 = (x \mid x).$$

The right side squared is

$$\left(C^{-1}\pi_{Hodge}(x) \mid C^{-1}\pi_{Hodge}(x)\right) = \left(\pi_{Hodge}(x) \mid \pi_{Hodge}(x)\right)^2 = (x \mid x)^2,$$

so the result is verified.

Recall that X/k is "strictly ordinary" iff for all q and all i, $H^q(X, B^i) = 0$, where $B^i =: Im[d: \Omega^{i-1}_{X/k} \to \Omega^i_{X/k}]$. It is known (and immediate) that this is equivalent to two conditions: the Hodge to DeRham spectral sequence should degenerate at E_1, and the filtrations $F^{\cdot}_{Hodge}$ and $F^{\cdot}_{con}$ should be transverse: $F^{i}_{Hodge}H^{N}_{DR} \cap F^{N-i+1}_{con}H^{N}_{DR} = 0$.

(4.3) *Remark.* If X/k is a smooth surface whose Hodge to DeRham spectral sequence degenerates at E_1, it is strictly ordinary iff the absolute Frobenius endomorphism induces isomorphisms on $H^i(X, \mathcal{O}_X)$ for $i = 0, 1, 2$.

If X/k is strictly ordinary. let $H^{q,p} =: F^{p}_{Hodge} \cap F^{q}_{con}$. Then the natural maps:

$$H^{q,p} \to gr^{p}_{F_{Hodge}} H^{q+p}_{DR}, \quad H^{q,p} \to gr^{q}_{con} H^{q+p}_{DR}, \quad \text{and } \bigoplus_{q+p=n} H^{q,p} \to H_{DR}$$

are isomorphisms. Moreover, the Cartier isomorphism

$$C^{-1}: gr^{q}_{F_{Hodge}} H^{q+p}_{DR} \longrightarrow gr^{q}_{F_{con}} H^{q+p}_{DR}$$

can be transported back to $H^{p,q}$, inducing a semi-linear automorphism ϕ of $H^{q,p}$, and the natural map $\{H^{q,p}\}^{\phi} \otimes_{\mathbb{F}_p} k \to H^{q,p}$ is an isomorphism.

(4.4) **Proposition.** *Assume that the smooth proper surface X/k is strictly ordinary and that there is a smooth proper scheme $\mathcal{X}$ over W lifting X. Then (4.1) is true for X.*

Proof. Let $\mathcal{X}_{\overline{\eta}}$ be a geometric generic fiber of $\mathcal{X}/W$. Without loss of generality we may assume that $k(\overline{\eta}) = \mathbb{C}$, so the classical Wu formula [15] tells us that $(a \mid a) = (c_1 \mid a)$ for all $a \in H^2_{\acute{e}t}(\mathcal{X}_{\overline{\eta}}, \mathbb{F}_2)$, where c_1 is the first Chern class of $\mathcal{X}_{\overline{\eta}}$ in $H^2_{et}(\mathcal{X}_{\overline{\eta}}, \mathbb{F}_2)$.

Let $j\colon \mathcal{X}_{\overline{\eta}} \longrightarrow \mathcal{X}$ be the natural map, let $i\colon X \longrightarrow \mathcal{X}$ be the natural inclusion, and let $\psi^q =: i^* R^q j_* \mathbb{F}_2$ (in the étale topology). We have the well-known spectral sequence: $E_2^{pq} = H^p(X, \psi^q) \Rightarrow H^n(\mathcal{X}_{\overline{\eta}}, \mathbb{F}_2)$; we shall denote the corresponding filtration on $H^n(\mathcal{X}_{\overline{\eta}}, \mathbb{F}_2)$ by $F^{\cdot}_{BGK}$. There is a natural multiplicative structure on the spectral sequence, and since $H^i(X, \psi^q) = 0$ for $i > 2$, it is clear that F^1_{BGK} annihilates F^2_{BGK}. Also, one can verify immediately that the map $Pic(\mathcal{X}) \to Pic(\mathcal{X}_{\overline{\eta}}) \to H^2(\mathcal{X}_{\overline{\eta}}, \mathbb{F}_2)$ factors through $F^1_{BGK} H^2_{et}(\mathcal{X}_{\overline{\eta}}, \mathbb{F}_2)$. According to [6], there are natural isomorphisms $\beta\colon H^p(X, \psi^q) \longrightarrow \{H^{p,q}\}^{\phi}$, and according to Bloch these maps are compatible with cup product, formation of Chern classes, and the trace map. Note that $gr^1_{F_{BGK}} H^2_{et}(\mathcal{X}_{\overline{\eta}}, \mathbb{F}_2) \cong \{H^{1,1}\}^{\phi}$, and $w_X \in \{H^{1,1}\}^{\phi}$ by (4.2).

It is clear that these properties imply that $w_X = c_1(X)$. Indeed, since β is an isomorphism and the form on $\{H^{1,1}\}^{\phi}$ is perfect, it suffices to prove that if ω is the canonical sheaf of $\mathcal{X}$, then for any $a \in F^1_{BGK}/F^2_{BGK}$, $(\beta(a) \mid c_{DR}(\omega)) = (\beta(a) \mid w_X)$ (with obvious abuse of notation). But

$$\begin{aligned}(\beta(a) \mid c_{DR}(\omega)) &= (\beta(a) \mid \beta c_{et}(\omega)) = (a \mid c_{et}(\omega)) = (a \mid a) \\ &= (\beta(a) \mid \beta(a)) = (\beta(a) \mid w_X)^2 = (\beta(a) \mid w_X)\end{aligned}$$

since $\beta(a)$ and w_X are fixed by ϕ.

(4.5) *Remark.* In the situation of (4.4), it can happen that $c_{et}(\omega) \neq 0$ in $H^2_{et}(\mathcal{X}_{\overline{\eta}}, \mathbb{F}_2)$ but $c_1(X) = 0$ in $H^2_{DR}(X/k)$. (When this happens, $c_{et}(\omega)$ lies in $F^2_{BGK}H^2_{et}(\mathcal{X}_{\overline{\eta}}, \mathbb{F}_2)$.) For example, let E be an elliptic curve over W with a W-valued point t of order 2, and let $\mathcal{X}$ by the quotient of $E \times E$ by the involution $(x, y) \mapsto (x + t, -y)$. In this case ω is a nontrivial 2-torsion element of $Pic(\mathcal{X})$, not divisible by 2 in $Pic(\mathcal{X}_{\overline{\eta}})$, and $\omega \mid_X$ is trivial. As a matter of fact, $\Omega^1_{\mathcal{X}} \mid_X$ is trivial, and this makes it easy to verify directly that $(a \mid a) = 0$ for all $a \in H^2_{DR}(X/k)$.

(4.6) **Lemma.** *Assume $f\colon X \longrightarrow S$ is a smooth proper morphism of relative dimension 2, and assume also that the sheaves $\mathbf{R}^i_* f_* \Omega^{\cdot}_{X/S}$ are locally free on S. If (4.1) is true over a dense open set of S, it is true over every point of S.*

Proof. We may assume without loss of generality that S is reduced. The map $x \to (x \mid x)$ defines an $\mathcal{O}_S$-linear map: $\sigma : \mathbf{R}^2 f_* \Omega^{\cdot}_{X|S} \to F_{S*}\mathcal{O}_S$, where F_S is the absolute Frobenius endomorhisms of S. We can also regard σ as a map $\tilde{\sigma} : F^*_S \mathbf{R}^2 f_* \Omega^{\cdot}_{X/S} \to \mathcal{O}_S$, and by Poincaré duality there is a unique section $\tilde{w}_{X/S}$ of $F^*_S \mathbf{R}^2 f_* \Omega^{\cdot}_{X/S}$ such that $\tilde{\sigma}(\tilde{x}) = (\tilde{w}_{X/S} \mid \tilde{x})$ for every section $\tilde{x}$ of $F^*_S \mathbf{R}^2 f_* \Omega^{\cdot}_{X/S}$. Since $\mathbf{R}^2 f_* \Omega^{\cdot}_{X/S}$ is locally free, its formation commutes with base change, and so it suffices to prove that $\tilde{w}_{X/S} = \tilde{c}_1$, where $\tilde{c}_1$ is the pull-back of the first Chern class of X/S to $F^*_S \mathbf{R}^2 f_* \Omega^{\cdot}_{X/S}$. Again using the fact that $\mathbf{R}^2 f_* \Omega^{\cdot}_{X/S}$ is locally free, we see that it suffices to verify equality at every point in a dense open set of S.

(4.7) **Theorem.** *If X_0/k is a $K3$ surface and char $k = 2$, $(a \mid a) = 0$ for $a \in H^2_{DR}(X_0/k)$.*

Proof. Let L_0 be an ample line bundle on X, which we may assume is not a square. Let $X/S, L$ be a versal k-deformation of (X_0, L_0); it is known that $\dim S = 19$ and that for all points lying in a dense open set of S, $X(s)$ is ordinary [19]. It is clear that an ordinary $K3$ surface is strictly ordinary and that the family X/S satisfies the local freeness assumption of (4.6). Moreover, an ordinary $K3$ surface can be lifted to W [19], so the theorem follows from (4.4) and (4.6).

(4.8) *Remark.* It is clear that a hypersurface in $\mathbb{P}^3$ is ordinary iff its Hesse-Witt matrix is invertible. Since this is true for a generic hypersurface [12], we can also apply the above reasoning to conclude that (4.1) holds for any hypersurface in $\mathbb{P}^3$

(4.9) Theorem. *Let k be an algebraically closed field of characteristic $p > 2$, and let X/k be a smooth projective surface without torsion in crystalline cohomology. Assume that X is a $K3$ surface or that X is strictly ordinary and liftable. Then the crystallline discriminant $\left(\frac{X}{p}\right) = \left(\frac{-1}{p}\right)^{\beta-1}$ where B is the second Betti number of X.*

Proof. We know that this formula is invariant under deformation [20], so by the arguments above, the result for $K3$ surfaces will follow for the result for ordinary sufaces.

Choose a lifting $\mathcal{X}$ of X, to W or to some ramified extension V of W, with fraction field K. Since X/k has no torsion in its crystalline cohomology, $\dim H^i_{DR}(X/k) = \operatorname{rank} H^i_{cris}(X/W)$, and by [5], this is the same as the rank of $H^i_{DR}(\mathcal{X}_K/K)$. The BGK spectral sequence then shows that

$$\begin{aligned}\dim H^i_{et}(\mathcal{X}_{\overline{\eta}}, \mathbb{F}_p) &\leq \sum_{a+b=i} \dim H^{a,b}(X/k)\\ &= \dim_k H^i_{DR}(X/k) = \dim_K H^i_{DR}(\mathcal{X}_K/K)\\ &= \operatorname{rank} H^i et(\mathcal{X}_{\overline{\eta}}, \mathbb{Z}_p) \leq \dim H^i_{et}(\mathcal{X}_{\overline{\eta}}, \mathbb{F}_p).\end{aligned}$$

It follows that there is no torsion in $H^*_{et}(\mathcal{X}_{\overline{\eta}}, \mathbb{Z}_p)$ and that the spectral sequence degenerates at E_2.

Since the form on $H^2_{et}(\mathcal{X}_{\overline{\eta}}, \mathbb{F}_p)$ is perfect and $p > 2$, the totally isotropic subspace F^2_{BGK} admits a hyperbolic complement U, and we see that the discriminant of $H^2(\mathcal{X}_{\overline{\eta}}, \mathbb{F}_p) = (-1)^{rk\, U} \operatorname{disc} gr^1_{F_{BGK}} H^2(\mathcal{X}_{\overline{\eta}}, \mathbb{F}_p)$ in $\mathbb{F}_p^*/\mathbb{F}_p^{*2}$. The Hodge index theorem implies that the discriminant of $H^2(\mathcal{X}^{an}_{\overline{\eta}}, \mathbb{Z}) =$

$(-1)^{\beta_2-1}$, hence of $H^2(\mathcal{X}_{\bar\eta}, \mathbb{F}_p)$, is also $(-1)^{\beta_2-1}$. According to [20, 3.2],[2]

$$\begin{aligned}\operatorname{disc} H^2_{cris}(X/W) &= (-1)^{rk\,F^2} \operatorname{disc}\{H^{1,1}\}^{\phi} \\ &= (-1)^{rk\,F^2} \operatorname{disc} gr^1_{F_{BGK}} H^2(\mathcal{X}_{\bar\eta}, \mathbb{F}_p) \\ &= (-1)^{rk\,F^2}(-1)^{rk\,U}(-1)^{\beta-1} = (-1)^{\beta-1}.\end{aligned}$$

(4.10) *Remark.* By using the $\mathbb{Q}_2$-adic form of the BGK spectral sequence, one can extend the above result to the case $p = 2$. In this case, one can conclude that $\operatorname{disc}(H) = (-1)^{\beta_2-1}$ in $\mathbb{Q}_2^*/\mathbb{Q}_2^{*2}$.

References

[1] Artin, M. "The implicit function theorem in algebraic geometry," *Algebraic Geometry* (Int. Coll. at Tasta, 1968), Oxford University Press, 1969, pp. 13–34.

[2] —— "Algebrization of formal moduli, I," in *Global Analysis* (Spencer and Iyanaga, eds.), Princeton Math. Studies No. 29, Princeton University Press, 1969.

[3] ——"Algebrization of formal moduli, II," *Annals of Math.* 91, 1970, pp. 88–135.

[4] —— "Supersingular $K3$ surfaces," *Ann. Sci. Ecole Norm. Sup.* 7, 1974, pp. 543–568.

[5] Berthelot, P. and Ogus, A. "F-crystals and DeRham cohomology, I," (to appear in *Inv. Math.*)

[6] Bloch, S., Gaber, O., and Kato, K. (to appear).

[7] Burns, D. and Rapoport M. "On the Torelli problem for Kählerian $K3$ surfaces," *Ann. Sci. Ecole Norm. Sup.* 8, 1975, pp. 235–273.

[8] Deligne, P. *Séminaire de Géométrie Algébrique* 4 1/2, Springer-Verlag LNM 569, 1977.

[9] Grothendieck, A. "Éléments de Géométrie Algébrique, III," *Publ. Math. de l'I.H.E.S.* 11, 1961.

[10] —— *Séminaire de Géométrie Algébrique 1*, Springer-Verlag LNM 224, 1971.

[11] Katz, N. "Slope filtration of F-crystals," *Astérisque* 63, 1979, pp. 113–164.

[2] There is a misprint in the formula [20, 3.2]: a minus sign appears as an "=". Even worse, this sign is inexplicably missing in [20, 3.3].

[12] Koblitz, N. "*P*-adic variation of the zeta function ..." *Comp. Math.* 31, 1975, pp. 119–218.

[13] Lang, W. and Nygaard, N. "A short proof of Rudakov Shafarevich theorem," *Math. Ann.* 251, 1980, pp. 171–173.

[14] Matsusaka, T. and Mumford, D. "Two fundamental theorems on deformations of polarized algebraic varieties," *Amer. J. Math.* 86, 1964, pp. 668–

[15] Milnor, J. and Stasheff, J. *Characteristic Classes*, Annals of Math. Study No. 76, Princeton University Press, 1974.

[16] Morrison, D. "Some remarks on the moduli of $K3$ surfaces," (to appear).

[17] Nygaard, N. "The Tate conjecture for ordinary $K3$ surfaces over finite fields," (to appear in Inv. Math.)

[18] —— "The Torelli theorem for ordinary $K3$ surfaces over finite fields," (this volume).

[19] Ogus, A. "Supersingular $K3$ crystals," *Asterisque* 64, 1979, pp. 3–86.

[20] —— "Hodge cycles and crystalline cohomology," in *Hodge Cycles, Motives and Shimura Varieties*, Springer-Verlag LNM 900, 1982, pp. 357–414.

[21] Piatetski-Shapiro, I. and Shafarevich, I. "A Torelli theorem for surfaces of type $K3$," *Math. USSR-Izv.* 5, 1971, pp. 547–588.

[22] Rudakov, A. and Shafarevich, I. "Inseparable morphisms of algebraic surfaces," *Math. USSR-Izv.* 10, 1976, pp. 1205–1237.

[23] —— "Degeneration of $K3$ surfaces over fields of finite characteristic," (preprint).

[24] —— "$K3$ surfaces over fields of finite characteristic," in *Modern Problems in Math.*, Vol. 18, 1981, Itogi Nauk Tekh., Acad. of Sci. USSR [Russian].

[25] Serre, J. P. *A Course in Arithmetic*, Springer-Verlag, 1973.

[26] Shioda, T. "On unirationality of supersingular varieties," *Math. Ann.* 225, 1977, pp. 155-159.

Received May 27, 1982.

Professor Arthur Ogus
Department of Mathematics
University of California
Berkeley, California 94720

Decomposition of Toric Morphisms

Miles Reid

To I.R. Shafarevich on his 60th birthday

§0. Statement and Discussion of Results

(0.1) This paper applies the ideas of Mori theory [4] to toric varieties. Let X be a projective toric variety (over any field) constructed from a simplicial fan F. The cone of effective 1-cycles $NE(X)$ is polyhedral (1.7), spanned by the 1-strata $l_w \subset X$; the condition that a 1-stratum l_w gives an extremal ray $R = \mathbb{Q}_+ l_w$ of $NE(X)$ has a nice interpretation (2.10) in terms of the geometry of F around the wall w.

If R is an extremal ray of $NE(X)$ then there is a toric morphism (2.4) $\varphi_R\colon X \longrightarrow Y$ with connected fibres, which is an elementary contraction in the sense of Mori theory: for a curve $C \subset X$,

$$\varphi_R C = pt. \in Y \Leftrightarrow C \in R.$$

Let

$$\begin{array}{ccc} A & \longrightarrow & B \\ \cap & & \cap \\ \varphi_R\colon X & \longrightarrow & Y \end{array}$$

be the loci where φ_R is not an isomorphism; A, B are irreducible. I say that R is an extremal ray of type (a, b) if $\dim A = a$, $\dim B = b$; clearly, $n \geq a > b \geq 0$. Any fibre of $\varphi_{R|A}\colon A \longrightarrow B$ is a weighted projective space of dimension $(a - b)$.

If $a = n$ or $n - 1$, the fan F_R of Y is again simplicial, but if $a \leq n - 2$ (when φ_R is an isomorphism in codimension 1), this is not the case. It turns out (3.4) that there is then another natural simplicial subdivision F_1 of F_R leading to an elementary transformation

$$\begin{array}{ccccc} X & & \xrightarrow{\psi_R} & & X_1 \\ & \searrow^{\varphi_R} & & \swarrow_{\varphi_{-R}} & \\ & & Y & & \end{array}$$

where φ_{-R} is also an elementary contraction which is an isomorphism in codimension 1, contracting an extremal ray of $NE(X_1)$ which is $-R$ when

$N_1(X)$ and $N_1(X_1)$ are identified. In particular, if $K_X R < 0$, then K_{X_1} is relatively ample for φ_{-R}.

The main result is the following.

(0.2) **Theorem.** *Let A be a projective toric variety, and let $f: V \longrightarrow A$ be a projective birational toric morphism, where V has $\mathbb{Q}$-factorial terminal singularities (see (1.8-10) for definitions; for example, V could be non-singular).*

Then f is a composite of toric maps:

$$V = V_0 - \to V_1 - \to \cdots - \to V_k = S \xrightarrow{\varphi} X$$

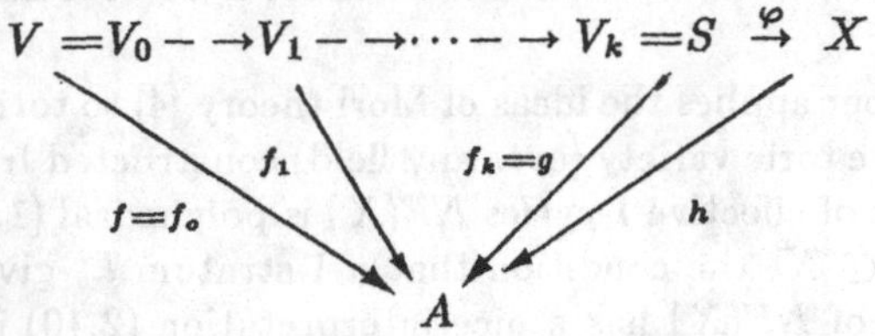

where

(1) "*X is a relative canonical model*": X has canonical singularities, $h: X \longrightarrow A$ is a projective birational morphism, and K_X is relatively ample for h;

(2) "*S is a relative minimal model*": S has $\mathbb{Q}$-factorial terminal singularities, $g: S \longrightarrow A$ is a projective morphism, and K_S is relatively nef for g (that is $K_S C \geq 0$ for every curve C contracted by g; it is known that this is equivalent to saying $K_S = \varphi^* K_X$ in the sense of $\mathbb{Q}$-Cartier divisors);

(3) *if A has $\mathbb{Q}$-factorial terminal singularities, then $S = X = A$;*

(4) *each V_i has $\mathbb{Q}$-factorial terminal singularities, and $f_i: V_i \longrightarrow A$ is a projective morphism, for $i = 0, \ldots, k$;*

(5) *for $i = 0, \ldots, k-1$, the rational map $V_i - \to V_{i+1}$ is either an elementary contraction φ_{R_i} (with R_i of type $(n-1, b)$), or an elementary transformation ψ_{R_i} (with R_i of type (a, b), $a \leq n-2$) specified by an extremal ray R_i of $NE(V_i/A)$ with $K_X R_i < 0$.*

(0.3) *Remarks.* The hypothesis that A is complete is not essential; it can be reduced to the projective case, or possibly eliminated by a careful (and rather tedious) rephrasing of the arguments of §§1 − 3. The projectivity hypothesis on f is needed in order for the statement of (0.2) to make sense, since without projectivity the cone $NE(V/A)$ will usually not have any extremal rays.

(0.4) The motivation for this result comes from Mori theory [4], and from a conjectural theory of minimal models and classification of 3-folds (see [6], §4). However, the result here is an easy exercise in manipulating fans and their decompositions – once the key idea of extremal rays has been recognised. While I believe that (0.2) holds without the toric hypothesis in dimension 3, over $\mathbf{C}$, it is certain that a wealth of new geometrical phenomena and technical difficulties are involved.

(0.5) In the surface case, $f: V \longrightarrow A$ is a resolution of a normal surface; S is the minimal resolution, and X the relative canonical model; the sequence $V_0 \rightarrow V_1 \rightarrow \cdots \rightarrow V_k = S$ is the sequence of contractions of (-1)-curves which gives the minimal model S in terms of any non-singular model.

In the higher-dimensional case, (0.2) provides a way of breaking down a birational toric morphism of varieties with $\mathbf{Q}$-factorial terminal singularities (in particular, non-singular) into elementary steps, which are however more complicated than blow-ups of smooth centres. Danilov [2] has given a more detailed treatment of the 3-fold case, showing that f can be further factorised as a composite of smooth blow-ups and blow-downs. I would like to thank Danilov for several invaluable conversations on toric topics.

(0.6) It has been a wonderful privilege for me to have studied and collaborated over the last 10 years with the great Moscow school of algebraic geometry founded by Professor I. R. Shafarevich, and it is an enormous pleasure to dedicate this paper to him.

§1. Toric Terminology and the Theorem on the Cone

(1.1) Toric ideas will be mainly as in [1]; let M be a lattice of rank n (that is, $M = \mathbf{Z}^n$). The elements $m \in M$ correspond to monomials

$$x^m \in k(M) = k(x_1, \ldots, x_n).$$

A cone σ in the dual vector space $N_{\mathbf{R}}$ (a cone is always assumed to be given by finitely many inequalities $m_i \geq 0$, with $m_i \in M_{\mathbf{Q}}$) gives rise to a finitely generated ring $k[\check{\sigma} \cap M]$, and hence to an affine variety X_σ.

In [1], §5, a *fan* F is defined as a finite collection F of cones $\sigma \subset N_{\mathbf{R}}$ such that

(i) σ and $\tau \in F \Rightarrow \sigma \cap \tau \in F$ and $\sigma \cap \tau$ is a face of σ, τ;

(ii) $\sigma \in F$ and τ a face of $\sigma \Rightarrow \tau \in F$;

(iii) every $\sigma \in F$ is a cone "with vertex", that is, σ does not contain any vector subspace of $N_{\mathbf{R}}$; equivalently, $\sigma \cap -\sigma = 0$.

(1.2) *Definition.* Let $U \subset N_{\mathbf{R}}$ be a rational vector subspace; a collection of cones G^* is a *degenerate fan with vertex* U if it satisfies the conditions for a fan, with (iii) replaced by

(iii′) for every $\sigma \in G^*$, $\sigma \cap -\sigma = U$.

Obviously this coincides with the usual notion of a fan $G = G^*/U$ in the quotient space $N^0_{\mathbf{R}} = N_{\mathbf{R}}/U$, and I will usually use the letters G^* and G in this way. Condition (iii′) is required in order that the affine ring corresponding to every cone $\sigma \in G^*$, $k[\check{\sigma} \cap M]$, has $k(U^{\perp} \cap M)$ as its quotient field; the corresponding X_σ overlap birationally, and can be glued together as usual to form a variety X_{G^*}, as in [1], §5.

(1.3) As in [1], §5, F is a *subdivision* of G^* if $|F| = |G^*|$, and every cone $\sigma \in G^*$ is a union of cones of F; in this situation there arises a proper morphism $f\colon X_F \longrightarrow X_{G^*}$, called a toric morphism. I will normally be assuming that the variety X_F is complete, which means that $|F| = N_{\mathbf{R}}$.

(1.4) *1-cycles and $(n-1)$-cycles.*

Let $f\colon X = X_F \to X_{G^*} = A$ be a proper toric morphism; a 1-cycle of X/A is a formal sum $\sum a_i C_i$, with C_i complete curves in fibres of f, and $a_i \in \mathbf{Q}$.

There is a pairing

$$\operatorname{Pic} X \times (\text{1-cycles of } X/A) \to \mathbf{Q}$$

defined by $(L, C) \mapsto \deg_C L$, extended by bilinearity. Define

$$N^1(X/A) = (\operatorname{Pic} X \otimes \mathbf{Q})/\equiv$$

and

$$N_1(X/A) = (\text{1-cycles of } X/A)/\equiv,$$

where numerical equivalence $\equiv$ is by definition the smallest equivalence relation which makes N^1 and N_1 into dual vector spaces.

Inside $N_1(X/A)$ there is a distinguished cone of effective 1-cycles,

$$NE(X/A) = \left\{ Z \mid Z \equiv \sum a_i C_i \text{ with } a_i \geq 0 \right\} \subset N_1(X/A).$$

(1.5) If $A = pt.$, write $NE(X)$ for $NE(X/A)$; assuming X is complete, $N_1(X/A) \subset N_1(X)$, and $N^1(X/A)$ is the corresponding quotient of $N^1(X)$. If A is projective, with ample $H \in \operatorname{Pic} A$, then $NE(X/A)$ is a face of $NE(X)$:

$$\text{for } Z \in NE(X), \quad (f^*H)Z \geq 0, \quad \text{and}$$

$$Z \in NE(X/A) \Leftrightarrow (f^*H)Z = 0.$$

(1.6) Proposition. *Let X be a complete toric variety; then every effective 1-cycle is rationally equivalent (and hence numerically equivalent) to an effective sum of 1-strata.*

Proof. Every irreducible curve $C \subset X$ belongs to a unique smallest closed stratum, say $Y \subset X$; if $\dim Y > 1$ then a suitable 1-parameter subgroup $\mathbf{G}_m$ of the big torus of X will move C inside Y. Considering the $\mathbf{G}_m$-action as a rational map $\mathbf{A}^1 \times Y - \rightarrow Y$ and resolving the indeterminacy leads to a diagram

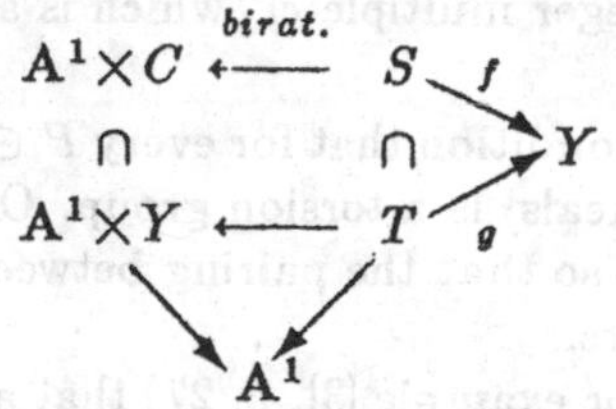

then $C = f(S_1) \sim f(S_0)$, and on the other hand $f(S_0) \subset g(T_0)$, which is a union of substrata of Y. Induction on $\dim Y$ completes the proof.

The degree of a line bundle L on the fibres of a flat family of curves is constant, so that rational equivalence as usual implies numerical equivalence.

(1.7) Corollary (the Theorem on the Cone). *Let $f\colon X \longrightarrow A$ be a proper toric morphism as in (1.3), and assume that X is proper. Then*

$$NE(X/A) = \sum \mathbf{Q}_+ l_w,$$

where l_w runs through the 1-dimensional strata of X in fibres of f, that is w runs through $F^{(n-1)} \setminus G^{*(n-1)}$.

Assuming that X is projective, the cone $NE(X)$ does not contain any vector subspace of $N_1(X)$, and it follows that $NE(X/A)$ is spanned as a convex cone by a finite number of *extremal rays* R of the form $\mathbf{Q}_+ l_w$, where

$$\text{if } Z_1 + Z_2 \in R, \text{ with } Z_1, Z_2 \in NE(X/A), \text{ then } Z_1, Z_2 \in R.$$

By (1.5), the extremal rays of $NE(X/A)$ are just the extremal rays of $NE(X)$ belonging to $N_1(X/A)$.

If $R = \mathbf{Q}_+ l_w$ is an extremal ray of $NE(X/A)$, l_w will be called an *extremal* 1-*stratum*, and $w \in F^{(n-1)}$ an *extremal wall*; I will write $w \in R$ in this case, and use the same letter R for the subset

$$\{w \in F^{(n-1)} \setminus G^{*(n-1)} \mid l_w \in R\}.$$

The condition that w is an extremal wall will turn out in (2.10) to have a nice interpretation in terms of the geometry of F near w.

(1.8) *Definition.* A normal variety X is $\mathbf{Q}$-*factorial* if every prime divisor $\Gamma \subset X$ has a positive integer multiple $c\Gamma$ which is a Cartier divisor.

This is really the local condition that for every $P \in X$, $Cl\, \mathcal{O}_{X,P}$ (divisorial ideals modulo principal ideals) is a torsion group. On a $\mathbf{Q}$-factorial variety $Cl\, X \otimes \mathbf{Q} = \operatorname{Pic} X \otimes \mathbf{Q}$, so that the pairing between divisors and 1-cycles extends to all Weil divisors.

(1.9) It is easy to see (for example [3], p. 27) that a toric variety $X = X_F$ is $\mathbf{Q}$-factorial if and only if the fan F is *simplicial*, that is, is made up of simplicial cones (I learned this from J. Fine). The singularities of X are then at worst Abelian quotient singularities (X is what Satake and Bailey call a V-manifold, and Steenbrink and Danilov call a quasi-smooth variety).

(1.10) *Traditional abuse of notation.* If F is a fan in $N_{\mathbf{R}}$, and $e \in F^{(1)}$ is a ray of F, then e contains a unique primitive lattice element $e \in N$, which I continue to denote e.

(1.11) *Definition.* A cone with vertex $\sigma \in N_{\mathbf{R}}$ is *terminal* (respectively *canonical*) if it satisfies the conditions (i) and (ii) (respectively (i) and (ii′)):

(i) the 1-faces $e_1, \ldots, e_k$ of σ are contained in an affine hyperplane $H\colon (m(n) = 1) \subset N_{\mathbf{R}}$, for some $m \in M_{\mathbf{Q}}$;

(ii) there are no other elements of the lattice N in the part of σ under or on H, that is $N \cap \sigma \cap (m(n) \leq 1) = \{0, e_1, \ldots, e_k\}$;

(ii′) there are no other lattice elements in the part of σ under H, that is $N \cap \sigma \cap (m(n) < 1) = \{0\}$.

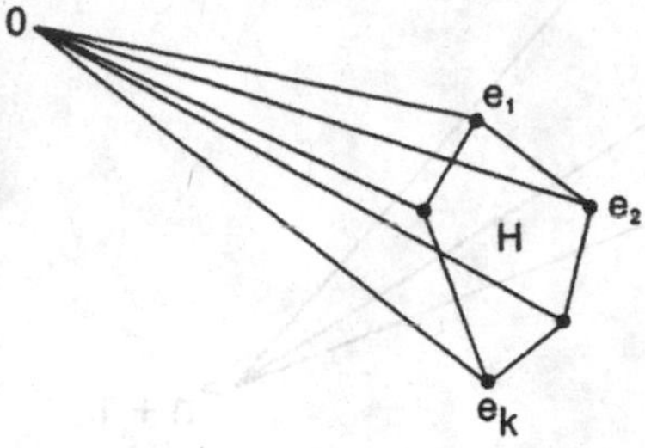

a terminal cone;

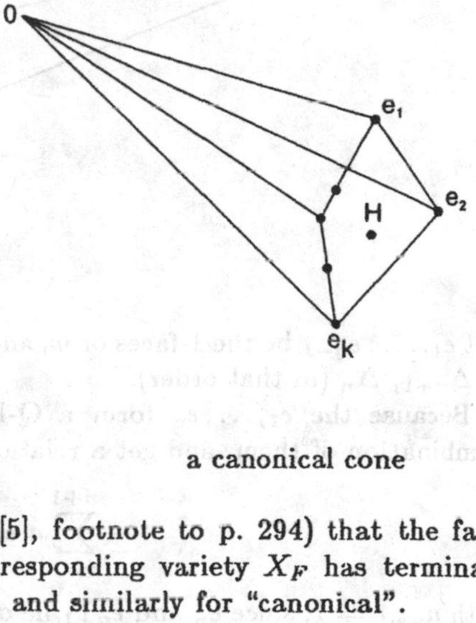

a canonical cone

(1.12) It is easy to see (compare [5], footnote to p. 294) that the fan F is terminal if and only if the corresponding variety X_F has terminal singularities, in the sense of [6], (0.2), and similarly for "canonical".

§2. The Contraction of Extremal Rays

(2.1) I now introduce some notation which will be used throughout §§2–4. Let F be a simplicial fan in $N_{\mathbf{R}}$, and $w \in F^{(n-1)}$ a wall; the corresponding closed stratum $l_w \subset X_F$ is a complete $\mathbf{P}^1$ if and only if w is an internal wall of F, that is w separates two cones Δ_n and $\Delta_{n+1} \in F^{(n)}$ (corresponding to the 0-strata 0 and $\infty \in \mathbf{P}^1$).

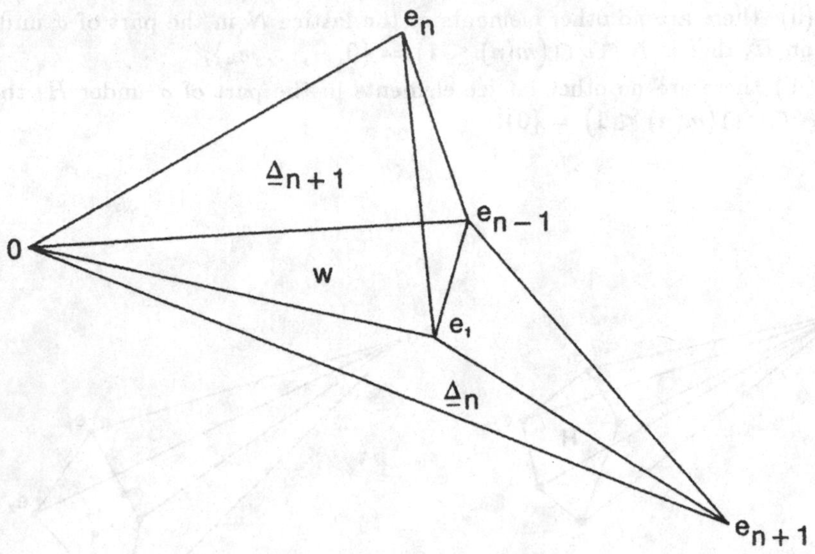

Let $e_1, \dots, e_{n-1}$ be the 1-faces of w, and let e_n, e_{n+1} be the opposite 1-faces of Δ_{n+1}, Δ_n (in that order).

Because the $e_1, \dots, e_n$ form a Q-basis of N, I can write e_{n+1} as a combination of them, and get a relation

$$\sum_{i=1}^{n+1} a_i e_i = 0 \qquad (*)$$

with $a_{n+1} = 1$; since e_n and e_{n+1} lie on opposite sides of w, it follows that $a_n > 0$.

(2.2) The sign of the a_i describes the shape of $\Delta_n \cup \Delta_{n+1}$ along the face of w opposite e_i: obviously $a_i < 0$ (respectively $= 0$, > 0) if and only if $\Delta_n \cup \Delta_{n+1}$ is strictly convex (respectively flat, concave) along

$$\langle e_1, \dots, \hat{e}_i, \dots, e_{n-1} \rangle.$$

By reordering the e_i, I can assume that

$$a_i \begin{cases} < 0 & \text{for } 1 \le i \le \alpha, \\ = 0 & \text{for } \alpha + 1 \le i \le \beta, \\ > 0 & \text{for } \beta + 1 \le i \le n+1; \end{cases}$$

here $0 \le \alpha \le \beta \le n-1$.

(2.3) Introduce the notation

$$\Delta = \Delta(w) = \Delta_n + \Delta_{n+1} = \langle e_1, \dots, e_{n+1} \rangle$$

and

$$U = U(w) = \langle e_1, \dots, e_\alpha, e_{\beta+1}, \dots, e_{n+1} \rangle.$$

Lemma. (i)

$$\Delta(w) = \bigcup_{j=\beta+1}^{n+1} \Delta_j = \bigcup_{j=1}^{\alpha} \Delta_j$$

(a union of n-simplices meeting along (n − 1)-faces), where

$$\Delta_j = \langle e_1, \dots, \hat{e}_j, \dots, e_{n+1} \rangle \qquad (\text{for } j \leq \alpha \text{ or } j \geq \beta + 1);$$

(ii) *$U(w)$ is a face of $\Delta(w)$, and is a vector subspace of $N_{\mathbb{R}}$ if and only if $\alpha = 0$.*

Proof. Let $x = \sum x_i e_i \in \Delta(w)$, where $x_i \geq 0$; then if j is the index $j \geq \beta + 1$ for which x_j/a_j is minimum, using (*) gives

$$x = \sum x_i e_i - (x_j/a_j) \sum a_i e_i \in \langle e_1, \dots, \hat{e}_j, \dots, e_{n+1} \rangle.$$

This proves the first equality, and the second is similar. For (ii) note that $\Delta(w)$ is in the region ($x_i \geq 0$ for $i = \alpha + 1, \dots, \beta$), and $U(w)$ is its intersection with the subspace ($x_i = 0$ for $i = \alpha + 1, \dots, \beta$). Finally, if $\alpha = 0$, (*) gives e_{n+1} as a linear combination of $e_{\beta+1}, \dots, e_n$ with strictly negative coefficients, so that

$$U(w) = \langle e_{\beta+1}, \dots, e_{n+1} \rangle = \langle \pm e_{\beta+1}, \dots, \pm e_n \rangle.$$

(2.4) **Theorem.** *Let F be a complete simplicial fan, R an extremal ray of $NE(X_F)$; then α, β and $U_R = U(w)$ are independent of $w \in R$, and*

$$F_R^{*(n-1)} = F^{(n-1)} \setminus R$$

form the walls of a fan F_R^*, degenerate with vertex U if $\alpha = 0$. Every $\Delta \in F_R^{*(n)} \setminus F^{(n)}$ is of the form $\Delta(w)$ for $w \in R$, and $\Delta_j \in F^{(n)}$ for $j = \beta + 1, \dots, n + 1$.

Suppose that $\alpha = 0$; then $F_R = F^*_R/U$ is simplicial.

Suppose that $\alpha = 1$; then $F_R = F^*_R$ is simplicial, and if F is terminal, with $K_X R < 0$, F_R is also terminal.

(2.5) **Corollary.** *There is a toric morphism $\varphi_R \colon X \longrightarrow Y = X_{F^*_R}$ which is an elementary contraction in the sense of Mori theory: $\varphi_{R*}\mathcal{O}_X = \mathcal{O}_Y$, and $\varphi_R C = pt. \Leftrightarrow C \in R$.*

(2.6) *Furthermore, let*

$$\begin{array}{rcc} & A & \rightarrow & B \\ & \cap & & \cap \\ \varphi_R : & X & \rightarrow & Y \end{array}$$

be the loci on which φ_R is not an isomorphism; A and B are the irreducible closed toric strata corresponding to the cones $\langle e_1, \dots, e_\alpha \rangle \in F^{(\alpha)}$ and $U_R \in F^{*(n-\beta)}_R$ respectively; $\dim A = a = n - \alpha$, $\dim B = b = \beta$, and $\varphi_{R|A} \colon A \longrightarrow B$ is a flat morphism, all of whose fibres are weighted projective spaces of dimension $(a - b)$.

Remark. It is not difficult to see that over a neighbourhood of any point $b \in B$, there are Abelian branched covers $\tilde{A}$ and $\tilde{B}$ such that $\varphi_{R|\tilde{A}} \colon \tilde{A} \longrightarrow \tilde{B}$ is a $\mathbf{P}^{n-\alpha-\beta}$-fibre bundle.

The proof of (2.4) will take the rest of §2.

(2.7) **Proposition.** *Let F be a complete simplicial fan, and let $w \in F^{(n-1)}$; for $e \in F^{(1)}$ let D_e denote the corresponding prime divisor of X_F. Then*

(i) $D_e l_w = 0$ *if* $e \neq e_1, \dots, e_{n+1}$;

(ii) $D_{e_i} l_w > 0$ *for* $i = n, n+1$;

(iii) $D_{e_i} l_w = a_i D_{e_{n+1}} l_w$ *for* $i = 1, \dots, n$.

Proof. The correspondence between cones of F and closed toric strata of X_F reverses inclusion, so that for $e \neq e_1, \dots, e_{n+1}$, D_e is disjoint from l_w, proving (i); D_{e_n} and $D_{e_{n+1}}$ are $(n-1)$-strata meeting l_w in its point strata, giving (ii). The remaining D_{e_i} contain l_w, so that to compute $D_{e_i} l_w$ I move D_{e_i} by linear equivalence: let $e^*_i \in M^*_{\mathbf{Q}}$ be the dual basis element, and let $m \in M$ be an integer multiple $m = ce^*_i$; the monomial x^m gives the linear equivalence

$$0 \sim \mathrm{Div}(x^m) = \sum ce_i^*(e)D_e = c(D_{e_i} + e_i^*(e_{n+1})D_{e_{n+1}} + \cdots),$$

where $\cdots$ are divisors disjoint from l_w. Since $a_i = -e_i^*(e_{n+1})$, the result follows.

The actual value of the intersection numbers can be expressed in terms of the volumes of the cones w, Δ_n, Δ_{n+1} in the sublattices of N which they span (compare [1], (10.9)), but fortunately I am only concerned with the signs. A similar remark applies to positive rational multiples appearing later.

(2.8) For smooth toric varieties, the numbers $D_{e_i}l_w$ represent the degrees of the normal bundle to l_w in X_F, and one would expect l_w to move in the directions corresponding to positive values of $D_{e_i}l_w$; the following trick achieves the same end in the Q-factorial toric case: let $\sigma \in F^{(n-2)}$ correspond to a complete surface stratum $S_\sigma \subset X_F$, and let

$$w_1, \ldots, w_s \in F^{(n-1)}$$

be the walls of F containing σ, giving the curve strata $l_{w_1}, \ldots, l_{w_s} \subset S_\sigma$:

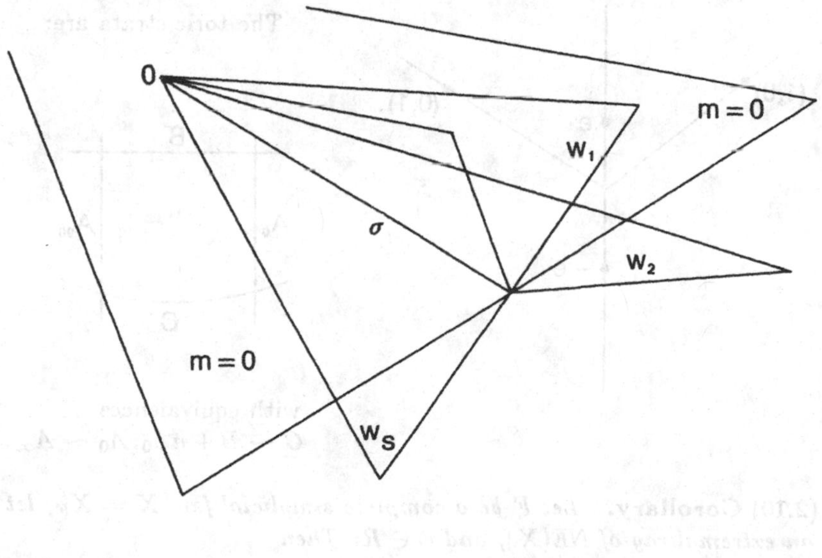

for $m \in M$, the condition $m \in (\sigma - \sigma)^{\perp}$ means that S_σ is not contained in the locus of zeros and poles of x^m, so that x^m restricts down to a rational monomial $x^m \mid_{S_\sigma} \in k(S_\sigma)$, the divisor of which provides a linear equivalence between the curves $l_{w_1}, \dots, l_{w_s} \subset S_\sigma$:

$$\mathrm{Div}(x^m \mid_{S_\sigma}) = \sum c_{m,w_i} l_{w_i} \sim 0,$$

with $c_{m,w_i} \in \mathbf{Z}$. By construction of S_σ, the monomial x^m has a zero (respectively is invertible, has a pole) along $l_{w_i} \subset S_\sigma$ if and only if w_i is contained in the $m > 0$ (respectively $m = 0$, $m < 0$) region of $N_{\mathbf{R}}$.

Proposition. *c_{m,w_i} has the same sign as m evaluated at an interior point of w_i.*

(2.9) *Example.* For $n > 0$ the surface $\mathbf{F}_n$ is given by the following fan in $N = \mathbf{Z}^2 + \mathbf{Z} \cdot e$, where $e = (\frac{1}{n}, \frac{1}{n})$:

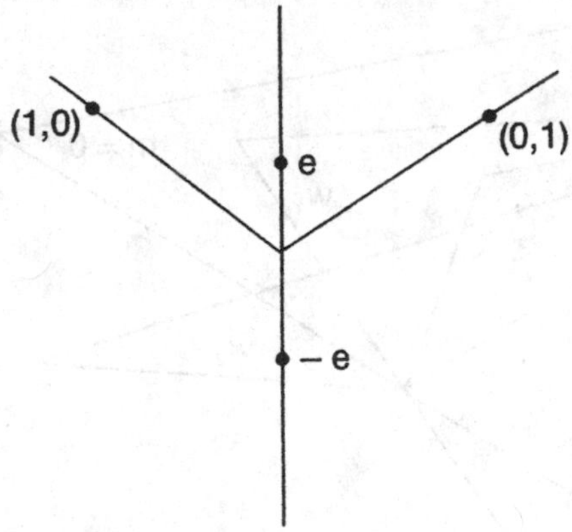

The toric strata are:

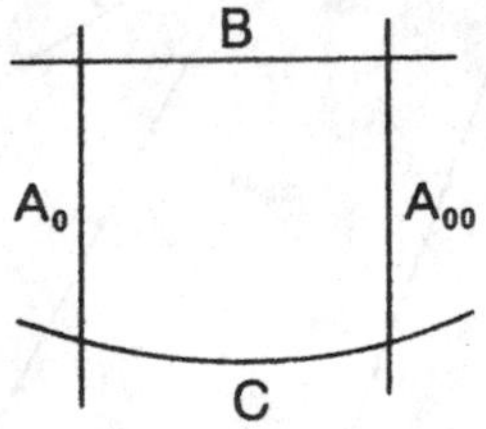

with equivalences $C \sim B + nA_0, A_0 \sim A_\infty$.

(2.10) **Corollary.** *Let F be a complete simplicial fan, $X = X_F$; let R be an extremal ray of $NE(X)$, and $w \in R$. Then*

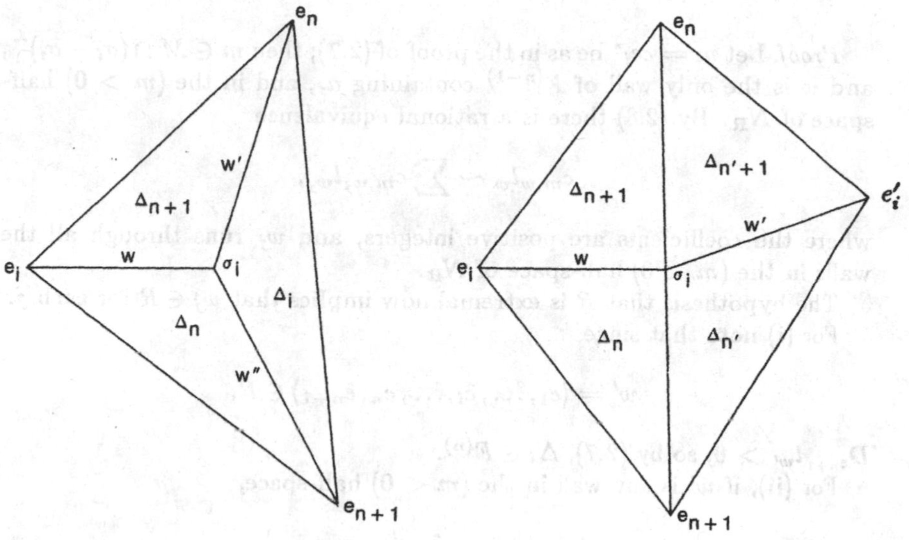

(the origin 0 is in the face σ_i, and is either behind or on the page);

(i) *for every i such that $D_{e_i} l_w > 0$,*

$$\Delta_i = \langle e_1, \ldots, \hat{e}_i, \ldots, e_{n+1} \rangle \in F^{(n)},$$

and the two walls

$$\left. \begin{aligned} w' &= \langle e_1, \ldots, \hat{e}_i, \ldots, e_n, \hat{e}_{n+1} \rangle \\ w'' &= \langle e_1, \ldots, \hat{e}_i, \ldots, \hat{e}_n, e_{n+1} \rangle \end{aligned} \right\} \in R;$$

(ii) *for every i such that $D_{e_i} l_w = 0$, there is a unique $e'_i \in F^{(1)}$ such that the star in F of $\sigma_i = \langle e_1, \ldots, \hat{e}_i, \ldots, e_{n-1} \rangle$ consists of the 4 cones*

$$\Delta_n, \Delta_{n+1}, \Delta'_n, \Delta'_{n+1} \in F^{(n)}$$

and their faces, and the wall

$$w' = \langle e_1, \ldots, \hat{e}_i, e'_i, \ldots, e_{n-1} \rangle \in R.$$

Proof. Let $m = ce_i^*$ be as in the proof of (2.7); then $m \in M \cap (\sigma_i - \sigma_i)^{\perp}$, and w is the only wall of $F^{(n-1)}$ containing a_i, and in the $(m > 0)$ half-space of $N_{\mathbf{R}}$. By (2.8) there is a rational equivalence

$$c_{m,w} l_w \sim \sum c_{m,w_j} l_{w_j},$$

where the coefficients are positive integers, and w_j runs through all the walls in the $(m < 0)$ half-space of $N_{\mathbf{R}}$.

The hypothesis that R is extremal now implies that $w_j \in R$ for each j.

For (i) note that since

$$w' = \langle e_1, \ldots, \hat{e}_i, \ldots, e_n, \hat{e}_{n+1} \rangle \in R,$$

$D_{e_{n+1}} l_{w'} > 0$, so by (2.7), $\Delta_i \in F^{(n)}$.

For (ii), if w' is any wall in the $(m < 0)$ half-space,

$$w' = \langle e_1, \ldots, \hat{e}_i, e'_i, \ldots, e_{n-1} \rangle$$

for some e'_i, and $w' \in R$ implies that $D_{e_n} l_{w'}$, $D_{e_{n+1}} l_{w'} > 0$, so by (2.7), Δ'_n and $\Delta'_{n+1} \in F^{(n)}$. Q.E.D.

(2.11) *Proof of (2.4).* For $w \in R$, (2.7) implies that

$$\{e_1, \ldots, e_\alpha, e_{\beta+1}, \ldots, e_{n+1}\} = \{e \in F^{(1)} \mid D_e l_w \neq 0\}.$$

It follows that for any $w' \in R$,

$$\Delta(w') = \langle e_1, \ldots, e_\alpha, f_{\alpha+1}, \ldots, f_\beta, e_{\beta+1}, \ldots, e_{n+1} \rangle$$

for some $f_i \in F^{(1)}$, and thus $U(w) = U(w')$.

I now claim that $\mathcal{V} = \bigcup_{w \in R} \Delta(w)$ is a convex neighbourhood of U_R in $N_{\mathbf{R}}$; indeed, pick a $w \in R$, a point $Q \in N_{\mathbf{R}}$ near $U(w)$, and a general point $P \in \Delta(w)$ near $U(w)$. Then the line interval PQ is entirely in $\mathcal{V}$: for $P \in \Delta(w)$, and the line interval can only escape from a $\Delta(w')$ by crossing a wall as in (2.10), (ii).

(2.10), (ii) also implies that the boundary of $\mathcal{V}$ is a union of faces in F; since $F^{(n-1)}$ and $F_R^{*(n-1)}$ agree outside $\mathcal{V}$ this implies that F_R^* is a fan.

If $\alpha = 0$, then every $\Delta \in F_R^{*(n)}$ is of the form

$$\Delta = \langle e_1, \ldots, e_\beta, \pm e_{\beta+1}, \ldots, \pm e_n \rangle = (\beta\text{-simplex}) \times U;$$

if $\alpha = 1$, then for every $w \in R$, the relation (2.1), (*) can be written $e_1 = \sum_{i=\beta+1}^{n+1} b_i e_i$, with $b_i > 0$, so that

$$\Delta(w) = \langle e_2, \ldots, e_{n+1} \rangle = (n\text{-simplex}).$$

The last sentence of (2.4) involves K_X, and its proof is deferred to (4.4).

§3. Elementary Transformation in the Case of a Contraction which is an Isomorphism in Codimension 1

(3.1) *Useful tautology.* Let $\psi\colon X \dashrightarrow X_1$ be a rational map which is an isomorphism in codimension 1 between two complete $\mathbf{Q}$-factorial varieties; then proper transform of Weil divisors induces an isomorphism

$$N^1(X) = N^1(X_1),$$

and hence a dual isomorphism $N_1(X) = N_1(X_1)$.

In more detail, the group of Weil divisors of X and X_1, and the notion of linear equivalence $\sim$ on them coincide. Also by hypothesis, $\operatorname{Pic} X$ is contained in

$$\{\text{Weil divisors of } X\}/\sim$$

as a subgroup of finite index, and

$$N^1(X) = (\operatorname{Pic} X/\operatorname{Pic}^0 X) \otimes \mathbf{Q};$$

hence $N^1(X) = N^1(X_1)$.

(3.2) For projective varieties, the cone $NE(X) \subset N_1(X)$ is characteristic for X: if $\psi\colon X \dashrightarrow X_1$ is as in (3.1), with both X and X_1 projective, then it follows from Kleiman's criterion that

$$\psi \text{ is an isomorphism} \Leftrightarrow NE(X) = NE(X_1).$$

(3.3) Let F be a complete simplicial fan, R an extremal ray of $NE(X)$ of type (a, b) with $a \leq n-2$. For any $w \in R$, I use the notation of §2 for the star of w in F. Let F_R be as in (2.4); every $\Delta \in F_R^{(n)} \setminus F^{(n)}$ is of the form $\Delta(w) = \langle e_1, \ldots, e_{n+1} \rangle$, and is not simplicial: the only relation (2.1), (*) between the e_i does not express one as a positive sum of the others. The key remark is that this relation, rewritten as

$$\sum_{i \leq \alpha} (-a_i) e_i = \sum_{i \geq \beta+1} a_i e_i \tag{$*$}$$

(all coefficients are now positive) is now symmetric in the vertices $e_1, \ldots, e_\alpha$ and $e_{\beta+1}, \ldots, e_{n+1}$. The two distinct simplicial subdivisions

$$\Delta = \bigcup_{j \geq \beta+1} \Delta_j = \bigcup_{j \leq \alpha} \Delta_j$$

seen in (2.3) are an expression of this symmetry. The first subdivision is what happens to $\Delta \in F_R^{(n)}$ on passing to F.

(3.4) Theorem. *Under the above hypotheses, let F_1 be the simplicial subdivision of F_R defined by*

$$F_1^{(n)} = F_R^{(n)} \setminus \{\Delta(w) \mid w \in R\} \cup \{\Delta_j \mid w \in R, \quad j = 1, \ldots, \alpha\},$$

or alternatively,

$$F_1^{(n-1)} = F_R^{(n-1)} \cup \{w_{jk} \mid w \in R, \quad j,\, k = 1, \ldots, \alpha\},$$

where $w_{jk} = \langle e_1, \ldots, \hat{e}_j, \hat{e}_k, \ldots, e_{n+1} \rangle$.

Let $\varphi_1 \colon X_1 = X_{F_1} \to Y$ be the toric morphism corresponding to the subdivision F_1 of F_R; then φ_1 is projective, and is an isomorphism in codimension 1. Write ψ for the composite

$$\begin{array}{ccc} X & \xrightarrow{\ \psi\ } & X_1 \\ & {\scriptstyle \varphi_R} \searrow \quad \swarrow {\scriptstyle \varphi_1} & \\ & Y & \end{array}$$

and use ψ to identify $N_1(X)$ and $N_1(X_1)$ as in (3.1). Then $-R$ is an extremal ray of $NE(X_1)$, and $\varphi_1 = \varphi_{-R}$ is the corresponding elementary contraction.

If F is terminal with $K_X R < 0$, then F_1 is also terminal.

(3.5) *Rremark.* Let $e_0 \in N$ be the unique primitive vector in the ray generated by

$$\sum_{i \leq \alpha} (-a_i) e_i = \sum_{i \geq \beta + 1} a_i e_i;$$

(by (2.4), this is independent of $w \in R$). Then barycentric subdivision at e_0 defines a common subdivision G of F and F_1, leading to a diagram

$$\begin{array}{ccc} & \tilde{X} & \\ \swarrow & & \searrow \\ X & & X_1 \\ \searrow & & \swarrow \\ & Y & \end{array}$$

in which all 4 arrows are elementary contractions. The exceptional locus $\tilde{A} \subset \tilde{X}$ is a divisor mapping $\tilde{A} \to B \subset Y$ with fibres a product of weighted projective spaces $q\mathbf{P}^{a-b} \times q\mathbf{P}^{n-a-1}$.

(3.6) *Proof.* Firstly I show that φ_1 is projective; for this it is sufficient to find an $e \in F_1^{(1)} = F^{(1)}$ such that $D_e l_v > 0$ for every new wall v of F_1 ("numerically ample $\Rightarrow$ ample" follows from [1], (6.7)). Let $w \in R$ and $e = e_1$; then $D_{e_1} R < 0$, so that e appears as e_1 for every $w \in R$. Each new wall v of F_1 belongs to $\Delta(w)$, and the argument of (2.7) shows that $D_{e_1} l_v > 0$.

Now $NE(X_1/Y)$ has at least one extremal ray, say $S = \mathbf{Q}_+ l_v$, where v is a new wall of F_1; I claim that $(F_1)_S = F_R$. It was shown in (2.11) that F_R is obtained from F by starting with some $w \in R$, knocking out the interior walls of $\Delta(w)$, and then proceeding to do the same for $\Delta(w')$ which can be reached by passing across a flat wall of $\Delta(w)$ in the sense of (2.10), (ii). Since for $v \in S$, $\Delta(v) = \Delta(w)$, and applying (2.10), (ii) to S, a moment's thought shows that every $\Delta(w)$ is of the form $\Delta(v)$, so that $(F_1)_S = F_R$.

The fact that $R = -S$ follows: if $H \in N^1(X)$ is relatively ample for φ_R then $N^1(X) = \varphi_R^* N^1 Y \oplus \mathbf{Q} H = \varphi_S^* N^1 Y \oplus \mathbf{Q} H$, and $-H$ is ample for φ_S. The final sentence is again deferred to (4.4). Q.E.D.

(3.7) **Conjecture.** *Let $P \in Y$ be an isolated 3-fold singularity, and $\varphi\colon X \longrightarrow Y$ a projective partial resolution which is an isomorphism outside P, and such that $\varphi^{-1} P$ is a union of curves. Suppose also that:*

(i) *the (algebraic) local class group of* $P \in Y$ *is finitely generated and of rank 1, that is* $\dim_{\mathbf{Q}} Cl\, \mathcal{O}_{Y,P} = 1$;

(ii) X *has terminal singularities (see* [6], (0.2));

(iii) $-K_X$ *is relatively ample for* φ.

Then there exists another projective partial resolution $\varphi_1 \colon X_1 \to Y$ *such that* K_{X_1} *is relatively ample for* φ_1.

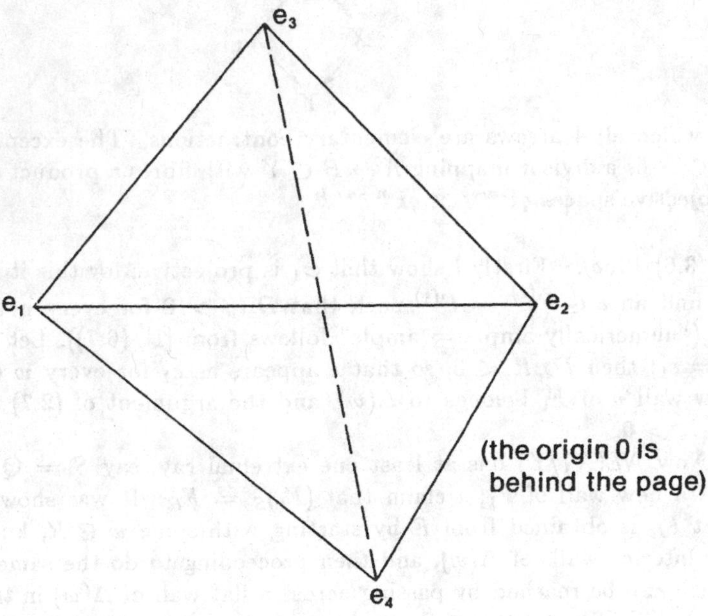

(3.8) *Example.* Let e_1, e_2, e_3 form the usual basis of $N = \mathbf{Z}^3$, and let e_4 be given by

$$ae_1 + (r-a)e_2 = re_3 + e_4,$$

where r and a are coprime integers with $0 < a < r$. The fan $F = \{\Delta_3, \Delta_4\}$ as drawn has only one internal wall, $w = \langle e_1, e_2 \rangle$, and the construction of (3.4) consists of replacing it by the dotted line $v = \langle e_3, e_4 \rangle$.

The contractions φ_R and φ_1 of (3.4) have in this case a copy of $\mathbf{P}^1$ as exceptional locus. The reader can check the following statements:

(1) X has a quotient singularity of type $\frac{1}{r}(a,-a,1)$ (that is, $\mathbf{A}^3/\mu_r$, where μ_r is the cyclic group of rth roots of 1, and

$$\mu_r \quad \epsilon : (x,y,z) \mapsto (\epsilon^a x, \epsilon^{-a} y, \epsilon z);$$

similarly below) at a point of l_w, and is non-singular otherwise;

(2) $K_X l_w = -1/r$;

(3) X_1 has two quotient singularities of type $\frac{1}{a}(r,-r,1)$ and $\frac{1}{r-a}(r,-r,1)$ at the 0-strata of l_v, and is otherwise non-singular;

(4) $K_{X_1} l_v = 1/(a(r-a))$.

The particular case when $a = 1$, $r = 2$ is a celebrated example due to P. Francia.

In the last paragraph of [2] it is shown how to describe this elementary transformation in terms of non-singular blow-ups and blow-downs on a "minimal" resolution; [2] also classifies explicitly the elementary transformations of toric 3-folds involved in Theorem (0.2).

(3.9) *Example (Mori).* Let $e_1, \ldots, e_4$ be the usual basis of $N = \mathbf{Z}^4$, and let e_5 be given by

$$e_1 + e_2 = e_3 + e_4 + e_5.$$

The toric variety corresponding to the cone $\Delta = \langle e_1, \ldots, e_5 \rangle$ is the cone on the Segre embedding of $\mathbf{P}^2 \times \mathbf{P}^1$, or in other words, is the determinental variety given by

$$rk \begin{vmatrix} x_{11} & x_{12} & x_{13} \\ x_{21} & x_{22} & x_{33} \end{vmatrix} \leq 1.$$

The reader can check that in the diagram

$$\begin{array}{ccc} & \tilde{X} & \\ \swarrow & & \searrow \\ X & & X_1 \\ \searrow & & \swarrow \\ & Y & \end{array}$$

obtained by the obvious simplicial decompositions of Δ, $\tilde{X} \to Y$ is the ordinary blow-up, X and X_1 are the contractions of the exceptional divisor

$\mathbf{P}^2 \times \mathbf{P}^1$ onto its two factors; both X and X_1 are smooth; and on the exceptional loci of φ and φ_1,

$$K_{X|\mathbf{P}^2} \cong \mathcal{O}_{\mathbf{P}^2}(-1); \quad K_{X_1|\mathbf{P}^1} \cong \mathcal{O}_{\mathbf{P}^1}(1).$$

It is easy to see that this is the only example of a smooth toric 4-fold, having an extremal ray of type (a, b) with $a \leq n-2$.

§4. Proof of (0.2)

(4.1) So far I haven't said anything about the canonical class; fortunately, there is a nice toric treatment of this too (see [1], (6.6)): if $X = X_F$ is a toric variety then for any $\mathbf{Z}$-basis $m_1, \ldots, m_n$ of M, the canonical differential

$$s = \frac{dx^{m_1}}{x^{m_1}} \wedge \cdots \wedge \frac{dx^{m_n}}{x^{m_n}}$$

is a global basis of $\mathcal{O}_X(K_X + D)$, where D is the complement of the big torus in X as a reduced divisor, that is

$$D = \sum_{e \in F^{(1)}} D_e; \quad K_X = -D.$$

(4.2) For vectors $v_1, \ldots, v_s \in N_{\mathbf{R}}$ write $[v_1, \ldots, v_s]$ for the convex hull

$$[v_1, \ldots, v_s] = \left\{ \sum \lambda_i v_i \mid \lambda_i \geq 0, \text{ and } \sum \lambda_i = 1 \right\};$$

(this is always a compact set, and is not to be confused with the cone $\langle v_1, \ldots, v_s \rangle$).

Definition. If σ is a cone with vertex in $N_{\mathbf{R}}$, write $e_1, \ldots, e_s$ for the 1-faces of σ (recall (1.10)), and define the *shed* of σ by

$$\operatorname{shed} \sigma = [0, e_1, \ldots, e_s];$$

if F is a fan of $N_{\mathbf{R}}$, with $|F| \subset N_{\mathbf{R}}$ a union to n-dimensional cones, define the *shed* of F by

$$\operatorname{shed} F = \bigcup_{\Delta \in F^{(n)}} \operatorname{shed} \Delta;$$

this is obviously a closed polyhedral region of $|F|$, and I define the *roof* of F to be the boundary of shed F in $|F|$.

(4.3) **Proposition.** (i) *Let F be a fan as above, and $X = X_F$; then K_X is a* **Q**-*Cartier divisor if and only if the roof of F is flat over every $\sigma \in F$; that is, for all $\sigma = \langle e_1, \dots, e_s \rangle \in F^{(k)}$, $[e_1, \dots, e_s]$ is a polyhedron contained in a $(k-1)$-dimensional affine subspace of $N_{\mathbf{R}}$.*

(ii) *Suppose that K_X is* **Q**-*Cartier, and let $w \in F^{(n-1)}$ be an interior wall. Then $K_X l_w < 0$ (respectively $= 0$, > 0) if and only if shed F is strictly convex (respectively flat, strictly concave) in a small neighbourhood of a point $P \in (\operatorname{int} w) \cap \operatorname{roof} F$:*

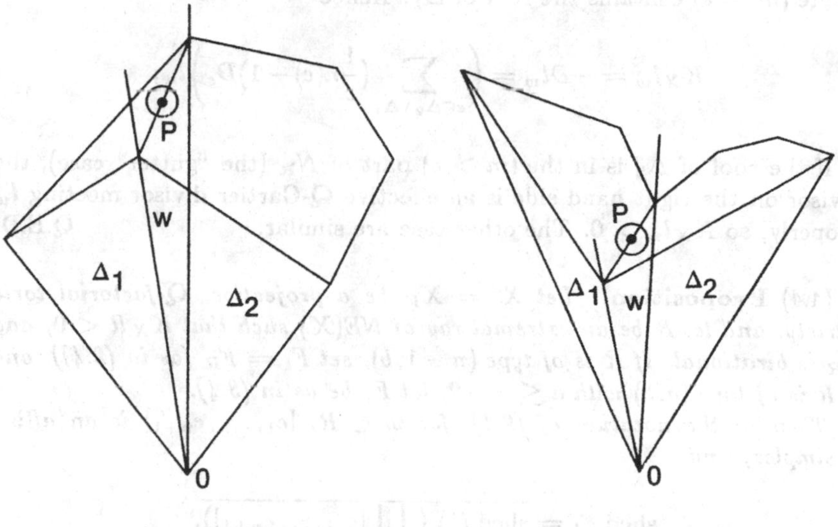

$K_x l_w < 0$: roof F has a ridge along w; $\qquad$ $K_X l_w > 0$: roof F has a gutter along w.

Proof. (i) If a multiple cD is locally principal at $P \in X$, it is not difficult to see (using the fact that the big torus $\mathbf{T} \subset X$ has $\operatorname{Pic} \mathbf{T} = 0$) that near P, cD is the divisor of a monomial x^m, for $m \in M$. On the other hand, if

$P \in X_\sigma$, where X_σ is the open stratum corresponding to $\sigma \in F^{(k)}$, locally near P,

$$\mathrm{Div}(x^m) = \sum m(e_i) D_{e_i};$$

hence $[e_1, \ldots, e_s]$ is contained in the affine hyperplane $(m = c) \subset N_{\mathbf{R}}$.

(ii) If w separates $\Delta_1, \Delta_2 \in F^{(n)}$, then in a neighbourhood of l_w,

$$D = \sum D_e; \quad e \in \{\text{1-faces of } \Delta_1 \text{ and of } \Delta_2\}.$$

I can move D away from l_w by using

$$0 \sim \frac{1}{c}\,\mathrm{Div}(x^m) = \sum_{e \in \Delta_1} D_e + \sum_{e \in \Delta_2 \setminus \Delta_1} \frac{1}{c} m(e) D_e,$$

where $(m = c)$ contains the roof of Δ_1. Hence

$$K_X l_w = -D l_w = \left(\sum_{e \in \Delta_2 \setminus \Delta_1} \left(\frac{1}{c} n(e) - 1\right) D_e \right) l_w;$$

If the roof of Δ_2 is in the $(m > c)$ part of $N_{\mathbf{R}}$ (the "gutter" case), the divisor on the right hand side is an effective $\mathbf{Q}$-Cartier divisor meeting l_w properly, so $K_X l_w > 0$. The other case are similar. Q.E.D.

(4.4) Proposition. *Let $X = X_F$ be a projective, $\mathbf{Q}$-factorial toric variety, and let R be an extremal ray of $NE(X)$ such that $K_X R < 0$, and φ_R is birational. If R is of type $(n-1, b)$, set $F_1 = F_R$ (as in (2.4)), and if R is of type (a, b) with $a \le n-2$, let F_1 be as in (3.4).*

Then in the notation of (2.1), for $w \in R$, $[e_1, \ldots, e_{n+1}]$ is an affine n-simplex, and

$$\mathrm{shed}\, F_1 = \overline{\mathrm{shed}\, F \setminus \Big(\bigcup_{w \in R} [e_1, \ldots, e_{n+1}] \Big)},$$

where $\overline{}$ denotes closure in $N_{\mathbf{R}}$.

Proof. In the notation of (2.3),

$$\Delta(w) \cap \mathrm{shed}\, F = \bigcup_{j \ge \beta+1} \mathrm{shed}\, \Delta_j,$$

and over $\Delta(w)$ the roof of F consists of the pieces

$$\Gamma_j = [e_1, \dots, \hat{e}_j, \dots, e_{n+1}].$$

The condition $K_X R < 0$ means that any two Γ_j meet in a ridge, so $\{e_1, \dots, e_{n+1}\}$ are not in any affine hypersurface, and

$$\Delta(w) \cap \operatorname{shed} F = [0, e_1, \dots, e_{n+1}];$$

hence $[e_1, \dots, e_{n+1}] \subset \operatorname{shed} F$.

An identical argument using the fact that shed F_1 has only gutters over $\Delta(w)$ shows that $[e_1, \dots, e_{n+1}]$ lies over the roof of F_1. Q.E.D.

(4.5) The last sentences of (2.4) and (3.4) are now obvious: the fan F is terminal if and only if $N \cap \operatorname{shed} F = \{0\} \cup F^{(1)}$; since $\operatorname{shed} F_1 \subset \operatorname{shed} F$, F_1 is again terminal.

Proof of (0.2). Let G be a complete fan corresponding to a projective toric variety A, and F a subdivision of G, corresponding to a projective birational morphism $f\colon V \to A$. If R is an extremal ray of $NE(V/A)$, $w \in R$ implies that $w \in F^{(n)} \setminus G^{(n)}$, so that F_R is also a subdivision of G, and all the varieties in the constructions of (2.4) and (3.4) have proper morphisms to A. As usual in Mori theory, on making an elementary contraction $\varphi_R\colon V \longrightarrow Y$, Y is again projective (if $H \in N^1(V)$ is such that the hyperplane $(HZ = 0) \subset N_1(V)$ intersects $NE(V)$ in R only, then $H \in \varphi_R^* H_Y$, where $H_Y \in N^1(Y)$ is ample).

Suppose that V has $\mathbf{Q}$-factorial terminal singularities, and that K_V is not relatively nef for f; then there exists an extremal ray R of $NE(V/A)$ with $K_V R < 0$. Let $\psi : V -\to V_1$ be the construction of (2.4) or (3.4); then V_1 satisfies all the same properties as V, and shed F_1 is strictly smaller (in volume, say) than shed F. It is obvious that $F_1^{(1)}$ is a subset of $F^{(1)}$, and that there are only finitely many fans with this property; hence, after a finite number of elementary transformations of this type, the variety $V_k = S$ has the properties claimed in (0.2), (1). Corresponding to the property that K_S is relatively nef for $f_k = g\colon S \longrightarrow A$, the shed of F_k has no ridges along walls $w \in F_k^{(n-1)} \setminus G^{(n-1)}$.

I claim that I can deltete all the walls $w \in F_k^{(n-1)} \setminus G^{(n-1)}$ for which $K_S l_w = 0$, and obtain a fan $\overline{F}_k$. The only thing that has to be proved is that the chambers of the corresponding subdivision of G are convex. This is clear from the following argument: above $\Delta \in G^{(n)}$, the roof of F_k is either flat or concave along every wall $w \in F_k^{(n-1)}$; hence the region of $N_{\mathbf{R}}$ given by $\overline{\Delta \setminus \text{shed}\, F_k}$ is a convex polyhedron. The chambers of $\overline{F}_k$ are the cones corresponding to its faces. Setting $X = X_{\overline{F}_k}$, X has the properties claimed in (0.2), (2).

Finally, for (0.2), (3), note that if G is terminal, the concavity of $\text{shed}\, F_k \cap \sigma$ for $\sigma \in G^{(r)}$ implies that $F_k^{(1)} = G^{(1)}$; if G is also simplicial, then $F_k = G$: any $\tau \in F_k$ is contained in some $\Delta = \langle e_1, \ldots, e_n \rangle \in G^{(n)}$, and its 1-faces are a subset of $\{e_1, \ldots, e_n\}$, hence τ is a face of Δ.

References

[1] V. Danilov, *The geometry of toric varieties*, Uspekhi Mat. Nauk *33* : 2(1978), 85–134 = Russian Math Surveys *33*:2(1978), 97–154.

[2] V. Danilov, *Birational geometry of toric 3-folds*, Izv. Akad. Nauk SSSR ser. Mat., *46* (1982), 971–982 = math USSR Izvestija, to appear.

[3] T. Oda, *Torus embeddings and applications*, Tata Inst. Lect. Notes, Springer 1978.

[4] S. Mori, Three-folds whose canonical bundles are not numerically effective, Ann. of Math. (2) *116* (1982), 133–176.

[5] M. Reid, *Canonical 3-folds*, in Proc. Journées de Géom. Alg. Angers, Ed. A. Beauville, Sijthoff and Noordhoff, Alphen, 1980, 273–310.

[6] M. Reid, *Minimal models of canonical 3-folds*, to appear in Advanced Studies in Pure Math. *1*, eds. S. Iitaka and H. Morikawa, Kinokuniya and North-Holland, 1982.

Received August 25, 1982

Miles Reid
Mathematics Institute
University of Warwick
Coventry CV4 7AL, England

A Solution to Hironaka's Polyhedra Game

Mark Spivakovsky

To I.R. Shafarevich

Abstract

In this paper we present a solution to Hironaka's polyhedra game. That is, we prove the existence of a winning strategy for the first player.

Introduction

In this paper, we present an affirmative solution to Hironaka's game in the original version which was proposed in 1970 (*Hironaka* [1], p. 5, and *Hironaka* [2], pp. 309–311) and has been left unsolved since then. The game is defined and our solution (a winning strategy to the game) is presented with a complete proof. Hironaka has also proposed a more complicated game whose affirmative solution would imply the local uniformization theorem (a local version of the resolution of singularities) for an algebraic variety over an algebraically closed field of *any* characteristic.[1] Our solution to the original and simple version of Hironaka's game is not enough for the local uniformization in its entirety but is effective in some special cases (such as local uniformization with respect to a valuation of large rational rank in the function field). We also hope that our solution will shed some light to the problem of the ultimate Hironaka's game and hence that of local uniformization in all characteristics.

My most grateful acknowledgements to Professor Hironaka, who has been of great help to me in this work. Among other things, he has put the original solution by induction on the dimension n into the form of an explicit winning strategy. I am also grateful to Professor Kyoji Saito and

---

[1] After this paper was written, I found a counterexample to the "hard" Hironaka's game. See [5].

all the other members of singularities seminar at RIMS at Kyoto University for very helpful discussions.

I. Formulation of the Problem

Let $\mathbb{R}^n_+$ denote the first quadrant of $\mathbb{R}^n$, i.e.

$$\mathbb{R}^n_+ = \{(x_1, \ldots, x_n) \in \mathbb{R}^n \mid \text{all } x_i \geq 0\}.$$

Definition. For a subset $M \subset \mathbb{R}^n_+$ the *positive convex hull* of M, denoted by $[M]$, is the convex hull of the set $\cup_{v \in M}(v + \mathbb{R}^n_+)$. A set is called *positively convex* if it coincides with its own positive convex hull.

We are given a non-empty subset $\Delta \subset \mathbb{R}^n_+$, which is the positive convex hull of a finite set of points, all of whose coordinates are rational numbers.

Initially, we assume $\sum_{j=1}^n x_j > 1$ for all $(x_1, \ldots, x_n) \in \Delta$.

Two people, A and B, are playing the following game: A chooses a non-empty subset $\Gamma \subset \{1, \ldots, n\}$ such that $\sum_{j \in \Gamma} x_j \geq 1$ for all $(x_1, \ldots, x_n) \in \Delta$ (such Γ is called permissible).

Then, B chooses an element $i \in \Gamma$. Let $\Delta' = [\sigma_{\Gamma,i}(\Delta)]$, where $\sigma_{\Gamma,i}$ is the transformation of $\mathbb{R}^n$ sending $(x_1, \ldots, x_n)$ to $(x'_1, \ldots, x'_n)$ defined by

$$x'_j = x_j \text{ if } j \neq i$$
$$x'_i = \sum_{j \in \Gamma} x_j - 1.$$

Δ is replaced by Δ', and this procedure is considered *one move* of our game. A wins, if after a finite number l of such moves the resulting set $\Delta^{(l)}$ contains a point $(x_1, \ldots, x_n)$ such that $\sum_{j=1}^n x_j \leq 1$.

By a *winning strategy of the player A for a set* Δ we shall mean a procedure of choosing a subset $\Gamma \subset \{1, \ldots, n\}$ at each stage of the game starting with $\Delta^{(0)} = \Delta$, such that by following it A is guaranteed to win in a finite number of moves, regardless of the responses of B.

Theorem. *There exists a winning strategy of A for any given* Δ.

Remark 1. In fact, our strategy is an algorithm of choosing Γ, depending only on Δ at the given stage of the game and *not* on the history of the preceding moves.

Remark 2. Our strategy does not always prescribe a unique choice of Γ, but an option of several possible Γ, any one of which ultimately guarantees victory.

Proof. First we shall define a certain finite sequence of positive rational numbers, associated with a set Δ.

Notations. (a) Let $\omega(\Delta)$ denote the vector in $\mathbb{R}_+^n$ defined by $(\omega_i)(\Delta) = (\omega_i)_{i=1,\ldots,n}$

$$\omega_i = \min\{x_i \mid (x_i, \ldots, x_n) \in \Delta\}, \quad i = 1, \ldots, n.$$

(b) $\tilde{\Delta}$ denotes the translation of Δ by $\omega(\Delta)$.

$$\tilde{\Delta} = \Delta - \omega(\Delta)$$

Note. By definition, $\tilde{\Delta} \subset \mathbb{R}_+^n$ and $\min\{\tilde{x}_i \mid (\tilde{x}_1, \ldots, \tilde{x}_n) \in \tilde{\Delta}\} = 0$ for all $i = 1, \ldots, n$.

(c) For a subset $\Gamma \subset \{1, \ldots, n\}$, $d_\Gamma(\Delta)$ will denote

$$d_\Gamma(\Delta) := \min\left\{\sum_{j\in\Gamma} x_j \mid (x_1, \ldots, x_n) \in \Delta\right\} \quad \text{and}$$

$$d(\Delta) := \min\left\{\sum_{j=i}^{n} x_j \mid (x_1, \ldots, x_n) \in \Delta\right\} = d_{\{1,\ldots,n\}}(\Delta)$$

Remark. $d(\tilde{\Delta})$ is the first important numerical character of Δ. As we shall see, A can insure that $d(\tilde{\Delta})$ does not increase under the moves of the game.

(d) Throughout this paper, if a is a point of Δ, a_j denotes the j-th coordinate of a, $\tilde{a} := a - \omega(\Delta)$, and $\tilde{a}_j := a_j - \omega_j$.

(e) For a point $y = (y_1, \dots, y_l) \in \mathbb{R}^l$, $|y|$ denotes the sum of the coordinates:

$$|y| := \sum_{j=1}^{l} y_j$$

(f) By $S(\Delta)$ we denote the following subset of $\{1, \dots, n\}$:

$$S(\Delta) := \{j \in \{1, \dots, n\} \mid \exists w \in \Delta \text{ such that } |w| = d(\Delta) \text{ and } \tilde{w}_j \neq 0\},$$

and by I_1 – the complement of $S(\Delta)$ in $\{1, \dots, n\}$, i.e.

$$I_1 := \{j \in \{1, \dots, n\} \mid \forall w \in \Delta \text{ with } |w| = d(\Delta) \text{ we have } \tilde{w}_j = 0\}.$$

(g) For $l \in \mathbb{Z}$ and two complementary subsets S and I of $\{1, \dots, l\}$, we define the subset $M_S^l \subset \mathbb{R}_+^l$ by

$$M_S^l := \left\{ (x_1, \dots, x_l)^{\in \mathbb{R}_+^l} \mid \sum_{j \in S} x_j < 1 \right\}$$

and the projection

$$P_S \colon M_S^l \longrightarrow \mathbb{R}_+^I, \text{ as follows.}$$

For $\alpha \in \mathbb{R}_+^I$, $\beta \in \mathbb{R}_+^S$ with $|\beta| < 1$, we set

$$P_S(\alpha, \beta) = \frac{\alpha}{1 - |\beta|}$$

With these notations for a positively convex set $\Delta \subset \mathbb{R}_+^n$, we define a convex set $\Delta_1 \subset \mathbb{R}_+^{I_1}$ by

$$\Delta_1 = P_{S(\Delta)}\left(M_{S(\Delta)}^n \cap \left[\frac{\tilde{\Delta}}{d(\tilde{\Delta})} \cup \Delta \right] \right)$$

It is easy to see that Δ_1 is again positively convex.

Now repeat the same procedure with the set Δ_1, namely, define $\tilde{\Delta}_1$, $I_2 = I_1 \setminus S(\Delta_1)$, etc.

Proceed to get a sequence of convex sets $\Delta_1, \Delta_2, \dots, \Delta_r$ and a decreasing sequence $I_1 \supset \dots \supset I_r$ of subsets of $\{1, \dots, n\}$ such that

$$\Delta_i \subset \mathbb{R}_+^{I_i} \quad \text{for } i = 1, \dots, r;$$
$$\omega(\Delta_i) = (\omega_j)_{j \in I_i}, \quad \omega_j = \min\{x_j \mid x \in \Delta_i\}, \quad j \in I_i$$

(For $a \in \Delta_i$, $\tilde{a} = a - \omega(\Delta_i)$, a_j and $\tilde{a}_j$ for $j \in I_i$ denote the j-th coordinate of a and $\tilde{a}$, respectively. The operation ~ depends on Δ_i, but this should cause no confusion).

$$\tilde{\Delta}_i = \Delta_i - \omega(\Delta_i)$$
$$S(\Delta_i) = \{j \in I_i \mid \exists w \in \Delta_i \text{ such that } |w| = d(\Delta_i) \text{ and } \tilde{w}_j \neq 0\}$$
$$I_{i+1} = I_i \setminus S(\Delta_i) \text{ for } i = 1, \ldots r-1$$
$$\Delta_{i+1} = Ps(\Delta_i)\left(M^{I_i}_{S(\Delta_i)} \cap \left[\frac{\tilde{\Delta}_i}{d(\tilde{\Delta}_i)} \cup \Delta \right] \right)$$

Stop when either $d(\tilde{\Delta}_r) = 0$ or $\Delta_r = 0$. (Adopt the convention that $\Delta = \Delta_0$, so that, in principle, r can be 0, if $d(\tilde{\Delta}) = 0$).

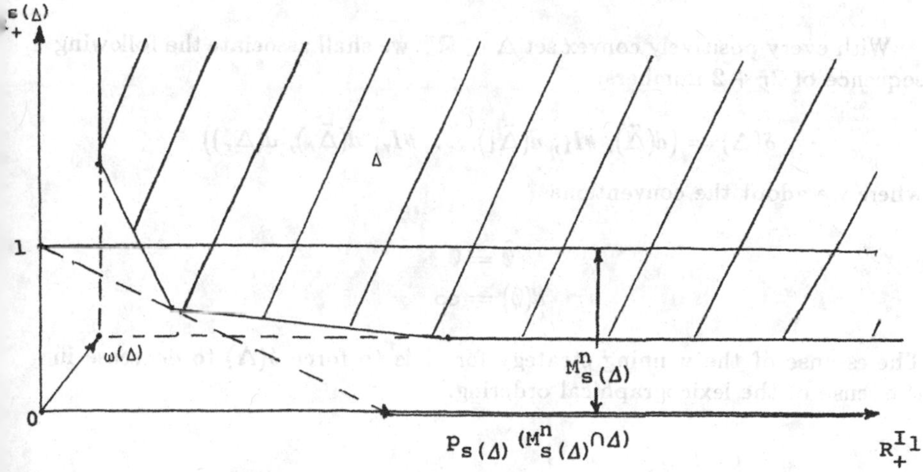

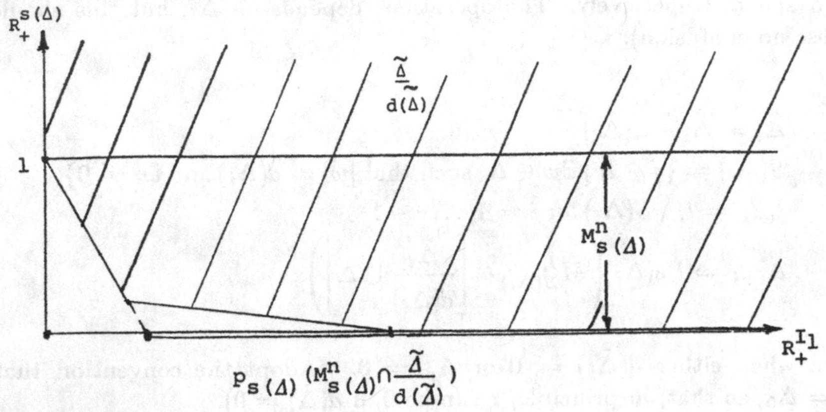

With every positively convex set $\Delta \subset \mathbb{R}^n_+$ we shall associate the following sequence of $2r+2$ numbers:

$$\delta(\Delta) = \big(d(\tilde{\Delta}),\ \#I_1,\ d(\tilde{\Delta}_1), \ldots, \#I_r,\ d(\tilde{\Delta}_r),\ d(\Delta_r)\big)$$

where we adopt the conventions

$$\tilde{\emptyset} = \emptyset$$
$$d(\emptyset) = \infty$$

The essense of the winning strategy for A is to force $\delta(\Delta)$ to decrease in the sense of the lexicographical ordering.

II. Description of the Winning Strategy

If $\Delta_r = \emptyset$, A chooses $\Gamma = \{1, \ldots, n\} \setminus I_r$.

If $\Delta_r \neq \emptyset$, A first chooses a subset $\Gamma_r \subset I_r$, which is a minimal one among those permissible for Δ_r (i.e. $\sum_{j \in \Gamma_r} x_j \geq 1$ for all $(x_1, \ldots, x_{\#I_r}) \in \Delta_r$ and no proper subset of Γ_r is permissible).

Then, he lets $\Gamma = (\{1, \ldots, n\} \setminus I_r) \cup \Gamma_r$.

III. Proof That the Above Strategy is Permissible and in Fact Guarantees Victory for A

Lemma 1. *The following two conditions on a subset $\Gamma_1 \subset I_1$ are equivalent:*

(a) Γ_1 *is permissible for* Δ_1
(b) $\Gamma = \Gamma_1 \cup S(\Delta)$ *is permissible for* Δ *and* $d_\Gamma(\tilde{\Delta}) = d(\tilde{\Delta})$.

Proof of Lemma 1.
Γ_1 is permissible for $\Delta_1 \Leftrightarrow \Gamma_1$ is permissible for both $P_{S(\Delta)}(M^l_{S(\Delta)} \cap \Delta)$ and $p_{S(\Delta)}(M^l_{S(\Delta)} \cap \frac{\tilde{\Delta}}{d(\tilde{\Delta})})$ (since Δ_1 is just the convex hull of the union of these two sets).

Now, Γ_1 is permissible for $p_{S(\Delta)}(M^l_{S(\Delta)} \cap \Delta)$ $\overset{\text{by definition}}{\Leftrightarrow}$
$\sum_{j\in\Gamma_1} x_j \geq 1$ for all $(x_1, \ldots, x_{\#I_1}) \in p_{S(\Delta)}(M^l_{S(\Delta)} \cap \Delta) \Leftrightarrow$
$\frac{1}{1-|\beta|}\sum_{j\in\Gamma_1} \alpha_j \geq 1$ for all $(\alpha, \beta) \in \Delta \cap M^l_{S(\Delta)}$,
where $\alpha \in \mathbb{R}^{I_1}_+$, $\beta \in \mathbb{R}^{S(\Delta)}_+$, $\Leftrightarrow$
$\sum_{j\in\Gamma_1} \alpha_j + |\beta| \geq 1$ for all $(\alpha,\beta) \in \Delta \cap M^l_{S(\Delta)} \Leftrightarrow$
$\sum_{j\in\Gamma} x_j \geq 1$ for all $(x_1, \ldots, x_n) \in \Delta \Leftrightarrow$
Γ is permissible for Δ.

Next, Γ_1 is permissible for $p_{S(\Delta)}(M^l_{S(\Delta)} \cap \frac{\tilde{\Delta}}{d(\tilde{\Delta})}) \Leftrightarrow$
$\frac{1}{1-|\beta|}\sum_{j\in\Gamma_1} \alpha_j \geq 1$ for all $(\alpha,\beta) \in M^l_{S(\Delta)} \cap \frac{\tilde{\Delta}}{d(\tilde{\Delta})}$,
where $\alpha \in \mathbb{R}^{I_1}_+$, $\beta \in \mathbb{R}^{S(\Delta)}_+$ $\Leftrightarrow$
$\frac{1}{d(\tilde{\Delta})-|b|}\sum_{j\in\Gamma_1} a_j \geq 1$ for all $(a, b) \in \tilde{\Delta}$,
where $a \in \mathbb{R}^{I_1}_+$, $b \in \mathbb{R}^{S(\Delta)}_+$ and $|b| < d(\tilde{\Delta}) \Leftrightarrow$
$\sum_{j\in\Gamma} \tilde{x}_j \geq d(\tilde{\Delta})$ for all $x \in \Delta \Leftrightarrow d_\Gamma(\tilde{\Delta}) \geq d(\tilde{\Delta})$. It is obvious from the definition that $d_\Gamma(\tilde{\Delta}) \leq d(\tilde{\Delta})$. Lemma 1 is proved.

Corollary. *The strategy described in II is permissible by the rules of the game.*

Proof of Corollary: If $\Delta_r \neq \emptyset$, then by applying Lemma 1 repeatedly r times, we get that Γ_r permissible for $\Delta_r \Rightarrow \Gamma_r \cup (I_{r-1} \setminus I_r)$ is permissible for $\Delta_{r-1} \Rightarrow$ etc... $\Rightarrow \Gamma_r \cup (I_1 \setminus I_r)$ is permissible for $\Delta_1 \Rightarrow \Gamma$ is permissible for Δ.

If $\Delta_r = \emptyset$, then by definition $M^{I_{r-1}}_{S(\Delta_{r-1})} \cap \Delta_{r-1} = \emptyset$, that is,

$$d_{S(\Delta_{r-1})}(\Delta_{r-1}) \geq 1$$

Hence, $S(\Delta_{r-1})$ is permissible for Δ_{r-1}, and applying Lemma 1 repeatedly $r-1$ times we get:

$S(\Delta_{r-1})$ permissible for $\Delta_{r-1} \Rightarrow S(\Delta_{r-1}) \cup (I_{r-2} \setminus I_{r-1})$ is permissible for $\Delta_{r-2} \Rightarrow \ldots \Rightarrow \Gamma = S(\Delta_{r-1}) \cup (\{1,\ldots,n\} \setminus I_{r-1})$ is permissible for Δ. Q.E.D.

Definition. The lexicographical ordering is the total ordering on the set of finite sequences of real numbers, in which $(a_1,\ldots,a_m)$ is less than $(b_1,\ldots,b_e)$ if and only if there exists a $k \in \mathbb{Z}$, $k \leq e$, $k \leq m$ such that $a_j = b_j$ for all $j \leq k$ and $a_{k+1} < b_{k+1}$, where we set the convention that $a_j = 0$ for $j > m$.

Proposition. *Let $\Delta \subset \mathbb{R}^n_+$ be as above and let $\Gamma \subset \{1,\ldots,n\}$ be chosen as described in II. Fix any $i \in \Gamma$, and let*

$$\Delta' = [\sigma_{\Gamma,i}(\Delta)].$$

Then $\delta(\Delta') < \delta(\Delta)$ in the lexicographical ordering.

The essence of the proof is contained in the following.

Lemma 2. *Suppose that $d(\tilde{\Delta}) \neq 0$, $\Delta_1 \neq 0$. Then*

(a) $d(\tilde{\Delta}') \leq d(\tilde{\Delta})$.

(b) $d(\tilde{\Delta}') = d(\tilde{\Delta}) \Rightarrow i \notin S(\Delta)$ *and* $S(\Delta) \subseteq S(\Delta')$

(c) $d(\tilde{\Delta}') = d(\tilde{\Delta})$ *and* $S(\Delta') = S(\Delta) \Rightarrow i \notin S(\Delta)$ *and* $\Delta'_1 = [\sigma_{\Gamma_1,i}(\Delta_1)]$, *where $\Gamma_1 = \Gamma \cap I_1$, and $\sigma_{\Gamma_1,i}$, naturally, is the transformation of $\mathbb{R}^{I_1}_+$, sending $(x_1,\ldots,x_{\#I_1})$ to $(x'_1,\ldots,x'_{\#I_1}) \geq$, with*

$$x'_j = x_j \text{ if } j \neq i \text{ and}$$
$$x'_i = \sum_{j\in\Gamma_1} x_j - 1.$$

Proof of Lemma 2.

(a) *Claim:* $\tilde{\Delta}' = [\tilde{\sigma}_{\Gamma,i}(\tilde{\Delta})]$, where $\tilde{\sigma}_{\Gamma,i}$ is the transformation of $\mathbb{R}^n_+$, sending $(\tilde{x}_1, \dots, \tilde{x}_n)$ to $(\tilde{x}'_1, \dots, \tilde{x}'_n)$, with $\tilde{x}'_j = \tilde{x}_j$ if $j \neq i$ and

$$(1) \qquad \tilde{x}'_i = \sum_{j\in\Gamma} \tilde{x}_j - d_\Gamma(\tilde{\Delta}) = \sum_{j\in\Gamma} \tilde{x}_j - d(\tilde{\Delta}).$$

Proof of Claim. Since $\tilde{\Delta}'$ and $\tilde{\Delta}$ are parallel shifts of Δ' and Δ respectively, and since for $j \neq i$ $\sigma_{\Gamma,i}$ leaves the j-th coordinates of all points invariant (in particular, $\omega'_j = \omega_j$ for all $j \neq i$, where ω'_j is the j-th coordinate of $\omega(\Delta')$), the transformation $\tilde{\sigma}^*_{\Gamma,i}$ sending $\tilde{\Delta}$ to $\tilde{\Delta}'$ must have the form:

$$\sigma^*_{\Gamma,i}(\tilde{x}_1, \dots, \tilde{x}_n) = (\tilde{x}'_1, \dots, \tilde{x}'_n)$$
$$\tilde{x}'_j = \tilde{x}_j \text{ for } j \neq i$$
$$\tilde{x}'_i = \sum_{j\in\Gamma} \tilde{x}_j - C,$$

where C depends only on Δ but not on the point x. It remains to determine C.

By definition of $d_\Gamma(\tilde{\Delta})$, $\min\{\tilde{x}'_i \mid x' \in \Delta'\} = d_\Gamma(\tilde{\Delta}) - C$. But, by definition of the operation $\sim$, $\min\{\tilde{x}'_i \mid x' \in \Delta'\} = 0$. Hence, $C = d_\Gamma(\tilde{\Delta})$.

Also, $d_\Gamma(\tilde{\Delta}) = d(\tilde{\Delta})$ by the assumptions on Γ and Lemma 1. (This is where we need the assumptions that $d(\tilde{\Delta}) \neq 0$ and $\Delta_1 \neq \emptyset$.)

The claim is proved.

Now, take any $v \in \Delta$ such that $|v| = d(\Delta)$, or equivalently, $|\tilde{v}| = d(\tilde{\Delta})$.

Let $v' = \sigma_{\Gamma,i}(v)$.

Then, by (1),

$$\tilde{v}'_j = \tilde{v}_j \text{ for } j \neq i, \text{ and}$$
$$\tilde{v}'_i = 0.$$

Hence,

$$(2) \qquad d(\tilde{\Delta}') \leq |\tilde{v}'| \leq |\tilde{v}| = d(\tilde{\Delta}).$$

(a) is proved.

(b) Assume that $d(\tilde{\Delta}') = d(\tilde{\Delta})$.

Suppose, $i \in S(\Delta)$. By definition of $S(\Delta)$, there exists a point $v \in \Delta$ such that $|v| = d(\Delta)$ and $\tilde{v}_j > 0$. Then, $0 = \tilde{v}'_i < \tilde{v}_i$, $|\tilde{v}'| < |\tilde{v}|$, and hence $d(\tilde{\Delta}') < d(\tilde{\Delta})$, which is a contradiction. Thus, $i \notin S(\Delta)$.

Next, take any $j \in S(\Delta)$. There exists a $v \in \Delta$ such that $|v| = d(\Delta)$ and $\tilde{v}_j \neq 0$. Since $i \notin S(\Delta)$, $i \neq j$. Hence, $\tilde{v}'_j = \tilde{v}_j \neq 0$. Moreover, $d(\tilde{\Delta}') \leq |v'| \leq |\tilde{v}| = d(\tilde{\Delta}) = d(\tilde{\Delta}')$ implies that $|\tilde{v}'| = d(\tilde{\Delta}')$. Thus, $j \in S(\Delta')$, as desired.

(c) Assume that $d(\tilde{\Delta}') = d(\tilde{\Delta})$ as before, and that $S(\Delta') = S(\Delta)$. It suffices to prove separately that

$$p_{S(\Delta)}(M^n_{S(\Delta)} \cap \Delta') = \left[\sigma_{\Gamma_1,i}(p_{S(\Delta)}(M^n_{S(\Delta)} \cap \Delta))\right] \tag{3}$$

and

$$p_{S(\Delta)}\left(M^n_{S(\Delta)} \cap \frac{\tilde{\Delta}'}{d(\tilde{\Delta})}\right) = \left[\sigma_{\Gamma_1,i}\left(p_{S(\Delta)}\left(M^n_{S(\Delta)} \cap \frac{\tilde{\Delta}}{d(\Delta)}\right)\right)\right]. \tag{4}$$

The proof is by a direct calculation:

Proof of (3). Since $i \notin S(\Delta)$, $\sigma_{\Gamma,i}$ leaves both $M^n_{S(\Delta)}$ and $\mathbb{R}^n_+ \backslash M^n_{S(\Delta)}$ invariant. Thus it suffices to prove for any $x \in \Delta \cap M^n_{S(\Delta)}$ that if $x' = \sigma_{\Gamma,i}(x)$ then

$$p_{S(\Delta)}(x') = \sigma_{\Gamma_1,i}(p_{S(\Delta)}(x))$$

Write $x = (\alpha, \beta)$ and $x' = (\alpha', \beta)$ with $\alpha, \alpha' \in \mathbb{R}^{I_1}_+$, $\beta \in \mathbb{R}^{S(\Delta)}_+$ and, by assumptions, $|\beta| < 1$. Then $p_{S(\Delta)}(x) = \frac{\alpha}{1-|\beta|}$ and $p_{S(\Delta)}(x') = \frac{\alpha'}{1-|\beta|}$. Compare $p_{S(\Delta)}(x')$ with $\sigma_{\Gamma_1,i}(p_{S(\Delta)}(x))$: For $j \in I_1 \setminus \{i\}$, $x'_j = x_j$ by definition of $\sigma_{\Gamma,i}$, hence

$$\frac{\alpha'_j}{1-|\beta|} = \frac{\alpha_j}{1-|\beta|} = \left(\sigma_{\Gamma_1,i}\left(\frac{\alpha}{1-|\beta|}\right)\right)_j$$

Finally,

$$(p_{S(\Delta)}(x'))_i = \frac{\alpha'_i}{1-|\beta|} = \frac{\sum_{k\in\Gamma} x_k - 1}{1-|\beta|} = \frac{\sum_{k\in\Gamma_1} \alpha_k + |\beta| - 1}{1-|\beta|}$$

$$= \sum_{k\in\Gamma_1} \frac{\alpha_k}{1-|\beta|} - 1 = \sigma_{\Gamma_1,i}(p_{S(\Delta)}(x))_i,$$

as desired.

Proof of (4). We have proved above that

$$\tilde{\Delta}' = [\tilde{\sigma}_{\Gamma,i}(\tilde{\Delta})],$$

where $\tilde{\sigma}_{\Gamma,i}$ is described by (1). Hence, from (1),

$$\frac{\tilde{\Delta}'}{d(\tilde{\Delta})} = \left[\sigma_{\Gamma,i}\left(\frac{\tilde{\Delta}}{d(\tilde{\Delta})}\right)\right].$$

Thus, (3) implies (4) (we only need to replace Δ in (3) by $\frac{\tilde{\Delta}}{d(\tilde{\Delta})}$). Q.E.D.

We now finish the proof of the proposition. Let $\delta_0, \ldots, \delta_{2r+1}$ denote the components of $\delta(\Delta)$, that is, $\delta_{2l} = d(\tilde{\Delta}_l)$, $\delta_{2l-1} = \#I_l$ for $l \leq r$ and $\delta_{2r+1} = d(\Delta_r)$. Let δ'_j denote the j-th component of $\delta(\Delta')$ and let r' be the smallest integer such that either $d(\tilde{\Delta}'_{r'}) = 0$ or $\Delta'_{r'} = \emptyset$. Now it follows easily from Lemma 2 that either $\delta(\Delta') < \delta(\Delta)$ in the lexicographical ordering or $r' \geq r$ and $\delta'_j = \delta_j$ for $0 \leq j \leq 2r$.

More precisely, we prove this statement by induction:

By Lemma 2(a) $d(\tilde{\Delta}') \leq d(\tilde{\Delta})$. If the inequality is strict, $\delta(\Delta') < \delta(\Delta)$, otherwise, we have $\delta'_0 = \delta_0$.

Suppose $\delta'_j = \delta_j$ for $0 \leq j \leq 2(k-1)$, where $1 \leq k \leq r$, and that $i \in \Gamma_{k-1} = \Gamma \cap I_{k-1}$ and $\Delta'_{k-1} = [\sigma_{\Gamma_{k-1},i}(\Delta_{k-1})]$.

Now, by Lemma 2(b), $\delta'_{2(k-1)} = \delta_{2(k-1)} \Rightarrow i \notin S(\Delta_{k-1})$ and

$$S(\Delta_{k-1}) \subset S(\Delta'_{k-1}) \Rightarrow \delta'_{2k-1} \leq \delta_{2k-1}.$$

If the inequality is strict, $\delta(\Delta') < \delta(\Delta)$, otherwise $\delta'_{2k-1} = \delta_{2k-1}$ and hence, $S(\Delta_{k-1}) = S(\Delta'_{k-1})$.

Then by Lemma 2(c), $\Delta'_k = [\sigma_{\Gamma_k,i}(\Delta_k)]$, hence, applying Lemma 2(a) now to Δ_k, we have $\delta'_{2k} \leq \delta_{2k}$. Strict inequality means $\delta(\Delta') < \delta(\Delta)$. Otherwise $\delta'_{2k} = \delta_{2k}$. The induction is completed.

Thus, we may assume $\delta'_j = \delta_j$ for $0 \leq j \leq 2r$. In particular, this implies that $\Delta_r \neq \emptyset$, $\Gamma_r \neq \emptyset$ and $i \in \Gamma_r$, since otherwise $\Gamma = \{1, \ldots, n\} \setminus I_r$ and $i \in S(\Delta_l)$ for some $l < r$ so that $d(\tilde{\Delta}'_l) < d(\tilde{\Delta}_l)$ and $\delta(\Delta') < \delta(\Delta)$.

By Lemma 2(c), we have

$$\Delta'_r = [\sigma_{\Gamma_r,i}(\Delta_r)].$$

$d(\tilde{\Delta}_r) = 0$ by definition of Δ_r, and hence

$$d(\tilde{\Delta}'_r) = 0 \quad \text{(so that } r' = r\text{)}$$

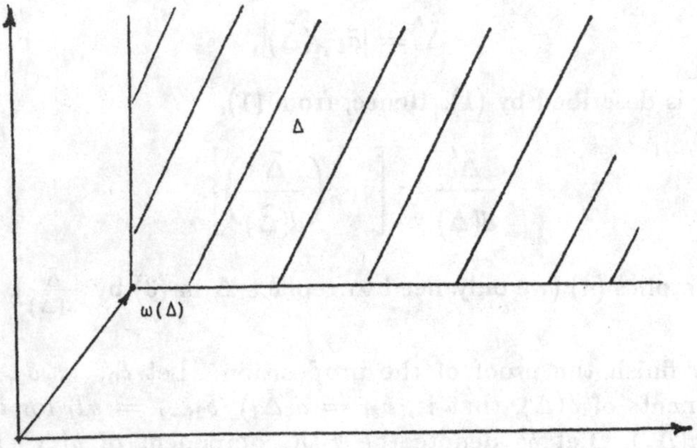

Δ is generated by one point, $w(\Delta)$.

It remains to prove that $d(\Delta'_r) < d(\Delta_r)$. We are reduced to

Lemma 3. *Suppose,* $\Delta = [\{v\}] = \text{hull}(v + \mathbb{R}^n_+)$ *is generated by one point, let* Γ *be a minimal subset of* $\{1, \dots, n\}$ *among those permissible for* Δ*, choose any* $i \in \Gamma$ *and let*

$$\Delta' = [\sigma_{\Gamma,i}(\Delta)].$$

Then, $d(\Delta') < d(\Delta)$.

Proof of Lemma 3. We have $\omega'_j = \omega_j$ for $j \neq i$ and

$$\omega'_i = \sum_{j \in \Gamma} \omega_j - 1 = \omega_i + \Big(\sum_{j \in \Gamma \setminus \{i\}} \omega_j - 1 \Big).$$

But since Γ was a *minimal* permissible subset, $\sum_{j \in \Gamma \setminus \{i\}} \omega_j < 1$ (otherwise $\Gamma \setminus \{i\}$ would also be permissible) and thus $\omega'_i < \omega_i$. Hence,

$$d(\Delta') = |\omega(\Delta')| < |\omega(\Delta)| = d(\Delta) \qquad \text{(Q.E.D.)}$$

The proof of the proposition is now completed.

The only way the game can end is if A wins, so it is sufficient to prove that $\delta(\Delta)$ cannot decrease forever.

Suppose, we have an infinite sequence $\Delta^{(0)}, \Delta^{(1)}, \ldots, \Delta^{(l)}, \ldots$ such that $\Delta^{(l+1)} = [\sigma_{\Gamma(l) i_l}(\Delta^{(l)})]$ and thus

$$\delta(\Delta^{(l+1)}) < \delta(\Delta^{(l)}).$$

Let $\delta_j^{(l)}$ denote the j-th component of $\delta(\Delta^{(l)})$. Let p be the greatest integer for which there exists $l_0 \in \mathbb{Z}$ such that $\delta_j^{(l)} = \delta_j^{(l_0)}$ for all $j < p$ and $l \geq l_0$ (possibly, $p = 0$).

Then $\delta_p^{(l_0)}, \delta_p^{(l_0+1)}, \ldots$ is a non-increasing infinite sequence of non-negative rational numbers, in which any number can occur only a finite number of times.

If p is odd, all the $\delta^{(l)}$ are non-negative integers which is absurd in view of the above condition.

If p is even, all the $\delta_p^{(l)}$ are rational number with bounded denominators for the following reason:

Let $p = 2k$, let $\Delta_j^{(l)}$ denote the j-th member of the recursive sequence $\Delta^{(l)}, \Delta_1^{(l)}, \ldots, \Delta_{r_l}^{(l)}$ for the set $\Delta^{(l)}$, defined in I. From Lemma 2 we deduce for each $l \geq l_0$ that the equality $\delta_j^{(l+1)} = \delta_j^{(l)}$, $j < p$ implies that $S(\Delta_j^{(l+1)}) = S(\Delta_j^{(l)})$ for $j < k$, $i_l \in \Gamma_k^{(l)} = I_k^{(l)} \cap \Gamma^{(l)}$ and $\Delta_j^{(l+1)} = [\sigma_{\Gamma_j^{(l)}, i_l}(\Delta_j^{(l)})]$ for $j \leq k$.

In other words, for $j < k$ and $l \geq l_0$ the sets $S(\Delta_j^{(l)})$ and $I_{j+1}^{(l)}$ are independent of l (and hence, so are the projections $p_S(\Delta_j^{(l)})$. Thus, for $l \geq l_0$ we can regard our game as the game on the set $\Delta_k^{(l)} \subset \mathbb{R}_+^{I_k}$, which is generated by a finite number of points, all of whose coordinates are rational (this follows from the analogous condition on Δ itself and the definition of the projections $p_{S(\Delta_j)}$). Let N be an upper bound on (say, the product of) the decominators of the vertices' coordinates. Clearly N remains a bound after the transformation $\sigma_{\Gamma_k, i}$. In particular, N is an upper bound for the denominator of

$$d(\tilde{\Delta}_k^{(l)}) = \delta_p^{(l)} \text{ for all } l \geq l_0.$$

Thus, the sequence $\{\delta_p^{(l)}\}_{l \geq l_0}$ must become constant for l sufficiently large, which contradicts the definition of p. Q.E.D.

References

[1] Hironaka, H., "Characteristic polyhedra of singularities," Journal of Mathematics of Kyoto University, Vol. 7, 1968, pp. 251–293.

[2] Hironaka, H., "Study of Algebraic Varieties" (in Japanese), Monthly Report, Japan Acad., Vol. 23, No. 5, 1970, pp. 1–5.

[3] Hironaka, H., "Schemes, ETC," Proc. 5th Nordic Summer School in Math., Oslo, 1970, pp. 291–313.

[4] Hironaka, H., "Certain numerical characters of singularities," Journal of Mathematics of Kyoto University, Vol. 10, 1970, pp. 151–187.

[5] Spivakovsky, M., "A counterexample to Hironaka's "hard" polyhedra game," to appear in the Journal of Mathematics of Kyoto University.

Received June 10, 1982

Mark Spivakovsky
Department of Mathematics
Harvard University
Cambridge, Massachusetts 02138

Research Institute of Mathematical Science
Kyoto, Japan

On the Superpositions of Mathematical Instantons

A. N. Tjurin

To I.R. Shafarevich

§0. Introduction

A *mathematical c-instanton* is, by definition, a vector bundle F on a projective space $\mathbf{P}^3 = \mathbf{P}(T)$, $T = \mathbf{C}^4$, with the following properties.

$$
\begin{aligned}
&1)\ \operatorname{rk} F = 2,\\
&2)\ c_1(F) = 0,\ c_2(F) = c,\\
&3)\ h^0(F) = 0,\\
&4)\ h^1(F(-2)) = 0.
\end{aligned}
\tag{0.1}
$$

(We refer to the papers [3], [4], [1] for the background and for a discussion of related topics).

For any c-instanton F, $\dim H^1(\mathbf{P}^3, F(-1)) = c$.

Let $H = \mathbf{C}^n$ be a vector space over $\mathbf{C}$ and $H^* = \operatorname{Hom}_{\mathbf{C}}(H, \mathbf{C})$.

Definition 1. A pair (F, i), where F is a mathematical c-instanton and $i\colon H^1(F(-1)) \longrightarrow H^*$ is a monomorphism, is called an H-marked c-instanton. A pair (F, i) is called an *exactly marked c-instanton* if i is an isomorphism.

Let M_c denote the moduli space of mathematical c-instantons, and let $M_c(H)$ denote the moduli space of H-marked c-instantons.

If we attach to every pair (F, i) its image $i(H^1(F(-1))) \subset H^*$, we have a fibering

$$\varphi_c\colon M_c(H) \longrightarrow G(c, n), \tag{0.2}$$

with the Grassmann variety of c-subspaces of H^* as a base and the moduli space $M_c(u)$ as the fiber over a point $u \subset H^*$.

The group $GL(n, \mathbf{C}) = \operatorname{Aut} H$ acts on $M_n(H)$ and defines a principal $GL(n, \mathbf{C})$-bundle:

$$\pi\colon M_n(H) \longrightarrow M_n \tag{0.3}$$

over the moduli space of n-instantons. Consequently these moduli spaces are birationally equivalent to direct products: $M_n(H) \overset{\text{bir}}{\sim} M_n \times GL(n, \mathbf{C})$; $M_c(H) \overset{\text{bir}}{\sim} M_c \times GL(c, n) \times GL(c, \mathbf{C})$, and the moduli space $M_c(H)$ has as many components as M_c has. Moreover, the unirationality of M_c is equivalent to the unirationality of $M_c(H)$.

There is a certain way to embed all these spaces $M_c(H)$ into the same vector space $\wedge^2 T^* \otimes S^2 H^*$ (§1), and any H-marked n-instanton is a superposition of marked 1-instantons (or half-instantons, see Definitions 2 and 3). In this paper we describe the properties of such superpositions.

§1. $M_n(H)$ as a Determinantal Locus (Determinantal Variety)

Let $(F, i) \in M_n(H)$ be any exactly marked n-instanton. A vector bundle F is a cohomology bundle of a complex of bundles over $\mathbf{P}^3 = \mathbf{P}(T)$, $(T = \mathbf{C}^4)$:

$$0 \to H \otimes \mathcal{O}_{\mathbf{P}(T)}(-1) \overset{a}{\to} H^* \otimes \Omega\mathbf{P}(T)(1) \overset{c}{\to} \mathbf{C}^{2n-2} \otimes \mathcal{O}_{\mathbf{P}(T)} \to 0, \tag{1.1}$$

where $\Omega\mathbf{P}(T) = \big(T\mathbf{P}(T)\big)^*$ is the cotangent bundle of $\mathbf{P}(T)$.

For the dual complex

$$0 \to \mathbf{C}^{2n-2} \otimes \mathcal{O}_{\mathbf{P}(T)} \overset{c^*}{\to} H \otimes T\mathbf{P}(T)(-1) \overset{a^*}{\to} H^* \otimes \mathcal{O}_{\mathbf{P}(T)}(1) \to 0, \tag{1.2}$$

the initial part of the corresponding cohomology sequence

$$0 \to \mathbf{C}^{2n-2} \overset{\Gamma(c^*)}{\to} H \otimes T \overset{\Gamma(a^*)}{\to} H^* \otimes T^* \tag{1.3}$$

is exact, and defines a homomorphism

$$\alpha(F, i)\colon H \otimes T \overset{\Gamma(a^*)}{\to} H^* \otimes T^* \tag{1.4}$$

of rank $2n+2$. This homomorphism can be considered as an element of the space $T^* \otimes T^* \otimes H^* \otimes H^*$. Hence we may identify such homomorphisms with the corresponding elements of the tensor space.

On the space $T^* \otimes T^* \otimes H^* \otimes H^*$ there are three involutions $*$, $*_H$ and $*_T$:

$$\begin{aligned}
&1) \quad *(t^1 \otimes t^2 \otimes h^1 \otimes h^2) \quad = t^2 \otimes t^1 \otimes h^2 \otimes h^1\\
&2) \quad *_H (t^1 \otimes t^2 \otimes h^1 \otimes h^2) = t^1 \otimes t^2 \otimes h^2 \otimes h^1\\
&3) \quad *_T (t^1 \otimes t^2 \otimes h^1 \otimes h^2) = t^2 \otimes t^1 \otimes h^1 \otimes h^2
\end{aligned}$$

These operators commute pairwise, and

$$* = *_H \cdot *_T. \tag{1.5}$$

The tensor $\alpha = \alpha(F, i)$, $(F, i) \in M_n(H)$ has additional symmetries:

1) $\alpha^* = -\alpha$ (1.6)
This follows from Serre duality and the existence of a skew-symmetric form $\gamma: F \longrightarrow F^*$, which induces an isomorphism $H^1(F(-1)) = H^1(F^*-1))$.
2) $\alpha^{*T} = -\alpha$ (1.7)
This follows from the representation of F as the cohomology bundle of the complex (1.1).

So we have $\alpha(M_n(H)) \subset S^2H^* \otimes \wedge^2T^*$, and $\alpha(F, i)$ is a hypernet of quadrics in H (see [7], §1, where some different interpretations of this notion are given).

Conversely, a tensor $\alpha \in S^2H^* \otimes \wedge^2T^*$ is of the form $\alpha(F, i)$ if and only if

$\alpha 0)$ rk $\alpha = 2n + 2$,

$\alpha 1)$ the homomorphism $\alpha : (t_1 \wedge t_2) \otimes H \to H^*$ is an isomorphism for some t_1, $t_2 \in T$, and

$\alpha 2)$ $\bigcap_{t \in T} \ker(\alpha : (t_0 \wedge t) \otimes H \to H^*) = 0$ for every vector $t^0 \in T$.

An arbitrary homomorphism α: $H \otimes T \longrightarrow H^* \otimes T^*$, tensored with $O_{\mathbf{P}(T)}$, can be included in a commutative diagram

$$\begin{array}{ccc} H\otimes T\mathbf{P}(T)(-1) & \overset{a^*}{\dashrightarrow} & H^*\otimes Q_{\mathbf{P}(T)}(1) \\ \uparrow & \overset{\alpha'}{\nearrow} & \uparrow \\ H\otimes T\otimes Q_{\mathbf{P}(T)} & \xrightarrow{\alpha} & H^*\otimes T^*\otimes Q_{\mathbf{P}(T)} \\ \uparrow & \nearrow & \uparrow \\ H\otimes Q_{\mathbf{P}(T)}(-1) & & H^*\otimes\Omega\mathbf{P}(1). \end{array} \tag{1.8}$$

This diagram can be completed by a homomorphism a^* if and only if the homomorphism $\alpha' : H\otimes \mathcal{O}_{\mathbf{P}(T)}(-1) \to H^*\otimes \mathcal{O}_{\mathbf{P}(T)}(1)$ is zero. If we consider this homomorphism as a tensor $\alpha' \in S^2T^* \otimes H^* \otimes H^*$, then we have:

$$2\alpha' = \alpha + \alpha^{*_T} \tag{1.9}$$

Therefore, α is $*_T$-skew-symmetric if and only if there exists a homomorphism a^* in (1.8).

We have, then, a complex

$$V\otimes \mathcal{O}_{\mathbf{P}(T)} \overset{c^*}{\to} H\otimes T\mathbf{P}(T)(-1) \overset{a^*}{\to} H^*\otimes \mathcal{O}_{\mathbf{P}(T)}(1), \tag{1.10}$$

where $V = H^0(\ker a^*)$, and c^* is the natural homomorphism of sections.

If F is the cohomology sheaf of the dual complex, then the property $\alpha 2)$ is equivalent to F being a bundle, and the property $\alpha 1)$ is equivalent to the existence of an exact H-marking of F, together with $\operatorname{rk} F = \operatorname{rk}\alpha - 2n$.

Consequently, any tensor $\alpha \in T^*\otimes T^*\otimes H^*\otimes H^*$ with $\alpha^{*_T} = -\alpha$ and the property

$$\alpha 2') : \left\{ \bigcap_{t\in T} \ker\bigl(\alpha : (t_0\wedge t)\otimes H \to H^*\bigr) \text{ doesn't depend on } t_0 \right\},$$

defines H-marked bundle F, which is the cohomology sheaf of the complex (1.1). Hence

Proposition 1.1. *α: $M_n(H) \longrightarrow S^2H^*\otimes \wedge^2 T^*$ is an imbedding.*

The variety $\alpha\bigl(M_n(H)\bigr)$ is a linear cone in $\wedge^2(H\otimes T)^*$, and we can pass to its projectivisation $\mathbf{P}\alpha\bigl(M_n(H)\bigr) \subset \mathbf{P}\wedge^2(H\otimes T)^*$.

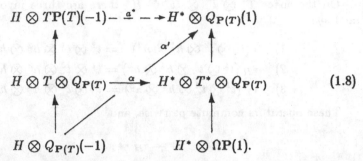

Proposition 1.2. $\mathbf{P}\alpha(M_n(H))$ *contains a Zariski open subset of the complete intersection*

$$M_{2n+2} = \Omega^{2n+2} \cap \mathbf{P}(S^2H^* \otimes \wedge^2 T^*) \tag{1.11}$$

in the projective space $\mathbf{P}\wedge^2(H \otimes T)^*$, *where*

$$\Omega^{2n+2} = \{\alpha \in \mathbf{P}\wedge^2(H \otimes T)^* \mid \mathrm{rk}\,\alpha \leq 2n+2\}.$$

To prove this fact, it is sufficient to show that the conditions $\alpha 1$) and $\alpha 2$) are open and that $5n(n-1)$ of the linear equations $\alpha - \alpha^{*T} = 0$ are independent on Ω^{2n+2}. This means that

$$\dim \mathbf{P}\alpha(M_n(H)) = \dim \Omega^{2n+2} - \dim \ker(1 - *_H). \tag{1.12}$$

But we have from (0.3) that

$$\dim \mathbf{P}\alpha(M_n(H)) = \dim M_n + \dim PGL(n, \mathbf{C}) = n^2 + 8n - 4.$$

On the other hand, $\dim \Omega^{2n+2} = 6n^2 + 3n - 4$ ([6], 10.4.3), and we obtain the equation (1.12).

So M_{2n+2} is a classical skew-symmetric determinantal variety of type $(W \mid 4n, 4n\,|_{2n+2}, [11n^2+n-1])$ ([6], ch. X). The ideal sheaf of this variety has the resolvent of Lascoux-Josefiak-Pragacz (see [5] and the reference in that paper). However the reducibility of this variety presents an obstacle to obtaining information about it from the general theory of determinantal varieties.

The projective variety $M_{2n+2} \subset \mathbf{P}(S^2H^* \otimes \wedge^2 T^*)$ (1.11) is a union of its components:

$$M_{2n+2} = \overline{M_n(H)} \cup M^0 \cup M^1 \cup M^2, \tag{1.13}$$

where $\overline{M_n(H)}$ is the closure of $\alpha(M_n(H))$, and where M^i is the union of those components on which the condition αi) is invalid.

The irreducibility of M_{2n+2} was proved for $n \leq 4$ by W. Barth ([1]). In the paper [7] the existence of a subvariety of dimension $n^2 - 1 + \frac{n^2+3n+6}{2}$ of M^0 is proved. Therefore M^0 is non-empty for $n \geq 12$. I.A. Artamkin has observed that M^1 is non-empty for $n \geq 9$, and so is M^2 for $n \geq 12$.

To check this it is sufficient to consider any tensor $\omega \in S^2H^* \otimes \wedge^2 T^*$ of rank 2 and all the superpositions $\alpha + \omega$, where α belongs to the component

M^0 of dimension $n^2 - 1 + \frac{n^2+3n+6}{2}$. We obtain a subvariety in M^2, whose general point satisfies the conditions $\alpha 0)$ and $\alpha 1)$ and whose dimension is greater than the dimension $n^2 + 8n - 4$ of $\overline{M_n(H)}$.

This shows that we have obtained a component of M^2.

§2. The Superpositions

The filtration

$$\mathbf{P}\wedge^2(H\otimes T)^* = \Omega^{4n} \supset \cdots \supset \Omega^{2n+2} \supset \Omega^{2n} \supset \cdots \supset \Omega^2 = G(2,4) \quad (2.1)$$

of the projective space $\mathbf{P}\wedge^2(H\otimes T)^*$, where

$$\Omega^{2k} = \{\alpha \in \mathbf{P}\wedge^2(H\otimes T)^* \mid \operatorname{rk}\alpha \leq 2k\},$$

induces the filtration

$$\begin{gathered} \mathbf{P}S^2H^*\otimes\wedge^2T^* = M_{4n} \supset \cdots \supset M_{2n+2} \supset M_{2n} \supset \cdots \supset M_2, \\ M_{2k} = \Omega^{2k} \cap \mathbf{P}S^2H^*\otimes\wedge^2T^*. \end{gathered} \quad (2.2)$$

Any $\alpha \in \Omega^{2k}$ is a superposition of k matrices of rank 2:

$$\alpha = \sum_{i=1}^{k} \omega_i, \quad \omega_i \in \Omega^2. \quad (2.3)$$

(The dimension of the variety of such decompositions is $2k(k-1)$).

Geometrically, Ω^{2k} is the union of k-chords of Ω^2. (A k-chord is a linear envelope of k points from Ω^2.) (See [6], 10.4.5).

Definition 2. A decomposition

$$\alpha(F,i) = \sum_{k=1}^{N} \alpha(F_k, i_k) \quad (2.4)$$

is called a ***representation of the marked instanton as superposition of marked instantons.***

Definition 3. A hypernet $\omega \in \Omega^2$ is called an *H-marked half-instanton.*

Any marked 1-instanton is a superposition of two marked half-instantons.

Proposition 2.1. 1) *The space of marked half-instantons is a direct product:* $M_2 = G \times \mathbf{P}(H^*)$, *where* $G = G(2,T)$ *is the grassmannian of lines in* $\mathbf{P}(T)$.

2) *The imbedding* $G \times \mathbf{P}(H^*) = M_2 \subset \mathbf{P}S^2H^* \otimes \wedge^2 T^*$ *is defined by all sections of the sheaf* $p_1^* \mathcal{O}_G(1) \otimes p_2^* \mathcal{O}_{\mathbf{P}(H^*)}(2)$, *where* p_1 *and* p_2 *are the projections of the direct product on its factors.*

3) *The moduli space of H-marked 1-instantons is a direct product:*

$$M_1(H) = (\mathbf{P} \wedge^2 T^* - G) \times \mathbf{P}(H^*) \subset \mathbf{P} \wedge^2 T^* \otimes S^2 H^*,$$

where the projection of $M_1(H)$ *on* $\mathbf{P}(H^*)$ *is the map* (0.2).

4) *The closure* $\overline{M_1(H)} \subset \mathbf{P} \wedge^2 T^* \otimes \mathbf{P}(H^*)$ *lies between two elements of the filtration* (2.2):

$$M_2 \subset \overline{M_1(H)} \subset M_4.$$

5) $\mathcal{O}_{\mathbf{P}S^2H^* \otimes \wedge^2 T^*}(1) \mid_{\overline{M_1(H)}} = p_1^* \mathcal{O}_{\mathbf{P}\wedge^2 T^*}(1) \otimes p_2^* \mathcal{O}_{\mathbf{P}(H^*)}(2)$.

To prove these facts, it is sufficient to point out that if $\alpha \in \mathrm{Hom}(\wedge^2 T, S^2 H^*)$, then $\mathrm{rk}\,\alpha \geq 2\,\mathrm{rk}\,\alpha(t_0 \wedge t_1)$ for every t_1, t_2 from T. Consequently, if $\mathrm{rk}\,\alpha = 2$, then $\mathrm{rk}\,\alpha(t_0 \wedge t_1) = 1$, and the projectivisation $\mathbf{P}(\alpha(t_0 \wedge t_1))$ does not depend on $t_0, t_1 \in T$. Hence $\alpha = \kappa \otimes h^2$, $\kappa \in \wedge^2 T^*$, $h \in H^*$. But $\mathrm{rk}\,\kappa \otimes h^2 = \mathrm{rk}\,\kappa$, and consequently, $\mathrm{rk}\,\kappa = 2$. From this the assertion 1) follows. The other assertions are proved along the same lines.

Corollary. *Any H-marked instanton F is a superposition of marked half-instantons and is a superposition of marked 1-instantons.*

Indeed, $\mathbf{P} \wedge^2 T^* \otimes S^2 H^*$ is the linear envelope of M_2 and $M_1(H)$.

Definition 4. If $\kappa \in \wedge^2 T^*$, $\kappa \wedge \kappa = 0$, $h^i \in H^*$, a hypernet $\gamma = \kappa \otimes h^1 \cdot h^2$ is called a *marked quasi-instanton.*

Any marked quasi-instanton $\gamma = \kappa \otimes h^1 \cdot h^2$ is a superposition of two half-instantons

$$2\kappa \otimes h^1 \cdot h^2 = \kappa \otimes h_+^2 - \kappa \otimes h_-^2, \quad h_\pm = h^1 \pm h^2 \tag{2.5}$$

Definition 5. Any mathematical instanton F defines three numbers $h(F)$, $d(F)$ and $q(F)$ as follows:

$h(F)$ is the minimal number of terms in a decomposition of the exactly marked instantons (F,i) into half-instantons. The numbers $d(F)$ and $q(F)$ have the corresponding meaning for decomposition into 1-instantons and quasi-instantons.

Remark. These numbers do not depend on the choice of the exact markings.

The functions $h(F)$, $d(F)$ and $q(F)$ are semi-continuous on M_n.

Evidently $2q(F) \geq h(F) \leq 2d(F)$, $h(F) \geq n+1$, and $d(F) \geq n+1$.

The behavior of these functions is rather complicated.

Proposition 2.2. $h(F) = c_2(F) + 1$ if an only if F is a 't Hooft-instanton, that is, $H^0(F(1)) \neq 0$.

Indeed, if (F,i) is an exactly marked n-instanton and

$$\alpha(F,i) = \sum_{i=1}^{n+1} \kappa_i \otimes h_i^2,\ \kappa_i \in \wedge^2 T^*, \quad \kappa_i \wedge \kappa_i = 0, \quad h_i \in H^*, \tag{2.6}$$

then every κ_i defines a projective line $L_i \subset \mathbf{P}(T)$. For any projective plane $\mathbf{P}^2 \supset L_i$, $F|_{\mathbf{P}^2}$ is not a stable bundle, that is, there exists $s \in H^0(F|_{\mathbf{P}^2})$, $s \neq 0$, and $(s)_0 = \bigcup_{i \neq j}(L_j \cap \mathbf{P}^2)$. Hence $F(1)$ has a section vanishing on each L_i, $i = 1, \ldots, n+1$.

Conversely, if $F(1)$ has a section vanishing on these lines L_i, then, fixing one L_1, we can define by the other lines $L_2, \ldots, L_{n+1}$ a β-commutative hypernet α_0 (see [7]), such that $\alpha(F,i) - \alpha_0 = \kappa_1 \otimes h_1^2$, where $\mathbf{P}\kappa_1 = L_1$ and $\alpha_0 = \sum_{i=2}^{n} \kappa_i \otimes h_i^2$.

Corollary. *If* $n \geq 3$, *then* $\max_{F \in M_n} h(F) \geq n+2$.

A trivial calculation shows that for $n \geq 13$,

$$\max_{F \in M_n} d(F) \geq n+2, \quad \max_{F \in M_n} q(F) \geq n+3,$$

and so on.

Any tensor α of the space $T^* \otimes T^* \otimes H^* \otimes H^*$ has the decomposition:

$$\begin{aligned} 4\alpha &= \alpha^+_+ + \alpha^-_- + \alpha^+_- + \alpha^-_+, \\ \alpha^+_+ &= \alpha + \alpha^{*_T} + \alpha^{*_H} + \alpha^* \in S^2T \otimes S^2H^*, \\ \alpha^-_- &= \alpha - \alpha^{*_T} - \alpha^{*_H} + \alpha^* \in \Lambda^2T^* \otimes \Lambda^2H^*, \\ \alpha^+_- &= \alpha + \alpha^{*_T} - \alpha^{*_H} - \alpha^* \in S^2T^* \otimes \Lambda^2H^*, \\ \alpha^-_+ &= \alpha - \alpha^{*_T} + \alpha^{*_H} - \alpha^* \in \Lambda^2T^* \otimes S^2H^*, \end{aligned} \tag{2.7}$$

where $*_T$, $*_H$ and $*$ are the operators (1.5).

Notice that for any $\alpha \in T^* \otimes T^* \otimes H^* \otimes H^*$, rank α is the rank of the corresponding homomorphism $T \otimes H \to T^* \otimes H^*$.

Any tensor ξ of rank 1 is defined by two homomorphisms

$$\begin{aligned} &\varphi_i : T \to H^*, \quad i = 1, 2, \text{ and} \\ &\xi_{\varphi_1,\varphi_2} = \varphi_1 \otimes \varphi_2 : T \otimes T \to H^* \otimes H^* \end{aligned} \tag{2.8}$$

If $\{t_i\}$, $i = 0, \dots, 3$, is a basis of T and $\{t^i\}$ is the dual basis of T^*, then such a tensor has the following components (2.7):

$$\begin{aligned} \xi^+_+ &= \sum_{i,j} t^i \cdot t^j \otimes [\varphi_1(t_i) \cdot \varphi_2(t_j) + \varphi_1(t_j)\varphi_2(t_i)], \\ \xi^-_- &= \sum_{i<j} t^i \wedge t^j \otimes [\varphi_1(t_i) \wedge \varphi_2(t_j) - \varphi_1(t_j) \wedge \varphi_2(t_i)], \\ \xi^+_- &= \sum_{i,j} t^i \cdot t^j \otimes [\varphi_1(t_i) \wedge \varphi_2(t_j) + \varphi_1(t_j) \wedge \varphi_2(t_i)], \\ \xi^-_+ &= \sum_{i<j} t^i \wedge t^j \otimes [\varphi_1(t_i)\varphi_2(t_j) - \varphi_1(t_j)\varphi_2(t_i)]. \end{aligned} \tag{2.9}$$

The last component ξ^-_+ is a superposition of 12 quasi-instantons, or of 24 half-instantons.

Any tensor $\alpha \in T^* \otimes T^* \otimes H^* \otimes H^*$ of rank $2n + 2$ is a superposition

$$\alpha = \sum_{k=1}^{2n+2} \xi_{\varphi_1^\kappa,\varphi_2^\kappa}, \quad \operatorname{rk} \xi_{\varphi_1^\kappa,\varphi_2^\kappa} = 1$$

of tensors of rank 1 and its $(\pm)$-component equals

$$\alpha^-_+ = \sum_{i<j} t^i \wedge t^j \otimes \left(\sum_{k=1}^{2n+2} \varphi_1^\kappa(t_i)\varphi_2^\kappa(t_j) - \varphi_1^\kappa(t_j) \cdot \varphi_2^\kappa(t_i) \right) \tag{2.10}$$

If $\alpha \in \wedge^2 T^* \otimes S^2 H^*$, then $\alpha = \alpha_+^-$ and (2.10) is a decomposition of it into $24(n+1)$ quasi-instantons, or $48(n+1)$ half-instantons.

Thus we obtain trivial estimates of the numbers $h(F)$ and $q(F)$:

$$q(F) \leq 24(n+1), \quad h(F) \leq 48(n+1)$$

But we have a more exact geometrical result:

The components ξ_-^- and ξ_+^- (2.9) of the general tensor ξ of rank 1 are $*_T$-skew-symmetric and have the property $\alpha 2'$) of §1. According to construction (1.8), they define H-marked bundles over $\mathbf{P}(T)$.

Let $\operatorname{ad} T\mathbf{P}(T)$ be the adjoint bundle to the tangent bundle $T\mathbf{P}(T)$ of the projective space $\mathbf{P}(T)$, i.e., the subbundle of $\operatorname{End} T\mathbf{P}(T)$ of endomorphisms with zero trace. We have an isomorphism

$$H^1(\operatorname{ad} T\mathbf{P}(T)(-1)) = T, \tag{2.11}$$

and any monomorphism $\varphi\colon T \longrightarrow H^*$ defines the H-marking of this bundle. By the well-known exact sequence for the tangent bundle of $\mathbf{P}^n$, we have

Proposition 2.3. *If $\varphi\colon T \longrightarrow H^*$ is a monomorphism and $\xi_{\varphi,\varphi}$ is the corresponding tensor of rank 1, then*

$$(\xi_{\varphi,\varphi})_-^- = \alpha(\operatorname{ad} T\mathbf{P}(T), \varphi) \tag{2.12}$$

Remark. The $*_H$-skew-symmetry of $\alpha(\operatorname{ad} T\mathbf{P}(T), \varphi)$ is a consequence of the fact that $\operatorname{ad} T\mathbf{P}(T)$ has the orthogonal structure defined by the Killing-form of $\operatorname{ad} T\mathbf{P}(T)$.

From now on we restrict our consideration to $*$-skew-symmetric tensors of $T^* \otimes T^* \otimes H^* \otimes H^*$, that is, to the subspace

$$S^2 T^* \otimes \wedge^2 H^* \oplus \wedge^2 T^* \otimes S^2 H^*.$$

Any $*$-skew-symmetric tensor ω of rank 2 is defined by a pencil of homomorphisms $T \to H^*$, or by a homomorphism $\psi : T \otimes W_0 \to H^*$, where $W_0 = \mathbf{C}^2$ is a fixed 2-vector space. Such a tensor has a decomposition

$$\begin{aligned} 2\omega &= \omega^+ + \omega^-, \\ \omega^+ &= \omega + \omega^{*_T} = \omega - \omega^{*_H} \in S^2 T^* \otimes \wedge^2 H^*, \\ \omega^- &= \omega - \omega^{*_T} = \omega + \omega^{*_H} \in \wedge^2 T^* \otimes S^2 H^*. \end{aligned} \tag{2.13}$$

If ψ is a monomorphism, then ω_ψ^- has the following geometric interpretation:

For the bundle ad $T\mathbf{P}(T) \otimes W_0$, we have $H^1(\mathrm{ad}\, T\mathbf{P}(T) \otimes W_0(-1)) = T \otimes W_0$, and a homomorphism $\psi : T \otimes W_0 \to H^*$ defines an H-marking of this bundle.

Proposition 2.4. *If $\psi : T \otimes W_0 \to H^*$ is a monomorphism and ω_ψ is the corresponding tensor of rank 2, then*

$$\omega_\psi^- = \alpha(\mathrm{ad}\, T\mathbf{P}(T) \otimes W_0, \psi). \tag{2.14}$$

Remark. The $*_H$-symmetry of ω_ψ^- is a consequence of the fact that ad $T\mathbf{P}(T) \otimes W_0$ has a symplectic structure, defined by the tensor product of the Killing-form and $\wedge^2 W_0$.

Any tensor α, $\alpha^* = -\alpha$, of rank $2n+2$, is a superposition $\alpha = \sum_{i=1}^{n+1} \omega_i$, $\omega_i^* = -\omega_i$, $\mathrm{rk}\, \omega_i = 2$. From this we have

Proposition 2.5. *Any exactly H-marked n-instanton (F, i) is a superposition of $n+1$ marked bundles ad $T\mathbf{P}(T) \otimes W_0$.*

A general homomorphism $\psi : T \otimes W_0 \to H^*$ can be specialized to a homomorphism of rank 1:

$$\psi : T \otimes W_0 \xrightarrow{s} \mathbf{C} \to H^*.$$

In that case,

$$T \xrightarrow{s} W_0^*$$

defines the composition

$$T \xrightarrow{s} W_0^* \xrightarrow{\kappa_0} W_0 \xrightarrow{s^*} T^*, \tag{2.16}$$

where κ_0 is a standard skew-symmetric correlation. We get that

$$\kappa_\psi = s \cdot \kappa_0 \cdot s^* : T \to T^*, \quad \kappa_\psi \in \wedge^2 T^*, \quad \kappa_\psi \wedge \kappa_\psi = 0.$$

If $\mathbf{C} \cdot h_\psi = \mathrm{im}\, \varphi \subset H^*$, the components (2.3) of the tensor ω_ψ are

$$\omega_\psi^+ = 0, \quad \omega_\psi^- = \kappa \otimes h_\psi^2.$$

The superposition $\alpha = \sum_{i=1}^{n+1} \omega_{\psi_i}$ for such ψ_i is an H-marked 't Hooft-instanton.

§3. The Special Superposition

The group $GL(n, \mathbf{C}) = \operatorname{Aut} H^*$ acts on the final element Ω^2 of the filtration (2.1). Consider certain orbits of this action:

I. $\omega \in M_2 \subset \Omega^2$, $\omega = \kappa \otimes h^2$ is a marked half-instanton, and $\dim M_2 = n+3$.

II. $t^i \in T^*$, $h^i \in H^*$, $i = 1, 2$,
$\omega = t^1 \otimes t^2 \otimes h^1 \otimes h^2 - t^2 \otimes t^1 \otimes h^2 \otimes h^1 \in \Omega^2$,
$\omega^+ = t^1 \cdot t^2 \otimes h^1 \wedge h^2$,
$\omega^- = t^1 \wedge t^2 \otimes h^1 \cdot h^2$ is a marked quasi-instanton.
Let $\Omega^2_{II} \subset \Omega^2$ be the subvariety of all such tensors ω. Then

$$\dim \Omega^2_{II} = 2(n+2).$$

III. If $\xi_{\varphi_1,\varphi_2}$, $\varphi_i : T \to H^*$, $\ker\varphi_1 = \ker\varphi_2$, $\dim\ker\varphi_i = 2$ is a tensor of rank 1 (see (2.8)), and $\omega = \xi_{\varphi_1,\varphi_2} - \xi^*_{\varphi_1,\varphi_2} = (\xi_{\varphi_1,\varphi_2})^+_- + (\xi_{\varphi_1,\varphi_2})^-_+$ (see (2.7)), then $T/\ker\varphi_i = W_0$, (φ_1, φ_2): $W_0 \otimes I_0 \longrightarrow H^*$, where $I_0 = \mathbf{C}^2$, and ω^+: $S^2T \xrightarrow{S^2(j)} S^2W_0 \to \wedge^2 H^*$, $\omega^- = \kappa \otimes q$, where $\ker\kappa = \ker\varphi_i$, and $q = \mathbf{P}(W_0) \times \mathbf{P}(I_0)$ is a quadric of rank 4 in $\mathbf{P}(H)$.
Let $\Omega^2_{III} \subset \Omega^2$ be the subvariety of all such tensors ω. Then

$$\dim \Omega^2_{III} = 4n.$$

IV. ω_ψ, ψ: $T \otimes W_0 \longrightarrow H^*$ (see (2.13)), $\dim \psi(T \otimes W_0) = 4$, and for some $\omega_0 \in W_0$ the homomorphism $\psi_{\omega_0} = \psi|_{T\times\omega_0}$ $T \to H^*$ is a monomorphism.

Let $\Omega^2_{IV} \subset \Omega$ be the subvariety of all such tensors. It has the following description: A homomorphism ψ can be interpreted as a pencil of homomorphisms:

$$0 \to T \otimes \mathcal{O}_{\mathbf{P}(W_0)}(-1) \xrightarrow{\overline{\psi}} H^* \otimes \mathcal{O}_{\mathbf{P}(W_0)}. \tag{3.1}$$

In the general case, among the homomorphisms $\overline{\psi}_p$, $p \in \mathbf{P}(W_0)$ there are four degenerate ones whose kernels define four points $\{p_i\}$ in $\mathbf{P}(T)$, $i = 0, \ldots, 3$. Any non-degenerate homomorphism $\overline{\psi}_p$ of the pencil (3.1) maps these points on the four points $\{q_i\} \in \mathbf{P}(H^*)$. In view of this, we have a map:

$$f\colon \Omega^2_{IV} \longrightarrow S^4(\mathbf{P}(T) \times \mathbf{P}(H)^*), \tag{3.2}$$

by which the points (p_i, q_i), $i = 0, \dots, 3$, are attached to a sheaf (3.1).

To describe the fiber of this map it is sufficient ot restrict oneself to the case $H^* = \operatorname{im} \psi$. The points $\{p_i\}$ in $\mathbf{P}(T)$ and $\{q_i\}$ in $\mathbf{P}(H^*)$ define the decompositions of the corresponding spaces into a direct sum of 1-spaces:

$$T = \bigoplus_{i=0}^{3} L_i, \quad H^* = \bigoplus_{i=0}^{3} M_i, \quad \dim L_i = \dim M_i = 1.$$

Using the components of these decompositions, we can construct a new 4-vector space:

$$T_{\{p_i,q_i\}} = \bigoplus_{i=0}^{3} L_i \otimes M_i^*. \tag{3.3}$$

Proposition 3.1. *The points of the fiber of the map (3.2) are in one-to-one correspondence with the projective lines $L \subset \mathbf{P}T_{\{p_i,q_i\}}$.*

Indeed, a pencil of homomorphisms (3.1) can be decomposed into a direct sum

$$\begin{gathered} 0 \to L_i \otimes \mathcal{O}_{\mathbf{P}(W_0)}(-1) \xrightarrow{\overline{\psi}_i} M_i, \\ \overline{\psi} = \bigoplus_{i=0}^{3} \psi_i \end{gathered} \tag{3.4}$$

Each 1-pencil is defined by a monomorphism $\psi_i \colon L_i \to W_0^* \otimes M_i$, which we can interpret as monomorphism $\overline{\psi}_i \colon L_i \otimes M_i^* \to W_0^*$.

Adding these monomorphisms, we obtain an epimorphism

$$\tilde{\psi} \colon \bigoplus_{i=0}^{3} L_i \otimes M_i^* \to W_0^* \to 0. \tag{3.5}$$

This epimorphism defines a skew-symmetric correlation

$$\begin{gathered} T_{\{p_i,q_i\}} \xrightarrow{\tilde{\psi}} W_0^* \xrightarrow{\kappa_0} W_0 \xrightarrow{\tilde{\psi}^*} T^*_{\{p_i,q_i\}}, \\ \kappa = \tilde{\psi} \cdot \kappa_0 \cdot \tilde{\psi}^*, \end{gathered} \tag{3.6}$$

where κ_0 is the standard skew-symmetric correlation of W_0 (see (2.16)).

Corollary. *The fiber of the map f (3.2) at $\{p_i, q_i\}$ is given by*

$$f^{-1}(\{p_i, q_i\}) = G \subset \mathbf{P} \wedge^2 T^*_{\{p_i,q_i\}},$$

where G is the grassmannian variety of lines in $\mathbf{P}T^*_{\{p_i,q_i\}}$.

Remark. A choice of basis $\{h^i\}$ in H^* such tht $\mathbf{P}(h_i) = q_i$ defines isomorphisms $M_i \cong \mathbf{C}$ and, by this, an isomorphism $T \cong T_{\{p_i,q_i\}}$. Analogously, a choice of a basis $\{t_i\}$ in T, such that $\mathbf{P}(t_i) = p_i$, defines an isomorphism $T^*_{\{p_i,q_i\}} \cong H^*$. Consequently, if the four points $q_i = q$ coincide, then the isomorphisms $M_i \cong \mathbf{C}$ define an isomorphism $\mathbf{P}(T) \cong \mathbf{P}(T_{\{p_i,q_i\}})$ and a line $L \subset \mathbf{P}(T)$, which is the line corresponding to a half-instanton (2.16).

The components of the correlation (3.6)

$$\begin{array}{ccc} T_{\{p_i,q_i\}} & & T^*_{\{p_i,q_i\}} \\ \| & \overset{\kappa}{\to} & \| \\ \bigoplus_{i=0}^{3} L_i \otimes M_i^* & & \bigoplus_{i=0}^{3} L_i^* \otimes M_i, \end{array}$$

$$\kappa_{ij}\colon L_i \otimes M_i^* \longrightarrow L_j^* \otimes M_j \tag{3.7}$$

can be interpreted as homomorphisms

$$\tilde{\kappa}_{ij}\colon L_i \otimes L_j \longrightarrow M_i \otimes M_j. \tag{3.8}$$

They can be collected together to form a homomorphism

$$\begin{array}{ccc} \left(\bigoplus_{i=0}^{3} L_i\right) \otimes \left(\bigoplus_{i=0}^{3} L_i\right) & & \left(\bigoplus_{i=0}^{3} M_i\right) \otimes \left(\bigoplus_{i=0}^{3} M_i\right) \\ \| & \overset{\tilde{\kappa}}{\to} & \| \\ T \otimes T & & H^* \otimes H^*. \end{array} \tag{3.9}$$

We obtain

$$\begin{gathered} \omega^- = \wedge^2 T = \bigoplus_{i<j} L_i \wedge L_j \overset{\tilde{\kappa}^-}{\to} \bigoplus_{i\le j} M_i \otimes M_j = S^2 H^*, \\ L_i \otimes L_j \overset{\tilde{\kappa}^-_{ij} = \tilde{\kappa}_{ij} - \tilde{\kappa}_{ji}}{\longrightarrow} M_i \otimes M_j, \\ \omega^+ = S^2 T = \bigoplus_{i\le j} L_i \otimes L_j \overset{\tilde{\kappa}^+}{\to} \bigoplus_{i<j} M_i \wedge M_j = \wedge^2 H^*, \\ L_i \otimes L_j \overset{\tilde{\kappa}^+_{ij} = \tilde{\kappa}_{ij} + \tilde{\kappa}_{ji}}{\longrightarrow} M_i \otimes M_j. \end{gathered} \tag{3.10}$$

Corollary. $\ker \omega^{+} = \bigoplus_{i=0}^{3} L_i^2$.

From this follows

Proposition 3.2. *The variety Ω^2_{IV} is birationally equivalent to a direct product:*

$$\Omega^2_{IV} \overset{\text{bir}}{\sim} G \times S^4(\mathbf{P}(T) \times \mathbf{P}(H^*)),$$

and

$$\dim \Omega^2_{IV} = 4(n+3).$$

If $\omega \in \Omega^2_{IV}$, then its parameters will be denoted by symbols, $\{p_i\}_\omega$, $\{q_i\}_\omega$, $T_\omega = T_{\{p_i, q_i\}}$, $L_\omega \subset \mathbf{P}(T_\omega)$.

Let us fix the quadruple of points $\{p_i\}$ in $\mathbf{P}(T)$ and consider the subvariety $\Omega^2_{\{p_i\}} \subset \Omega^2_{IV}$ defined by

$$\Omega^2_{\{p_i\}} = \{\omega \in \Omega^2_{IV} \mid \{p_i\}_\omega = \{p_i\}\}. \tag{3.11}$$

Then

$$\dim \Omega^2_{\{p_i\}} = 4n,$$

Definition 6. If $D \subset \Omega^2$ is any subvariety of Ω^2, then a formal sum

$$\sum_{i=1}^{n+1} \omega_i, \quad \omega_i \in D, \tag{3.12}$$

defined up to a multiplicative constant, is called a *D-superposition.*

By the symbol $S(D)$ we denote the variety of the all D-superpositions.

From this definition it follows that $S(D)$ is birationally equivalent to a direct product

$$S(D) \overset{\text{bir}}{\sim} \mathbf{P}^n \times S^{n+1}(D), \tag{3.13}$$

where S^{n+1} is $(n+1)$-th symmetric power of our variety.

Associating the tensor $\alpha = \sum_{i=1}^{n+1} \omega_i \in \mathbf{P}(\wedge^2(T \otimes H)^*)$ to the formal sum (3.12), we define a map

$$\alpha\colon S(D) \longrightarrow \mathbf{P}\wedge^2(T \otimes H)^*, \tag{3.14}$$

and the projections on components α^{+}_{-} and α^{-}_{+} (2.7) provide maps:

$$\begin{aligned} \alpha^{+}&: S(D) \longrightarrow \mathbf{P}(S^2T^* \otimes \wedge^2 H^*),\\ \alpha^{-}&: S(D) \longrightarrow \mathbf{P}(\wedge^2 T^* \otimes S^2 H^*). \end{aligned} \tag{3.15}$$

These projections define a subvariety $S^{-}(D) \subset S(D)$ by

$$S^{-}(D) = \left\{ \sum_{i=1}^{n+1} \omega_i \in S(D) \mid \alpha^{+}\left(\sum_{i=1}^{n+1} \omega_i\right) = 0 \right\} \tag{3.16}$$

Finally, the image

$$\alpha^{-}(S^{-}(D)) \subset M_{2n+2}$$

belongs to the $(2n+2)$-th element of the filtration (2.2).

Applying this construction to the orbits I-IV in Ω^2, we obtain the following:

I. $\alpha^{-}(S^{-}(M_2)) \subset \overline{M_n(H)}$ is a subvariety of exact marked 't Hooft-instantons.

II. Since $\alpha^{-}(S^{-}(\Omega^2_{II})) \supset \alpha^{-}(S^{-}(M_2))$,

$$\alpha^{-}(S^{-}(\Omega^2_{II})) \subset \overline{M_n(H)} \tag{3.17}$$

is a subvariety of exact marked instantons which are the superpositions of $(n+1)$ quasi-instantons.

Analogously

$$\begin{aligned} \alpha^{-}(S^{-}(\Omega^2_{III})) &\subset \overline{M_n(H)},\\ \alpha^{-}(S^{-}(\Omega^2_{IV})) &\subset \overline{M_n(H)},\\ \alpha^{-}(S^{-}(\Omega^2_{\{P_i\}})) &\subset \overline{M_n(H)}. \end{aligned}$$

Proposition 3.3 *The variety $\alpha^{-}(S^{-}(\Omega^2_{\{P_i\}}))$ is the component of $\overline{M_n(H)}$ containing 't Hooft-instantons.*

The dimension of the fiber of the map α^{-} over a point of $\alpha^{-}(S^{-}(\Omega^2_{\{P_i\}}))$ is not more than 4.

Indeed, by (3.13), $\dim S(\Omega^2_{\{p_i\}}) = 4n^2 + 5n$. By the corollary to (3.10), a subvariety $S^{-}(\Omega^2_{\{p_i\}} \subset S(\Omega^2_{\{p_i\}})$ (3.16) is defined by no more than $3n(n-1)$

equations. From this we get

$$\dim S^-(\Omega^2_{\{p_i\}}) \geq n^2 + 8n.$$

Careful checking of the second assertion of our proposition concludes the proof.

For small value of n the conditions

$$\sum_{i=1}^{n+1} \omega_i^+ = 0, \quad \omega_i \in \Omega^2_{\{p_i\}}$$

have a simple geometrical meaning. In this situation the direct geometrical constructions provide

Proposition 3.4. The variety $S^-(\Omega^2_{\{p_i\}})$ is unirational for $n < 6$.

This proves the unirationality of M_4 and also of the component of M_5 containing the 't Hooft-instantons.

References

[1] Barth W., Irreducibility of the space of mathematical instanton bundles with rank 2 and $c_2 = 4$, Math. Ann. 258, 1981, 81–106.

[2] Barth W., Hulek K., Monads and moduli of vector bundles. Manuscripta Math., 25, 1978, 323–347.

[3] Hartshorne R., Stable vector bundles of rank 2 on $\mathbf{P}^3$, Math. Ann., 238, 1978, 229–280.

[4] Hartshorne R., Algebraic vector bundles on projective spaces. A problem list. Topology, 18, 1979, 117–128.

[5] Jozefiak T., Lascoux A., Pragacz P. Klassu, determinantnux mnogoobrasij, associirovannux s simmetricheskoy i kososimmetricheskoy matricami, Izv. AN SSSR, ser. math., t.45, N3, 1981, 662–673 (in Russian).

[6] Room T. G., The geometry of determinantal loci, Cambridge University Press, 1938.

[7] Tjurin A. N., Struktura mnogoobrazia par kommutirujushin puchkov simmetricheskih matric, Izv. AN SSSR, ser. math., t.46, N2, 1982, 409–430 (in Russian).

Received August 30, 1982

Professor Andrei Nikolaevic Tjurin
Steklov Institute of Mathematics
ul. Vavilova, 42
Moscow 117966 GSP-1, USSR

How Many Kähler Metrics Has a K-3 Surface?

A. N. Todorov

To my teacher Igor Rostislavovich Shafarevich on the occasion of his 60th birthday

§ 0 Introduction

The aim of this article is to prove the following theorem:

Theorem 1. *Let X be a Kähler K-3 surface and let u be a two form representing a class in $H^{1,1}(X,\mathbb{R})$. The form u is cohomologous to the imaginary part of a Kähler metric on X iff a) $\int_X u\wedge u > 0$, and b) for any effective divisor D on X, $\int_D u > 0$.*

This theorem is an important step in the description of the moduli space of Einstein metrics on a compact 4-dimensional manifold diffeomorphic to a K-3 surface. The description of the moduli space of Einstein metrics on a K-3 surface is given in [T2], assuming theorem 1. There we use the following results in order to prove that the moduli space of all Einstein metrics is isomorphic to an open and everywhere dense subset of

$$SO(3,19)/SO(3)\times SO(19)/\Gamma,$$

where $\Gamma = Aut(L)$ and L is a Euclidean lattice defined in §1:

1) Yau's proof of the Calabi conjecture, 2) Shafarevich and Piatetski-Shapiro's proof of the global Torelli theorem for K-3 surfaces, 3) the uniqeness of a Kähler-Einstein metric in a fixed class $u\in H^{1,1}(X,\mathbb{R})$ fulfilling conditions a) and b) of theorem 1, 4) The fact that each Einstein metric is isometrically equivalent to a Kähler-Einstein metric with respect to a one-complex parameter family of complex structures, compatible with the given Einstein metric.

The author wishes to thank Professors B. Moishezon, M. Kuranishi, H. Pinkham, I. Morrison and D. Morrison for their helpful remarks, when the author lectured on theorem 1 at Columbia University.

§ 1 Some Standard Facts about K-3 Surfaces

Definition 1.1. A two-dimensional complex compact manifold is a K-3 *surface* if a) $\pi_1(X) = 0$, and b) the canonical class of X is zero, (i.e., trivial).

It is a well-known fact that $rk\, H^2(X,\mathbb{Z}) = 22$. Since $\pi_1(X) = 0$, $H^2(X,\mathbb{Z})$ is a free abelian group. Cup product defines on $H^2(X,\mathbb{Z})$ a scalar product with values in $\mathbb{Z}$, with the following properties:

a) $\langle u,u\rangle \equiv 0 \pmod 2$,
b) If $e_1,\ldots,e_{22}$ is a basis of $H^2(X,\mathbb{Z})$, then $\det(\langle e_i,e_j\rangle) = -1$,
c) The scalar product defined by cup product has signature $(3,19)$.

It is a well-known fact that all Euclidean lattices, i.e., free abelian groups with a scalar product, having properties a), b) and c), are isometric. Let us fix one of these lattices and denote it by L.

Definition 1.2. A pair (X,φ) is called a *marked* K-3 *surface* if X is a K-3 surface and φ is an isomorphism of lattices

$$\varphi : H^2(X,\mathbb{Z}) \simeq L.$$

Definition 1.3. An *admissible Hodge structure of type* $(1,20,1)$ on L is a filtration $H^{2,0} \subset H^{2,0} + H^{1,1} \subset L \otimes \mathbb{C}$ such that a) $\dim H^{2,0} = 1$, b) $H^{1,1} = (H^{2,0} + H^{0,2})^{\perp}$, and c) if $w \in H^{2,0}$ and $w \neq 0$, then $\langle w,w\rangle = 0$ and $\langle w,w\rangle > 0$.

NOTATION. Let $M \subset \mathbf{P}(L\otimes\mathbb{C})$ be defined by the following formulas:

$$\langle z,z\rangle = Q(z,z) = 0 \tag{1.4}$$

$$\langle z,\overline{z}\rangle = Q(z,\overline{z}) > 0, \tag{1.5}$$

We also denote by Q the quadric in $\mathbf{P}(L\otimes\mathbb{C})$ defined by the scalar product in L.

Proposition 1.6. *There is a one to one map between points of M and oriented two dimensional subspaces E in $L\otimes\mathbb{C}$ such that $\langle\ ,\ \rangle|_E > 0$.*

This map is given in the following way: a point $w \in M$ defines a line L_w in $L \otimes \mathbb{C}$, and w corresponds to $E_w = L_w + \overline{L}_w$ with the following orientation: let $w \neq 0$, then $(Re\, w, Im\, w)$ will be an oriented orthonormal basis of E_w.

For the proof see [T].

Definition 1.7. Let (X, φ) be a marked K-3 surface. The map

$$p\colon (X, \varphi) \longrightarrow m \in M$$

where m is defined by the line $\varphi(H^{2,0}(X, \mathbb{C}))L \otimes \mathbb{C}$ is called the *period map*.

We consider M as a complex manifold embedded in $\mathbf{P}(L \otimes \mathbb{C})$ and M is defined by (1.4.) and (1.5.).

Remark 1.8. Let $\pi\colon \mathcal{X} \longrightarrow \mathcal{D}$ be a family of K-3 surfaces and suppose that $\pi_1(\mathcal{D}) = 0$, and that $\pi^{-1}(o) = X_o$ is a marked K-3 surface. Then since $\pi_1(\mathcal{D}) = 0$, we may suppose that all K-3 surfaces in the family $\pi\colon \mathcal{X} \longrightarrow \mathcal{D}$ are marked K-3 surfaces, because $R^2\pi_*\mathbb{Z}$ will be a trivial local system on $\mathcal{D}$. So the period map $p\colon \mathcal{D} \longrightarrow M$ is a well-defined map. Furthermore it is an etale holomorphic map if $\mathcal{D}$ is the Kuranishi family of X_o. (See [Gr]).

Definition 1.9. Let X be a K-3 surface with a fixed Kähler metric $(g_{i\bar{j}})$. Let

$$V(X) = \{v \in H^{1,1}(X, \mathbb{R}) \mid \langle v, v\rangle > 0\}.$$

Clearly $V(X) = V^+(X) \cup (V^-(X))$. Let

$$\Delta(X) = \{b \in H^{1,1}(X, \mathbb{R}) \cap H^2(X, \mathbb{Z}) \mid \langle b, b\rangle = -2\}.$$

Let G be the subgroup of $Aut(L)$ generated by the reflections $s_b(v) = v + \langle v, b\rangle b$ for $b \in \Delta(X)$. It is a well-known fact that G acts properly and discontinuously on $V^+(X)$. $V^+(X)$ is the part of $V(X)$ which contains the imaginary part of the Kähler metric $(g_{i\bar{j}})$, i.e. $i/2 \sum g_{i\bar{j}} dz_i \wedge d\bar{z}_j$. The fundamental domain of G in $V^+(X)$ will be a convex polyhedron, bounded by the hyperplanes $H_b = \{v \in H^{1,1}(X, \mathbb{R}) \mid \langle v, b\rangle = 0\}$. Let us denote the fundamental domain of G which contains $i/2 \sum_1 g_{i\bar{j}} dz_i \wedge \overline{dz_j}$ by $V_p^+(X)$. *We will call this fundamental domain the Kählerian cone of X.*

It is a well-known fact that $V_p^+(X)$ defines a partition

$$P : \Delta = \Delta^+ \cup (-\Delta^+),$$

where $\Delta^+ = \{b \in \Delta \mid \langle v, b\rangle > 0 \text{ for all } v \in V_p^+(X)\}$, and Δ^+ has the following property:

$$(*) \qquad \text{If } b_1, \ldots, b_k \in \Delta^+ \text{ and } \sum n_i b_i \text{ for } n_i \in \mathbb{Z} \text{ and } n_i > 0, \text{ then } \sum n_i b_i \in \Delta^+.$$

Each partition P of $\Delta = \Delta^+ \cup (-\Delta^+)$ with the property $(*)$ defines a fundamental domain $V_P^+(X) = \{v \in V_P^+(X) \mid \langle v, b\rangle > 0$ for $b \in \Delta^+\}$. From the way we define the Kählerian cone $V_P(X)$, it follows that $\Delta^+ = \{$effective divisors D, such that $\langle D, D\rangle = -2\}$. For the proof of this fact see [B & R].

§ 2 How Many Kähler Metrics has a Kählerian K-3 Surface X?

Theorem 1. *Let X be a Kähler K-3 surface and let $V_P^+(X)$ be the Kählerian cone of X, then for every $v \in V_P^+(X)$, there exists a Kähler metric $(g_{i\bar{j}})$ such that $v = i/2 \sum g_{i\bar{j}} dz_i \wedge d\bar{z}_j$ in $H^{1,1}(X, \mathbb{R})$.*

Proof. The proof is based on several lemmas.

Lemma 2.1. There is an open and everywhere dense subset $M' \subset M$ such that:

a) each point of M' corresponds to a marked K-3 surface.

b) $M \setminus M' = \cup V_i$, where V_i are complex analytic subspaces in M. ($M \subset \mathbf{P}(L \otimes \mathbb{C})$ is defined by (1.4.) and (1.5.).)

Proof. Let $q \in L \otimes \mathbb{Q}$ and $\langle q, q\rangle > 0$. Let

$$\mathbf{P}(H_q) = \{v \in \mathbf{P}(L \otimes \mathbb{C} \mid \langle v, q\rangle = 0\} \text{ and } M_q = M \cap \mathbf{P}(H_q).$$

It is a well-known fact that:

i) $M_q \cong SO(2,19)/SO(2)\times SO(19)$

ii) On M_q there is an open and everywhere dense subset of algebraic K-3 surfaces. (See [SP] and [Sh], ch. X.)

Let $M'_q = \{m \in M_q \mid m$ correspond to a marked K-3 surface (X,φ) such that $Nq = \varphi(H)$, where H is a Poincaré dual of a very ample divisor on X.$\}$.

Proposition 2.1.1. *$M_q \setminus M'_q = \cup V^i_q$, where V^i_q are countably many analytic subsets of M_q.*

Proof. Let N_o be the minimal number such that $N_o q = \varphi H$, where H is the Poincaré dual of a hyperplane section on some marked K-3 surface (X,φ). That such a number N_o exists follows from the definition of M'_q. Of course we suppose that $p(X,\varphi) = m \in M'_q$. Let $|H|: X \longrightarrow \mathbb{P}^n$ be an embedding. We will denote by $Hilb^q_{\mathbb{P}^n}$ the Hilbert scheme of all nonsingular K-3 surface in $\mathbb{P}^n$ which contain X. In [SP] it is proved that $Hilb^q_{\mathbb{P}^n}$ is a nonsingular scheme. From the general results of Grothendieck it follows that $Hilb^q_{\mathbb{P}^n}$ is quasi-projective. Let $\hat{Hilb}^q_{\mathbb{P}^n}$ be a projective variety which is non singular, contains $Hilb^q_{\mathbb{P}^n}$ and such that $\hat{Hilb}^q_{\mathbb{P}^n} \setminus Hilb^q_{\mathbb{P}^n}$ is a divisor with normal crossings. The existence of $\hat{Hilb}^q_{\mathbb{P}^n}$ follows from Hironaka's resolution of singularities theorem. Let

$$\Gamma_q = \{\varphi \in Aut(L) \mid \varphi(N_o q) = N_o q\}.$$

$M_q = SO(2,19)/SO(2)\times SO(19)$ is a Siegel domain of type IV and Γ_q acts on M_q as a discrete group. From the Baily-Borel compactification theory we know that M_q/Γ_q is an open Zariski set in a projective variety $\widehat{N_q/\Gamma_q}$. Over $Hilb^q_{\mathbb{P}^n}$ we have a universal family $\pi: \mathcal{X} \longrightarrow Hilb^q_{\mathbb{P}^n}$, so we have a well-defined map $p: Hilb^q_{\mathbb{P}^n} \longrightarrow M_q/\Gamma_q$, where p is the period map. Borel proved in [B] that p can be prolonged to a holomorphic map $\hat{p}: \hat{Hilb}^q_{\mathbb{P}^n} \longrightarrow \widehat{M_q/\Gamma_q}$. From the local Torelli theorem it follows that the map $\hat{p}$ is a surjective map. From the global Torelli theorem, and the fact that $PGL(n)$ acts on $Hilb^q_{\mathbb{P}^n}$ in such a way that two points z_1 and z_2 of $Hilb^q_{\mathbb{P}^n}$ correspond to isomorphic K-3 surfaces iff z_1 and z_2 are in the same orbit of $PGL(n)$, it follows that $p(\hat{Hilb}^q_{\mathbb{P}^n} \setminus Hilb^q_{\mathbb{P}^n}) \cap p(Hilb^q_{\mathbb{P}^n}) = 0$. Now it is clear that

$p(Hilb^q_{\mathbb{P}^n})$ is a Zariski-open subset in $M_q\hat{/}\Gamma_q$, hence in M_q/Γ_q. So $V_q = M_q/\Gamma_q \setminus p(Hilb^q_{\mathbb{P}^n})$ is an algebraic subspace of M_q/Γ_q. If $f\colon M_q \longrightarrow M_q/\Gamma_q$ is the natural projection, then it is clear that $M_q \setminus M'_q = f^{-1}(V_q) = \cup V^i_q$. Q.E.D.

Remark. It is not too difficult to prove this lemma without using the global Torelli theorem.

Definition 2.1.2. A Kähler metric $g_{i\bar{j}}$ on a K-3 surface X will be called a *Kähler-Einstein metric* if $Ricci(g_{i\bar{j}}) = 0$.

Let X be a K-3 surface with a Kähler metric $(g_{i\bar{j}})$. It follows from Yau's result that we can find a Kähler-Einstein metric $(g_{i\bar{j}})$ such that $[i/2\sum g_{ij}dz_i \wedge d\bar{z}_j] = [i/2\sum g_{i\bar{j}}dz_i \wedge d\bar{z}_j]$ in $H^{1,1}(X,\mathbb{R})$ (See [Y]). The metric $g_{i\bar{j}}$ is uniquely determined up to a constant. Let us fix this Kähler-Einstein metric on X. Denote by Λ^2 the bundle of real two forms on X. Since the real dimension of X is four, the Hodge operator $*$ maps Λ^2 onto Λ^2. From the fact that $*^2 = id$ we have the following decomposition: $\Lambda^2 = \Lambda_+ \oplus \Lambda$, where $\Lambda_+ = \{w \in \Lambda^2 \mid *w = w\}$ and $\Lambda_- = \{w \in \Lambda^2 \mid *w - w\}$. The Kähler-Einstein metric induces a connection D on Λ^2 compatible with the metric $(g_{i\bar{j}})$. So D induces a connection D_+ on Λ_+.

Lemma 2.1.3. *The curvature of D_+ is zero, i.e., Λ_+ is a flat bundle with respect to D.*

Corollary 2.1.4. a) *Λ_+ is a trivial bundle on X.*

b) *A form $w \in \Gamma(X,\Lambda_+)$ is a parallel form iff $dw = 0$.*

c) *The dimension of the space of parallel forms in $\Gamma(X,\Lambda_+)$ is three.*

For the proof of these facts see [T] or [H].

2.1.5. A construction of an isometric deformation with respect to the Kähler-Einstein metric $(g_{i\bar{j}})$.

The construction is based on the following remark due to Andreotti and Weil: Let W be a complex valued two-form with the following properties: a) at each point $x \in X$, $W \wedge W \equiv 0$, b) at each point $x \in X$, $W \wedge \overline{W} > 0$, & c) $dW \equiv 0$.

a) is equivalent to the fact that W defines a two-dimensional subspace $E_x \subset T_x^* \otimes \mathbb{C}$ at each point $x \in X$, where T_x^* is the real cotangent space at x.

b) is equivalent to the fact that $E_x \overline{E}_x = T_x^* \otimes \mathbb{C}$.

c) is equivalent to the fact that the almost complex structure defined by a) & b) is integrable. (See [W]).

2.1.5.1. On each fibre $(\Lambda^2)_x$, the metric $(g_{i\bar{j}})$ defines in a natural manner a scalar product. Let e_1, e_2 & e_3 be an orthonormal basis of $(\Lambda^2)_x$ with respect to the scalar product on Λ^2 induced from $(g_{i\bar{j}})$. It follows from the definition of the scalar product that:

$$(*) \qquad e_i \wedge e_j = \delta_{ij} vol(g_{k\bar{l}}).$$

Let $\tilde{e}_1, \tilde{e}_2, \tilde{e}_e$ be elements of $\Gamma(x, \Lambda_+)$ obtained from e_1, e_2, e_3 by parallel displacement. Then

$$(**) \qquad \tilde{e}_i \wedge \tilde{e}_j = \delta_{ij} vol(g_{k\bar{l}})$$

Proposition 2.1.5.2. *Let* $W = \tilde{e}_1 + i\tilde{e}_2 \in H^2(X, \mathbb{C})$. *Then* a) $W \wedge W \equiv 0$, b) $W \wedge \overline{W} \equiv vol(g_{i\bar{j}})$, & c) $dW \equiv 0$.

Proof. Proposition 2.1.5.2. follows immediately from (**). Q.E.D.

Let $H(g_{i\bar{j}})$ be the subspace of $\Gamma(X, \Lambda_+)$ which consists of all parallel sections with respect to D_+. Since each parallel section of $\Gamma(X, \Lambda_+)$ is a harmonic form, we get from Hodge theory that $H(g_{i\bar{j}}) \subset H^2(X, \mathbb{R})$. From Proposition 2.1.5.2. and the Androtti-Weil remark, we get:

Proposition 2.1.5.3. *Every oriented two plane in* $H(g_{i\bar{j}})$ *defines a complex structure on* X.

Remark 2.1.5.4. Let $w_X(2,0)$ be the holomorphic two form on X. It is not difficult to prove that $Re\, w_X(2,0)$, $Im\, w_X(2,0)$ and $Im(g_{i\bar{j}})$ are parallel forms and form an orthonormal basis of $H(g_{ij})$. On the other hand all oriented two planes in $H(g_{ij})$ are parametrized by $\mathbb{P}^1(\mathbb{C})$. So we get a family $\mathcal{X} \to \mathbb{P}^1(\mathbb{C})$.

Remark 2.1.5.5. Let (X, φ) be a marked K-3 surface with a fixed Kähler-Einstein metric $(g_{i\bar{j}})$ such that $\varphi(Im(g_{i\bar{j}})) = q \in L \otimes \mathbb{Q}$. We know that the point $p(X,\varphi) = x$ corresponds to an oriented two dimensional subspace $E_x = \big((Re\, w_X(2,0),\ Im\, w_X(2,0)\big)$. See 1.6. Let us denote $E_x(q) = \varphi\big(H(g_{i\bar{j}})\big)$. Clearly $E_x \subset E_x(q)$. Notice that the form $\langle\ ,\ \rangle$ restricted to $E_x(q)$ is positive definite. Let us denote by $\mathbb{P}^1_x(q)$ the plane quadric in $\mathbf{P}(E_x(q) \otimes \mathbb{C})$ defined by $Q(z,z) = 0$.

Proposition 2.1.5.6. *The image via P of the family of K-3 surfaces obtained from the isometric deformation with respect to* $(g_{i\bar{j}})$ *is* $\mathbb{P}^1_x(q)$, *where* $\varphi\big(Im(g_{i\bar{j}})\big) = q$.

Proof. See [T]. Q.E.D.

Definition 2.1.5.7.

$$M' \overset{\text{def}}{=} \bigcup_{\substack{q \in L \otimes \mathbb{Q} \\ \langle q,q\rangle > 0}} \Big(\bigcup_{x \in M'_q} \mathbb{P}^1_x(q) \Big),$$

where $\mathbb{P}^1_x(q) = M \cap \mathbf{P}\,(E_x(q) \otimes \mathbb{C})$, and $E_x(q)$ is the three-dimensional subspace in $L \otimes \mathbb{R}$ spanned by E_x and q.

Notice that the form $\langle\ ,\ \rangle$ on $E_x(q)$ is positive definite. Also, M'_q consists of those points of $M_q \overset{\text{def}}{=} M \cap \mathbf{P}\,(H_q)$ which correspond to marked K-3 surface having the following property: $N = \varphi^{-1}(q)$ is the Poincaré dual of an ample divisor on the K-3 surface (X,φ).

From the way we define M', lemma 2.1.1, and the isometric deformations coming from the Hodge metrics $\varphi^{-1}(q)$, lemma 2.1. follows. Q.E.D.

Let (X,φ) be any marked Kählerian K-3 surface and $p(X,\varphi) = x$. Let $w \in V^+_p(X)$, where $V^+_p(X)$ is the Kählerian cone of X[1] (See def. 1.9.). Clearly $\langle w,w\rangle > 0$ and $\langle w, E_x\rangle = 0$. Let us denote by $E_x(w)$ the three-dimensional subspace of $L \otimes \mathbb{R}$ spanned by E_x and w. Clearly $\langle\ ,\ \rangle$ restricted to $E_x(w)$ is positive definite and it follows that $\mathbf{P}\,(E_x(w) \otimes \mathbb{C})$ is contained in the open set U of $\mathbf{P}\,(L \otimes \mathbb{C})$ defined by the inequality 1.5.. From this fact, it follows that the plane quadric $\mathbb{P}^1_x(w)$ in $\mathbf{P}\big(E_x(w) \otimes \mathbb{C}\big)$ defined by $Q(z,z) = 0$ is contained in M.

---

[1]There we denote by $V^+_p(X)$ also the image $\varphi\big(V^+_p(X)\big)$ in $L \otimes \mathbb{R}$.

Definition 2.3. $K_x(\mathbb{R}) \stackrel{\text{def}}{=} \bigcup_{w \in V_p^+(X)} \mathbb{P}_x^1(w)$.

Remark 2.4. $K_x(\mathbb{R})$ is a real analytic subset of dimension 21 in M.

Definition 2.5. $K'_x(\mathbb{R}) = \{\mathbb{P}_x^1(w) \subset K_x(\mathbb{R}) \mid \mathbb{P}_x^1(w) \subset M'\}$, where M' is defined by 2.1.5.7.

Lemma 2.6. *$K_x(\mathbb{R}) \backslash K'_x(\mathbb{R})$ is a finite union of real analytic subspaces of real codimension ≥ 2. $K'_x(\mathbb{R})$ is an open and everywhere dense subset in $K_x(\mathbb{R})$.*

Proof. Let us denote by $Grass(3, 19; \mathbb{R})$ the set of subspaces $E \subset L \otimes \mathbb{R}$ such that a) $dim_R E = 3$, and b) $\langle\ ,\ \rangle|_E$ is positive definite. Clearly

$$(2.6.1.) \qquad Grass(3, 19; \mathbb{R}) \simeq SO(3, 19)/SO(3) \times SO(19)$$

Let $Grass(3, 19; \mathbb{C}) = \{E \subset L \otimes \mathbb{C} \mid \text{a) } \dim E = 3, \text{ b) } \langle u, \overline{u}\rangle_E \text{ is positive definite}\}$. Then

$$(2.6.2.) \qquad Grass(3, 19; \mathbb{C}) \simeq SO(3, 19; \mathbb{C})/SO(3; \mathbb{C}) \times SO(19; \mathbb{C}).$$

Notice that $Grass(3, 19; \mathbb{R}) = Grass(3, 19; \mathbb{C})^{\tilde{c}}$, and so $Grass(3, 19; \mathbb{R})$ consists of the fixed points of the real involution $\tilde{c}$, where $\tilde{c}(E) = \overline{E}$, for $E \subset L \otimes \mathbb{C}$.

From the definition of $\mathbb{P}_x^1(w)$ it follows that $\mathbb{P}_x^1(w)$ is uniquely determined by $E_x(w)$. All $\mathbb{P}_x^1(w)$ in $K_x(\mathbb{R})$ are in a one-to-one correspondence with $E_x(w)$, where $w \in V_p^+(X)$. $\{E_x(w) \mid w \in V_p^+(X)\}$ is a real analytic subvariety of $Grass(3, 19; \mathbb{R})$.

2.6.3. Let us denote this analytic subvariety by $\mathbf{P}\left(V_p^+(X)\right)$, where $\mathbf{P}\left(V_p^+(X)\right)$ is the projectivization of $V_p^+(X)$ in $\mathbf{P}\left(H^{1,1}(X, \mathbb{R})\right)$. Clearly $dim_{\mathbb{R}} \mathbf{P}\left(V_p^+(X)\right) = 19$.

2.6.4. Let $V_p^+(X)(\mathbb{C})$ be $V_p^+(X) + iV_p^+(X) \subset H^{1,1}(X, \mathbb{C})$. If $w \in V_p^+(X)(\mathbb{C})$, then $\langle w, w\rangle > 0$. Let $E_x(w)$ be the subspace of $H^{1,1}(X, \mathbb{C})$ generated by E_x and w. It is clear that a) $dim_{\mathbb{C}} E_x = 3$, and b) $\langle\ ,\ \rangle|_{E_x(w)}$ is positive definite. From b) it follows that

$$\mathbf{P}\left(E_x(w)\right) \subset U = \{z \in \mathbf{P}\left(L \otimes \mathbb{C}\right) \mid \langle z, \overline{z}\rangle > 0\}.$$

So $\mathbb{P}^1_x(w) \stackrel{\text{def}}{=} \mathbf{P}\left(E_x(w)\right) \cap M$ will be a plane quadric contained in M.

2.6.5. Let $K_x(\mathbb{C}) \stackrel{\text{def}}{=} \bigcup_{w \in V_p^+(X)(\mathbb{C})} \mathbb{P}^1(w)$. Then $dim_{\mathbb{C}} K_x(\mathbb{C}) = 20$.

2.6.6. The union of the spaces $E_x(w)$ for all $w \in V_p(X)(\mathbb{C})$ forms a complex submanifold of $Grass(3, 19; \mathbb{C})$. Let us denote this complex submanifold by $\mathbf{P}(V_p^+(X)(\mathbb{C}))$.

2.6.7. $\mathbf{P}(V_p^+(X)(\mathbb{C})) = \left(\mathbf{P}\left(V_p^+(X)\right)\right)^{\tilde{c}}$, where $\tilde{c}(E) = \overline{E}$, $E \in L \otimes \mathbb{C}$.

2.6.8. We know from 2.1. that 1) M' is an open and everywhere dense subset of M 2) $M \setminus M' = \bigcup V_i$, where V_i are complex analytic subspaces of M.

Let $\tilde{V}_i = \{\mathbb{P}^1_x(w) \subset K_x(\mathbb{C}) \mid \mathbb{P}^1_x(w) \cap V_i \neq 0\}$. $\tilde{V}_i$ defines a complex analytic subset $\mathbf{P}(\tilde{V}_i) \subset \mathbf{P}\left(V_p^+(X)(\mathbb{C})\right) \subset Grass(3, 19; \mathbb{C})$.

Proposition 2.6.9. $\mathbf{P}(\tilde{V}_i)$ *does not contain* $\mathbf{P}\left(V_p^+(X)\right)$ *for any* i.

Proof. Suppose $\mathbf{P}\left(V_p^+(X)\right) \subset \mathbf{P}(\tilde{V}_i)$ for some i. Since $\mathbf{P}(\tilde{V}_i)$ are complex analytic subsets in $\mathbf{P}\left(V_p^+(X)(\mathbb{C})\right)$, are given locally by a system equations $f_j(z_1, \ldots, z_k) = 0$, where f_j are complex analytic functions in $\mathbb{C}^k$. Since $\mathbf{P}\left(V_p^+(X)(\mathbb{C})\right)^{\tilde{c}} = \mathbf{P}\left(V_p^+(X)\right) \subset \mathbf{P}(\tilde{V}_i)$, it follows that $f_j = 0$. Indeed, if f is a complex analytic function and $f(Re\, z_1, \ldots, Re\, z_k) \equiv 0$, then $f \equiv 0$. This shows that $\mathbf{P}\left(V_p^+(X)\right)$ is not contained in any of the $\mathbf{P}(\tilde{V}_i)$. Furthermore $\mathbf{P}\left(V_p^+(X)\right) \cap \mathbf{P}(V_i)$ is a real analytic subset of $\mathbf{P}\left(V_p^+(X)\right)$ of codimension 2. Q.E.D.

2.6.10. Let $\overline{V_p^+(X)(\mathbb{C})} = \overline{V_p^+(X)} + i\overline{V_p(X)}$, denote the closure of $V_p^+(X)(\mathbb{C})$ in $H^{1,1}(X, \mathbb{C})$. Clearly

$$\overline{K_x(\mathbb{C})} = \bigcup_{w \in V_p^+(X)(\mathbb{C})} \mathbb{P}^1_x(w)$$

is a compact subset of M, and $\overline{K_x(\mathbb{C})}$ is the closure of $K_x(\mathbb{C})$ in M. Since the closure of $K_x(\mathbb{C})$ in M is a compact set, it follows from the way we constructed V_i that those V_i for which $K_x(\mathbb{C}) \cap V_i \neq 0$ are finite in number. Q.E.D.

2.6.11. Let $K'_x(\mathbb{R}) = \{\mathbb{P}^1_x(w) \subset K_x(\mathbb{R}) \mid \mathbb{P}^1_x(w) \subset M'\}$, then from 2.1.

and 2.6.9. it follows that:

a) $K'_x(\mathbb{R})$ is an open and everywhere dense subset in $K_x(\mathbb{R})$.

b) $K_x(\mathbb{R}) \setminus K'_x(\mathbb{R}) = W_i$, where W_i are real analytic subsets in $K_x(\mathbb{R})$ of real codimension ≥ 2

This proves lemma 2.6. Q.E.D.

2.7. Let $U(X) = \bigcup\{E_x(w) \mid w \in V_p^+(X)\}$. Clearly $U(X) \subset L \otimes \mathbb{R}$.

Proposition 2.7.1. *$U(X)$ is an open set in $L \otimes \mathbb{R}$, diffeomorphic to $E_x \times V_p^+(X)$.*

Proof. It is clear that if $u \in U(X)$, then $u = u' + u''$, where $u' \in E_x$ & $u'' \in V_p^+(X)$, and this decomposition is unique. From this it follows that $U(X)$ is diffeomorphic to $E_x \times V_p(X)$. This proves proposition 2.7.1.. Q.E.D.

2.7.2. Let

$$U'(X) = \bigcup\{E_x(w) \subset U(X) \mid \mathbb{P}^1_x(w) \subset M', \quad w \in V_p^+(X)(\mathbb{R})\}.$$

It[2] follows from lemma 2.6. that $U'(X)$ is an open and everywhere dense subset of $U(X)$.

Proposition 2.7.3. *Let*

$$W(X) \stackrel{\text{def}}{=} \bigcup\{E_x(w) \subset U'(X) \mid E_x(w) \cap (L \otimes \mathbb{Q}) \neq 0\}.$$

Then $W(X)$ is an everywhere dense subset in $U(X)$.

Proof. Since $U'(X)$ is an open set in $L \otimes \mathbb{R}$,

$$W(X) \subset U'(X) \cap (L \otimes \mathbb{R})$$

is an everywhere dense subset of $U'(X)$. Since $U'(X)$ is an open and everywhere dense subset of $U(X)$, $W(X)$ is an everywhere subset of $U(X)$. Q.E.D.

---

[2]Here $\mathbb{P}^1_x(w)$ means $M \cap \mathbb{P}\left(E_x(w)\right)$, where $w \in V_p^+(X)(\mathbb{C})$.

2.7.4. Let $V_p^+(X)(\mathbb{Q}) \stackrel{\text{def}}{=} V_p^+(X) \cap W(X)$. It follows from proposition 2.7.3. that $V_p^+(X)(\mathbb{Q})$ is an everywhere dense subset in $V_p^+(X)$.

Proposition 2.7.5. *Suppose that (X, φ) is a marked Kähler K-3 surface and $p(X, \varphi) = x$. If $w \in V_p^+(X)(\mathbb{Q})$, then $\varphi^{-1}(w)$ is the imaginary part of a Kähler metric on X.*

Proof. This lemma is proved in [T] (see the proof of lemma 3.5. on p. 261). The same proof can be found in the paper by [Siu]. Q.E.D.

It follows from proposition 2.7.5. that there is an everywhere dense subset $V_p(X)(\mathbb{Q})$ of $V_p^+(X)$ such that every element of this subset is the imaginary part of a Kähler metric on X. Theorem 1 follows from the following remark:

Remark 2.7.6. Suppose that $v_1, \ldots, v_{20}$ are linearly independent elements of $V_p^+(X)$ and that all of them are imaginary parts of Kähler metrics on X, then $v = \sum a_i v_i$ is the imaginary part of a Kähler metric g on X if all a_i are positive real numbers.

Proof. If v_i corresponds to the imaginary part of a Kähler metric (g_i), i.e. $v_i = Im(g_i)$ in $H^{1,1}(X, \mathbb{R})$, then v will correspond to the imaginary part of the metric $g = \sum a_i g_i$. Q.E.D.

Thoerem 1 follows immediately from proposition 2.7.4. and remark 2.7.5. Q.E.D.

References

[B R] D. Burns and M. Rapoport, On the Torelli problem for Kählerian K-3 surfaces, Ann. Sci. Ecole Norm. Sup. 4, ser. 8 (1975).

[H] N. Hitchin, Compact four-dimensional Einstein manifolds, Journal of Differential Geometry, 435–441 (1974).

[Sh] I. R. Shafarevich, Algebraic surfaces, Proc. Steklov Inst., vol. 75 (1965).

[S P] Shafarevich and Piatetski-Shapiro, A Torelli theorem for algebraic surfaces of type K-3, Izv. Akad. Nauk 35, 530–365 (1971).

[T] A. Todorov, Application of Kähler-Einstein-Calabi-Yau metric to Moduli of K-3 surfaces, Inventiones Math., 61, 251–265 (1980).

[T2] A. Todorov, The moduli space of Einstein metrics on a K-3 surface, To appear in a Proc. of a conference in math. physics Promorsko.

[Yau] S. T. Yau, On the Ricci curvature of a compact Kähler manifold and the Monge-Ampere equation 1, Comm. Pure Appl. Math. XXXI, 339–411 (1978).

[W] A. Weil, Collected papers, vol. 2, 365–395, Springer-Verlag 1979.

[Siu] Y. T. Siu, A simple proof of the surjectivity of the period map for K-3 surfaces, Manuscripta Math. 35, 311–321 (1981).

Received July 2, 1982

Professor A. N. Todorov
Bulgarian Academy of Sciences
BAN Institute of Math.
Sofia 1090, P.B. 373
Sofia
Bulgaria

On the Problem of Irreducibility of the Algebraic System of Irreducible Plane Curves of a Given Order and Having a Given Number of Nodes

Oscar Zariski

To I.R. Shafarevich

Introduction. Let k be an algebraically closed group field of characteristic zero. The following result is well known (see Severi [3], Anhang F):

If $\sigma_{n,d}$ is a maximal irreducible algebraic system, defined over k, of plane algebraic (not necessarily irreducible) curves of a given order n, and if the general curve C^* of $\sigma_{n,d}/k$ has d nodes (and no other singularities), then the dimension of $\sigma_{n,d}$ is equal to $3n+p-1$, where $p = \frac{(n-1)(n-2)}{2} - d$ is the "effective" genus of C^*.

We note that $3n+p-1$ is equal to $N-d$. Here $N = \frac{n(n+3)}{2}$ is the dimension of the (linear) system of all plane curves of order n. This is in agreement with the intuitive expectation that the requirement that a curve of order n possess d nodes (in *non-assigned* position) imposes d independent algebraic (non-linear) conditions on the curve.

The above result is easily proved and was included by us in a more general theorem which we have proved in our recent paper [6] which deals with maximal irreducible algebraic systems σ of curves of order n in which the general curve C^* has arbitrary singularities (but is free from multiple components; equivalently: C^* is a *reduced* curve). This theorem asserts that *if p is the "effective" genus of C^** [as defined below in formula (2)], then $\dim \sigma \leq 3n+p-1$, *with equality if and only if C^* has only nodes as singularities.*

It is easily seen (see Severi, loc. cit.) that if $0 \leq d \leq \frac{n(n-1)}{2}$, then there always exist curves of order n having d nodes (the maximum $d = \frac{n(n-1)}{2}$ being reached only for n-gons, i.e., for curves consisting of n distinct lines, no three of which have a common point). For *irreducible* curves of order n with d nodes to exist it is necessary and sufficient that d satisfy the stronger inequality $d \leq (n-1)(n-2)/2$.

We shall denote by $\sigma_{n,d}$ any *maximal* irreducible algebraic system of plane curves of order n in which the general curve C^* has d nodes and no other singularities. Thus, we always have $d \leq n(n-1)/2$, and if C^* is irreducible, then $d \leq (n-1)((n-2)/2$.

In our paper [6] we have pointed out that Severi's "proof" of the following important statement (see Severi [3], Anhang F) is trivially erroneous:

Basic Conjecture I. *For any positive integer n and any non-negative integer d such that $d \leq (n-1)(n-2)/2$, there exists only one system $\sigma_{n,d}$ in which the general curve C^* is irreducible.*

Since it is easily proved (see Severi [3], Anhang F, pp. 316-317) that every plane curve C of order n which has d nodes (and no other singularities) belongs to a unique system $\sigma_{n,d}$, it would follow (if the above conjecture is proved) that *all the irreducible plane curves of a given order n and with a given number d of nodes belong to a single irreducible algebraic system $\sigma_{n,d}$.*

It can be shown by examples that if $d \geq n-1$ and $n \geq 4$, then the set of all curves of order n with d nodes (including *reducible curves*) does not belong to one and the same maximal irreducible system $\sigma_{n,d}$. For instance, if $n = 4$ and $d = 3$, then there exist *two* maximal irredicible systems $\sigma_{4,3}$: in one of the systems the general curve is an irreducible (rational) quartic curve with 3 nodes, while in the other system the general curve is the union of a non-singular (elliptic) cubic curve and a line meeting that cubic in 3 distinct points. However, a modified *irredicibility* theorem would follow trivially from the Basic Conjecture I also for reducible curves.

This modified theorem is the following:

For any finite set of positive integers $n_1, n_2, \ldots, n_q$ and for any finite set of q non-negative integers $d_1, d_2, \ldots, d_q$ such that $d_i \leq (n_i - 1)(n_i - 2)/2$, there exists only one system $\sigma_{n,d}$, where $n = n_1 + n_2 + \cdots + n_q$ and $d = d_1 + d_2 + \cdots + d_q + \sum\sum_{i \neq j} n_i n_j$, in which the general curve C^* has q irreducible components $C_1^*, C_2^*, \ldots, C_q^*$ of orders $n_1, n_2, \ldots, n_q$, and where each C_i^* has d_i nodes. (Note that for any i, j the two curves C_i^* and C_j^* must have only simple intersections, and no three of the curves $C_1^*, C_2^*, \ldots, C_q^*$ can have a common point. Thus the nodes of C^* consist of the $d_1 + d_2 + \cdots + d_q$ nodes of the q curves C_i^* and of the $\sum\sum_{i \neq j} n_i n_j$

intersections of the $\binom{q}{2}$ pairs of curves $C_i^*, C_j^*, i \neq j$).

Proof. We consider *the* q systems Σ_{n_i,d_i} in which the general curve C_i^* is irreducible (and which are therefore uniquely determined by the Basic Conjecture I). We consider, as we did in [6], the quasi-direct sum of the q systems Σ_{n_i,d_i} i.e., the set of all plane curves of the form $C_1+C_2+\cdots+C_q$, where C_i is any curve in Σ_{n_i,d_i}. This is an irreducible algebraic system of plane curves of order $n = n_1 + n_2 + \cdots + n_q$ and of dimension equal to the sum of the dimensions of q systems Σ_{n_i,d_i}. We have proved in [6] that this system is a system $\Sigma_{n,d}$. This system satisfies the conditions stated in the above theorem. Now, if $\Sigma'_{n,d}$ is any other system having the properties stated in that theorem, and $C'^* = C_1'^* + C_2'^* + \cdots + C_q'^*$ is the general curve of $\Sigma'_{n,d}$, then it follows from the Basic Conjecture I that $C_i'^*$ belongs to the above system Σ_{n_i,d_i}. Hence C'^* belongs to the above system $\Sigma_{n,d}$, and thus $\Sigma'_{n,d}$ is contained in $\Sigma_{n,d}$. Since $\Sigma'_{n,d}$ is a maximal system with the properties stated in the above theorem, it follows that $\Sigma'_{n,d}$ *coincides* with the system $\Sigma_{n,d}$ constructed in the beginning of the proof. This completes the proof of the above theorem.

Our contribution to a possible proof of the Basic Conjecture I will be based exclusively on another conjecture which we shall refer to as Basic Conjecture II and which should be easier to prove (or to disprove). We now proceed to state the Basic Conjecture II.

Let $\Sigma_{n,d}$ be a system in which the general curve C^* is *irreducible*. Let

$$f(X,Y) = \sum_{0 \leq i+j \leq n} A_{ij} X^i Y^j = 0 \tag{1}$$

be the equation of C^*. The ring R generated over k by the ratios of the coefficients A_{ij} is the non-homogeneous coördinate ring of the representative variety $\pi(\Sigma_{n,d})$ of $\Sigma_{n,d}$ in $\mathbf{P}_N(N = \frac{n(n+3)}{2})$. Let $(-\alpha_\nu, -\beta_\nu)$ be the coördinates of the d nodes $Q_1^*, Q_2^*, \ldots, Q_d^*$ of $C^*(\nu = 1, 2, \ldots, d)$. The α_ν's and β_ν's are algebraic over the quotient field K of R, since α_ν and β_ν must satisfy the 3 equations $f(-\alpha_\nu, -\beta_\nu) = 0$, $f'_X(-\alpha_\nu, -\beta_\nu) = 0$, $f'_Y(-\alpha_\nu, -\beta_\nu) = 0$, and since by our assumption concerning the curve C^*, these 3 equations have exactly d solutions $(-\alpha_\nu, -\beta_\nu)$.

Basic Conjecture II: *If the general curve C^* of a $\Sigma_{n,d}$ is irreducible, then the d nodes $Q_1^*, Q_2^*, \ldots, Q_d^*$ of C^* from a* complete *set of d conjugate* *algebraic points over the function field K of the representative variety $\pi(\Sigma_{n,d})$ of $\Sigma_{n,d}$ in the projective space $\mathbf{P}_N(N = n(n+3)/2)$.*

Note. The Basic Conjecture II is not generally true if C^* is reducible. For instance, if $\Sigma_{n,d}$ is a quasi-direct sum $\Sigma_{n-1,d_1} \oplus \Sigma_{1,0}$, the general curve C^* is the union $l^* + C^*_{n-1}$, where l^* is a generic line and C^*_{n-1} is an irreducible curve of order $n-1$ with d_1 nodes. Here $d = n - 1 + d_1$. The $n-1$ intersections of l^* and C^*_{n-1} are nodes of C^*, and they form by themselves a complete system of conjugate points over K, the remaining d_1 nodes of C^* being the nodes of C^*_{n-1}. On the other hand, the basic conjecture is true also in one (and only one) special case in which C^* is reducible. This is the case in which C^* is the union of two irreducible *non-singular* curves C^*_1, C^*_2 (having, of course, only simple intersections). In this case, we have $d = n_1 n_2$ where n_i is the order of $C^*_i (i = 1, 2)$.

We note that if the coördinate axes X, Y are chosen to be lines rational over k, then no line parallel to the X-axis can have more than one multiple intersection with C^*, and that any such intersection has necessarily multiplicity 2. Equivalently: any line parallel to the X-axis is neither a flex tangent, nor a double (or a multiple) tangent, nor is tangent to C^* at any of the d nodes $Q^*_1, Q^*_2, \ldots, Q^*_d$. In particular, the abscissas $-\alpha_1, -\alpha_2, \ldots, -\alpha_d$ of the nodes are distinct (and are, of course, algebraic over K). It follows that if $D(X)$ is the Y-discriminant of $f(X, Y)$ [see (1)] then $D(X) = [D_1(X)]^2 D_2(X)$, where

$$D_1(X) = \sum_{\nu=1}^{d} (X + \alpha_\nu)$$

and where $D_2(X)$ is a polynomial of degree $2n + 2p - 2$, free from multiple roots. The coefficients of $D_1(X)$ and $D_2(X)$ belong to the field K. Our Basic Conjecture II states that *the polynomial $D_1(X)$ is irreducible over K.*

The Basic Conjecture I is trivial for low values of n, such as $n = 2, 3, 4$. This conjecture is also trivially true for any positive integer n and for $d = 0, 1$. We shall therefore use, in our proof, induction with respect to n and d, i.e., we shall assume that the Basic Conjecture I is true for $n - 1$ (and any $d \leq \frac{(n-2)(n-3)}{2}$) and is also true for any n and any non-negative integer d' less than d.

Our derivation of the Basic Conjecture I from the Basic Conjecture II will require a good deal of preliminary considerations and several auxiliary results. Of special importance will be a "Basic Lemma" which we now state.

Basic Lemma. *Given any system $\Sigma_{n,d}$ in which the general curve of the system is irreducible, there exist systems $\Sigma_{n-1,d}$ such that $\Sigma_{n,d}$ contains all the curves $l + C_{n-1}$, where l is any line in the plane and C_{n-1} is any curve in $\Sigma_{n-1,d}$.*

The basic lemma has the following immediate consequence (derived by induction from $n-1$ to n, the statement being trivial for $n = 2$):

Corollary of the basic lemma. Any system $\Sigma_{n,d}$ *contains all the n-gons of the plane (i.e., all the curves which consist of n lines).*

From this corollary the Basic Conjecture I follows without difficulty (see Severi [3], Anhang F). However, it is Severi's "proof" of this corollary that is triviallly erroneous.

We observe incidentally that the above corollary of the Basic Lemma was used by us in [4] in order to prove that the Poincaré group $G = \pi(\mathbf{P}_2 - C)$ of the residual space of any plane curve C having only nodes is abelian (and, in particular, is cyclic of order n if C is an irreducible curve of order n). The present paper would therefore eliminate the basic weak point of the cited paper [4], *if* the Basic Conjecture II is proved. In this connection we call the reader's attention to the paper [1] of Deligne where it is proved that G is indeed an abelian group. Deligne's proof is a topological adaptation of a purely algebraic proof by Fulton of the following result (see Fulton [2]): *if H is any invariant subgroup of G, of finite index, then the quotient group G/H is abelian.*

§1. Let $\Sigma_{n,d}$ be an irreducible algebraic system the general curve C^* of which is irreducible. We consider a line l in the plane. To simplify the algebraic considerations given below, we temporarily choose a coördinate system X, Y in which the line l is the Y-axis ($X = 0$).

We consider the subsystem of $\Sigma_{n,d}$ which consists of all the curves in $\Sigma_{n,d}$ which contain l as component. This subsystem is defined algebraically by the condition that the polynomical $f(0, Y)$ (see (1), Introduction) be identically zero. Let $(-\alpha_\nu, -\beta_\nu)$ be any of the d nodes Q_ν of C^*. Then we can write

$$f(X,Y)=\sum_{2\leq i+j\leq n} a_{ij}^{(\nu)}(X+\alpha_\nu)^i(Y+\beta_\nu)^j \qquad (\nu=1,2\ldots,d).$$

Setting $X=0$ we get

$$f(0,Y)=\sum_{2\leq i+j\leq n} a_{ij}^{(\nu)}\alpha_\nu^i(Y+\beta_\nu)^j.$$

The coefficients of this polynomial in $(Y+\beta_\nu)$ are the following $n+1$ quantities:

$$\begin{aligned}
&\varphi_0(\{a_{ij}^{(\nu)}\};\alpha_\nu)=\sum_{i=2}^{n} a_{i0}^{(\nu)}\alpha_\nu^i,\\
(1)\qquad &\varphi_1(\{a_{ij}^{(\nu)}\};\alpha_\nu)=\sum_{i=1}^{n-1} a_{i1}^{(\nu)}\alpha_\nu^i,\\
&\varphi_\mu(\{a_{ij}^{(\nu)}\};\alpha_\nu)=\sum_{i=0}^{n-\mu} a_{i\mu}^{(\nu)}\alpha_\nu^i \qquad (\mu=2,3,\ldots,n).
\end{aligned}$$

Thus the above subsystem of $\Sigma_{n,d}$ is defined by the following $n+1$ equations (where we note in particular, $\varphi_n=a_{0n}^{(\nu)}$):

$$(2)\qquad \varphi_\mu(\{a_{ij}^{(\nu)}\};\alpha_\nu)=0 \qquad (\mu=0,1,\ldots,n).$$

The subsystem of $\Sigma_{n,d}$ defined by the equations (2) may not be irreducible, but each of its irreducible components has dimension $\geq r-n-1$, where $r=\dim\Sigma_{n,d}=3n+p-1$. One general solution of the equations (2) is $\alpha_\nu=a_{02}^{(\nu)}=a_{03}^{(\nu)}=\ldots=a_{0n}^{(\nu)}=0$. We shall be interested primarily in the possible solutions of (2) in which $\alpha_\nu\neq 0$.

We observe that $\varphi_0(\{a_{ij}^{(\nu)}\};\alpha_\nu)$ and $\varphi_1(\{a_{ij}^{(\nu)}\};\alpha_\nu)$ are divisible by $\alpha^{(\nu)2}$ and $\alpha^{(\nu)}$, respectively. We shall therefore replace φ_0 and φ_1 by $\varphi_0/\alpha^{(\nu)2}$ and $\varphi_1/\alpha^{(\nu)}$, respectively, and we shall be interesed primarily in the subsystem of $\Sigma_{n,d}$ which is defined by the following $n+1$ equations:

$$\begin{aligned}
&\psi_0(\{a_{ij}^{(\nu)}\};\alpha_\nu)\ (=\ \varphi_0(\{a_{ij}^{(\nu)}\};\alpha_\nu)/\alpha_\nu^2)=0,\\
(3)\qquad &\psi_1(\{a_{ij}^{(\nu)}\};\alpha_\nu)\ (=\ \varphi_1(\{a_{ij}^{(\nu)}\};\alpha_\nu)/\alpha)=0,\\
&\psi_\mu(\{a_{ij}^{(\nu)}\};\alpha_\nu)\ (=\ \varphi_\mu(\{a_{ij}^{(\nu)}\};\alpha_\nu))=0 \qquad (\mu=2,3,\ldots,n),
\end{aligned}$$

or, more explicitly:

$$\text{(3a)}\qquad \begin{aligned} a_{20}^{(\nu)} &= -(a_{30}^{(\nu)}\alpha_\nu + \cdots + a_{n0}^{(\nu)}\alpha_\nu^{n-2}),\\ a_{11}^{(\nu)} &= -(a_{21}^{(\nu)}\alpha_\nu + \cdots + a_{n-1,1}^{(\nu)}\alpha_\nu^{n-2}),\\ a_{0\mu}^{(\nu)} &= -(a_{1\nu}^{(\nu)}\alpha_\nu + \cdots + a_{n-\mu,\mu}^{(\nu)}\alpha_\nu^{n-\mu}) \qquad (\mu = 2, 3, \ldots, n). \end{aligned}$$

We shall show, first of all, that the *algebraic subsystem of* $\Sigma_{n,d}$ defined by the equations (3) (and which *a priori* may depend on the choice of the node $(-\alpha_\nu, -\beta_\nu)$) is non-empty (equivalently: the ideal generated by $\psi_0, \psi_1, \ldots, \psi_n$ in the ring $R[\alpha_\nu, \beta_\nu]$ is not the unit ideal).

To show this we recall (see our paper [6], §3, Lemma 1) that the maximality of $\Sigma_{n,d}$ implies that $\Sigma_{n,d}$ *is invariant under all collineations of the plane.* We recall also that as a consequence of this lemma we have shown (loc, cit., §3, Lemma 2) the following:

Let C be any curve in $\Sigma_{n,d}$, let O be a point of C and let l_1 be a line which is not a component of C and does not contain the point O. Let $P_1, P_2, \ldots, P_n$ be the intersections of C with l_1. (The n points P_i need not be distinct; each point P_i is to be counted to a suitable multiplicity.) Then the system $\Sigma_{n,d}$ contains the curve consisting of the n lines $OP_1, OP_2, \ldots, OP_n$ (each line to be counted to a suitable multiplicity).

We now apply this result to proving that the system defined by the equations (3) is non-empty. We consider a *reduced* curve C in $\Sigma_{n,d}$ which has a node at the origin $X = Y = 0$:

$$C\colon \sum_{2\le i+j\le n} \bar{a}_{ij}^{(\nu)} X^i Y^j = 0.$$

Let l be the line $X = 0$. If we take for l_1 any line not containing the origin, which is not a component of C, which passes through an intersection of C with l and meets C in n distinct points, and if we assume, as we may, that l_1 is the line at infinity, the above-stated result signifies that $\Sigma_{n,d}$ contains the curve consisting of n distinct lines through the origin, namely the curve defined by the equation

$$\text{(4)}\qquad \sum_{i+j=n} \bar{a}_{ij}^{(\nu)} X^i Y^j = 0,$$

and $X = 0$ is one of these n lines (i.e., we have $\bar{a}_{o,n}^{(\nu)} = 0$, since l_1 contains an intersection of C with l). It follows that the ideal in $R[\alpha_\nu, \beta_\nu]$ generated by the elements $a_{ij}^{(\nu)}, i+j < n$, the element $a_{o,n}^{(\nu)}$ and the elements α_ν, β_ν, is not the unit ideal and that the set of n distinct concurrent lines defined by (4) (or, more precisely, the point which represents this set of n lines in the projective space $\mathbf{P}_N, N = \frac{n(n+3)}{2}$) is a zero of that ideal. Now, from the definition (1) of the φ_μ's and the definition (3) of the ψ_μ's, follows that *this ideal contains the ideal generated by the $n+1$ elements ψ_μ*. This proves our assertion that the subsystem of $\Sigma_{n,d}$ defined by the equations (3) is non-empty.

We point out explicitly another consequence of the above argument, namely that *the subsystem of $\Sigma_{n,d}$ defined by the equations (3) contains curves which have no multiple components*, since it contains, in particular, the set of n distinct concurrent lines defined by (4). This consequence will be of prime importance in the sequel. It implies that the *subsystem of $\Sigma_{n,d}$ defined by the equations (3) has at least one irreducible component the general curve of which is reduced.* We shall denote by $\Sigma_{n,d}^{(\nu)}(l)$ the union of those irreducible components of that subsystem, the general curves of which are reduced. The subsystem of $\Sigma_{n,d}$ defined by (3) will be denoted in the sequel by $S_{n,d}^{(\nu)}(l)$. Both systems $\Sigma_{n,d}^{(\nu)}(l)$ and $S_{n,d}^{(\nu)}(l)$ have l as fixed component. We shall write $\Sigma_{n,d}^{(\nu)}(l) = l + \Sigma_{n-1}^{(\nu)}(l)$, where $\Sigma_{n-1}^{(\nu)}(l)$ is an algebraic system of curves of order $n-1$. The systems $\Sigma_{n,d}^{(\nu)}(l)$ obtained as l varies will be the main object of the two propositions which we shall prove below (Proposition 1 and Propsition 2).[1]

We now go back to the equations (3). We denote by $\mathfrak{A}^{(\nu)}$ the ideal generated in $R[\alpha_\nu, \beta_\nu]$ by the $n+1$ elements $\psi_\mu(\{a_{ij}^{(\nu)}\}; \alpha_\nu)(\mu = 0, 1, \ldots, n)$. We have shown above that $\mathfrak{A}^{(\nu)}$ is not the unit ideal. Hence every minimal prime ideal of $\mathfrak{A}^{(\nu)}$ has dimension $\geq r - n - 1 = 2n + p - 2$. Here r is the transcendence degree of R over k ($r =$ dimension of the system $\Sigma_{n,d}$). If $\mathfrak{P}^{(\nu)}$ is any minimal prime ideal of $\mathfrak{A}^{(\nu)}$, then the intersection ideal $\mathfrak{P}^{(\nu)} \cap R$ is a prime ideal in R and hence defines an irreducible algebraic subsystem $\Sigma(\mathfrak{P}^{(\nu)})$ of $\Sigma_{n,d}/k$. We have shown earlier that there exist minimal prime ideals $\mathfrak{P}^{(\nu)}$ of $\mathfrak{A}^{(\nu)}$ such that the algebraic subsystem $\Sigma(\mathfrak{P}^{(\nu)})$

[1] It is highly probable that, as a consequence of the maximality of the system $\Sigma_{n,d}$, every irreducible component of the system defined by the equaations (3) has the property that its general curve is reduced and that therefore $\Sigma_{n,d}^{(\nu)}(l)$ coincides with the system $S_{n,d}^{(\nu)}(l)$ defined by (3).

of $\Sigma_{n,d}$ contains *reduced* curves. The general curve $C^*_{(\nu)}$ of this subsystem has therefore only a finite number of singular points. This implies that if $\alpha^*_\nu, \beta^*_\nu, A^{*\nu}_{ij}/A^{*(\nu)}_{n,o}$ are the $\mathfrak{P}^{(\nu)}$-residues of $\alpha_\nu, \beta_\nu, A_{ij}/A_{(n,o)}$, then $\alpha^*_\nu, \beta^*_\nu$ are algebraic over $k[A^{*(\nu)}_{ij}/A^{*(\nu)}_{n,o}]$. This implies that *the dimension of the prime ideal* $\mathfrak{P}^{(\nu)} \cap R$ is equal to *the dimension of the prime ideal* $\mathfrak{P}^{(\nu)}$, and hence is $\geq 2n+p-2$. Thus, *the algebraic subsystem* $l + \Sigma^{(\nu)}_{n-1}(l)$ *of* $\Sigma_{n,d}$, *which we have introduced above, has the property that all the irreducible components of that subsystem have dimension* $\geq 2n+p-2$. Hence $\Sigma^{(\nu)}_{n-1}(l)$ *is an algebraic system of curves of order* $n-1$, *all irreducible components of which have dimension* $\geq 2n+p-2$.

We have here d algebraic systems $\Sigma^{(\nu)}_{n-1}(l)$ (l being a given line) which depend *a priori* on the choice of the node $(-\alpha_\nu, -\beta_\nu)$. If, however, we assume the Basic Conjecture II stated in the Introduction, then it is clear that every minimal prime ideal $\mathfrak{P}^{(1)}$ of $\mathfrak{A}^{(1)}$ determines a set of K-conjugate minimal prime ideals $\mathfrak{P}^{(1)}, \mathfrak{P}^{(2)}, \ldots, \mathfrak{P}^{(d)}$ of the ideals $\mathfrak{A}^{(1)}, \mathfrak{A}^{(2)}, \ldots, \mathfrak{A}^{(d)}$ in the rings $R[\alpha_1, \beta_1], R[\alpha_2, \beta_2], \ldots, R[\alpha_d, \beta_d]$, and that consequently $\mathfrak{P}^{(1)} \cap R = \mathfrak{P}^{(2)} \cap R = \ldots = \mathfrak{P}^{(d)} \cap R$. It follows that *the algebraic system* $\Sigma^{(\nu)}_{n-1}(l)$ *does not depend on the choice of the node* $Q^*_\nu(-\alpha_\nu, -\beta_\nu)$. We shall denote this system and the system $\Sigma^{(\nu)}_{n,d}(l) = l + \Sigma^{(\nu)}_{n-1}(l)$ by $\Sigma^d_{n-1}(l)$ and $\Sigma^d_n(l)$, respectively. In a similar manner it follows that also the system $S^{(\nu)}_{n,d}(l)$, defined by the equations (3), is independent of ν. We shall denote this system by $S^d_n(l)$, or, equivalently, by $l + S^d_{n-1}(l)$.

In the above derivation of the equations (3) of the system $S^d_n(l)$ we have used a special coördinate system, namely one in which l is the Y-axis. We wish, however, to fix once and for all our coördinate system X, Y, taking as coördinate axes two lines defined over k, and take for l the general line l^* of the plane, over k. The equation of l^* is then of the form $X = uY + v$, where u and v are indeterminates. We set $X' = X - uY - v$ and apply the equations (3) to the coördinate system X', Y, which now replace the X, Y coördinate used in (3). The equation of the general curve C^* of $\Sigma_{n,d/k}$ in the X', Y-coördinaes is the following:

$$(5) \qquad C^*: \sum_{2 \leq i+j \leq n} a^{(\nu)}_{ij}(X' + uY + v + \alpha_\nu)^i (Y + \beta_\nu)^j = 0.$$

The X', Y coördinates of the double point $X = -\alpha_\nu, Y = -\beta_\nu$ of C^* are

$-\alpha'_\nu, -\beta_\nu$ where $\alpha'_\nu = \alpha_\nu - u\beta_\nu + v$. Thus, the equation (5) can be written as follows:

$$C^* : \sum_{2 \le i+j \le n} a_{ij}^{(\nu)} [X' + \alpha'_\nu + u(Y + \beta_\nu)]^i (Y + \beta_\nu)^j = 0,$$

and this equation is of the form

$$C^* : \sum_{2 \le i+j \le n} a_{ij}'^{(\nu)} (X' + \alpha'_\nu)^i (Y + \beta_\nu)^j = 0,$$

where

$$\text{(6)} \qquad \begin{aligned} a_{ij}'^{(\nu)} &= a_{ij}^{(\nu)} + (i+1) u a_{i+1,j-1}^{(\nu)} \\ &+ \binom{i+2}{2} u^2 a_{i+2,j-2}^{(\nu)} + \cdots + \binom{i+j}{j} u^j a_{i+j,0}^{(\nu)}. \end{aligned}$$

The equations which define the system $S_{n,d}^{(\nu)}(l^*)$ are therefore obtainable by replacing in the equations (3) the $a_{ij}^{(\nu)}$ by the $a_{ij}'^{(\nu)}$ and α_ν by α'_ν:

$$\text{(7)} \qquad \psi_\mu(\{a_{ij}'^{(\nu)}\}; \alpha'_\nu) = 0, \qquad (\mu = 0, 1, \ldots, n)$$

where the $a_{ij}'^{(\nu)}$ are given by (6) and where

$$\text{(8)} \qquad \alpha'_\nu = \alpha_\nu - u\beta_\nu + v.$$

The equations (7) are relations between the quantities $a_{ij}^{(\nu)}, \alpha_\nu, \beta_\nu$ and the quantities u, v:

$$\text{(9)} \qquad \psi_\mu^*(\{a_{ij}^{(\nu)}\}, \alpha_\nu, \beta_\nu, u, v) = 0. \qquad (\mu = 0, 1, \ldots, n)$$

The equations (9) can be viewed as the equations of an algebraic correspondence ψ^*, defined over k, between the dual of the (X, Y)-plane and some subvariety V of $\pi(\Sigma_{n,d})$: a line $l : X = uY + v$ and a point $\pi(C)$ of V, where the curve C is defined by the equation

$$C : \sum_{2 \le i+j \le n} \bar{a}_{ij}^{(\nu)} (X + \bar{\alpha}_\nu)^i (Y + \bar{\beta}_\nu)^j = 0,$$

are corresponding elements under ψ^* if and only if the quantities $\bar{a}_{ij}^{(\nu)}, \bar{\alpha}_\nu, \bar{\beta}_\nu, \bar{u}, \bar{v}$ satisfy the $(n+1)$ equations (9). The total ψ^*-transform of the line

l is the subsystem $S_{n,d}(l)$ (independent of ν), and as before, there is a non-empty subsystem $\Sigma_{n,d}(l)$ of $S_{n,d}(l)$ consisting of the irreducible components of $S_{n,d}(l)$ in which the general curve is reduced. The curves of the system $S_{n,d}(l)$ are all the form $l + C_{n-1}$, where C_{n-1} ranges over an algebraic system of plane curves of order $n-1$, this system being defined over the field $k(l)(= k(u,v))$. We shall denote this system by $S^d_{n-1}(l)$. Similarly, we shall denote by $\Sigma^d_{n-1}(l)$ the algebraic subsystem of $S^d_{n-1}(l)$ consisting of those curves C_{n-1} of $S^d_{n-1}(l)$ for which $l + C_{n-1} \in \Sigma_{n,d}(l)$.

The correspondence ψ^* may very well be reducible, in fact is probably, in general, reducible. For instance, in the case of the system $\Sigma_{5,3}$ of irreducible quintics having 3 nodes, the system $S^3_4(l)$ is always the union of the two irreducible systems $\Sigma_{4,3}$ mentioned in the introduction, both systems having dimension 11. We note that in this case the system $S^3_4(l)$ *is independent of the line* l and coincides with the system $\Sigma^3_4(l)$ (equivalently: in each irreducible component of the system $S_{5,3}(l)$ the general curve is reduced). The independence of the system $\Sigma^d_{n-1}(l)$ on the line l is a general fact which we shall prove below (see proposition 2) and which is an essential ingredient of our proof of the Basic Conjecture I (*always assuming the validity of the Basic Conjecture II*).

Proposition 1. *Let C_{n-1} be a curve in the system $S^d_{n-1}(l)$, where l is any line in the plane, and let C^* be the general (hence irreducible) curve of the system $\Sigma_{n,d}/k$ of the Basic Lemma.*

(a) *If in the specialization $C^* \to l + C_{n-1}$ a node Q^* of C^* specializes to a point Q of l, then Q is necessaarily at least a triple point of $l + C_{n-1}$ (hence at least a double point of C_{n-1}).*

(b) *Furthermore, if in the above specialization the point Q of l is a node of C_{n-1}, then Q^* is the only node of C^* which specializes to Q.*

Proof. Part (a) of the proposition follows directly from the equations (3a), for if the node $Q^*_\nu(-\alpha_\nu, -\beta_\nu)$ specializes to a point $(0, -\overline{\beta})$ of the line l (i.e.,of the line $X = 0$), then upon setting $\alpha = 0$ in the equations (3a) we find that $a^{(\nu)}_{20}, a^{(\nu)}_{11}$ and $a^{(\nu)}_{02}$ specialize to zero, and that consequently the point $(0, -\overline{\beta})$ is at least a triple point of $l + C_{n-1}$.

Let $\overline{a}_{ij}$ be the specialization of $a^{(\nu)}_{ij}$ as C^* specializes to $l + C_{n-1}$, while (α_ν, β_ν) specializes, say, to the origin $Q(0,0)$. Using the first 3 of the equations (3a) we find that the quotients $-\frac{a_{20}^{(\nu)}}{\alpha_\nu}, -\frac{a_{11}^{(\nu)}}{\alpha_\nu}, -\frac{a_{02}^{(\nu)}}{\alpha_\nu}$ specialize to $\overline{a}_{30}, \overline{a}_{21}$ *and* $\overline{a}_{12}$, respectively. Hence the node Q^* of C^* specializes to the point Q of the curve Γ given by an equation of the form

$$0 = \overline{a}_{30}X^2 + \overline{a}_{21}XY + \overline{a}_{12}Y^2 + \text{terms of degree} > 2.$$

On the other hand, the specialization $l + C_{n-1}$ of the curve C^* is given by the equation

$$X(\overline{a}_{30}X^2 + \overline{a}_{21}XY + \overline{a}_{12}Y^2 + \text{terms of degree} \geq 2) = 0.$$

If Q is a node of C_{n-1}, then Q is also a node of Γ, and hence no other node Q_1^* of C^* can specialize to the same point Q of C_{n-1}, for in the contrary case Q would be at least a tacnode of Γ, hence also of C_{n-1}.

This completes the proof of Proposition 1.

The next proposition is the decisive step in our proof of the Basic Lemma.

Proposition 2.

(a) *The algebraic system $\Sigma_{n-1}^d(l)$ is independent of the line l (we shall therefore denote this system by Σ_{n-1}^d).*

(b) *The system Σ_{n-1}^d is a union of systems $\Sigma_{n-1,d}$.*

Proof. (a) We shall use induction on d by assuming that part (a) of the proposition is true for n and $d-1$. It is trivially true for $d = 1$, *since $\Sigma_{n,1}(l)$ is equal to $l + \Sigma_{n-1,1}$.* This can be seen as follows:

If $d = 1$ then the elements $a_{ij} (= a_{ij}^{(1)})$ in (1) are indeterminates. Substituting the expressions (3a) of $a_{20}^{(1)}, a_{11}^{(1)}$ and $a_{0\mu}^{(1)} (\mu = 2, 3, \ldots, n)$ into the equation $f(X, Y) = 0$ of C^*, we obtain an equation $\tilde{f}(X, Y) = 0$, which defines a curve having the line $X = 0$ as component but still has the point $(-\alpha, -\beta)$ as double point. Therefore, $\Sigma_{n-1}^1 \subset \Sigma_{n-1,1}$. On the other hand, we have $\dim \Sigma_{n-1}^1 \geq \frac{n(n+3)}{2} - 1 - (n+1)$, i.e., $\dim \Sigma_{n-1}^1 \geq \frac{(n-1)(n+2)}{2} - 1$, while we have the $\dim \Sigma_{n-1,1} = \frac{(n-1)(n+2)}{2} - 1$. Hence Σ_{n-1}^1 must coincide with the irreducible system $\Sigma_{n-1,1}$.

To prove (a) it is sufficient to prove that if l^* is the general line $X = uY + v$, then the system $\Sigma_{n-1}^d(l^*)$ is defined over k, for if that is proved then it will follow that any specialization, over k, of the representative variety $\pi(\Sigma_{n-1}^d(l^*))$ coincides with $\pi(\Sigma_{n-1}^d(l))$ and consequently $\Sigma_{n-1}^d(l^*) = \Sigma_{n-1}^d(l)$, where l is any line of the plane.

We shall show that the assumption that $\Sigma_{n-1}^d(l^*)$ is not defined over k leads to a contradiction. The union of all the systems $\Sigma_{n-1}^d(l)$, as l ranges over the set of all the lines in the plane, is an algebraic system defined over k, for if $\Sigma_{n-1}^d(l^*)$ has ρ irreducible components (over $k(l^*)$) and if $C_{n-1,1}^*, C_{n-1,2}^*, \ldots, C_{n-1,\rho}^*$ are the general curves of these components, then the union of the systems $\Sigma_{n-1}^d(l)$ is the union of the ρ irreducible algebraic systems having as general curves, over k, the ρ curves $C_{n-1,1}^*, C_{n-1,2}^*, \ldots, C_{n-1,\rho}^*$. We shall denote the union of the systems $\Sigma_{n-1}^d(l)$ by Σ_{n-1}^d. If $\Sigma_{n-1}^d(l^*)$ is not defined over k, then it is clear that the dimension of Σ_{n-1}^d is greater than the dimension of $\Sigma_{n-1}^d(l^*)$. We fix an irreducible component $\overline{\Sigma}_{n-1}^d$ of Σ_{n-1}^d whose dimension is greater than $\dim \Sigma_{n-1}^d(l^*)$. We have therefore

$$\dim \overline{\Sigma}_{n-1}^d \geq 2n + p - 1, \tag{10}$$

since, as we have pointed out earlier, the dimension of Σ_{n-1}^d is greater than or equal to $2n + 2p - 2$, for every line l.

The system $\Sigma_{n,d}$ is contained in some system $\Sigma_{n,d-1}$. For the proof, consider the curves of order n which have $d-1$ nodes which specialize to a given set of $d-1$ nodes of a generic C of the system $\Sigma_{n,d}$ and apply the same argument used by us in [6] in the course of the proof that C is contained in a (unique) irreducible system $\Sigma_{n,d}$. The general curve of $\Sigma_{n,d-1}$ is necessarily irreducible, since $\Sigma_{n,d-1}$ contains $\Sigma_{n,d}$ and since we hae assumed that the general curve of $\Sigma_{n,d}$ is irreducible. The equations (3), as applied to the system $\Sigma_{n,d-1}$, define a system $\Sigma_{n,d-1}(l)$ and a system $\Sigma_{n-1}^{d-1}(l)$, and we have obviously the inclusion $\Sigma_{n-1}^d(l) \subset \Sigma_{n-1}^{d-1}(l)$, for any line l. By our induction hypothesis, the proposition is true for n and $d-1$. Therefore the system $\Sigma_{n-1}^{d-1}(l)$ is independent of l and is a union of systems $\Sigma_{n-1,d-1}$. The irreducible system $\overline{\Sigma}_{n-1}^d$ (see (10)) must be contained in some irreducible component of $\Sigma_{n-1}^{d-1}(l)$, i.e., in some system $\Sigma_{n-1,d-1}$. We fix one such system. We have

$$\begin{aligned}\dim \Sigma_{n-1,d-1} &= \frac{(n-1)(n+2)}{2} - d + 1 = \frac{n(n+3)}{2} - d - n \\ &= \dim \Sigma_{n,d} - n = 2n + p - 1.\end{aligned}$$

Hence, by (10), it follows *that* $\overline{\Sigma}_{n-1}$ coincides with $\Sigma_{n-1,d-1}$. Let C^*_{n-1} be the general curve of $\Sigma_{n-1,d-1}/k$, hence also of $\overline{\Sigma}^d_{n-1}/k$. Then $l^* + C^*_{n-1}$ is a specialization of the general curve C^* of $\Sigma_{n,d}$. Since C^*_{n-1} has only nodes, it follows Proposition 1, part (a), that if q of the nodes of C^*_{n-1} lie on l^*, then q and only q of the nodes of C^* are specialized to the nodes of C^*_{n-1} which lie on l^*. Hence the remaining $d - q$ nodes of C^* must specialize to the remaining $d-1-q$ nodes of C^*_{n-1}, and this is impossible. This completes the proof of part (a) of Proposition 2.

We now proceed to the proof of part (*b*) of Proposition 2. By part (*a*) of that proposition we know that for any line l in the plane the system $l + \Sigma^d_{n-1}$ (of curves of order n) is a subsystem of $\Sigma_{n,d}$. We know also (by our definition of the system $\Sigma^d_{n-1}(l)$ and by part (*a*) of the proposition) that Σ^d_{n-1} is contained in some system Σ^{d-1}_{n-1}. We have, by our induction hypothesis, $\dim \Sigma^{d-1}_{n-1} = 2n + p - 1$, while we know that

$$\dim \Sigma^d_{n-1} \geq 2n + p - 2,$$

and this inequality holds for each irreducible component Σ' of the system Σ^d_{n-1}. We have therefore $\dim \Sigma' \geq \dim \Sigma_{n-1,d-1} - 1$. We know that the general curve of any such system Σ' is reduced. If then C'^*_{n-1} is the general curve of Σ'/k, then a generic line l meets C'^*_{n-1} in $n-1$ distinct (hence simple) points of C'^*_{n-1}. Consequently, by part (*a*) of Proposition 1, in the specializaton $C^* \to l + C'^*_{n-1}$ of the general curve C^* of $\Sigma_{n,d}$, no node of C^* is specialized to a point of l. *Therefore, the d nodes of C^* specialize to singular points of C'^*_{n-1}* (which, by our choice of l, do not belong to l). It follows that the effective genus p'^* of the *reduced* curve C'^*_{n-1} must be less than or equal to $\frac{(n-2)(n-3)}{2} - d$. We have

$$\frac{(n-2)(n-3)}{2} - d = \frac{(n-1)(n-2)}{2} - d - n + 2 = p - n + 2.$$

Hence $p'^* \leq p - n + 2$. We have

$$\dim \Sigma' \geq \dim \Sigma_{n,d} - n - 1 = 3(n-1) + (p - n + 2) - 1.$$

Hence $\dim \Sigma' \geq 3(n-1) + p'^* - 1$. This implies, by Theorem 2 of our paper [6], that $\dim \Sigma' = 3(n-1) + p'^* - 1$, that $p'^* = p - n + 2$ and that the general curve of Σ' must have precisely d nodes (since $3(n-1) + p'^* - 1 =$

$3n+p-1-(n+1)=\frac{n(n+3)}{2}-d-(n+1)=\frac{(n-1)(n+2)}{2}-d)$. Thus every irreducible component Σ' of Σ^d_{n-1} is a system $\Sigma_{n-1,d}$. This completes the proof of part (b) of the proposition.

The Basic Lemma (see Introduction) is an immediate consequence of Proposition 2. The corollary of the Basic Lemma, stated in the Introduction, is an immediate consequence of the Basic Lemma since, by induction on n, we may assume that $\Sigma_{n-1,d}$ contains all $(n-1)$-gons of the plane and since $\Sigma_{n,d}$ contains all the curves of the form $l+C_{n-1}$, where $C_{n-1}\in\Sigma_{n-1,d}$ and where l is an *arbitrary* line of the plane.

Note. We add a few words about a possible approach to a proof of the conjecture started in footnote (1), i.e., that our subsystem $\Sigma_{n,d}(l)$ of $S_{n,d}(l)$ coincides with $S_{n,d}(l)$ (or, equivalently, the general curve of every irreducible component of the system $S_{n,d}(l)$ is reduced). One could, namely, try to prove a stronger form of Proposition 2, by replacing in the statement of that proposition the system $\Sigma^d_{n-1}(l)$ by the possibly bigger system $S^d_{n-1}(l)$. It is easily seen that the proof of part (a) of that proposition remains valid without any changes. However, when it comes to the proof of part (b) of the proposition, the following difficulty arises.

Our conclusion, in that proof, that Σ^d_{n-1} is contained in some system $\Sigma_{n-1,d-1}$, is now to be replaced by the (correct) statement that S^d_{n-1} is contained in some $\Sigma_{n-1,d-1}$, and that consequently every irreducible component S' of S^d_{n-1} is contained in a $\Sigma_{n-1,d-1}$. But now we cannot assert (as we did in the case when we dealt with Σ^d_{n-1} and Σ') that the general curve of S'/k is reduced. However, the inequality $\dim\Sigma'\geq\dim\Sigma_{n-1,d-1}$ remains valid if we replace Σ' by S'. Hence, it would be necessary to prove the following (highly plausible) conjecture:

If S' is an irreducible subsystem of a system $\Sigma_{n,d}$ and if $\dim S'=\dim\Sigma_{n,d}-1$, then the general curve of S' is reduced.

§2. We shall now derive the corollary of the basic lemma, stated in the Introduction. We shall use induction on n, since the corollary is trivial for $n=2$. We consider first the case in which the general curve C^* of $\Sigma_{n,d}$ is reducible. In this case we make use of the result proved in our paper [6] (§3, proof of the Theorem 2 in the case of $\Sigma_{n,d}$'s with reducible general curve C^*), namely that $\Sigma_{n,d}$ is then a quasi-direct sum $\Sigma_{n_1,d_1}+\Sigma_{n_2,d_2}+\cdots+\Sigma_{n_q,d_q}$ of a certain number $q>1$ of complete systems Σ_{n_i,d_i}, the general curves of which are irreducible. Here $n=n_1+n_2+\cdots+n_q$, while $d = d_1 + d_2 + \ldots + d_q + \sum \sum_{i \neq j} n_i n_j$. By our induction hypothesis each system Σ_{n_i, d_i} has the property that it contains all n_i-gons of the plane. Since $\Sigma_{n,d}$ is the set of all curves of the form $C_1 + C_2 + \cdots + C_q$, $C_i \in \Sigma_{n_i, d_i}$, it follows that $\Sigma_{n,d}$ contains all the n-gons of the plane.

We now consider the case in which the general curve C^* of $\Sigma_{n,d}$ is irreducible. By the basic lemma, there exists then a system $\Sigma_{n-1,d}$ such that $\Sigma_{n,d}$ contains all the curves of the form $l + C_{n-1}$, where l *is any line* and C_{n-1} *is any curve* in $\Sigma_{n-1,d}$. Since, by our induction hypothesis, $\Sigma_{n-1,d}$ contains all the $(n-1)$-gons of the plane, it follows that $\Sigma_{n,d}$ contains all the n-gons of the plane.

This completes the proof of the corollary of the basic lemma.

References

[1] P. Deligne, "Le groupe fondamental du complément d'une courbe plane n'ayant que des points doubles ordinaires est abélien," *Séminaire Bourbaki* (November, 1979).

[2] W. Fulton, "On the fundamental group of the complement of a node curve," Ann. of Math., *111* (1980) pp. 407-409.

[3] F. Severi, *Vorlesungen über algebraische Geometrie*, B. G. Teubner, Leipzig (1921).

[4] O. Zariski, "On the problem of existence of algebraic functions of two variables possessing a given branch curve," Amer. J. Math., *51* (1929) pp. 305-328.

[5] ________, *Algebraic Surfaces*, second edition (1971), Springer-Verlag, Berlin, Heidelberg, and New York.

[6] ________, "Dimension-theoretic characterization of maximal irreducible algebraic systems of plane nodal curves of a given order n and with a given number d of nodes," Amer. J. Math., *104* (1982) pp. 209-226.

[7] O. Zariski and P. Samuel, *Commutative Algebra*, v. 1 (1958) and v. 2 (1960), D. Van Nostrand Company, Princeton, also Springer-Verlag, Berlin, Heidelberg, and New York.

Received February 1, 1982

Professor Oscar Zariski
Department of Mathematics
Harvard University
Cambridge, Massachusetts 02138

Progress in Mathematics
Edited by J. Coates and S. Helgason

Progress in Physics
Edited by A. Jaffe and D. Ruelle

- A collection of research-oriented monographs, reports, notes arising from lectures or seminars
- Quickly published concurrent with research
- Easily accessible through international distribution facilities
- Reasonably priced
- Reporting research developments combining original results with an expository treatment of the particular subject area
- A contribution to the international scientific community: for colleagues and for graduate students who are seeking current information and directions in their graduate and post-graduate work.

Manuscripts

Manuscripts should be no less than 100 and preferably no more than 500 pages in length.

They are reproduced by a photographic process and therefore must be typed with extreme care. Symbols not on the typewriter should be inserted by hand in indelible black ink. Corrections to the typescript should be made by pasting in the new text or painting out errors with white correction fluid.

The typescript is reduced slightly (75%) in size during reproduction; best results will not be obtained unless the text on any one page is kept within the overall limit of 6x9½ in (16x24 cm). On request, the publisher will supply special paper with the typing area outlined.

Manuscripts should be sent to the editors or directly to:
Birkhäuser Boston, Inc., P.O. Box 2007, Cambridge, Massachusetts 02139

PROGRESS IN MATHEMATICS

Already published

PM 1 Quadratic Forms in Infinite-Dimensional Vector Spaces
Herbert Gross
ISBN 3-7643-1111-8, 432 pages, paperback

PM 2 Singularités des systemes différentiels de Gauss-Manin
Frédéric Pham
ISBN 3-7643-3002-3, 346 pages, paperback

PM 3 Vector Bundles on Complex Projective Spaces
C. Okonek, M. Schneider, H. Spindler
ISBN 3-7643-3000-7, 396 pages, paperback

PM 4 Complex Approximation, Proceedings, Quebec, Canada, July 3-8, 1978
Edited by Bernard Aupetit
ISBN 3-7643-3004-X, 128 pages, paperback

PM 5 The Radon Transform
Sigurdur Helgason
ISBN 3-7643-3006-6, 202 pages, paperback

PM 6 The Weil Representation, Maslov Index and Theta Series
Gérard Lion, Michèle Vergne
ISBN 3-7643-3007-4, 348 pages, paperback

PM 7 Vector Bundles and Differential Equations Proceedings, Nice, France, June 12-17, 1979
Edited by André Hirschowitz
ISBN 3-7643-3022-8, 256 pages, paperback

PM 8 Dynamical Systems, C.I.M.E. Lectures, Bressanone, Italy, June 1978
John Guckenheimer, Jürgen Moser, Sheldon E. Newhouse
ISBN 3-7643-3024-4, 298 pages, paperback

PM 9 Linear Algebraic Groups
T. A. Springer
ISBN 3-7643-3029-5, 314 pages, hardcover

PM10 Ergodic Theory and Dynamical Systems I
A. Katok
ISBN 3-7643-3036-8, 346 pages, hardcover

PM11 18th Scandinavian Congress of Mathematicians, Aarhus, Denmark, 1980
Edited by Erik Balslev
ISBN 3-7643-3040-6, 526 pages, hardcover

PM12 Séminaire de Théorie des Nombres, Paris 1979-80
Edited by Marie-José Bertin
ISBN 3-7643-3035-X, 404 pages, hardcover

PM13 Topics in Harmonic Analysis on Homogeneous Spaces
Sigurdur Helgason
ISBN 3-7643-3051-1, 152 pages, hardcover

PM14 Manifolds and Lie Groups, Papers in Honor of Yozô Matsushima
Edited by J. Hano, A. Marimoto, S. Murakami, K. Okamoto, and H. Ozeki
ISBN 3-7643-3053-8, 476 pages, hardcover

PM15 Representations of Real Reductive Lie Groups
David A. Vogan, Jr.
ISBN 3-7643-3037-6, 776 pages, hardcover

PM16 Rational Homotopy Theory and Differential Forms
Phillip A. Griffiths, John W. Morgan
ISBN 3-7643-3041-4, 258 pages, hardcover

PM17 Triangular Products of Group Representations and their Applications
S.M. Vovsi
ISBN 3-7643-3062-7, 142 pages, hardcover

PM18 Géométrie Analytique Rigide et Applications
Jean Fresnel, Marius van der Put
ISBN 3-7643-3069-4, 232 pages, hardcover

PM19 Periods of Hilbert Modular Surfaces
Takayuki Oda
ISBN 3-7643-3084-8, 144 pages, hardcover

PM20 Arithmetic on Modular Curves
Glenn Stevens
ISBN 3-7643-3088-0, 236 pages, hardcover

PM21 Ergodic Theory and Dynamical Systems II
A. Katok, editor
ISBN 3-7643-3096-1, 226 pages, hardcover

PM22 Séminaire de Théorie des Nombres, Paris 1980-81
Marie-José Bertin, editor
ISBN 3-7643-3066-X, 374 pages, hardcover

PM23 Adeles and Algebraic Groups
A. Weil
ISBN 3-7643-3092-9, 138 pages, hardcover

PM24 Enumerative Geometry and Classical Algebraic Geometry
Patrick Le Barz, Yves Hervier, editors
ISBN 3-7643-3106-2, 260 pages, hardcover

PM25 Exterior Differential Systems and the Calculus of Variations
Phillip A. Griffiths
ISBN 3-7643-3103-8, 349 pages, hardcover

PM26 Number Theory Related to Fermat's Last Theorem
Neal Koblitz, editor
ISBN 3-7643-3104-6, 376 pages, hardcover

PM27 Differential Geometric Control Theory
Roger W. Brockett, Richard S. Millman, Hector J. Sussmann, editors
ISBN 3-7643-3091-0, 349 pages, hardcover

PM28 Tata Lectures on Theta I
David Mumford
ISBN 3-7643-3109-7, 254 pages, hardcover

PM29 Birational Geometry of Degenerations
Robert Friedman and David R. Morrison, editors
ISBN 3-7643-3111-9, 410 pages, hardcover

PM30 CR Submanifolds of Kaehlerian and Sasakian Manifolds
Kentaro Yano, Masahiro Kon
ISBN 3-7643-3119-4, 223 pages, hardcover

PM31 Approximations Diophantiennes et Nombres Transcendants
D. Bertrand and M. Waldschmidt, editors
ISBN 3-7643-3120-8, 349 pages, hardcover

PM32 Differential Geometry
Robert Brooks, Alfred Gray, Bruce L. Reinhart, editors
ISBN 3-7643-3134-8, 267 pages, hardcover

PM 33 Uniqueness and Non-uniqueness in the Cauchy Problem
Claude Zuily
ISBN 3-7643-3121-6, 185 pages, hardcover

PM 34 Systems of Microdifferential Equations
Masaki Kashiwara
ISBN 3-7643-3138-0, 182 pages, hardcover

PROGRESS IN PHYSICS

Already published

PPh1 Iterated Maps on the Interval as Dynamical Systems
Pierre Collet and Jean-Pierre Eckmann
ISBN 3-7643-3026-0, 256 pages, hardcover

PPh2 Vortices and Monopoles, Structure of Static Gauge Theories
Arthur Jaffe and Clifford Taubes
ISBN 3-7643-3025-2, 294 pages, hardcover

PPh3 Mathematics and Physics
Yu. I. Manin
ISBN 3-7643-3027-9, 112 pages, hardcover

PPh4 Lectures on Lepton Nucleon Scattering and Quantum Chromodynamics
W.B. Atwood, J.D. Bjorken, S.J. Brodsky, and R. Stroynowski
ISBN 3-7643-3079-1, 574 pages, hardcover

PPh5 Gauge Theories: Fundamental Interactions and Rigorous Results
P. Dita, V. Georgescu, R. Purice, editors
ISBN 3-7643-3095-3, 406 pages, hardcover

PPh6 Third Workshop on Grand Unification
University of North Carolina, Chapel Hill, April 15-17, 1982
P.H. Frampton, S.L. Glashow, and H. van Dam, editors
ISBN 3-7643-3105-4, 382 pages, hardcover

Printed in the United States
By Bookmasters